T0137019

Advances in Intelligent Systems and Computing

Volume 584

Series editor

Janusz Kacprzyk, Polish Academy of Sciences, Warsaw, Poland
e-mail: kacprzyk@ibspan.waw.pl

The series "Advances in Intelligent Systems and Computing" contains publications on theory, applications, and design methods of Intelligent Systems and Intelligent Computing. Virtually all disciplines such as engineering, natural sciences, computer and information science, ICT, economics, business, e-commerce, environment, healthcare, life science are covered. The list of topics spans all the areas of modern intelligent systems and computing.

The publications within "Advances in Intelligent Systems and Computing" are primarily textbooks and proceedings of important conferences, symposia and congresses. They cover significant recent developments in the field, both of a foundational and applicable character. An important characteristic feature of the series is the short publication time and world-wide distribution. This permits a rapid and broad dissemination of research results.

More information about this series at http://www.springer.com/series/11156

Millie Pant · Kanad Ray
Tarun K. Sharma · Sanyog Rawat
Anirban Bandyopadhyay
Editors

Soft Computing: Theories and Applications

Proceedings of SoCTA 2016, Volume 2

 Springer

Editors

Millie Pant
Department of Applied Science
  and Engineering
IIT Roorkee
Saharanpur
India

Kanad Ray
Department of Physics
Amity School of Applied Sciences, Amity
  University Rajasthan
Jaipur, Rajasthan
India

Tarun K. Sharma
Department of Computer Science
  and Engineering
Amity School of Engineering
  and Technology, Amity University
  Rajasthan
Jaipur, Rajasthan
India

Sanyog Rawat
Department of Electronics
  and Communication Engineering
SEEC, Manipal University Jaipur
Jaipur, Rajasthan
India

Anirban Bandyopadhyay
Surface Characterization Group, NIMS
Nano Characterization Unit, Advanced Key
  Technologies Division
Tsukuba, Ibaraki
Japan

ISSN 2194-5357   ISSN 2194-5365 (electronic)
Advances in Intelligent Systems and Computing
ISBN 978-981-10-5698-7   ISBN 978-981-10-5699-4 (eBook)
https://doi.org/10.1007/978-981-10-5699-4

Library of Congress Control Number: 2017947482

Printed on acid-free paper

This Springer imprint is published by Springer Nature
The registered company is Springer Nature Singapore Pte Ltd.
The registered company address is: 152 Beach Road, #21-01/04 Gateway East, Singapore 189721, Singapore

Preface

It is a matter of pride to introduce the first international conference in the series of "Soft Computing: Theories and Applications (SoCTA)", which is a joint effort of researchers from Machine Intelligence Lab, USA, and the researchers and Faculty members from Indian Institute of Technology, Roorkee; Amity University Rajasthan.

The maiden conference took place in the historic city of Jaipur at the campus of research-driven university, Amity University Rajasthan. The conference stimulated discussions on various emerging trends, innovation, practices, and applications in the field of Soft Computing.

This book that we wish to bring forth with great pleasure is an encapsulation of 149 research papers, presented during the three-day international conference. We hope that the initiative will be found informative and interesting to those who are keen to learn on technologies that address to the challenges of the exponentially growing information in the core and allied fields of Soft Computing.

We are thankful to the authors of the research papers for their valuable contribution in the conference and for bringing forth significant research and literature across the field of Soft Computing.

The editors also express their sincere gratitude to SoCTA 2016 Patron, Plenary Speakers, Keynote Speakers, Reviewers, Programme Committee Members, International Advisory Committee and Local Organizing Committee, Sponsors without whose support the support and quality of the conference could not be maintained.

We would like to express our sincere gratitude to Prof. Sanghamitra Bandyopadhyay, Director, ISI Kolkata, for gracing the occasion as the Chief Guest for the Inaugural Session and delivering a Plenary talk.

We would like to express our sincere gratitude to Dr. Anuj Saxena, Officer on Special Duty, Chief Minister's Advisory Council, Govt. of Rajasthan, for gracing the occasion as the Chief Guest for the Valedictory Session.

We would like to extend our heartfelt gratitude to Prof. Nirupam Chakraborti, Indian Institute of Technology, Kharagpur; Prof. Ujjwal Maulik, Jadavpur University; Prof. Kumkum Garg, Manipal University Jaipur; Dr. Eduardo Lugo,

Université de Montréal; Prof. Lalit Garg, University of Malta for delivering invited lectures.

We express our special thanks to Prof. Ajith Abraham, Director, MIR Labs, USA, for being a General Chair and finding time to come to Jaipur amid his very busy schedule.

We are grateful to Prof. W. Selvamurthy and Ambassador (Retd.) R.M. Aggarwal for their benign cooperation and support.

A special mention of thanks is due to our student volunteers for the spirit and enthusiasm they had shown throughout the duration of the event.

We express special thanks to Springer and its team for the valuable support in the publication of the proceedings.

With great fervor, we wish to bring together researchers and practitioners in the field of Soft Computing year after year to explore new avenues in the field.

Saharanpur, India   Dr. Millie Pant
Jaipur, India   Dr. Kanad Ray
Jaipur, India   Dr. Tarun K. Sharma
Jaipur, India   Dr. Sanyog Rawat
Tsukuba, Japan   Dr. Anirban Bandyopadhyay

Organizing Committee

Patrons-in-Chief

Dr. Ashok K. Chauhan, Founder President, Ritnand Balved Education Foundation (RBEF)
Dr. Aseem Chauhan, Chancellor, Amity University Rajasthan

Patron

Prof. (Dr.) S.K. Dube, Vice Chancellor, AUR

Co-patron

Prof. (Dr.) S.L. Kothari, Pro Vice Chancellor, AUR

General Chair

Prof. (Dr.) Ajith Abraham, Director, MIR Labs, USA
Dr. Millie Pant, Indian Institute of Technology, Roorkee
Prof. (Dr.) Kanad Ray, Amity University Rajasthan

Program Chairs

Dr. Tarun K. Sharma, Amity University Rajasthan
Dr. Sanyog Rawat, Manipal University Jaipur

Organizing Chair

Prof. D.D. Shukla, Amity University Rajasthan
Prof. Jagdish Prasad, Amity University Rajasthan

Finance Chair

Brig. (Retd) S.K. Sareen, Amity University Rajasthan
Mr. Sunil Bhargawa, Amity University Rajasthan

Conference Proceedings and Printing & Publication Chair

Dr. Millie Pant, Indian Institute of Technology, Roorkee
Prof. (Dr.) Kanad Ray, Amity University Rajasthan
Dr. Tarun K. Sharma, Amity University Rajasthan
Dr. Sanyog Rawat, Manipal University Jaipur
Dr. Anirban Bandoypadhyay, NIMS, Japan

Best Paper and Best Ph.D. Thesis Chair

Prof. S.C. Sharma, Indian Institute of Technology, Roorkee
Prof. K.K. Sharma, MNIT, Jaipur
Dr. Anirban Bandoypadhyay, NIMS, Japan

Technical and Special Sessions Chair

Dr. Musrrat Ali, Glocal University Saharanpur
Dr. Sushil Kumar, Amity University Uttar Pradesh

Publicity Chair

Dr. Pravesh Kumar, Jaypee University Noida
Mr. Jitendra Rajpurohit, Amity University Rajasthan
Mr. Anil Saroliya, Amity University Rajasthan
Dr. Divya Prakash, Amity University Rajasthan
Mr. Anurag Tripathi, Arya, Jaipur

Registration Chair

Mr. Amit Hirawat, Amity University Rajasthan
Mr. Jitendra Rajpurohit, Amity University Rajasthan

Outcome Committee

Prof. Kanad Ray, Amity University Rajasthan
Dr. Tarun K. Sharma, Amity University Rajasthan
Prof. (Dr.) P.V.S. Raju, Amity University Rajasthan

Web Administrator

Mr. Chitreshh Banerjee, Amity University Rajasthan

Reporting

Dr. Ratnadeep Roy, Amity University Rajasthan
Ms. Pooja Parnami, Amity University Rajasthan

Hospitality Chair

Mr. Vikas Chauhan, Amity University Rajasthan
Ms. Preeti Gupta, Amity University Rajasthan
Dr. Divya Prakash, Amity University Rajasthan

Mr. Jitendra Rajpurohit, Amity University Rajasthan
Mr. Amit Chaurasia, Amity University Rajasthan
Mr. Deepak Panwar, Amity University Rajasthan

Local Organizing Committee

Prof. Upendra Mishra, Amity University Rajasthan
Dr. Swapnesh Taterth, Amity University Rajasthan
Ms. Parul Pathak, JECRC Jaipur
Mr. Anurag Tripathi, Arya, Jaipur
Mr. Ashwani Yadav, Amity University Rajasthan
Ms. Vaishali Yadav, Amity University Rajasthan
Ms. Bhawana Sharma, Amity University Rajasthan
Ms. Pallavi Sharma, Amity University Rajasthan
Mr. Abhay Sharma, Amity University Rajasthan
Dr. Irshad Ansari, Indian Institute of Technology, Roorkee
Mr. Bilal, Indian Institute of Technology, Roorkee
Mr. Nathan Singh, Indian Institute of Technology, Roorkee
Mr. Sunil K. Jauhar, Indian Institute of Technology, Roorkee
Ms. Meenu Singh, Indian Institute of Technology, Roorkee
Mr. Het, Indian Institute of Technology, Roorkee

International Advisory Board

Aboul Ella Hassanien, University of Cairo, Egypt
Adel Alimi, University of Sfax, Tunisia
Aditya Ghose, University of Wollongong, Australia
André Ponce de Leon F de Carvalho, University of São Paulo, Brazil
Ashley Paupiah, Amity Mauritius
Bruno Apolloni, University of Milano, Italy
Francesco Marcelloni, University of Pisa, Italy
Francisco Herrera, University of Granada, Spain
Imre J. Rudas, Obuda University, Hungary
Javier Montero, Complutense University of Madrid, Spain
Jun Wang, Chinese University of Hong Kong, Hong Kong
Naren Sukurdeep, Amity Mauritius
Mo Jamshidi, University of Texas at San Antonio, USA
Sang-Yong Han, Chung-Ang University, Korea
Sebastián Ventura, University of Cordoba, Spain

Sebastian Basterrech, Technical University of Ostrava, Czech Republic
Witold Pedrycz, University of Alberta, Canada

National Advisory Committee

Ankush Mittal, Dehradun
Aruna Tiwari, IIT Indore
Ashish Verma, IIT Guwahati
Ashok Deshpande, Pune
Ashok Kumar Singh, DST, New Delhi
B.K. Das, Delhi University, India
C. Thangaraj, IIT Roorkee, India
D. Gnanaraj Thomas, Madras Christian College, Chennai
D. Nagesh Kumar, IISc., Bangalore
Debasish Ghose, IISc., Bangalore
Deepti, AHEC, IIT Roorkee
Dharmdutt, Saharanpur Campus, IIT Roorkee
Ghanshyam Singh Thakur, MANET Bhopal
Himani Gupta, IIFT, Delhi
Kanti S. Swarup, IIT Chennai
M.P. Biswal, IIT Kharagpur
Manoj Kumar Tiwari, IIT Kharagpur
N.R. Pal, ISI, Kolkata
Nirupam Chakraborti, IIT Kharagpur
P.C. Jha, Delhi University
Pankaj Gupta, Delhi University
Punam Bedi, University of Delhi
Raju George, (IIST), Trivandrum
Rama Mehta, NIH IIT Roorkee
Rama Sushil, Dehradun
Ravindra Gudi, IIT Bombay
Steven Fernendes, Sahyadri College of Engineering & Management, Mangaluru,
Karnataka

Contents

Designing ANFIS Model to Predict the Reliability of
Component-Based System. 1
Rajni Sehgal, Deepti Mehrotra and Manju Bala

Bone Fracture Detection Using Edge Detection Technique 11
Nancy Johari and Natthan Singh

Virtual Experimental Analysis of Redundant Robot Manipulators
Using Neural Networks. 21
H.P. Singh, Surendra Kumar, Pravesh Kumar and Akanshu Mahajan

Selection of Energy-Efficient Material: An Entropy–TOPSIS
Approach. 31
Chiranjib Bhowmik, Sachin Gangwar, Sumit Bhowmik and Amitava Ray

Analyses and Detection of Health Insurance Fraud Using Data Mining
and Predictive Modeling Techniques. 41
Pallavi Pandey, Anil Saroliya and Raushan Kumar

Software Cost Estimation Using Artificial Neural Network. 51
Shaina Arora and Nidhi Mishra

SteganoCrypt: An App for Secure Communication 59
Neha Mudgal, Pallavi Singh and Shweta Saxena

Step-Stress Partially Accelerated Life Testing Plan for Rayleigh
Distribution Using Adaptive Type-II Progressive Hybrid
Censoring. 67
Showkat Ahmad Lone, Ahmadur Rahman and Arif-Ul-Islam

A Three-Layer Approach for Overlay Text Extraction
in Video Stream 79
Lalita Kumari, Vidyut Dey and J. L. Raheja

A Comprehensive Review and Open Challenges of Stream
Big Data.. 89
Bharat Tidke and Rupa Mehta

Biomass Estimation at ICESat/GLAS Footprints Using Support
Vector Regression Algorithm for Optimization of Parameters 101
Sonika and Priya Rathi

Novel Miniaturized Microstrip Patch Antenna for Body Centric
Wireless Communication in ISM Band.......................... 113
Raghvendra Singh, Pinki Kumari, Pushpendra Singh, Sanyog Rawat
and Kanad Ray

Recognition of Noisy Numbers Using Neural Network.............. 123
Chanda Thapliyal Nautiyal, Sunita Singh and U.S. Rana

Pollution Check Control Using License Plate Extraction via
Image Processing.. 133
Shivani Garg and Nidhi Mishra

Watermarking Technology in QR Code with Various Attacks 147
Azeem Mohammed Abdul, Srikanth Cherukuvada, Annaram Soujanya,
G. Sridevi and Syed Umar

Cognitive Networked Redemption Operational Conception and
Devising (CNROCD).. 157
Azeem Mohammed Abdul, Srikanth Cherukuvada, Annaram Soujanya,
G. Sridevi and Syed Umar

Secure Group Authentication Scheme for LTE-Advanced............ 167
M. Prasad and R. Manoharan

Modeling the Alterations in Calcium Homeostasis in the Presence
of Protein and VGCC for Alzheimeric Cell 181
Devanshi D. Dave and Brajesh Kumar Jha

Solution of Multi-objective Portfolio Optimization Problem Using
Multi-objective Synergetic Differential Evolution (MO-SDE).......... 191
Hira Zaheer and Millie Pant

Frequency Fractal Behavior in the Retina Nano-Center-Fed Dipole
Antenna Network of a Human Eye............................. 201
P. Singh, R. Doti, J.E. Lugo, J. Faubert, S. Rawat, S. Ghosh, K. Ray
and A. Bandyopadhyay

DNA as an Electromagnetic Fractal Cavity Resonator: Its Universal
Sensing and Fractal Antenna Behavior.......................... 213
P. Singh, R. Doti, J.E. Lugo, J. Faubert, S. Rawat, S. Ghosh, K. Ray
and A. Bandyopadhyay

Compact Half-Hexagonal Monopole Planar Antenna for UWB Applications......... 225
Ushaben Keshwala, Sanyog Rawat and Kanad Ray

Effective Data Acquisition for Machine Learning Algorithm in EEG Signal Processing......... 233
James Bonello, Lalit Garg, Gaurav Garg and Eliazar Elisha Audu

Solving Nonlinear Optimization Problems Using IUMDE Algorithm 245
Pravesh Kumar, Millie Pant and H.P. Singh

Health Recommender System and Its Applicability with MapReduce Framework 255
Ritika Bateja, Sanjay Kumar Dubey and Ashutosh Bhatt

Trigonometric Probability Tuning in Asynchronous Differential Evolution 267
Vaishali, Tarun Kumar Sharma, Ajith Abraham and Jitendra Rajpurohit

Relevance Index for Inferred Knowledge in Higher Education Domain Using Data Mining 279
Preeti Gupta, Deepti Mehrotra and Tarun Kumar Sharma

Analytical Study on Cardiovascular Health Issues Prediction Using Decision Model-Based Predictive Analytic Techniques......... 289
Anurag Bhatt, Sanjay Kumar Dubey and Ashutosh Kumar Bhatt

Automated Sizing Methodology for CMOS Miller Operational Transconductance Amplifier......... 301
Pankaj P. Prajapati and Mihir V. Shah

Analytical Review on Image Compression Using Fractal Image Coding......... 309
Sobia Amin, Richa Gupta and Deepti Mehrotra

Biological Infrared Antenna and Radar 323
P. Singh, R. Doti, J.E. Lugo, J. Faubert, S. Rawat, S. Ghosh, K. Ray and A. Bandyopadhyay

The Fulcrum Principle Between Parasympathetic and Sympathetic Peripheral Systems: Auditory Noise Can Modulate Body's Peripheral Temperature 333
J.E. Lugo, R. Doti and J. Faubert

Fractal Information Theory (FIT)-Derived Geometric Musical Language (GML) for Brain-Inspired Hypercomputing 343
Lokesh Agrawal, Rutuja Chhajed, Subrata Ghosh, Batu Ghosh, Kanad Ray, Satyajit Sahu, Daisuke Fujita and Anirban Bandyopadhyay

Design and Analysis of Fabricated Rectangular Microstrip Antenna with Defected Ground Structure for UWB Applications 373
Sandeep Toshniwal, Tanushri Mukherjee, Prashant Bijawat, Sanyog Rawat and Kanad Ray

Improved Clustering Algorithm for Wireless Sensor Network 379
Santar Pal Singh and S.C. Sharma

Performance Measurement of Academic Departments: Case of a Private Institution 387
Sandeep Kumar Mogha, Alok Kumar, Amit Kumar and Mohd Hussain Kunroo

Application of Shuffled Frog-Leaping Algorithm in Regional Air Pollution Control 397
Divya Prakash, Anurag Tripathi and Tarun Kumar Sharma

Estimating Technical Efficiency of Academic Departments: Case of Government PG College 405
Imran Ali, U.S. Rana, Millie Pant, Sunil Kumar Jauhar and Sandeep Kumar Mogha

Fuzzy-based Probabilistic Ecological Risk Assessment Approach: A Case Study of Heavy Metal Contaminated Soil 419
Vivek Kumar Gaurav, Chhaya Sharma, Rakesh Buhlan and Sushanta K. Sethi

Performance Evaluation of DV-HOP Localization Algorithm in Wireless Sensor Networks 433
Vikas Gupta and Brahmjit Singh

Performance Comparisons of Four Modified Structures of Log Periodic Three Element Microstrip Antenna Arrays 441
Abhishek Soni and Sandeep Toshniwal

Optimization of Compressive Strength of Polymer Composite Brick Using Taguchi Method 453
Nitesh Singh Rajput, Dipesh Dilipbhai Shukla, Lav Ishan and Tarun Kumar Sharma

Application of Unnormalized and Phase Correlation Techniques on Infrared Images 461
Himanshu Singh, Millie Pant, Sudhir Khare and Yogita Saklani

Modified Least Significant Bit Algorithm of Digital Watermarking for Information Security 473
Devendra Somwanshi, Indu Chhipa, Trapti Singhal and Ashwani Yadav

Megh: A Private Cloud Provisioning Various IaaS and SaaS 485
Tushar Bhardwaj, Mohit Kumar and S.C. Sharma

Predicting the Calorific Value of Municipal Solid Waste of
Ghaziabad City, Uttar Pradesh, India, Using Artificial Neural
Network Approach 495
Dipti Singh, Ajay Satija and Athar Hussain

Advertisement Scheduling Models in Television Media:
A Review 505
Meenu Singh, Millie Pant, Arshia Kaul and P.C. Jha

Preliminary Study of E-commerce Adoption in Indian Handicraft
SME: A Case Study 515
Rohit Yadav and Tripti Mahara

Optimization of End Milling Process for Al2024-T4 Aluminum
by Combined Taguchi and Artificial Neural Network Process 525
Shilpa B. Sahare, Sachin P. Untawale, Sharad S. Chaudhari,
R.L. Shrivastava and Prashant D. Kamble

Dynamic Classification Mining Techniques for Predicting
Phishing URL 537
Surbhi Gupta and Abhishek Singhal

Protection from Spoofing Attacks Using Honeypot in Wireless
Ad Hoc Environment 547
Palak Khurana, Anshika Sharma, Sushil Kumar
and Shailendra Narayan Singh

Threat Detection for Software Vulnerability Using DE-Based
Adaptive Approach 555
Anshika Sharma, Palak Khurana, Sushil Kumar
and Shailendra Narayan Singh

An Implementation Case Study on Ant-based Energy Efficient
Routing in WSNs 567
Kavitha Kadarla, S.C. Sharma and K. Uday Kanth Reddy

Comparison Between Gaming Consoles and Their Effects
on Children 577
Devansh Chopra, Ekta Sharma, Anchal Garg and Sushil Kumar

Design of Chamfered H-bend in Rectangular Substrate Integrated
Waveguide for K-band Applications 585
Anamika Banwari, Shailza Gotra, Zain Hashim and Sanjeev Saxena

An Approach to Vendor Selection on Usability Basis by AHP
and Fuzzy Topsis Method 595
Kirti Sharawat and Sanjay Kumar Dubey

Analysis and Comparative Exploration of Elastic Search,
MongoDB and Hadoop Big Data Processing. 605
Praveen Kumar, Parveen Kumar, Nabeel Zaidi and Vijay Singh Rathore

Methods to Choose the 'Best-Fit' Patch in Patch-Based Texture
Synthesis Algorithm 617
Arti Tiwari, Kamanasish Bhattacharjee, Sushil Kumar and Millie Pant

Block Matching Algorithm Based on Hybridization of Harmony
Search and Differential Evolution for Motion Estimation in Video
Compression 625
Kamanasish Bhattacharjee, Arti Tiwari and Sushil Kumar

A Rigorous Investigation on Big Data Analytics. 637
Kajal Rani and Raj Kumar Sagar

Proposed Algorithm for Identification of Vulnerabilities and
Associated Misuse Cases Using CVSS, CVE Standards During
Security Requirements Elicitation Phase. 651
C. Banerjee, Arpita Banerjee, Ajeet Singh Poonia and S.K. Sharma

Vulnerability Identification and Misuse Case Classification
Framework 659
Ajeet Singh Poonia, C. Banerjee, Arpita Banerjee and S.K. Sharma

Revisiting Requirement Analysis Techniques and Challenges 667
Shreta Sharma and S.K. Pandey

Test Data Generation Using Optimization Algorithm:
An Empirical Evaluation 679
Mukesh Mann, Pradeep Tomar and Om Prakash Sangwan

Telugu Speech Recognition Using Combined MFCC, MODGDF
Feature Extraction Techniques and MLP, TLRN Classifiers. 687
Archek Praveen Kumar, Ratnadeep Roy, Sanyog Rawat,
Ashwani Kumar Yadav, Amit Chaurasia and Raj Kumar Gupta

Speech Recognition with Combined MFCC, MODGDF and ZCPA
Features Extraction Techniques Using NTN and MNTN Conventional
Classifiers for Telugu Language. 697
Archek Praveen Kumar, Ratnadeep Roy, Sanyog Rawat,
Rekha Chaturvedi, Abhay Sharma and Cheruku Sandesh Kumar

Classification Model for Prediction of Heart Disease 707
Ritu Chauhan, Rajesh Jangade and Ruchita Rekapally

Determination and Segmentation of Brain Tumor Using Threshold
Segmentation with Morphological Operations 715
Natthan Singh and Shivani Goyal

A Brief Overview of Firefly Algorithm 727
Bilal and Millie Pant

**Analysis of Indian and Indian Politicians News in the New York
Times** 739
Irshad Ahmad Ansari and Suryakant

**Comparative Analysis of Clustering Techniques for Customer
Behaviour** 753
Shalini and Deepika Singh

**Real-time Sentiment Analysis of Big Data Applications Using
Twitter Data with Hadoop Framework** 765
Divya Sehgal and Ambuj Kumar Agarwal

**Addressing Security Concerns for Infrastructure of Cloud
Computing** 773
Shweta Gaur Sharma and Lakshmi Ahuja

Author Index 781

About the Editors

Dr. Millie Pant is an associate professor in the Department of Paper Technology, Indian Institute of Technology, Roorkee (IIT Roorkee) in India. A well-known figure in the field of swarm intelligence and evolutionary algorithms, she has published several research papers in respected national and international journals.

Dr. Kanad Ray is a professor of Physics at the Department of Physics at the Amity School of Applied Sciences, Amity University Rajasthan (AUR), Jaipur. In an academic career spanning over 19 years, he has published and presented research papers in several national and international journals and conferences in India and abroad. He has authored a book on the Electromagnetic Field Theory. Dr. Ray's current research areas of interest include cognition, communication, electromagnetic field theory, antenna and wave propagation, microwave, computational biology, and applied physics.

Dr. Tarun K. Sharma has a Ph.D. in artificial intelligence as well as MCA and MBA degrees and is currently associated with the Amity University Rajasthan (AUR) in Jaipur. His research interests encompass swarm intelligence, nature-inspired algorithms, and their applications in software engineering, inventory systems, and image processing. He has published more than 60 research papers in international journals and conferences. He has over 13 years of teaching experience and has also been involved in organizing international conferences. He is a certified internal auditor and a member of the Machine Intelligence Research (MIR) Labs, WA, USA and Soft Computing Research Society, India.

Dr. Sanyog Rawat is presently associated with the Department of Electronics and Communication Engineering, SEEC, Manipal University Jaipur, Jaipur, India. He holds a B.E. in Electronics and Communication, an M.Tech. in Microwave Engineering and Ph.D. in Planar Antennas. Dr. Rawat has been involved in organizing various workshops on "LabVIEW" and antenna designs and simulations using FEKO. He has taught various subjects, including electrical science, circuits and system, communication system, microprocessor systems, microwave devices, antenna theory and design, advanced microwave engineering and digital circuits.

Dr. Anirban Bandyopadhyay is a Senior Scientist in the National Institute for Materials Science (NIMS), Tsukuba, Japan. Ph.D. from Indian Association for the Cultivation of Science (IACS), Kolkata 2005, on supramolecular electronics. During 2005–2008, he was selected as Independent Researcher, ICYS Research Fellow in the International Center for Young Scientists (ICYS), NIMS, Japan, he worked on brain-like bio-processor building. In 2007, he started as permanent Scientist in NIMS, working on the cavity resonator model of human brain and brain-like organic jelly. During 2013–2014, he was a visiting professor in Massachusetts Institute of Technology (MIT), USA. He has received many honors such as Hitachi Science and Technology award 2010, Inamori Foundation award 2011–2012, Kurata Foundation Award, Inamori Foundation Fellow (2011–), Sewa Society International member, Japan, etc.

Geometric phase space model of a human brain argues to replace Turing tape with a fractome tape and built a new geometric-musical language which is made to operate that tape. Built prime metric has to replace space–time metric. Designed and built multiple machines and technologies include: (1) angstrom probe for neuron signals, (2) dielectric imaging of neuron firing, (3) single protein and its complex structure's resonant imaging, and (4) fourth circuit element Hinductor. A new frequency fractal model is built to represent biological machines. His group has designed and synthesized several forms of organic brain jelly (programmable matter) that learns, programs, and solves problems by itself for futuristic robots during 2000–2014, also several software simulators that write complex codes by itself.

Designing ANFIS Model to Predict the Reliability of Component-Based System

Rajni Sehgal, Deepti Mehrotra and Manju Bala

Abstract Predicting reliability of any product is always a desire of quality-oriented industry. The fault-free working of products depends on large number of parameters, and designing a machine learning model that can predict the reliability considering these as input parameters will help to plan testing and maintenance of the product. In this paper, ANFIS approach is adopted to train a model that can predict reliability of component-based software system. The parameters considered for designing the model are standard design metrics which are evaluated for quality benchmarking during software development process.

Keywords FIS · ANFIS · Component-based system · Halstead fault

1 Introduction

There is a close dependency between the software reliability and the quality of a software system. Software reliability is defined as "the probability of failure-free operation of a computer program for a specified period occurring in a specified environment." Development of Software requires new technologies to deal with size and complexity that is increasing day by day [1]. The component-based software development approach has been established as viable solution to the above problem. The biggest advantage of component-based software engineering is that it supports the reuse of components. Reliability of a component-based system

R. Sehgal (✉) · D. Mehrotra
Amity School of Engineering and Technology, Amity University,
Noida, Uttar Pradesh, India
e-mail: rsehgal@amity.edu

D. Mehrotra
e-mail: dmehrotra@amity.edu

M. Bala
I.P College for Women, University of Delhi, New Delhi, India
e-mail: manjugpm@gmail.com

© Springer Nature Singapore Pte Ltd. 2018
M. Pant et al. (eds.), *Soft Computing: Theories and Applications*,
Advances in Intelligent Systems and Computing 584,
https://doi.org/10.1007/978-981-10-5699-4_1

increases if the reliable component is reused. These emerged techniques can predict the reliability of component-based applications. The reliability of a system can be predicted either at the system level or component level. System-level reliability prediction is for the application as a whole, and component-based reliability prediction is by the reliability of the individual components and their interconnection mechanisms.

Many researchers have used traditional approaches like software testing to estimate the reliability. During the software test phase, only test data is used to model the software's interactions with the outside world neglecting the structure of software constructed from components as well as the reliability of individual components and is therefore unsuitable for modeling CBSS applications. In recent literature, soft computing techniques deal with uncertainty making it ideal for predicting CBSS reliability. Neural networks and fuzzy logic are the two primary soft computing techniques. This paper outlines an adaptive neuro-fuzzy inference system (ANFIS) model for the prediction of the reliability of CBSS. However, this is a relatively slow and prolonged process. The ANFIS model uses the combination of fuzzy logic and learning capabilities of neural networks to solve the problem making it more advantageous than an FIS model. The ANFIS yields an improvement over an FIS. This paper is divided into five sections. Existing literature is reviewed in Sect. 2. Section 3 describes the framework of proposed ANFIS approach, parameters taken in this study number of faults, cyclomatic complexity, coupling, and cohesion. In Sect. 4, experiment conducted is discussed, followed by result and conclusion in Sect. 5.

2 Related Work

Predicting the reliability of any system is a critical task. Researchers consider different approaches for predicting the reliability. In this study, the focus is on prediction of reliability based on design metric and use of soft computing techniques for the prediction. Many authors conclude that design metrics can be suitably used for predicting the reliability of software [1]. Sehgal and Mehrotra [2] predict the faults before testing phase using the Halstead metric to improve the reliability of software. Tripathi [3] purposed a model for early reliability prediction based on reliability block diagram and found that the coupling is a parameter which affects the reliability of the system. Shatnawi and Li [4] found that software metrics such as CBO, RFC, WMC, DIT, and NOC metrics are efficient to find out the classes predisposed to error. Shin and Willams [5] perform the statistical analysis on different complexity metrics to find the impact of software complexity on security. Graylin [6] presents the high correlation between lines of code and cyclomatic complexity. Subramanyam and Krishan [7] utilize CK metrics suit to find out the fault in an early phase of software development in an object-oriented system. Lee et al. [8] identify that coupling and cohesion as two parameters for component identification for reusability. In a good component, the coupling should be weak, and cohesion should be high. Briand et al. [9] state that cohesion is a parameter to measure the quality of software system and is a degree which tells how firmly two elements belong to each other. Kumar et al. [10] suggest a new measure for the cohesion of class and called it the conceptual cohesion of classes (CCC). It is used to evaluate the strength of class relation to each other conceptually and capturing conceptual aspects of cohesion of classes. This new method measures the quality of the system. System having high cohesion implies high quality. Binkley and Schach [11] state that coupling is suitable quality measure. Yadav and Khan [12] state that higher cohesion decreases the complexity of the software making the system less fault-prone and more reliable. Chowdhury [13] considers complexity, coupling, and cohesion metrics to locate the code vulnerabilities.

Many researchers applied soft computing techniques to various domains such as hydraulic engineering, electrical engineering, flood forecasting [14–17]. ANFIS has gained importance in software engineering field. Nagpal et al. [18] identify relevant parameters of an educational Web site for evaluating the usability of the site using ANFIS. Kaur et al. [19] predict the software maintenance efforts and compare various soft computing techniques. Ardil and Sandhu [20] considered NASA data set and applied ANFIS to discover the severity of faults. Reliability is an important aspect of any software system. Tyagi and Sharma [21] used ANFIS to estimate the reliability by taking four parameters namely component dependency, operational profile, application complexity, and reusability of the component.

3 Framework for Proposed Model

In this paper, an automated reliability prediction model for component-based software system (CBSS) is proposed using ANFIS, which was proposed in 1992 by Jang [22]. Optimization of parameters of a given FIS can be done by ANFIS by mapping the relation between input and output data through a learning algorithm. In an ANFIS network, nodes and directional links and learning rules are associated. In this study, four important parameters of reliability of component-based software are considered, namely number of faults, cyclomatic complexity, cohesion, and coupling (Fig. 1).

3.1 Number of Faults

A fault is an incorrect step, human mistake in typing, in correct syntax, which causes a software program to work in an intended manner. When a fault executes, it leads to decrease in reliability of the software system. So reliability is inversely proportional to the number of faults. Halstead metric [23] is used to predict the fault before the system undergoes testing phase by evaluating number of operator and operand in a software program.

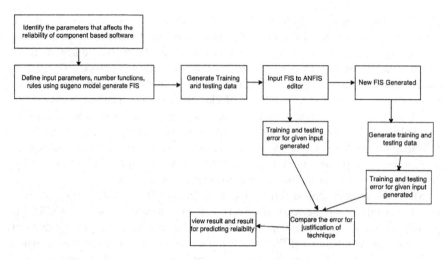

Fig. 1 Framework for proposed model

3.2 Cyclomatic Complexity

Mccabe cyclomatic complexity was developed by Thomas J. McCabe in 1976 to measure the complexity of a program by measuring the number of its decision point. It is a technique which works on the software code by measuring the number of linearly independent path in control graph. Control graph is made up of nodes and edges, whereas nodes represent the group of statements and edges represent the flow of that statement. Complexity of the program is closely related to the reliability of software system. A program number with high decision point will have higher complexity and will have higher probability of occurrence of error in the system.

3.3 Cohesion

Cohesion is a metric which is directly related to the reliability, i.e., higher the cohesion, higher the value of reliability. It measures the strength of the function within a module. Modules with high cohesion are not complex, and they are easy to maintain and can be reusable.

3.4 Coupling

Reliability of a software system is strongly dependent on coupling. Coupling is a metric which measures the interdependency of one module on to another module. If two modules are strongly coupled, then error in one module is propagated to another model thereby decreasing the reliability of software system. Modules which are strongly coupled are more complex, tough to maintain, and difficult to reuse.

4   Experimental Setup

In this paper, ANFIS approach is proposed for the prediction of reliability which depends on the four factors, namely number of faults, cyclomatic complexity, cohesion, and coupling. Intermediate value between conventional evaluations like true/false, yes/no, high/low can be defined with fuzzy logic which is a multi-valued logic [24]. Knowledge base is formed with the number of fuzzy rules (IF-THEN) in a fuzzy reasoning system. Decisions are made by decision-making unit based on this rule. An ANFIS approach combines the benefits of artificial neural network (ANN) and fuzzy logic. The model proposed in this paper is additive in nature.

To create the ANFIS model, fuzzy toolbox of MATLAB is considered. Following are the steps for implementing the ANFIS approach for predicting the reliability:

(i)   Crisp values of parameters are provided as input.
(ii) For each parameter, membership functions (low, medium, high) are determined.
(iii) Rules are fired based on the input parameter and membership function.
(iv) Training and testing data for ANFIS editor are evaluated based on fuzzy rules.
(v)   Error is evaluated using training and testing data in ANFIS editor.
(vi) To evaluate the network error, a new, trained FIS is created, and same procedure is applied again.

In this study, each input variable has three fuzzy linguistic set (low, medium, high). Fuzzification processes are done by using the triangular membership function (TMF). Values of the input parameter number of faults (NF), cyclomatic complexity (CC), coupling (CP), cohesion (CH) are normalized between [0, 3] for the fuzzification. Defuzzified values give the crisp values between the range [0, 1]. The ANFIS system provides 81 rules and is given in Fig. 2.

Rule set based on above-mentioned parameter for first-order Sugeno fuzzy model is as follows:

Number of faults = NF
Cyclomatic complexity = CC
Cohesion = CH
Coupling = CP
Rule 1
If NF is A_i and CC is B_i and CH is C_i and CP is $D_{i,}$ then

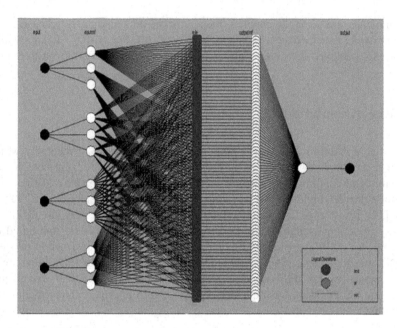

Fig. 2 Proposed architecture

$$F = p_i\text{NF} + q_i\text{CC} + r_i\text{CH} + n_j\text{CP} + m_i \tag{1}$$

Fuzzy sets are represented by A_i, B_i, C_i, and D_i, respectively, and during the training process, parameters determined are p_i, q_i, r_i, n_i, and m_i.

5   Results and Conclusion

Data generated is used to train and test the ANFIS. A total of 80% of the data is used for training and rest 20% for testing the network. A total of 81 rules were formed on the basis of the available data. The rules divide the input factors into three variables: low, medium, and high. The training data set was loaded to evaluate the proposed model. ANFIS was trained using the data set. The model is validated using the testing data set, and testing error is plotted as shown in Figs. 3 and 4.

Reliability values obtained by FIS and ANFIS for different input sets are compared after creating the ANFIS model. There is a significant difference in output obtained by FIS and ANFIS which is calculated by root mean square value (RMSE). Error obtained in results using FIS is 10.4%, while it is 1.4% using ANFIS, which shows that the ANFIS gives the better results than FIS as shown in Figs. 3 and 4. FIS is trained using the ANFIS on the basis of training data rules that are formed to produce the output of the trained model.

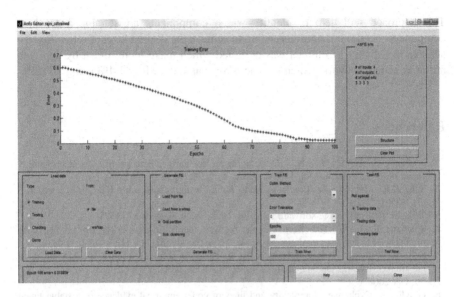

Fig. 3 Mapping original Sugeno FIS to ANFIS training error (1.2%)

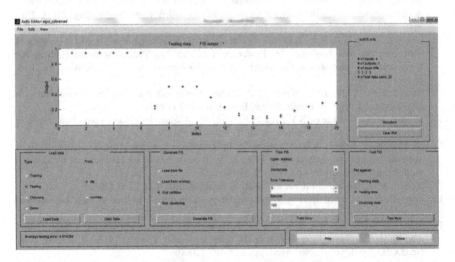

Fig. 4 Mapping original Sugeno FIS to ANFIS testing error (1.4%)

The reliability of component-based system depends on the constitute components and their interaction with each other. In this paper, the only reliability of
individual component is predicted before testing phase. Halstead metric for fault
prediction, McCabe complexity metrics for predicting complexity of design, coupling and cohesion metrics are used as input for ANFIS model is designed. These
metrics can be measured during design and coding stage of software development.

This approach will help to identify which components are reliable and which components need more rigorous testing to achieve the desired level of reliability. To achieve this, firstly FIS model was designed which make more robust by applying artificial neural network principle. The error was reduced by 1.4% using ANFIS.

References

1. Mohanta, S., Vinod, G., Mall, R.: A technique for early prediction of software reliability based on design metrics. Int. J. Syst. Assur. Eng. Manag. 2(4), 261–281 (2011)
2. Sehgal, R., Mehrotra, D.: Predicting faults before testing phase using Halstead' s metrics. 9 (7), 135–142 (2015)
3. Tripathi, R.: Early stage software reliability and design assessment (2005)
4. Shatnawi, R., Li, W.: The effectiveness of software metrics in identifying error-prone classes in post-release software evolution process. J. Syst. Softw. 81(11), 1868–1882 (2008)
5. Shin, Y., Williams, L.: Is complexity really the enemy of software security. In: Proceedings of the 4th ACM Workshop on Quality of protection, Alexandria, Virginia, USA, vol. 27, pp. 47–50, Oct 2008
6. Graylin, J.: Cyclomatic complexity and lines of code: empirical evidence of a stable linear relationship. J. Softw. Eng. Appl. 2(3), 137–143 (2009)
7. Subramanyam, R., Krishnan, M.S.: Empirical analysis of CK metrics for object-oriented design complexity: implications for software defects. IEEE Trans. Softw. Eng. 29(4), 297–310 (2003)
8. Lee, J.K., Jung, S.J., Kim, S.D., Jang, W.H., Ham, D.H.: Component identification method with coupling and cohesion. In: Proceedings of the Asia-Pacific Software Engineering Conference in International Computer Science Conference on APSEC ICSC, Feb 2016, pp. 79–86 (2001)
9. Briand, L.C., Daly, J.W., Wust, J.: A unified framework for cohesion measurement in object-oriented systems. Empir. Softw. Eng. 3, 65–117 (1998)
10. Kumar, M.S., Achutarao, S.V., Ali, S., Shaik, A.: A class level fault prediction in object oriented systems: cohesion approach. 2(2), 918–922 (2012)
11. Binkley, A.B., Schach, S.R.: Validation of the coupling dependency metric as a predictor of run-time failures and maintenance measures. In: Proceedings of the International Conference on Software Engineering, pp. 452–455 (1998)
12. Yadav, A., Khan, R.A.: Impact of cohesion on reliability. 3(1), 7762 (2012)
13. Chowdhury, I.: Can complexity, coupling, and cohesion metrics be used as early indicators of vulnerabilities? 1963–1969 (2010)
14. Schurter, K.C., Roschke, P.N.: Fuzzy modeling of a magnetorheological damper using ANFIS. In: The Ninth IEEE International Conference Fuzzy Systems. FUZZ-IEEE 2000 (Cat. No.00CH37063), vol. 1, No. April, pp. 122–127 (2000)
15. Samandar, A.: A model of adaptive neural-based fuzzy inference system (ANFIS) for prediction of friction coefficient in open channel flow. Sci Res Essay 6(5), 1020–1027 (2011)
16. Geethanjali, M., Raja, S.M.: A combined adaptive network and fuzzy inference system (ANFIS) approach for over current relay system. Neurocomputing 71(4–6), 895–903 (2008)
17. Mukerji, A., Chatterjee, C., Raghuwanshi, N.S.: Flood forecasting using ANN, neuro-fuzzy, and neuro-GA models. June (2009)
18. Nagpal, R., Mehrotra, D., Sharma, A., Bhatia, P.: ANFIS method for usability assessment of website of an educational institute. World Appl. Sci. J. 23(11), 1489–1498 (2013)
19. Kaur, D.A., Kaur, K., Malhotra, D.R.: Soft computing approaches for prediction of software maintenance effort. Int. J. Comput. Appl. 1(16), 80–86 (2010)

20. Ardil, E., Sandhu, P.S.: A soft computing approach for modeling of severity of faults in software systems. Int. J. Phys. Sci. **5**(2), 74–85 (2010)
21. Tyagi, K., Sharma, A.: An adaptive neuro fuzzy model for estimating the reliability of component-based software systems. Appl. Comput. Informatics **10**(1–2), 38–51 (2014)
22. Jang, J.S.R.: ANFIS: adaptive-network-based fuzzy inference system. IEEE Trans. Syst. Man Cybern. **23**(3), 665–685 (1993)
23. Halstead, M.H.: Elements of software science, vol. 7, p. 127. Elsevier, New York (1977)
24. Kamel, T., Hassan, M.: Adaptive neuro fuzzy inference system (ANFIS) for fault classification in the transmission lines. Online J. Electron. Electr. Eng. **2**(2), 164–169 (2009)

Bone Fracture Detection Using Edge Detection Technique

Nancy Johari and Natthan Singh

Abstract Diagnosis through computer-based techniques is nowadays is tremendously growing. Highly efficient system that incorporates modern techniques and fewer resources is required to speed up the diagnosis process and also to increase the level of accuracy. Fracture in a bone occurs when the external force exercised upon the bone is more than what the bone can tolerate. A disassociation between two cartilages is also referred as a fracture. The purpose of this paper is to find out the accuracy of an X-ray bone fracture detection using Canny Edge Detection method. Edge detection through Canny's algorithm is proven to be an ideal edge identification approach in determining the end of line with impulsive threshold and less error rate.

Keywords Canny edge detection · Gaussian filter · Pattern recognition
Sobel operator · Threshold value

1 Introduction

Medical image processing is the integration of various fields that include computer science, data science, biological science and medical science. Medical image computation is an effectual and also an economic fashion that aids in the generation of visual representations of the internal part of the body for clinical inspection and medical diagnosis. To extricate the information which is clinically important from the medical image is the principal objective of medical image processing. The field of medical imaging has been witnessing advances not only in acquisition of medical images but also in its techniques and expertise of interpretation. The most common

N. Johari (✉)
Information Technology, Banasthali University, Jaipur, India
e-mail: nancysrms@gmail.com

N. Singh
IIT Roorkee, Roorkee, India
e-mail: ns_pipil@rediffmail.com

© Springer Nature Singapore Pte Ltd. 2018
M. Pant et al. (eds.), *Soft Computing: Theories and Applications*,
Advances in Intelligent Systems and Computing 584,
https://doi.org/10.1007/978-981-10-5699-4_2

ailment of human bone is fracture. Bone fractures are nothing but the ruptures which occur due to accidents. There are many types of bone fractures such as normal, transverse, comminute, oblique, spiral, segmented, avulsed, impacted, torus and greenstick. Computer-aided diagnosis is exceedingly active field of research in which computer systems are developed to provide a quick and accurate diagnosis. A bone fracture is the outcome of intense force exertion, thrust or a slight trauma or a shock. A damaged bone is referred as a fracture. It can range from a slight crack to a completely collapsed bone, referred as hairline fracture. Generally, a bone gets fractured, if the force thrust upon it is more than what it can tolerate. The force frail the bone and breaks it. The asperity of the fracture is determined by the strength of the force. Falls, direct strikes on the body surface, traumatic events, etc., can cause bone fractures. Computerized diagnosis of fracture has helped the physician to diagnose the fracture with higher accuracy and in less time. The main objective of this work is to use the advancement of computer processing techniques to automatically diagnose the fracture. This paper focuses on the utilization of canny edge detection algorithm that helps the radiologist in automated diagnosis of image.

2 Motivation

The main reasons behind bringing up this system are:

To reduce the human faults as it is very obvious that diagnosis done by human experts can be ambiguous and can drop down beyond admissible levels. Computerized processing can be proved helpful in such case as the efficiency of software does not degrade due to overuse. Computerized system also helps to cut down the time and effort that is required to train the practitioners, and this system can be incorporated with the software of the X-ray imaging device that allows the physicians to make quick and effective diagnosis [1].

Additional inspiration behind building this system is to comfort the physicians, patients and researchers to look forward various cases for exploration.

3 Existing Methodology

Nowadays, X-ray image is widely used for fracture diagnosis and bone treatment as it provides accurate results and quick treatment. Several researches have been previously done for the same purpose. Numerous softwares had been implemented to diagnose fracture in human body. In this paper, fracture diagnosis is done using canny edge detection method. Software using canny edge detection has been implemented, and it also provide quite accurate results, but the work can be modified and more refined results can be obtained. This increases the efficiency of software. In the previous work, the edge detection operator, Sobel operator, is used for the edge identification which does not uses any sigma parameter, so the image

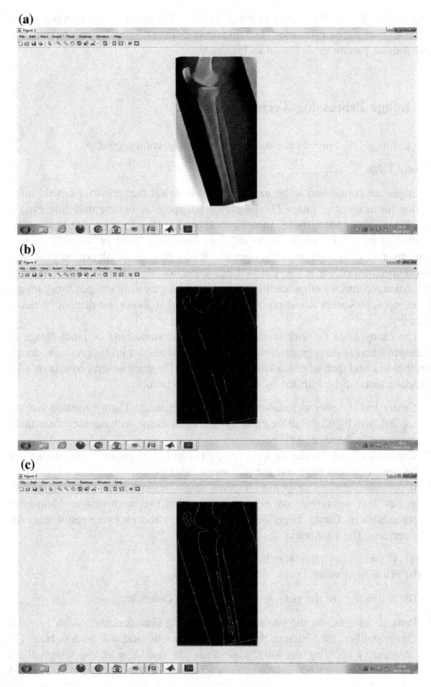

Fig. 1 **a** Original image, **b** edge detection using Sobel operator without sigma value, **c** edge detection using Sobel operator with sigma value of 4.75

obtained by this can be more enhanced. This work uses the Sobel operator on the value 4.75 as sigma parameter. The difference between using Sobel operator with and without parameter is shown in Fig. 1a–c.

4 Image Processing Technique

The technique developed here uses canny edge detection algorithm.

Canny Edge

Edges are considered to be most significant aspect that provides crucial information for analysis of image [2]. Edges are basically the outline that differentiates the object with its background. Identification of edge is quite a complex process that is affected by deterioration due to different fluctuating level of noise [2]. Canny edge detector is an edge detection operator. It uses multi-stage algorithm to identify broad spectrum of edges [3]. Canny Edge Detection algorithm is proven to be an optimal algorithm for edge identification. It takes a greyscale image (X-ray image) as an input, processes it and produces the output that shows the discontinuities of intensity.

The canny edge detector works in five phases: smoothing of input image by Gaussian filter, finding gradients of the image, non-maximum suppression, double thresholding and then edge tracking by hysteresis. There are several criteria on edge detecting that can be fulfilled by canny edge detection:

1. Canny has a better detection (for detection criteria). Canny method has the capability to highlight all the existing edges that match with the user-determined parameter's threshold [4].
2. Canny has better localizing way. Canny is capable of producing minimum gap between detected edge and the real image edge [4].
3. It provides obvious response. It gives single response for every edge. This makes less confusion on edge detection for the next image. Identifying parameters on Canny Edge Detection will give effect on every result and edge detection. The parameters are [4]:

   (a) Gaussian deviation standard value.
   (b) Threshold value.

The following are the steps to do Canny Edge Detection:

1. Remove all noise on the image by implementing Gaussian filter. After applying Gaussian filter, the resulting image that will be obtained will be less blur. The intention of applying the filter is to obtain the real edge of the image. If the Gaussian filter is not applied, sometimes noise itself will be detected as an edge [4].

Fig. 2 Application of Sobel
operator for edge detection in
horizontal and vertical
directions (*Gx*) and (*Gy*)

-1	0	+1
-2	0	+2
-1	0	+1

+1	+2	+1
0	0	0
-1	-1	-1

2. Detect the edge with Sobel operator on the value "4.75", and this will give the clear view of the edges and distortions. Following is an example of an edge detection using Sobel operator (Fig. 2):

   The results from both the operators are combined to fetch the combined result by the equation given below [4]:

$$[G] = [Gx] + [Gy]$$

3. Determine the direction by using the formula:

$$G = \sqrt{G2x + G2y}$$

   Detection uses two thresholds (maximum threshold value and minimum threshold value). If the pixel gradient is higher than maximum threshold, the pixel will be denied as background image. If the pixel gradient between maximum threshold and minimum threshold, the pixel will be accepted as an edge if it is connected with other edge pixel that is higher than the maximum threshold [4].

4. Minimize the emerging edge line by applying non-maximum suppression. The process gives slimmer edge line [4].
5. The last step is to fetch the binary value of the image pixels by applying two threshold values.

5 System Overview

System overview of this project is discussed through flow chart. The flow chart explains the process flow from detecting X-ray image until producing the bone fracture detection on the X-ray image (Fig. 3).

The realization of the system is demonstrated below:

1. Firstly user fed the X-ray as an input into the system. The input gets processed.
2. The previous step is then carried further by filtering the input image by using filter. Here, in this system, Gaussian filter is used.

Fig. 3 Flow chart of the
system

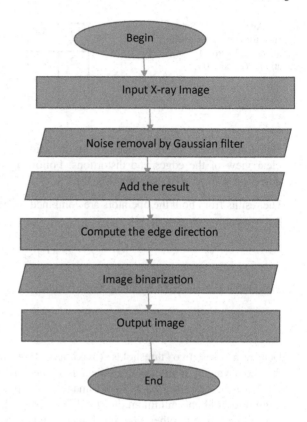

3. The next step is performed after the filtered image is obtained. The image will be
   then processed using Canny Edge Detection method. As a result, it will produce
   more visible lines on the X-ray image.
4. The system then couples the result of previous step with the original image.
   After this step, the edges of the bone get clearly visible and then further this
   image is processed by the system.
5. To detect the location of fracture in the image, shape detection with multiple
   parameters is used by the system. A broken bone is expressed when the line has
   an end and does not have any connection with other line [4].
6. The various parameters on which the system specifies the location of broken
   bone are as follows:

   (a) The red colour indicates the position where the line ends after it gets pro-
       cessed by canny edge detection. The line obtained is a single line.
   (b) The blue colour indicates the position of the line that is next to one another;
       such types of lines generally indicate the hairline fracture.
   (c) The green colour depicts the location of the end of the line with multiple
       ends.

6 Results

This system processes the input X-ray with canny edge detection method. In Fig. 4a, the user inputs the image which is an X-ray image of the bone. Then, the system will undergo canny edge detection that includes Gaussian filter and edge detection Sobel operator followed by non-maximum suppression, and the output obtained through this step is shown in Fig. 4b. In Fig. 4c, the system demonstrates the output by combining the output of the result obtained in Fig. 4b and by inverting the original image that is uploaded by the user. After this, the system detects the location where it highlights the end of line, which is shown in Fig. 4d. In Fig. 4e, the system identifies the fractured bone and location of fracture (Fig. 5).

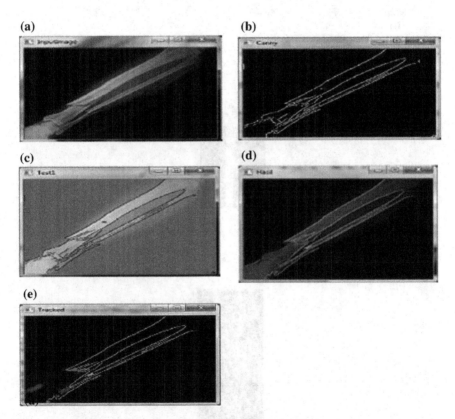

Fig. 4 **a** Input image, **b** output image through canny edge detection, **c** inverted output image by canny edge detection, **d** output image with canny edge detection at every edge, **e** output image with fracture detection

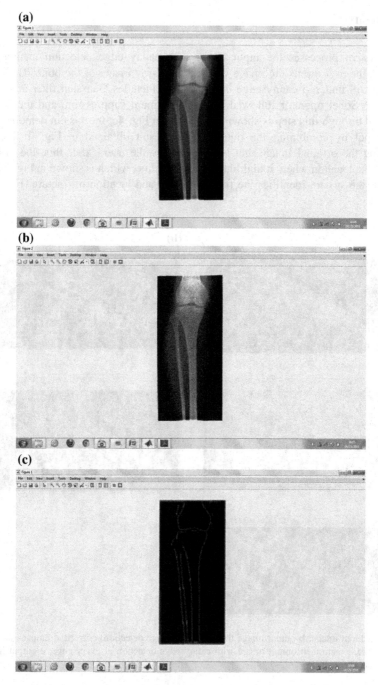

Fig. 5 **a** Original image, **b** image after undergoing Gaussian filter, **c** fracture detection using Sobel operator

7 Conclusion

In this paper, fracture identification is done using Canny Edge Detection operator. This framework will help the physicians to obtain more accurate results with less effort and also in less time. The system has been tested upon real data. Using Sobel operator with the parameter sigma 4.75 helps to enhance the efficiency of the system and also helps to diagnose the hairline fracture more effectively. Because hairline fracture is basically the fracture that has multiple fractures combined together, in this fracture the complete distortion of bone occurs. So, at this value, edges can be diagnosed in such a way that all the distortions and joints are clearly visible that help to increase the success rate of the system. Along with that, soft computing techniques [5–9] and formal methods [10] will also be investigated to improve the clustering performance.

References

1. Dhanabal, R.: Digital image processing using sobel edge detection algorithm in FPGA. J. Theor. Appl. Inf. **58**(1) (2013)
2. Mahajan, S.R., et al.: Review of an enhanced fracture detection algorithm design using X-ray image processing. IJIRSET J. **1**(2) (2012)
3. Fazal-E-Malik: Mean and standard deviation features of color histogram using laplacian filter for content-based image retrieval. J. Theor. Appl. Inf. Technol. **34**(1) (2011)
4. Kurniawan, S.F.: Bone fracture detection using openCV. J. Theor. Appl. Inf. Technol. **64** (2015)
5. Gonzalez, R.C., Woods, R.E.: Digital Image Processing, 3rd edn. (2009)
6. Ansari, I.A., Pant, M., Ahn, C.W.: Robust and false positive free watermarking in IWT domain using SVD and ABC. Eng. Appl. Artif. Intell. **49**, 114–125 (2016)
7. Jauhar, S.K., Pant, M., Deep, A.: An approach to solve multi-criteria supplier selection while considering environmental aspects using differential evolution. In: International Conference on Swarm, Evolutionary, and Memetic Computing, pp. 199–208. Springer International Publishing (2013)
8. Ahmad, A.I., Pant, M.: SVD watermarking: particle swarm optimization of scaling factors to increase the quality of watermark. In: Proceedings of Fourth International Conference on Soft Computing for Problem Solving. NIT Silchar, Assam, India, Springer India (2015)
9. Zaheer, H., Pant, M., Monakhov, O., Monakhova, E.: A portfolio analysis of ten national banks through differential evolution. In: Proceedings of Fifth International Conference on Soft Computing for Problem Solving, pp. 851–862. Springer, Singapore (2016)
10. Jauha, S. K., Pant, M.: Recent trends in supply chain management: a soft computing approach. In: Proceedings of Seventh International Conference on Bio-Inspired Computing: Theories and Applications (BIC-TA 2012), pp. 465–478. Springer, India (2013)

Virtual Experimental Analysis of Redundant Robot Manipulators Using Neural Networks

H.P. Singh, Surendra Kumar, Pravesh Kumar and Akanshu Mahajan

Abstract This study presents a theoretical–experimental scheme to control a redundant robot manipulator in the presence of unmodeled dynamics and discontinuous friction. The proposed control scheme does not require a priori knowledge of upper bounds, robot's parameters, and external disturbance. The advantage of a feed-forward neural network (FFNN) controller is its robustness and ability to handle the model uncertainties. The virtual experimental results are carried out for a three-link planar redundant manipulator to show the effectiveness of the controller.

Keywords Feed-forward neural network · Redundant robot manipulator Uncertainties · Self-motion criteria

1 Introduction

The important role of redundant robot manipulators, in space, undersea, etc., has promoted the research on these manipulators. Many schemes have been proposed to design the controller for manipulators with dynamic uncertainties [1]. From the literature survey, dynamical controller for redundant robots is very limited [2–4]. Li et al. [5] deigned a robust control scheme in Cartesian space with redundancy

H.P. Singh (✉)
Cluster Innovation Centre, University of Delhi, New Delhi, India
e-mail: harendramaths@gmail.com

S. Kumar
Department of Mathematics, University of Delhi, New Delhi, India
e-mail: surendraiitr8@gmail.com

P. Kumar
Jaypee Institute of Information Technology, Noida, India
e-mail: praveshtomariitr@gmail.com

A. Mahajan
SVC, University of Delhi, New Delhi, India
e-mail: akimahajan01@gmail.com

© Springer Nature Singapore Pte Ltd. 2018
M. Pant et al. (eds.), *Soft Computing: Theories and Applications*,
Advances in Intelligent Systems and Computing 584,
https://doi.org/10.1007/978-981-10-5699-4_3

21

utilization measures. The drawback of this scheme is that the knowledge of the bounds of parameter vibrations and unmodeled dynamics is required, and there is no friction and external disturbance in robot dynamics. Zergeroglu et al. [6] and later on Ozbay et al. [7] have designed a robust controller to perform multiple subtasks. These control schemes are based on linear-parameterization property. An adaptive control scheme in operational space for redundant manipulators is proposed in [8]. Maaroof et al. [9] designed a subtask controller for redundant robot manipulators using self-motion criteria. Soto and Campa [10] proposed a two-loop control scheme for redundant robots. Madania et al. [11] designed a control scheme for redundant manipulators constrained by moving obstacles. A terminal sliding manifold controllers for n-degree-of-freedom (DOF) rigid robotic manipulator is proposed in [12]. There are many FFNN-based controllers for manipulators available in the literature [1, 13, 14]. Chien et al. [15] constructed a NN control algorithm for a nonlinear system such as ball and beam control system. Kumar et al. [16] proposed a nonlinear tracking controller without any disturbance term in the model. Singh and Sukavanam [17] developed a trajectory tracking controller for redundant manipulators with continuous friction. A novel control scheme for online path tracking and obstacle avoidance is proposed by Jasour and Farrokhi [18]. Shoushtari et al. [19] designed an innovative control algorithm for redundant robots. A three-degree-of-freedom model is presented to handle kinematic redundancy. This study presents a theoretical–experimental scheme to control a redundant manipulator with unmodeled dynamics and discontinuous fraction. The proposed control scheme does not require a priori knowledge of upper bounds, robot's parameters, and external disturbance. The advantage of a FFNN controller is its robustness and ability to handle the model uncertainties. The outlay of this study is as follows. Section 2 provides the kinematics and dynamics model of a robot manipulator. The error system formulation and controller are given in Sect. 3. Section 4 illustrates and discusses the experimental results for performance of the proposed controller. Section 5 provides the final conclusions.

2   Kinematics and Dynamics Model

The manipulator end-effector position and orientation in the Cartesian space denoted by $x = [x_1, x_2, \ldots, x_m]^T$, is defined as

$$x = f(q) \tag{1}$$

where $f(q) \in R^m$ denotes the direct kinematics and $q = [q_1, q_2, \ldots, q_n]^T$ represents the $n \times 1$ $(m < n)$ position vector of an n-link robot manipulator. Differentiating (1) with respect to time yields

$$\dot{x} = J(q)\dot{q} \tag{2}$$

where $J(q) = \partial f(q)/\partial q \in R^{m \times n}$ denotes the manipulator Jacobian. The general solution of Eq. (2) is given as

$$\dot{q} = J^+(q)\dot{x} + k(I_n - J^+ J) \sum_{i=1}^{s} w_i \nabla h_i(q) \tag{3}$$

where $h_i(q)$ is the ith performance criteria, $s \in N$ is the maximum number of self-motion (subtask), and $w_i's$ are positive weights of the corresponding criteria subjected to the following constraint.

$$\sum_{i=1}^{s} w_i = C \tag{4}$$

where C is a real-valued constant and is used in conjunction with the self-motion control parameter k [6]. The pseudoinverse $J^+ = J^T(JJ^T)^{-1}$ of J satisfies the following conditions.

$$JJ^T J = J \quad J^+ JJ^+ = J^+ \quad (J^+ J)^T = J^+ J \quad (JJ^+)^T = JJ^+ \tag{5}$$

$$(I_n - J^+ J)(I_n - J^+ J) = (I_n - J^+ J) \quad J(I_n - J^+ J) = 0$$
$$(I_n - J^+ J)^T = (I_n - J^+ J) \quad (I_n - J^+ J)J^+ = 0 \tag{6}$$

The dynamics model for an n-link robot manipulator in joint space has the following form [1]

$$M(q)\ddot{q} + N(q, \dot{q}) + \tau_d = \tau \tag{7}$$

where $N(q, \dot{q}) = V_m(q, \dot{q})\dot{q} + G(q) + F(\dot{q}).M(q) \in R^{n \times n}, V_m(q, \dot{q}) \in R^{n \times n}, G(q) \in R^n$, and $F(\dot{q}) \in R^n$ are mass matrix, centripetal-coriolis matrix, gravity effects, and friction effects, respectively. $\tau_d \in R^n$ denotes a bounded external disturbance, and $\tau \in R^n$ represents the control input exerted on joints.

3 Controller Design Using Neural Network

The tracking error in Cartesian space is defined as

$$e = x - x_d \tag{8}$$

where $x_d \in R^m$ the desired trajectory. Let us consider the following controller

$$\tau = \hat{M}[J^+(\ddot{x}_d - \dot{J}\dot{q} - K_v\dot{e} - K_p e + u_0) + k(I_n - J^+J)(\nabla h(q) - \dot{q})] + \hat{N} \quad (9)$$

where $\hat{M}$ and $\hat{N}$ are estimated versions of M and N. K_v and K_p are the positive definite diagonal gain matrices. For convenience, we write

$$v = \ddot{x}_d - \dot{J}\dot{q} - Ky \quad (10)$$

where $K = [K_p K_v]$ and $y = [e\dot{e}]^T$. Substituting (9) into (7), we get

$$\begin{aligned}
\ddot{q} &= M^{-1}\hat{M}J^+v + M^{-1}\hat{M}(k(I_n - J^+J)(\nabla h(q) - \dot{q})) + M^{-1}\hat{M}J^+u_0 + M^{-1}\Delta N - M^{-1}\tau_d \\
&= J^+v + J^+u_0 + (M^{-1}\hat{M} - I)J^+v + (M^{-1}\hat{M} - I)J^+u_0 \\
&\quad + M^{-1}\hat{M}(k(I_n - J^+J)(\nabla h(q) - \dot{q})) + M^{-1}\Delta N - M^{-1}\tau_d \\
&= J^+v + J^+u_0 + EJ^+v + EJ^+u_0 + M^{-1}\hat{M}(k(I_n - J^+J)(\nabla h(q) - \dot{q})) + M^{-1}\Delta N - M^{-1}\tau_d
\end{aligned} \quad (11)$$

where $\Delta N = \hat{N} - N, E = M^{-1}\hat{M} - I$. From (10), by substituting v into the first term of (11) and multiplying both sides by J with the properties $JJ^+ = I$ and $kJ(I_n - J^+J)(\nabla h(q) - \dot{q}) = 0$, we get

$$\begin{aligned}
J\ddot{q} &= \ddot{x}_d - \dot{J}\dot{q} - Ky + u_0 + JEJ^+v + JEJ^+u_0 + JE(k(I_n - J^+J)(\nabla h(q) - \dot{q})) \\
&\quad + JM^{-1}\Delta N - JM^{-1}\tau_d
\end{aligned} \quad (12)$$

Now, using (12), $\dot{e} = \dot{x} - \dot{x}_d, \ddot{x} = \dot{J}(q)\dot{q} + J(q)\ddot{q}, \ddot{e} = \ddot{x} - \ddot{x}_d$, and $y = [e\dot{e}]^T$, we get the error dynamics system

$$\dot{y} = Ay + B(u_0 + \eta) \quad (13)$$

where $A = \begin{bmatrix} 0 & I \\ -K_p & -K_v \end{bmatrix}, B = \begin{bmatrix} 0 \\ I \end{bmatrix}$ and

$$\eta = JEJ^+v + JEJ^+u_0 + JE(k(I_n - J^+J)(\nabla h(q) - \dot{q})) + JM^{-1}\Delta N - JM^{-1}\tau_d \quad (14)$$

Due to the mismatch between M, N, and their estimated version $\hat{M}$, $\hat{N}$, and unmodeled dynamics, η in (14) is an uncertain nonlinear function. Let us assume that the singularities are always avoided and all terms in (14) are assumed to be bounded [20], then there will always exist a continuous function $\varphi(\cdot)$ such that $\|\eta\| < \varphi(t)$. Mathematical model of a one-layer neural network in terms of uncertain bound φ has the following form [21]

$$\varphi = Z^{*T}\phi(s)\Delta\varphi$$

and the estimate value of φ is given as $\varphi = \hat{Z}^T\phi(s)$ where $\Delta\varphi$ denotes NN approximation error satisfying $\|\varphi(t)\| = \|Z^{*T}\phi(s) - \varphi\| < \xi$, $\phi(t)$ denotes a basis vector and $\hat{Z}$ is the estimate value of the optimal matrix Z^*. Let us assume that the norm of η and φ satisfies the condition $\varphi - \|\eta\| > \xi$. The term u_0 represents the compensator used to stabilize the error dynamics system and enhances the system robustness against system uncertainties and external disturbance. The compensator u_0 in neural network is given as

$$u_0 = -\frac{\hat{\rho}\omega}{\hat{\rho}\|\omega\| + \varepsilon} \tag{15}$$

where $\dot{\varepsilon} = -\gamma\varepsilon$, $\varepsilon(0) > 0$ and γ is a positive constant. The adaptive neural network law is designed as $\dot{\hat{Z}} = F\|\omega\|\phi(s)$ where $\omega = B^T Py$, P is a symmetric matrix with positive eigenvalues such that

$$A^T P + PA + Q = 0 \tag{16}$$

with a positive definite symmetric matrix Q defined as

$$Q = \begin{bmatrix} 2K_p^2 & 0 \\ 0 & 2K_v^2 - 2K_p \end{bmatrix} \tag{17}$$

such that $K_v^2 > K_p$, F is a positive definite design parameter.

4 Experimental Results

To confirm the validity of the controller, experiments are performed on the three-link planar redundant manipulator. The model terms are given as follows

$$M(q) = \begin{bmatrix} \beta_1 + 2p_1c_2 + p_2c_{23} + p_3c_3 & \beta_2 + p_1c_2 + p_2c_{23} & \beta_3 \\ \beta_2 + p_1c_2 + p_2c_{23} & \beta_2 + 2p_3c_3 & \beta_3 + p_3c_3 \\ \beta_3 & \beta_3 + p_3c_3 & \beta_3 \end{bmatrix}$$

$$V_m(q,\dot{q}) = \begin{bmatrix} V_{m11} & V_{m12} & V_{m13} \\ V_{m21} & V_{m22} & V_{m23} \\ V_{m31} & V_{m32} & V_{m33} \end{bmatrix}$$

where

$$V_{m11} = -(p_1 s_2 + p_2 s_{12})\dot{q}_2 - (p_2 s_{12} + p_3 s_{12})\dot{q}_3,$$
$$V_{m12} = -(p_1 s_2 + p_2 s_{12})(\dot{q}_1 + \dot{q}_2) - (p_2 s_{12} + p_3 s_{12})\dot{q}_3,$$
$$V_{m13} = (p_2 s_{12} + p_3 s_{12})(-\dot{q}_1 + \dot{q}_2 + \dot{q}_3), V_{m21} = (p_1 s_2 + p_2 s_{12})\dot{q}_1 + p_3 s_3 \dot{q}_3,$$
$$V_{m22} = -(p_2 s_{12} + p_3 s_{12})\dot{q}_3, V_{m23} = -p_3 s_3(3\dot{q}_1 + \dot{q}_2 + \dot{q}_3),$$
$$V_{m31} = (p_1 s_2 + p_2 s_{12})\dot{q}_1 - p_3 s_3 \dot{q}_2, V_{m32} = p_3 s_3(\dot{q}_1 + \dot{q}_2), V_{m33} = 0,$$

and $\beta_1, \beta_2, \beta_3, p_1, p_2, p_3$ represent the inertia parameters and defined as

$$\beta_1 = 1.1956 \text{ kg m}^2, \quad \beta_2 = 0.3946 \text{ kg m}^2, \quad \beta_3 = 0.0512 \text{ kg m}^2$$
$$p_1 = 0.4752 \text{ kg m}^2, \quad p_2 = 0.1280 \text{ kg m}^2, \quad p_3 = 0.1152 \text{ kg m}^2$$

and c_i denotes $\cos(q_i)$, s_i denotes $\sin(q_i)$, c_{ij} represents $\cos(q_i + q_j)$, and s_{ij} represents $\sin(q_i + q_j)$. The external disturbance is defined as $\tau_d = [4\cos 2(t) \quad \sin(t) + \cos(2t)2\sin(t)]^T$. The masses of the link and corresponding lengths are taken as 3.60, 2.60, 2.00 kg and 0.40, 0.36, 0.30 m, respectively. For controller design, estimated values are given as

$$\hat{M} = 14M,$$

$$\hat{V}_{m11} = -(\hat{p}_1 s_2 + \hat{p}_2 s_{12})\dot{q}_2 - (\hat{p}_2 s_{12} + \hat{p}_3 s_{12})\dot{q}_3,$$
$$\hat{V}_{m12} = -(\hat{p}_1 s_2 + \hat{p}_2 s_{12})(\dot{q}_1 + \dot{q}_2) - (\hat{p}_2 s_{12} + \hat{p}_3 s_{12})\dot{q}_3,$$
$$\hat{V}_{m13} = (\hat{p}_2 s_{12} + \hat{p}_3 s_{12})(-\dot{q}_1 + \dot{q}_2 + \dot{q}_3), \hat{V}_{m21} = (\hat{p}_1 s_2 + \hat{p}_2 s_{12})\dot{q}_1 + \hat{p}_3 s_3 \dot{q}_3,$$
$$\hat{V}_{m22} = -(\hat{p}_2 s_{12} + \hat{p}_3 s_{12})\dot{q}_3, \hat{V}_{m23} = -\hat{p}_3 s_3(3\dot{q}_1 + \dot{q}_2 + \dot{q}_3),$$
$$\hat{V}_{m31} = (\hat{p}_1 s_2 + \hat{p}_2 s_{12})\dot{q}_1 - \hat{p}_3 s_3 \dot{q}_2, \hat{V}_{m32} = \hat{p}_3 s_3(\dot{q}_1 + \dot{q}_2), \hat{V}_{m33} = 0,$$
$$\hat{p}_1 = 0.4 \text{ kg m}^2, \quad \hat{p}_2 = 0.1 \text{ kg m}^2, \quad \hat{p}_3 = 0.1 \text{ kg.m}^2$$

Discontinuous friction is given as follows

$$F(\dot{q}) = \begin{bmatrix} f_{d1} & 0 & 0 \\ 0 & f_{d2} & 0 \\ 0 & 0 & f_{d3} \end{bmatrix} \begin{bmatrix} \dot{q}_1 \\ \dot{q}_2 \\ \dot{q}_3 \end{bmatrix} + \begin{bmatrix} \text{sgn}(\dot{q}_1) \\ \text{sgn}(\dot{q}_2) \\ \text{sgn}(\dot{q}_3) \end{bmatrix}$$

where

$$f_{d1} = 4.3, \quad f_{d2} = 1.4, \quad f_{d3} = 0.4$$
$$\hat{f}_{d1} = 4, \quad \hat{f}_{d2} = 1, \quad \hat{f}_{d3} = 0.2$$

The desired trajectory of the end-effector and subtask are given as

$$x_d = \begin{bmatrix} 0.30 + 0.2\cos{(t)} \\ 0.40 + 0.1\sin{(t)} \end{bmatrix} \tag{18}$$

and

$$h(q) = 0.5\det(JJ^T) - 0.5((q_3 - 0.5q_2) - 0.5(q_2 - q_1))^2, \tag{19}$$

respectively. The self-motion parameters k and C in (4) are selected to be 10 and 1, respectively. For simulation, we select

$$B = \begin{bmatrix} 0 & 0 \\ 0 & 0 \\ 1 & 0 \\ 0 & 1 \end{bmatrix} \quad \text{and} \quad P = \begin{bmatrix} 2K_pK_v & K_p \\ K_p & K_v \end{bmatrix}$$

where $K_p = \text{diag}(10, 10), K_v = \text{diag}(4, 4)$. To confirm the validity of the controller, we have the following two cases.

(a) **Desired end-effector trajectory is taken as a straight line:**

In this case, the end-effector trajectory is taken as $x_d = [\,0.30 + 0.2\cos(t) \quad 0.5\,]^T$. The experimental results are shown in Figs. 1, 2 and 3.

(b) **Desired end-effector trajectory is taken as an elliptical path (18):**

In this case, we choose an elliptical trajectory for the end-effector. The experimental results are shown in Figs. 4, 5, and 6.

Fig. 1 Desired end-effector trajectory

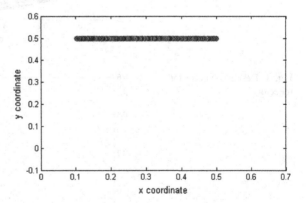

Fig. 2 Performance of robot and end-effector at different time level

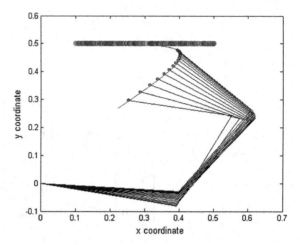

Fig. 3 Performance of robot and end-effector at different time level

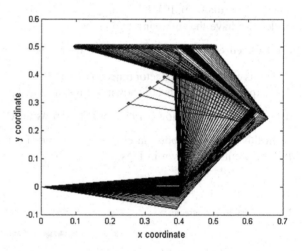

Fig. 4 Desired end-effector trajectory

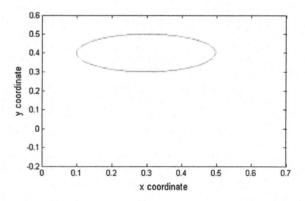

Fig. 5 Performance of robot and end-effector at different time level

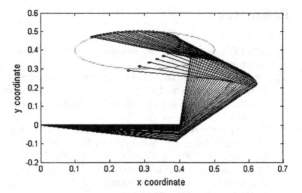

Fig. 6 Performance of robot and end-effector at different time level

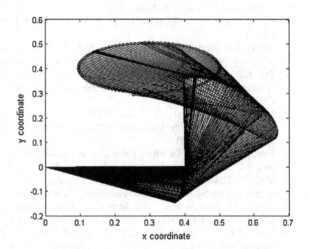

5 Conclusions

A robust adaptive scheme is designed for redundant manipulators with uncertainties such as model uncertainties, external disturbances, and discontinuous friction. The extra degrees of freedom are used to avoid singularity, maintain good manipulability, etc., without affecting the end-effector Cartesian space trajectory. A FFNN is employed to learn the nonlinear uncertainty bound. The results show that the NN-based compensator eliminates the effects of system uncertainties and external disturbance.

Acknowledgements This work is financially supported by research and development grant, University of Delhi, New Delhi, India.

References

1. Lewis, F.L., Jagannathan, S., Yesildirek, A.: Neural Network Control of Robot Manipulators. Taylor and Francis (1999)
2. Khatib, O.: Dynamic control of manipulators in operational space. In: Proceedings of the 6th IFTOMM Congress on Theory of Machines and Mechanisms, pp. 1–10 (1983)
3. Hsu, P., Hauser, J., Sastry, S.: Dynamic control of redundant manipulators. J. Robotic Syst. **6**, 133–148 (1989)
4. Xian, B., de Quieroz, M.S., Dawson, D., Walkar, I.: Task space tracking control of redundant robot manipulators via quaternion feedback. In: Proceedings of the IEEE International conference on control applications, pp. 363–368 (2001)
5. Li, Z., Sinha, N.K., Elbestawi, M.: Simulation of robust controller for redundant manipulators. In: IEEE 34th International Midwest Symposium on Circuits and Systems (MWSCAS), pp. 521–524 (1991)
6. Zergeroglu, E., Dawson, D.M., Walker, I.W., Setlur, P.: Nonlinear tracking control of kinematically redundant robot manipulators. IEEE/ASME Trans. Mechatron. **9**, 129–132 (2004)
7. Ozbay, U., Sahin, H.T., Zergeroglu, E.: Robust tracking control of kinematically redundant robot manipulators subject to multiple self-motion criteria. Robotica **26**, 711–728 (2008)
8. Tee, K.P., Yan, R.: Adaptive operational space control of redundant robot manipulators. In: American Control Conference on O'Farrell Street, San Francisco, CA, USA, pp. 1742–1747 (2011)
9. Maaroof, O.W., Gezgin, E., Dede, M.I.C.: General subtask controller for redundant robot manipulators. In: 12th International Conference on Control, Automation and Systems, pp. 1352–1357 (2012)
10. Soto, I., Campa, R.: Two-loop control of redundant manipulators: analysis and experiments on a 3-DOF planar arm. Int. J. Adv. Rob. Syst. **10**, 1–8 (2013)
11. Madania, T., Daachib, B., Benalleguea, A.: Adaptive variable structure controller of redundant robots with mobile/fixed obstacles avoidance. Robot. Auton. Syst. **61**, 55–564 (2013)
12. Galicki, M.: Robust task space finite-time chattering-free control of robotic manipulators. J. Intell. Rob. Syst. (2016). doi:10.1007/s10846-016-0387-3
13. Lewis, F.L., Yesildirek, A., Liu, K.: Multilayer neural-net robot controller with guaranteed tracking performance. IEEE Trans. Neural Networks **7**, 388–399 (1996)
14. Wai, R.J.: Tracking control based on neural network strategy for robot manipulator. Neurocomputing **51**, 425–445 (2003)
15. Chien, T.L., Chen, C.C., Huang, Y.C., Lin, W.J.: Stability and almost disturbance decoupling analysis of nonlinear system subject to feedback linearization and feedforward neural network controller. IEEE Trans. Neural Networks **19**, 1220–1230 (2008)
16. Kumar, N., Panwar, V., Sukavanam, N., Sharma, S.P., Borm, J.H.: Neural network based nonlinear tracking control of kinematically redundant robot manipulators. Math. Comput. Model. **53**, 1889–1901 (2011)
17. Singh, H.P., Sukavanam, N.: Neural network based control scheme for redundant robot manipulators subject to multiple self-motion criteria. Math. Comput. Model. **55**, 1275–1300 (2012)
18. Jasour, A.M., Farrokhi, M.: Adaptive neuro-predictive control for redundant robot manipulators in presence of static and dynamic obstacles: a Lyapunov-based approach. Adapt. Control Signal Process. **28**, 386–411 (2014)
19. Shoushtari, A.L., Mazzoleni, S., Dario, P.: Bio-inspired kinematical control of redundant robotic manipulators. Assembly Autom. **36**, 200–215 (2016)
20. Waibel, B.J.E., Kazerooni, H.: Theory and experiments on the stability of robot compliance control. IEEE J. Robot. Autom. **7**, 95–104 (1991)
21. Singh, H.P., Sukavanm, N.: Simulation and stability analysis of neural network based control scheme for switched linear systems. ISA Trans. **51**, 105–110 (2012)

Selection of Energy-Efficient Material: An Entropy–TOPSIS Approach

Chiranjib Bhowmik, Sachin Gangwar, Sumit Bhowmik
and Amitava Ray

Abstract Reduction of environmental effect from material uses is a big challenge on the view of material selection as per recent scientific reports. This paper aims to identify the major challenges of material selection for the future use and suggest an appropriate energy-efficient material based on the performance by the decision maker on the various issues. The proposed work presents a multi-criteria decision-making (MCDM) analysis—Technique for order of preference by similarity to ideal solution (TOPSIS) to find the appropriate selection and evaluate the best energy-efficient material on the basis of criteria. In this paper, different energy-efficient materials have been taken into consideration under multiple uncertainties. Firstly, materials are figure out acquiesce to one and the other approximate and perceptible precedent ensue from entropy analysis. Entropy is a convenient technique in critical administration and takes advantage of this crucial executive strategy for criteria weight to deal with material selection in environment-friendly way. Thereafter, the results are used as an input for TOPSIS to designate the array of materials. To our observation, this is the first analysis in the energy-efficient material selection which considers environmental threats. The effect of weighting factors has also been deliberated.

Keywords Entropy · TOPSIS · Energy-efficient material

C. Bhowmik (✉) · S. Bhowmik
Department of Mechanical Engineering, NIT Silchar, Silchar, Assam, India
e-mail: chiranjibbhowmik18@gmail.com

S. Bhowmik
e-mail: bhowmiksumit04@yahoo.co.in

S. Gangwar
Department of Mechanical Engineering, JKLU, Jaipur, Rajasthan, India
e-mail: sachingangwar@jklu.edu.in

A. Ray
Department of Training and Placement Cell, JGEC, Jalpaiguri, West Bengal, India
e-mail: amitavaray.siliguri@gmail.com

© Springer Nature Singapore Pte Ltd. 2018   31
M. Pant et al. (eds.), *Soft Computing: Theories and Applications*,
Advances in Intelligent Systems and Computing 584,
https://doi.org/10.1007/978-981-10-5699-4_4

Nomenclature

A1   Alkaline earth lead glass
A2   Silicon
A3   Cast Magnesium
A4   Wrought Magnesium
A5   Cast Nickel–Iron alloy
A6   Lanthanum commercial purity min 99%
A7   Magnesium commercial purity
A8   Nickel–Iron Chromium alloy HW grade; aged
A9   Cerium commercial purity
M1   Density
M2   Bulk modulus
M3   Compressive strength
M4   Thermal conductivity
M5   Thermal expansion
M6   Resistivity
M7   Cost
M8   Energy production
M9 CO_2 Emission
MCDM Multi-criteria decision making
TOPSIS Technique for order of preference by similarity to ideal solution

1   Introduction

Energy-efficient materials have been exploited in immense scientific and technological applications [1, 5, 12] including radiation windows, florescent lamp
envelopes, television bulbs, color TV necks, electronic component, photovoltaic
cell, aerospace, sports goods, automobiles, lining of furnace, cyanide pots, hearth
plates, and flints for cigarette lighters, electrical heating elements, carbon arc
lighting etc. However, development of new energy-efficient materials has been
imposed by technical and industrial demands [5]. The imminent application of
energy-efficient materials is depend upon the environmental impact of their
mechanical, electrical, and thermal properties. In the recent past, abundance of
energy-efficient materials are fabricated and reported for their promising physical
properties [5, 12]. These materials are usually classified in commercial purity based
on the values of density, bulk modulus, compressive strength, thermal conductivity,
thermal expansion, and resistivity. The selection of energy-efficient materials of
desired properties for technological applications from the large number of materials
pool available in the literature is a difficult task. Material selection for real engineering process is based on several stipulations which include utilitarian precon-
dition, process-ability requirements, cost, energy production, and CO_2 emission.
However, simultaneous consideration of all these criteria in material selection is

also a tedious task. A systematic approach is further required to select the optimum material for eco-friendly application [5]. Assorted accesses have been suggested for excerpt of materials. Ashby [2] has discussed the method for material design and selection by using Pareto-optimal theory which is used in initial screening [3]. Other methods (which are common in material selection) are artificial neural network, and multi-criteria decision-making approaches (AHP, VIKOR, Grey relation analysis, ELECTRE, etc.). These recipes are elucidated in detail for materials selection viewpoint [5, 12].

It would be impressive to compose perceptible analogy of energy-efficient materials for their scholarly utilization attitude. Material selection for industrial application also has been investigated by interval weighting MULTIMOORA technique [19]. Through this paper, an endeavor has been shaped for initial screening and ranking imperative energy-efficient materials using MCDM approach [8, 12, 18]. In the present study TOPSIS method is used to rank the energy-efficient materials for their best utilization [7], considering environmental constructs. A risk-based fuzzy axiomatic design approach has been put forward for gas turbine material selection problem [18]. Relative weights of the material properties (criteria) are approximated using Shannon's entropy method [9, 12, 14, 16]. The paper is standardized as follows. Problem formulation has been depicted in region 2. Methodology is illustrated in piece 3. Discussion and an algorithmic precedent supervised in Sect. 4 to view the practicality and potency of the planned model. The paper halts in segment 5.

2 Problem Formulation

Material selection is the outmost crucial parameter for development society. An unsound excerpt of materials favors enormous cost and extravagant CO_2 emission to the environment. Hence, availability of distinct materials and the selection of peculiar energy-efficient materials become difficult. A superior procedure is more and more needed for the election of energy-efficient materials [12].

In real scenario, energy-efficient materials compared to auxiliary materials are the superlative choice on the grounds of lessened emission to the environmental. The quandary embroil testimony of distinct energy-efficient materials that are passed down in the manufacturing of [12] radiation windows, solar cells, fluorescent lamp envelop and to select the best among them discussed by Kumar and Suman [12]. A survey has been made on energy-efficient materials and their properties and similar properties of others are tabulated. Nine energy-efficient materials important properties ("Density; kg/m^3 (M1), bulk modulus; Gpa (M2), compressive strength; Mpa (M3)—physical properties", "thermal conductivity; W/mk (M4), thermal expansion; μstrain/°C (M5)—thermal properties", "resistivity; μohm.cm (M6)—electrical property", "cost; GBP/kg (M7), energy production; MJ/kg (M8)—economical property" and "CO_2 emission; kg/kg (M9)—environmental property") are considered in this study and shown in Table 1.

Table 1 Energy-efficient materials and their properties

S. No.	Materials	M1	M2	M3	M4	M5	M6	M7	M8	M9
1	A1	3175	34.45	261.5	0.84	9.965	5.05	2.5	25.05	1.35
2	A2	2330	100	3330	160	3	5	5.85	59.9	3.23
3	A3	1815	37	140	87.5	27.3	10.18	3	485	30.5
4	A4	0.066	5.584	38.07	59.51	14.62	9.575	2.565	5.254	30.5
5	A5	0.296	21.03	44.96	7.223	6.945	114	5.558	1.446	8.4
6	A6	0.222	4.2	18.13	7.742	2.778	58	10.26	3.727	21.65
7	A7	0.063	5.07	11.96	88.12	14.46	4.6	2.56	4.171	34.5
8	A8	0.294	21.03	51.85	7.223	6.945	112	5.558	1.24	7.2
9	A9	0.241	2.9	14.48	6.355	4.167	87.5	20.52	6.923	40.2

Source CES Edu pack (2005) [17]

From Table 1, it is very tough enough to select the best energy-efficient material from a given set of alternative materials like Alkaline Earth Lead Glass (A1), Silicon (A2), Cast Magnesium (A3), Wrought Magnesium (A4), Cast Nickel Iron Alloy (A5), Lanthanum Commercial Purity min 99% (A6), Magnesium Commercial Purity (A7), Nickel Iron Chromium Alloy HW grade; aged (A8), Cerium Commercial Purity (A9) etc. Because most of the energy-efficient material are mixtures of rare earth elements which makes them very pricey for their better utilization are excellent in fortitude but encompass rare earth inclusion made them very pricey. Similarly, each and every material is acquiring its own positive and negative properties, respectively. Thence, the outcome judgment has to correlate all the energy-efficient materials in regard to particular aspects and has to choose the outstanding one. So this study allows entropy-TOPSIS based methodology for the selection of best energy-efficient material. The next section briefly describes the methodology for selection purpose.

3 Methodology

The following steps satisfy the methodology of MCDM methods to rank the best energy-efficient material.

3.1 Identifying the Weights of Each Criterion Based on Entropy Method

Among numerous criterion, find the weightage of each criterion i.e. the relative importance of the framework with admiration to other. The weightage importance made by the decision maker plays a crucial role in selecting the fortune material. The dilemma recommended in this research, consists of entropy based equal weightage factors to each criterion and to decide the selection process more precisely [12]. This method is considered as a convenient tool to distinguish the weights of each criterion in minimal time.

Table 2 Equal weightings given to each criterion using entropy method

	M1	M2	M3	M4	M5	M6	M7	M8	M9
Weights	0.594	0.594	0.594	0.594	0.594	0.594	0.594	0.594	0.594

In entropy practice, the criteria including execution assessment contradict with each other have more advanced consequence in the crunch; that is, the criteria is treated with curtailed predilection, if all the alternative materials acquire analogous consequence appraisal to the distinct criteria. The preeminent assessment of entropy comparable to the precise criteria signifies the petite criteria weight and fewer influences of that criterion in decision-making process [12]. Entropy method consists the following steps to justify the weights of each criterion suggested by Lotfi and Fallahnejad [13] Dashore et al. [6]. The weighting components were reckoned by rubric and systematized below as shown in Table 2.

Normalizing the decision matrix to disregard irregularities with distinctive appraisal entity and proportions is the original decision matrix;

$$P_{ij} = \frac{X_{ij}}{\sum_{i=1}^{m} X_{ij}} \tag{1}$$

After normalization, appraise the entropy values as;

$$e_j = -k \sum_{j=1}^{n} p_{ij} \ln p_{ij} \tag{2}$$

where k is a uninterrupted, let $k = (\ln(m))^{-1}$
The intensity of discrepancy can be calculated as;

$$d_j = 1 - e_j \tag{3}$$

The value d_j speak for the genetic contradiction of C_j.

3.2 Implementing TOPSIS Methodologies

After identifying the weights of each criterion by entropy method, a chic of TOPSIS methodology can be adopted. TOPSIS model was first put forward by Hwang and Yoon [8], and their application history also (2012) described by them. According to Chauhan and Vaish [5]; Kumar et al. [11], TOPSIS implies that a decision matrix having 'm' alternatives and 'n' criteria can be pretended to be dilemma of 'n' dimensional hyperplane having 'm' points whose whereabouts is obsessed by the value of their [5] criteria. Akyene [1] believed that TOPSIS find the finest surrogate by play down the distance to the ideal solution and widen the span to the nadir or negative—ideal solution. Generally, TOPSIS method comprises the following steps:

Calculate the normalized decision matrix a_{ij}. The normalized value is calculated as

$$a_{ij} = x_{ij} / \sqrt{\sum_{i=1}^{m} (x_{ij})^2} \quad (1 \le i \le m, 1 \le j \le n) \tag{4}$$

After normalization, calculate the weighted normalized matrix;

$$V = \left(a_{ij}^* w_j\right) \quad (1 \le i \le m, 1 \le n) \tag{5}$$

where w_j is the weight of the ith criterion and $\sum_{i=1}^{n} w_j = 1$.

Calculate the ideal solution V^+ and the negative ideal solution V^- [11];

$$\begin{aligned} V^+ &= \{v_1^*, v_2^*, \ldots, v_n^*\} = \{(\text{Max} v_{ij} | j \in J), (\text{Min} v_{ij} | j \in J)\} \\ V^- &= \{v_1^-, v_2^-, \ldots, v_n^-\} = \{(\text{Min} v_{ij} | j \in J), (\text{Max} v_{ij} | j \in J)\} \end{aligned} \tag{6}$$

Calculate the separation measures [4];

$$S_i^+ = \sqrt{\sum_{j=1}^{n} (V_{ij} - V^*)^2} \quad (1 \le i \le m, 1 \le j \le n) \tag{7}$$

$$S_i^- = \sqrt{\sum_{j=1}^{n} (V_{ij} - V^-)^2} \quad (1 \le i \le m, 1 \le j \le n) \tag{8}$$

Calculate the relative closeness to the ideal solution;

$$C_i = \frac{S_i^-}{S_i^+ + S_i^-} \quad (1 \le i \le m) \tag{9}$$

where $C_i \in (0, 1)$. The larger C_i is the closer alternative to the ideal solution [15].

4 Result and Discussion

Data occupied in Tables 1 and 2, respectively, prepare Table 3 with normalized weighted data of criteria. This segment deliberates all the equity of the perceptible material with identical relevance, so that the decision maker contemplates on each individual tract amidst the equivalent emphasis. Thereafter, classify the convenient weights [5] by entropy method and apply TOPSIS method by using the data of Table 3 that calculated positive ideal solution and negative ideal solution. Thereafter, calculate the [10] separation measure as described by TOPSIS method shown in Table 4a, b, respectively. Finally, consummation outcome seizes by the technique, each alternative is graded and encapsulate, is depicted in Table 5. Lanthanum commercial purity min 99% is the preferred choice of energy-efficient

Table 3 Weighted normalized data by TOPSIS method

S. No.	Materials	M1	M2	M3	M4	M5	M6	M7	M8	M9
1	A1	0.435	0.008	0.015	0.000	0.003	0.001	0.000	0.001	0.000
2	A2	0.319	0.025	0.192	0.038	0.001	0.001	0.001	0.003	0.001
3	A3	0.248	0.009	0.008	0.021	0.009	0.002	0.001	0.032	0.010
4	A4	9.135	0.001	0.002	0.014	0.005	0.002	0.000	0.000	0.010
5	A5	4.025	0.005	0.002	0.001	0.002	0.030	0.001	9.551	0.002
6	A6	3.047	0.001	0.001	0.001	0.000	0.015	0.003	0.000	0.007
7	A7	8.613	0.001	0.000	0.021	0.005	0.001	0.000	0.000	0.012
8	A8	4.025	0.005	0.003	0.001	0.002	0.029	0.001	8.190	0.002
9	A9	3.306	0.000	0.000	0.001	0.001	0.023	0.006	0.000	0.014

Table 4 Separation measured data

S. No.	Materials	M1	M2	M3	M4	M5	M6	M7	M8	M9
(a)										
1	A1	0	0.000	0.031	0.001	4.010	0.000	3.700	0.000	0.000
2	A2	0.102	0.000	0.036	0.001	6.377	1.136	1.278	1.501	4.423
3	A3	0.034	0.000	0.034	0.000	0	0.000	3.497	0	1.177
4	A4	0.189	0.000	0.036	0.000	2.082	0.000	3.673	0.001	1.177
5	A5	0.189	0.000	0.036	0.001	5.361	0	2.551	0.001	0.000
6	A6	0.189	0.000	0.036	0.001	7.781	0.000	1.199	0.001	4.306
7	A7	0.189	0.000	0.036	0.000	2.138	0.000	3.675	0.001	4.066
8	A8	0.189	0.000	0.036	0.001	5.361	2.841	2.551	0.001	0.000
9	A9	0.189	0.000	0.036	0.001	6.924	4.988	0	0.000	0
(b)										
1	A1	0.189	6.400	0.000	0	6.191	1.438	0	2.473	0
2	A2	0.102	0.000	0.036	0.001	6.377	1.136	1.278	1.501	4.423
3	A3	0.061	7.476	5.493	0.000	7.781	2.207	2.848	0.001	0.000
4	A4	2.716	4.631	2.284	0.000	1.813	1.758	4.814	7.030	0.000
5	A5	1.001	2.113	3.649	2.362	2.246	0.000	1.065	1.851	6.220
6	A6	4.780	1.086	1.275	2.761	0	0.000	6.862	2.698	5.157
7	A7	0	3.027	0	0.000	1.761	0	4.102	3.748	0.000
8	A8	1.001	2.113	5.332	2.362	2.246	0.000	1.065	0	4.283
9	A9	5.980	0	2.128	1.764	2.496	0.000	3.700	1.409	0.000

Table 5 Performance score and ranking by TOPSIS

Materials	A1	A2	A3	A4	A5	A6	A7	A8	A9
Score	0.3013	0.5	0.5121	0.9634	0.9413	0.9672	0.9514	0.9423	0.9470
Rank	9	8	7	2	6	1	3	5	4

material in dispersion of the correlated other materials with the accustomed equity to wardrobe for environment when commensurate weight was accustomed to all criteria [5].

5 Conclusion and Future Scope

In this paper, we contemplate a speculative case as an arithmetical illustration. This paper comes up with the utilization of MCDM analysis to energy-efficient material election dilemma. The application of these methods gives an orderly and plausible explanation to the decision makers. Weighting factors influence has also been discussed. Each approach suggests its own excerpt criteria, but completely it is preferred by the ruling maker to go with. Lanthanum commercial purity min 99% is an acceptable choice of energy-efficient material for the accustomed dilemma. The main features of the prospective miniature can be cataloged below:

- Recommending an exemplary which simultaneously brings about energy-efficient material selection and procedure appropriation,
- Determining the precedent from critical glimpse.

Further investigation may also assimilate a practice to contemplate new restraints in prospective entropy–TOPSIS method. Alternative optimization techniques such as artificial neural network, bee-colony optimization seems to be practical to appraise and putrid the energy-efficient materials properly.

References

1. Akyene, T.: Cell phone evaluation base on entropy and TOPSIS. Interdisc. J. Res. Bus. **1**(12), 9–15 (2012)
2. Ashby, M.F.: Multi-objective optimization in material design and selection. Acta Mater. **48** (1), 359–369 (2000)
3. Bligaard, T., Jóhannesson, G.H., Ruban, A.V., Skriver, H.L., Jacobsen, K.W., Nørskov, J.K.: Pareto-optimal alloys. Appl. Phys. Lett. **83**(22), 4527–4529 (2003)
4. Chang, C.W.: Collaborative decision making algorithm for selection of optimal wire saw in photovoltaic wafer manufacture. J. Intell. Manuf. **23**(3), 533–539 (2012)
5. Chauhan, A., Vaish, R.: Magnetic material selection using multiple attribute decision making approach. Mater. Des. **36**, 1–5 (2012)
6. Dashore, K., Pawar, S.S., Sohani, N., Verma, D.S.: Product evaluation using entropy and multi criteria decision making methods. Int. J. Eng. Trends Tech. **4**(5), 2183–2187 (2013)
7. Hwang, C.L., Yoon, K.: Multiple attribute decision making: methods and applications a state-of-the-art survey, p. 186. Springer, Berlin (2012)
8. Hwang, C.L., Yoon, K.: Multiple attribute decision making: methods and applications. Springer, New York (1981)
9. Jahan, A., Ismail, M.Y., Sapuan, S.M., Mustapha, F.: Material screening and choosing methods–a review. Mat. Design. **31**(2), 696–705 (2010)

10. Kou, G., Wu, W., Zhao, Y., Peng, Y., Yaw, N.E., Shi, Y.: A dynamic assessment method for urban eco-environmental quality evaluation. J Multi-Criteria Decis. Anal. **18**(1–2), 23–38 (2011)
11. Kumar, R., Bhomik, C., Ray, A.: Selection of cutting tool material by TOPSIS method. In: National Conference on Recent Advancement in Mechanical Engineering. NERIST, Itanagar, India. pp. 978–993 (2013)
12. Kumar, D.S., Suman, K.N.S.: Selection of magnesium alloy by MADM methods for automobile wheels. Int. J. Eng. Manuf. **4**(2), 31 (2014)
13. Lotfi, F.H., Fallahnejad, R.: Imprecise Shannon's entropy and multi attribute decision making. Entropy **12**(1), 53–62 (2010)
14. Milani, A.S., Shanian, A., Madoliat, R., Nemes, J.A.: The effect of normalization norms in multiple attribute decision making models: a case study in gear material selection. Strut. Multi. Optim. **29**(4), 312–318 (2005)
15. Senan, S., Arik, S.: Global robust stability of bidirectional associative memory neural networks with multiple time delays. IEEE Trans. Syst. Man Cybern. B **37**(5), 1375–1381 (2007)
16. Vaish, R.: Piezoelectric and pyroelectric materials selection. Int. J. App. Ceramic Tech. **10**(4), 682–689 (2013)

CES Edu pack 2005

17. Govindan, K., Shankar, K.M., Kannan, D.: Sustainable material selection for construction industry–a hybrid multi criteria decision making approach. Renew. Sustain. Energy Rev. **55**, 1274–1288 (2016)
18. Hafezalkotob, A., Hafezalkotob, A.: Risk-based material selection process supported on information theory: a case study on industrial gas turbine. Appl. Soft Comp. J. http://dx.doi.org/10.1016/j.asoc.2016.09.018
19. Hafezalkotob, A., Hafezalkotob, A., Sayadi, K.M.: Extension of MULTIMOORA method with interval numbers: an application in materials selection. Appl. Math. Mod. **40**, 1372–1386 (2016)

Analyses and Detection of Health Insurance Fraud Using Data Mining and Predictive Modeling Techniques

Pallavi Pandey, Anil Saroliya and Raushan Kumar

Abstract As establishing fraud detection mechanism in the healthcare industry is an evolving challenge, this research work proposes a comprehensive approach for predicting the most probable fraudulent claims by the help of traditional and advanced statistical data analysis tools as well as algorithms. Keeping in mind the business value of the fraudulent claims, the research work seeks to develop a systemic methodology of modeling the information in such a way that the defined business goals in healthcare fraud management are achieved. Given the nature of insurance industry, the currently available mechanisms lack a standardized approach for encountering the events of fraudulent activities. So the research area focuses on the use of a distinguished methodology of "**Rules-based scoring system**" and creating a well-defined statistical model to assess the correlating factors which describe the probability and propensity of fraudulent claims in the industry. The research work leads to develop various data models using algorithms such as logistic regression, neural network, and decision tree in order to understand the cohesion between probable factors and also build up the most accurate model which can address or predict the claims based on the prevailing conditions and transactional activities.

Keywords Predictive analytics · Data mining · Rule-based scoring system
Logistic regression · Neural networks · Decision trees

P. Pandey (✉)
Amity School of Engineering and Technology, Jaipur, India
e-mail: pandeypallavi027@gmail.com

A. Saroliya
Amity School of Engineering and Technology, Noida, India
e-mail: asaroliya@jpr.amity.edu

R. Kumar
CapGemini, Noida, India
e-mail: rkumar71.sas@gmail.com

© Springer Nature Singapore Pte Ltd. 2018
M. Pant et al. (eds.), *Soft Computing: Theories and Applications*,
Advances in Intelligent Systems and Computing 584,
https://doi.org/10.1007/978-981-10-5699-4_5

1 Introduction

Fraud can be witnessed in all insurance types including health insurance. In health insurance industry, fraud is committed by intentional dissimulation or misrepresentation of data in order to gain some benefit from the insurance companies in the form of medical health expenditures [1]. The alarming rate at which the incidence of abuse and fraud in health insurance is increasing is a matter of great concern. Since there was no provision of punishment earlier, it led to the increment in the occurrence of such claims. Recently, IRDA quoted the definition provided by the International Association of Insurance Supervisors (IAIS) and identified "Insurance Fraud" as a punishable offense. The IRDA defines fraud as "an act or omission intended to gain dishonest or unlawful advantage for a party committing the fraud or for other related parties."

The major goal of the insurance companies is to manage the risks of the consumers. Insurance companies offer to pay its customers a definite amount of money upon the occurrence of a predetermined event such as a car accident, doctor's visit in exchange for a constant stream of premiums. The main functionality of insurance companies is to pool and redistribute various types of risks. This is done by collecting premiums from every customer that the company insures and then paying them out to a few customers who really need the money. It is a common practice to use credit scores by banks and credit card companies which help them in evaluating the potential risk involved in lending the money to the consumers and to mitigate losses due to bad debt. These credit scores are used by the lenders to determine to whom the loan can be given, at what interest rate, and what will be the credit limits. These credit scores also help in determining which customers can bring in the most revenue. The credit scores used by the banks and credit card companies are not much prevalent in health insurance domain. Thus, there is no direct parameter to decide the worthiness of the customer to whom the claimed amount should be given. Among millions of claims, the fraudulent claims need to be segregated from the non-fraudulent ones. Thus, some rules need to be followed which will eventually help in discriminating the fraudulent and non-fraudulent claims. These set of business rules have been identified and discussed later.

2 Literature Review

Supervised Learning Techniques: It is a type of machine learning technique in which a supervised learning algorithm is used to train the model by using training data. Later, the learning from training data can be used to generalize the unseen situations in a logical way [2–5].

Hybrid Model: The hybrid model is designed by using the concepts of supervised as well as unsupervised learning techniques [6, 7].

Rule-Based Method: In this system, fraud patterns are identified as rules. Rules may consist of one or more conditions. When all conditions are met, an alert is raised [8].

Cost Matrix: Since all the misclassifications have equal weights, to improve the performance of the classification decision trees a model needs to be generated which can deal with such misclassified data. For this purpose, the cost matrix is used [9].

Social Network Analysis: Using social network analysis, the investigators can look after the data patterns across the complete product line to detect the existence of any potential fraudsters [10].

Computational Intelligence: Computational intelligence comprises of nature-inspired techniques for computation which are used to deal with the real-world problems [11].

3 Proposed Methodology

The main aim of this research was to identify a fraud indicator in the health insurance claims dataset [12] which could successfully distinguish between the fraudulent and the non-fraudulent claims. Since there was no apparent fraud indicator in the dataset, a rule-based scoring system was used. In this approach, a set of certain rules, with the predefined weights (scores), are implemented on the dataset. After the implementation, each score has a corresponding frequency. The claim frequencies lying in the third quartile are considered to be the probable fraudulent cases of claims as in insurance industry the frequency of such claims is estimated to be very less. After the identification of the scores lying in the third quartile, a threshold value is set for the score to segregate the fraudulent claims with the non-fraudulent ones. After this, various supervised learning techniques were used to train the model in order to identify and predict the fraudulent claims. Finally, these models have been compared to arrive at the best prediction model for insurance claims dataset.

3.1 About the Dataset

The formal structure of any insurance business has many variables which actually define the basic components of real-time events. These data have very generic component in the form of variables, and some of them are as follows:

(a) Date of claim,
(b) Date of birth,
(c) Policy end date,

(d) Policy start date,
(e) Discharge date,
(f) Admission date,
(g) Age of insured person,
(h) Amount of sum insured,
(i) Claim number,
(j) Gender,
(k) Medical history,
(l) Hospital name.

3.2 Implementation Methodology

The implementation was done on the health insurance claims dataset which initially had 65,435 records in total with 56 attributes.

1. *Data Preprocessing*
   Before implementing any modeling techniques, the dataset needs to be pre-processed. For this, all the numeric values in the dataset were imputed with zero. The missing numeric values that could have been imputed with the help of other attributes were calculated. After performing imputation of the missing values, certain conditions were checked on the claims dataset and the records that did not match with these conditions were deleted from the dataset.

2. *Selection of Variables for Scoring and Model Building*
   The claims dataset had 56 variables *which were categorized, as per their roles, in four different categories, that is, Variables for Modelling Only, Variables for Business Rules of the Scoring Model, Variables Selected for Both* and the Unselected Variables.

3. *Scoring Model for Creating Fraud Indicator*
   As there was no direct fraud indicator present in the dataset, a collection of business rules [13] was examined and implemented on the dataset (Fig. 1).
   The above graph depicts the frequency of the corresponding scores generated after the implementation of the business rules on claims dataset. Based on the

Fig. 1 Distribution of scores

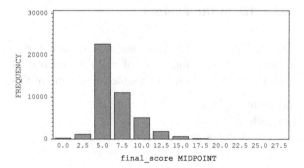

inferences drawn from the above histogram showing the frequency of the scores, the threshold value should lie in the third quartile, as it is assumed in health insurance industry that the number of fraudulent claims is very less. Hence, the threshold value for fraud indicator is set as 10.

4. *Scoring Model Validation*
   For scoring model validation, multiple linear regression has been used.

   Multiple Linear Regression: Multiple linear regression depicts the relationship between one dependent variable and two or more independent variables. Each value of the dependent variable x is associated with the value of independent variable y. The regression line p for independent variables $x_1, x_2,...,x_p$ is defined to be

$$\mu y = \beta 0 + \beta 1 x i 1 + \beta 2 x i 2 + \cdots + \beta p x i p$$
   for $i = 1, 2, 3, ... n$.

5. *Predictive Modeling Techniques*
   Predictive analytics comprises of a variety of statistical techniques ranging from predictive modeling, machine learning, and data mining that can be used to scrutinize the historical as well as current data to make predictions about future or otherwise unknown events. The algorithms that were used for building prediction model are logistic regression, decision trees, and neural networks.

   Logistic Regression: When the dependent variable is in binary form, then the regression analysis is done using logistic regression. The logistic regression generates the coefficient of the formula:

$$\text{logit}(p) = b_0 + b_1 X_1 + b_2 X_2 + \cdots + b_k X_k$$

   where p is the probability of the presence of the characteristic of interest. The logit transformation is defined as the logged odds:

$$\text{odds} = \frac{p}{1-p} = \frac{\text{probability of presence of characteristic}}{\text{probability of absence of characteristic}}$$

   and

$$\text{logit}(p) = \ln\left(\frac{p}{1-p}\right)$$

   Decision Trees: For multiple variable analysis, decision trees are powerful yet simple tool. The algorithm used in decision trees identifies various ways in which a dataset can be split into branch like segments. At the top of the tree is a root node, and the segments originating from the root node form an inverted decision tree.

Neural Networks: Artificial neural networks are used for supervised prediction problems and are a flexible nonlinear model in data analysis. Hidden units are known to be the basic building blocks of an artificial neural network. In data analysis, the multi-layer perceptron (MLP) is the most popular and widely used type of neural network. An MLP is composed of an input layer, hidden layers, and an output layer and is a feed-forward network. The MLP and many other neural networks learn using an algorithm called **backpropagation**.

4 Results

4.1 Scoring Model Validation

The validation of scoring model was done by multiple linear regression. For this purpose, two models were built with total score being the dependent variable and some selected attributes being the independent ones. The following table compares the results obtained from these two models (Table 1).

4.2 Models for Fraud Detection

After cleaning the dataset, the raw dataset which contained 65,535 records was reduced down to 46,805 records. In order to train the models, the cleaned dataset was broken down to train and test datasets in the ratio of 70:30, respectively.

Model Building Using Logistic Regression

As a part of model building, logistic regression was applied with fraud_indicator as dependent variable and other variables as independent variables (Table 2).

Model Building Using Decision Trees

For model building, the decision tree was incorporated with fraud_indicator as a dependent variable with other covariates being the independent variables (Table 3).

Model Building Using Neural Networks

For model building, the neural network has been incorporated with fraud_indicator as a dependent variable (Table 4).

Table 1 Comparative study of results obtained from multiple linear regression models

Values	Model 1	Model 2
R-square (%)	16.37	14.10
Adjusted R-square	0.1635	0.1409

Table 2 Comparison table for logistic regression

	Training			Validation			Test
ROC index	0.814			0.807			0.798
Gini coefficient	0.628			0.614			0.596
Confusion matrix		**1**	**0**		**1**	**0**	–
	1	402	1979	**1**	181	839	
	0	280	22541	**0**	137	9644	
Sensitivity	0.1688			0.1774			–
Specificity	0.9877			0.9859			–
Accuracy	0.9104			0.9096			–

Table 3 Comparison table for decision tree

	Training			Validation		
Confusion matrix		**1**	**0**		**1**	**0**
	1	484	1897	**1**	197	823
	0	37	22784	**0**	27	9754
Specificity	0.9983			0.9972		
Sensitivity	0.2032			0.1931		
Accuracy	0.9233			0.9213		

Table 4 Comparison table for neural network

	Training			Validation		
Confusion matrix		**1**	**0**		**1**	**0**
	1	483	1898	**1**	194	826
	0	215	22606	**0**	113	9668
Specificity	0.9905			0.9884		
Sensitivity	0.2028			0.1902		
Accuracy	0.9161			0.9130		

4.3 Comparison of Models

The best performing variant of each type of model is picked and compared against each other in Table 5.

As ROC curve represents relationship between true positive rate and false positive rate and the area under the ROC curve represents the trade-off between these two measures. Hence, area under the ROC curve has been selected as the criteria for selecting the most suitable model from multiple models.

Table 5 Comparative analysis of algorithms

Model name	Logistic		Decision tree		Neural network	
	Train	Validation	Train	Validation	Train	Validation
ROC area	**0.814**	**0.807**	**0.831**	**0.816**	**0.845**	**0.834**
Sensitivity	0.1688	0.1774	0.2032	0.1931	0.2028	0.1902
Specificity	0.9877	0.9859	0.9983	0.9972	0.9905	0.9884
Accuracy	0.9104	0.9096	0.9233	0.9213	0.9161	0.9130
Misclassification rate	0.0896	0.0904	0.0.0767	0.0786	0.0896	0.0869

Thus, by comparing the area under the ROC curve, neural network is slightly better than the decision tree. Hence, either of them can be used in scoring any new claim data.

5 Conclusion and Future Scope

In this research work, the concepts of data mining have been implemented for building a prediction model which can help in detecting any fraudulent claims made by the customers by training the model using supervised learning technique. For this purpose, a scoring model has been implemented for creating a fraud indicator which helped in identifying the fraudulent claims and the classification of claims into fraudulent and non-fraudulent claims has been done on the basis of fraud_indicator. This scoring model is validated using linear regression. After this, the fraud detection model was made. Logistic regression, decision tree, and neural network were used for the fraud detection. By comparing different statistics derived from these three models, neural network was found to be slightly better than decision tree for the purpose of making predictions and for scoring any new claim data.

In future, research can be done to determine which traditional fraud indicators carry good claim sorting information. In this work, the supervised learning methods have been used to build the model. In future, unsupervised methods or the hybrid of supervised and unsupervised methods can be used by extracting the best features from both the approaches and building a hybrid model. Also, a model can be built which processes dynamic data and detects frauds and flags them for further investigation dynamically.

References

1. Kirlidog, M., Asuk, C.: A fraud detection approach with data mining in health insurance. In: World Conference on Business, Turkey, vol. 62, pp. 989–994 (2012)
2. He, H., Wang, J., Graco, W., Hawkins, S.: Application of neural networks to detection of medical fraud. Expert Syst. Appl. **13**, 329–336

3. Wilson, J.H.: An analytical approach to detecting insurance fruad using logistic regression. J. Financ. Acc.
4. Swamy, N.S., Lingareddy, S.C.: Fruad detection using data mining technique. Int. J. Innov. Eng. Technol. **4**(1), (2014)
5. Wipro White Paper: Comparative Analysis of Machine Learning Techniques for Detecting Insurance Claims Fraud
6. Rawte, V., Anuradha, G.: Fraud detection in health insurnace using data mining techniques. In: International Conference on Communication, Information and Computing Techniques, Mumbai (2015)
7. Thornton, D., Capelleveen, G.V., Poel, M., Hillegersberg, J.V., Muller, R.: Outlier based health insurance fraud detection for U.S. medicaid data. In: International Conference on Enterprise Information Systems, pp. 684–694 (2014)
8. Rajani, S., Padmavathamma, M.: A model for rule based fraud detection in telecommunications. Int. J. Eng. Res. Tehnol. **1**(5), 570–575 (2012)
9. Rodrigues, L.A., Omar, N.: Auto claim fraud detection using multi classifier system. Department of Electrical Engineering, Mackenzie Presbiterian University, Brazil, pp. 37–44 (2014)
10. CGI White Paper; "Implementing Social Network Analysis for Fraud Prevention" 2011
11. West, J., Bhattacharya, M., Islam, R.: Intelligent financial fraud detection practices: an investigation. School of Computing and Mathematics, Charles Sturt University, Australia
12. IIB. Available at: https://iib.gov.in/IIB/About.htm
13. Dogra, A.: Trigger based scoring system for health insurance claims

Software Cost Estimation Using Artificial Neural Network

Shaina Arora and Nidhi Mishra

Abstract Cost estimation is something we can call the most challenging task of software project management. Cost estimation is to precisely assess required assets and schedules for software improvement ventures, and it includes a number of things under its wide umbrella, for example, estimation of the effort required, estimation of the size of the software product to be produced, and last but not the least estimating the cost of the project. The overall project life cycle is impacted by the accurate prediction of the software development cost. Lots of models for software cost estimations are proposed by researchers. The COCOMO model makes employments of multilayer feedforward neural system while being actualized and prepared to utilize the perceptron learning algorithm. To test and prepare the system, the COCOMO dataset is actualized. Whatever result is generated from multilayer neural system is then compared with Kaushik [12]. This paper has the goal of creating the quantitative measure not only in the current model but also in our proposed model.

Keywords Software cost estimation · Artificial neural network · COCOMO Feedforward neural network · Magnitude relative error

1 Introduction

The estimation of required resources and schedule can be done through accurate cost estimation. Moreover, cost estimation can also relate as well as organize advancement of tasks apportion what an assets to perform the undertaking and by what technique these assets will be utilized. Talking about the parameters, the exactness of the product advancement and the precision of the administration

S. Arora (✉) · N. Mishra
Department of Computer Science & Engineering, Poornima University, Jaipur, India
e-mail: shaina.arora22@gmail.com

N. Mishra
e-mail: nidhi.mishra@poorinma.edu.in

© Springer Nature Singapore Pte Ltd. 2018   51
M. Pant et al. (eds.), *Soft Computing: Theories and Applications*,
Advances in Intelligent Systems and Computing 584,
https://doi.org/10.1007/978-981-10-5699-4_6

choices both are interrelated. The exactness of the previous is dependent on the precision of the last mentioned in terms of relying; that is, the former will rely on the latter. There are number of parameters, for example, improvement time, group and effort estimation, and for the calculation of each one of the models is required. COCOMO model is said to be the most popular models for the estimation of software cost effort estimation technique. What is unique about COCOMO model is that it makes use of mathematical formula to analyze project cost effort estimation. The paper based on COCOMO model is making use of multilayer neural network technique using perception algorithm.

2 COCOMO

The Constructive Cost Model or better known as the COCOMO model, first presented by Dr. Barry Boehm in 1981, surpassed all the software development practices that took place in those days. Software development techniques have been undergoing many changes and evolving since those days. The COCOMO model can be subdivided in the following models based on the type of application [1].

2.1 Basic COCOMO

Project quality details are not needed to implement parameterized equation of basic COCOMO model.

$$\text{Person Month} = a(\text{KLOC})^b \tag{1}$$

$$\text{Development Time} = 2.5 * \text{PM}^c \tag{2}$$

Three modes of progress of projects are there, on which all the three parameters depend, namely a, b, and c.

2.2 Intermediate COCOMO

According to basic COCOMO model, there is no provision to do software development. To add the accuracy in basic COCOMO model, there are 15 cost drivers provided by Boehm. Cost driver can be classified into

1. Product attributes

   - Required software reliability or better known as RELY
   - Database size or better known as DATA
   - Product complexity or better known as CPLX

2 Computer attributes

- Execution time constraint or better known as TIME
- Main storage constraint or better known as STOR
- Virtual machine volatility or better known as VIRT
- Computer turnaround time or better known as TURN

3 Personnel attributes

- Analyst capability or better known as ACAP
- Application experience or better known as AEXP
- Programmer capability or better known as PCAP
- Virtual machine experience or better known as VEXP
- Programming language experience or better known as LEXP

4. Project attributes

- Modern programming practices or better known as MODP
- Use of software tools or better known as TOOLS
- Required development schedule or better known as SCED [1].

3 Artificial Neural Network

Neurons or the interconnected basic processing devices are used to build artificial neural network (ANN). Much like the human brain, an ANN is stimulated by the way biological nervous systems are. What plays a key role in this paradigm is the novel structure of the input data processing system. It is constructed from number of highly interconnected processing elements (neurons) working in unification to resolve specific tribulations [2].

A simple artificial neural network is shown in Fig. 1 with two inputs neurons ($I1$, $I2$) and one output neuron (O). The inter connected weights are given by $w1$ and $w2$. Represented by the mathematical symbol $I(n)$, various inputs to the network are shown in Fig. 1. Each of these inputs is multiplied by a connection weight. These weights are represented by $w(n)$. In the simplest case, these products are simply summed, fed through a convey function to generate a result, and then delivered as output. This process lends itself to physical achievement on a large scale in a small

Fig. 1 A simple artificial neural net

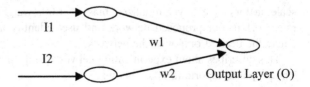

I1

w1

I2

w2   Output Layer (O)

package. This electronic implementation is still possible with other network structures, which utilize different calculation functions as wells as different transfer functions [2].

4 Related Work

Researchers in effort estimation models have developed multiple software. Artificial neural network is capable of generating good information and modeling complex nonlinear relationships. For the calculation of software effort estimation, researchers all across the world have used the artificial neural network approach. Moreover, Boehm's COCOMO dataset is also used. Tadayon [3] reports the use of neural network with a backpropagation. Kaushik [1] also research on multilayer neural network using perceptron learning algorithm. COCOMO [4] is the most effective and widely used software for effort estimation model, which arranges beneficial expert knowledge. For getting appropriate calculations, COCOMO model is one of the most imperative tools that produce capacity for developing effort estimation models with better analytical accuracy. In this paper, multilayer feedforward neural network using perceptron learning algorithm and COCOMO dataset has been used. Using this approach, an effort estimation model for software cost evaluation has been proposed.

5 Proposed Neural Network Model

Figure 1 shows the simple artificial neural network [5]. Many of the network architectures have been developed for different applications and performance. Neural network configuration depends on its design and their performance constraint settings.

In neural network, several parameters are used such as:

- Number of nodes in input layer and hidden layer,
- Training algorithm parameters,
- Weights–Neurons are connected to each other by directed communication links, which are associated with weights.

In proposed model, 20 inputs are taken in which 15 are effort multipliers and 5 scale factors. We have initialized bias as 1.00, weight $w = 1.08$, and learning rate $= 0.1$ in our proposed network and used identity activation function to calculate the desired output of the network.

This describes about experimenting networks and calculating new set of weight. The proposed algorithm is as follow:

Step1. Initialize the bias (b = 1), weights (Wi = 1 for i = 1 to 15), set learning rate (lrnrate = 0.1) and Threshold theta (Θ) value = 4. Set Learning Rate (0 < N <= 1).

Step 2. Execute steps 3-8 until condition is false.

Step 3. Execute step 5-7 for each training pair.

Step 4. The input layer gets information flag and transmit it to concealed layer by executing identity activation function on the inputs from x = 1 to 20.

Step 5. Calculate the response of each response per unit is as follow:

   5.1 Calculate effort multiplier hidden layer Using Hiddenem = b + input * weightem for input(x) = 1 to 15.

   5.2 Calculate scale factor hidden layer using Hiddensf = b + input* (weightsf + size) for input (x) = 16 to 20.

Step 6. Calculate estimated output i.e. effort using Estimated Output = Hiddenem *Weightem + Hiddensf*Weightsf. where Weightem = effort multiplier weight and Weightsf = scale factor weight.

Step 7. Weights are not updated. Go to step 10.

Step 8. If the difference is acceptable limit the output is considered else weights are modified using weight (new) = weight (old) +learning rate*input(x).

Step 9. Repeat step 5 to 6.

Step 10. Calculate Magnitude Relative Error (MRE) MRE = ((Oact-Oest)/Oact) *100

Step 11. Stop.

6   Evaluation Criteria and Results

In this area, we depict the procedure utilized for figuring endeavors and the outcomes get when actualizing proposed neural system model to the COCOMO information set [6]. COCOMO information set is open-source cost evaluating apparatus which comprises of 63 undertakings. The tool does the logical appraisal among the exactness of the assessed exertion with the genuine exertion. For examining programming exertion estimation, we have calculated error using Magnitude of Relative Error (MRE) which is characterized as follows:

$$MRE = (actual\,effort - estimated\,effort/actual\,effort) * 100 \qquad (3)$$

14 experimental values are shown in Table 1 which were tested. Actual effort of the model has been compared with these values. The comparison reflects us about the efficiency of our network. Table 2 contains the estimated effort, actual effort, and Mean Magnitude of Relative Error (MRE) values for 14 experimented projects (Table 3).

Table 1 Cost drivers

Cost drivers			
Product attributes	Computer attributes	Personnel attributes	Project attributes
RELY	TIME	ACAP	MODP
DATA	STOR	AEXP	TOOLS
CPLX	VIRT	PCAP	SCED
	TURN	VEXP	
		LEXP	

Table 2 Assessment of calculated effort

Project No.	Actual effort	Kaushik [1] model	Our proposed model
P1	120	104.17	116.33
P2	60	53.64	56.3
P3	18	18.69	14.33
P4	239	240	235.58
P5	170	154.87	166.52
P6	480	439.17	476.33
P7	300	245.39	296.33
P8	50	38.42	46.33
P9	210	194.64	206.32
P10	62	54.22	58.58
P11	70	66.74	66.58
P12	82	71.13	78.6
P13	31.2	30.14	27.3
P14	25.2	20.55	21.33

Table 3 Assessment of MRE (magnitude relative error)

Project No.	Actual effort	Calculated effort our proposed model	MRE using Kaushik [1] model	MRE using our proposed model
P1	120	116.33	13.19	3.1
P2	60	56.3	10.6	6.1
P3	18	14.33	3.83	20.4
P4	239	235.58	0.41	1.6
P5	170	166.52	8.9	2.3
P6	480	476.33	8.5	0.8
P7	300	296.33	18.20	1.2
P8	50	46.33	23.16	7.3
P9	210	206.32	7.31	1.8
P10	62	58.58	12.54	6.1
P11	70	66.58	4.65	5.4
P12	82	78.6	13.25	4.6
P13	31.2	27.3	3.39	12.8
P14	25.2	21.33	18.45	15.4

Fig. 2 Graphical representation of calculated effort

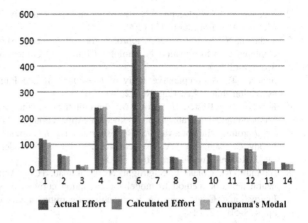

Tables 1, 2 and Fig. 2 show that the described neural network model gives the most proficient effort estimation results as compared to other models.

7   Conclusion

Having a dependable and precise estimate of software development has never been an easy task, and this is where has always lied the problem for many scholarly and industrial conglomerates since ages. Talking about anticipating the future programming improvement exertion, there are a number of product exertion estimating models that can be used for the purpose. This paper shows how a cost estimation model is built based on artificial neural network. The neural network that is used to estimate the software improvement effort is multilayer feedforward neural network with identity activation function. Accurate value is attained through neural network. In future, for software cost estimation, we will put our focus on neurofuzzy approach.

Acknowledgements I would like to express my deep gratitude and thanks to Dr Nidhi Mishra, Associate Professor, Department of Computer Engineering, Poornima University, IT Developer Devesh Arora and Prof. Pramod Choudhry for giving me an opportunity to work under his guidance for preparing the paper. Finally thanks to my family members Mr. Gopal Das (Father), Anju Arora (Mother), Abhishek (Brother), Shivani, Neeraj Munjal and my friend Hridya Narang for their constant encouragement and support throughout the research.

References

1. Kaushik, A., Chauhan, A., Mittal, D.: COCOMO estimates using neural networks. Int. J. Intell. Syst. Appl. **4**(9), (2012)

 2. Sivanandam, S.N., Deepa, S.N.: Introduction to Neural Networks using MATLAB 6.0. Tata McGraw-Hill Education, US (2006)
 3. Tadayon, N.: Neural network approach for software cost estimation. In: IEEE International Conference on Information Technology: Coding and Computing (ITCC'05), vol. 2, pp. 815–818,4–6 (2005)
 4. Sharma, T.: A comparative study of COCOMO II and Putnam models of software cost estimation. Int. J. Sci. Eng. Res. 2(11), (2011)
 5. Krenke, A., Bešter, J., Kos, A.: Introduction to the artificial neural networks. In: Methodological Advances and Biomedical Applications (2011)
 6. http://promise.site.uottawa.ca/SERepository/datasets/cocomo81.arff
 7. Reddy, C., Raju, K.V.S.V.N.: A concise neural network model for estimating software effort. Int. J. Recent. Trends. Eng. 1(1), (2009)
 8. Bayindir, R., Colak, I., Sagiroglu, S., Kahraman, H.T.: Application of adaptive artificial neural network method to model the excitation currents of synchronous motors. In: 11th International Conference on in Machine Learning and Applications (ICMLA), vol. 2, pp. 498–502 (2012)
 9. Kumari, S.: Performance analysis of the software cost estimation methods: a review. Int. J. Adv. Res. Comput. Sci. Softw. Eng. 3(7), (2013)
10. Sharma, T.: Statistical analysis of various models of software cost estimation. Int. J. Eng. Res. Appl. 2(3), 683–685 (2012)
11. Kaushik, A., Soni, A.K., Soni, R.: A simple neural network approach to software cost estimation. Glob. J. Comput. Sci. Technol. Neural Artif. Intell. 13(1), (2013) (Version 1.0)
12. Hamza, H., Kamel, A., Shams, K.: Software effort estimation using artificial neural networks. A Survey of the Current Practices. In: Tenth International Conference on Information Technology: New Generations (ITNG), pp. 731–733, 15–17 (2013)
13. Mukherjee, S.: Optimization of project effort estimate using neural network. In: IEEE International Conference on Advanced Communication Control and Computing Technologies, pp. 406–410 (2014)
14. Jodpimai, P., Lursinsap, P.C.: Estimating software effort with minimum features using neural functional approximation. In: International Conference on Computational Science and Its Applications (ICCSA), pp. 266–273, 23–26 (2010)
15. Ghose, M.K., Bhatnagar, R., Bhattacharjee, V.: Comparing some neural network models for software development effort prediction. In: National Conference on Emerging Trends and Applications in Computer Science (NCETACS), pp. 1–4, 4–5 (2011)

SteganoCrypt: An App for Secure Communication

Neha Mudgal, Pallavi Singh and Shweta Saxena

Abstract In this technical era, information sharing and transfer has increased explosively as well as security of information has also become prime concern to inoculate information from intruders. Many cryptography techniques have been discovered to ensure security of information. To secure the information more robustly, this paper depicts cryptography along with steganography techniques. Cryptography renders data in unreadable form, whereas steganography is the subfield of cryptography and aims at hiding clandestine information in the image. In this paper, using both the techniques successively, an Android application is developed on the proposed method. In the proposed methodology, clandestine information is encrypted using blowfish algorithm, and then, it is embedded into cover image using PVD and LSB algorithms to ensure high level of data security. This methodology increases the robustness and security and capacity of data as compared to the existing methods.

Keywords Cryptography · Steganography · Blowfish · PVD · LSB

1 Introduction

With the advancement in electronic communication, information security becomes more important. From decades, research has been done on securing the information from intruders. And for this purpose, cryptography technique comes in the lime-light. In our research, cryptography along with its subfield steganography has been used to secure the information in more robust way. A three-tier system has been

N. Mudgal (✉) · P. Singh · S. Saxena
Jaipur Engineering College and Research Center, Jaipur, India
e-mail: nehamudgal100@gmail.com

P. Singh
e-mail: pallavisingh2911@gmail.com

S. Saxena
e-mail: shweta.saxena20oct@gmail.com

© Springer Nature Singapore Pte Ltd. 2018   59
M. Pant et al. (eds.), *Soft Computing: Theories and Applications*,
Advances in Intelligent Systems and Computing 584,
https://doi.org/10.1007/978-981-10-5699-4_7

implemented so that paramount secrecy level could be achieved. At first tier, encryption of information using blowfish algorithm is done. At second tier and third tier, secrecy is maintained by hiding the encrypted information in the image using steganography technique via combination of PVD (pixel vector differencing) and LSB (least significant bit). Now, if the intruder tries to breach the information then all he will get is garbage value.

2 Related Work

2.1 Cryptography

In the year 1993, for the first time, the blowfish block cipher was introduced by Schneier [1]. In the research paper [2], a comparative study is done among common symmetric key cryptography algorithms, namely DES, AES, and blowfish based on speed (decryption time), block size, and key size and the study shows that blowfish algorithm is superior among three in terms of processing time.

Various encryption algorithms have been combined with various steganography algorithms for more security. It is observed in the survey of cryptography that dual encryption process is done, i.e., RSA and blowfish, [3] that provides better security and gives more ideas for using higher level cryptographic algorithms. The text data is used for encryption till now, but in further research, some other types of data have been used for encryption. In [4, 5], an image is encrypted using blowfish; i.e., different types of data are used. After that, a new idea has emerged for encryption of text and hides it into audio [6]. The authors of [7] used blowfish and RIJANDAEL algorithm for encryption and steganography using LSB, i.e., triple security, and they also compared their results with existing results. The next new idea is to combine blowfish and advanced encryption standard, and the use of DCT transform domain for hiding message is used [8].

2.2 Steganography

With the combination of cryptography, steganography places most important role in security. There always occurs a slight confusion between cryptography and steganography as both techniques are used to hide information, but the basic difference is between them is—cryptography is the process for converting a text language into another text language by using secret key, but steganography is to hide exact data into another data type or medium, for example, image, audio, video. Hence, the researchers have found that combination of two is not easy to breach.

So, in the field of steganography, there are many algorithms which contribute for security purpose. The author of [9] used PVD (pixel value difference) and LSB (least significant bit) for steganography; i.e., the encrypted message is hidden into true color RGB image. The two major techniques observed are LSB and PVD.

2.2.1 LSB

LSB stands for least significant bit is an insertion method, which uses fixed K-LSB in each pixel to hide the encrypted message. LSB is very easy for hiding the one bit of message into the least bit place of pixel value of an image. Less distracted image has sent with the encrypted message to the receiver. But this is very common technique which can be easily cracked by hacker. There is some variation into LSB technique, i.e., Dynamic LSB [10] and Hash LSB [3].

2.2.2 PVD

PVD stands for pixel value difference is a method with high-potential embedding capacity and amazingly undetectable for the stego-object, proposed by Wu and Tsai [11]. In this method, the difference value is calculated in each non-overlapping block of two consecutive pixels into number of ranges. Then, the difference value is replaced by a new value for embedding the value of subpart of secret messages [12]. The number of bits which can be embedded between two pixels lies in the range of difference value belongs. Leading to further research in the field of PVD [13], there is an embedding algorithm and an extraction algorithm which have detailed explanation.

Considering all the scenarios of cryptography and steganography techniques, a method is proposed in the next section by choosing better technique of cryptography, i.e., blowfish, and steganography, i.e., LSB and PVD.

3 Proposed Methodology

In proposed method, the clandestine information is encrypted using blowfish algorithm, and then, it is embedded into cover image using PVD and LSB algorithms to ensure high level of data security.

3.1 While Embedding the Information

Input:

- Text message to be sent (M)
- Key (K)

- Two cover image behind which we hide information (C1, C2)

  Output:

- Stego-image (S)

  Procedure:

1. Apply cryptography technique by using blowfish on text message (M) for encryption.
2. Develop encrypted text (E_M) by using key (K) and blowfish algorithm.
3. Apply steganography technique on the encrypted text message (E_M), we get in Step 2.

   a. Hide encrypted message (E_M) into cover image (C1) using LSB method.
   b. Hide the image received from Step (a), into another cover image (C2) using PVD method to acquire final stego-image (S).

4. Transmit the final stego-image (S)

3.2 While Extracting the Information

Input:

- Stego-image (S)
- Key (K)

  Output:

- Text message (M)

  Procedure:

1. Apply steganography technique on stego-image (S) to retrieve the hidden information.

   (a) Remove the cover image (C2) by using PVD method to acquire the hidden cover image (C1).
   (b) Remove the cover image (C1) by using LSB method to acquire the hidden information.

2. Apply cryptography technique to decrypt the acquired hidden information in Step 1(b).
3. Decrypt the hidden encrypted message (E_M) using key (K) and blowfish algorithm
4. Retrieve original text message (M)

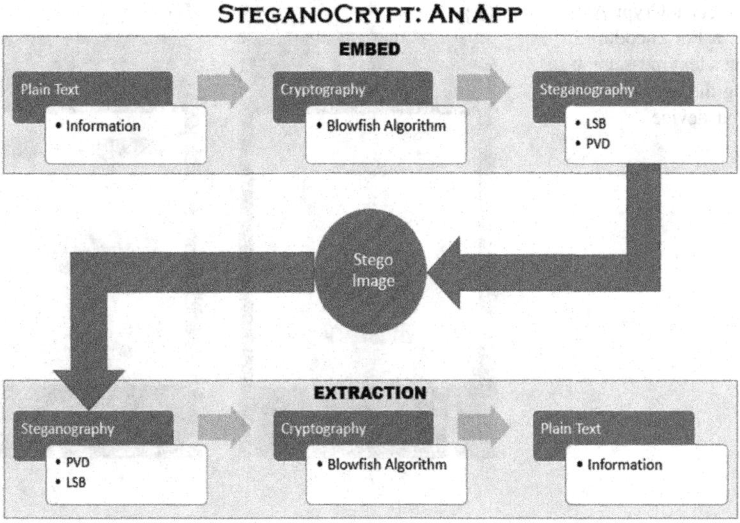

Fig. 1 SteganoCrypt: An app

This methodology has been embedded into an Android app where single activity app is created. This app will remove all the unnecessary effort done by the user for transferring information securely. Block diagram Fig. 1 shows the process.

4 Experimental Results

The experiment is done by simulating the combination of cryptography and steganography through Android app. As in the proposed method described in Sect. 3, three-tier security combining blowfish as encryption technique and steganography combination of LSB and PVD is used. The blowfish algorithm is implemented in Java, and MATLAB framework is used for implementation of LSB and PVD. The algorithm through Android application is implemented on many images, and results are analyzed on the basis of the quality of image we are getting after applying steganography technique. An app is developed, Fig. 2a, b, which will encrypt the message and then hide that message into images, by embedding the codes into Android Studio.

For a cover image, at first without using cryptography, only steganography is performed to hide text message and PSNR of the resultant image is calculated. Then, on the similar cover image, after applying both techniques, PSNR of image is calculated.

Table 1 gives the comparison of the result of image's PSNR using the existing algorithm AES and our proposed method using blowfish. Figures 3 and 4 show the results of the images after embedding the encrypted message using app.

Fig. 2 SteganoCrypt App location. **a** For encoding or decoding a text message, **b** for browsing the text and image from your device

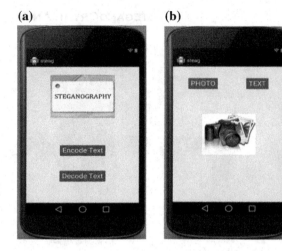

Table 1 Comparision between AES algorithm and blowfish algorithm PSNR values of images

S. No.	Images	Message	PSNR using plain text (M)	PSNR using AES algorithm	PSNR using blowfish algorithm
1		Hoo	58.3266	58.8145	58.5112
2		Hello	63.4151	63.2567	63.4395
3		Hello world	56.6170	56.5873	56.6783
4		I love India	57.5000	58.1046	58.1122
5		This message is secured	58.3260	59.2543	59.7239

Fig. 3 Original images
a dog.jpg. **b** flower.jpg

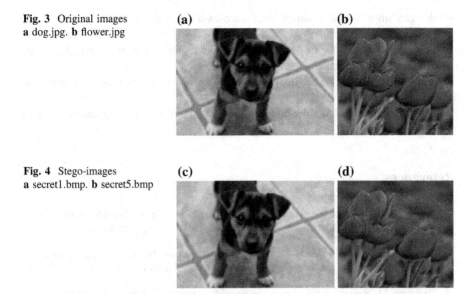

(a) (b)

Fig. 4 Stego-images
a secret1.bmp. **b** secret5.bmp

(c) (d)

The tools used for implementing the algorithms are Java JDK 1.7, MATLAB 2010, Android Studio 2.1.

Here, it is also observed that the bigger images give better result than smaller one. But this fact does not affect the working of the algorithm.

5 Conclusion

This paper concludes that there can be a secured app "SteganoCrypt" which will encrypt your text like your bank details or passwords by using blowfish, and then, it will ask you to choose any image from your device and then hide that encrypted message into image and then send that stego-image to receiver.

At receiver side, this SteganoCrypt App should be installed on device for decryption of that message sent by sender. Hence, this research paper emphasizes on working of secured app which uses both concepts of cryptography and steganography algorithms.

6 Future Scope

The limitation of this application is that it can only work for text messages. Some enhancements can be made further to extend and improve the working of this application:

- We can improve the features for secure exchange of our private images and audios using this app.
- This app only works for Android framework. There can be a possibility to extend it for IOS framework.
- Large images, i.e., DSLR quality images can also be encrypted by using this app.
- We can also extend this methodology to detect and filter the malicious images.

References

1. Schnier, B.: Description of a new variable-length key, 64-bit block Cipher (Blowfish). In: Fast Software Encryption, Cambridge Security Workshop Proceedings, Springer, pp. 191–204, December 1993, (1994)
2. Thakur, J., Kumar, N.: DES, AES and blowfish: symmetric key cryptography algorithms simulation based performance analysis. Int. J. Emerg. Technol. Adv. Eng. 1(2) (2011). ISSN-2250-2459
3. Rajkamal, M., Zoraida, B.S.E.: Image and text hiding using RSA & blowfish algorithms with Hash LSB technique. Int. J. Innov. Sci. Eng. Technol. 1(6), (2014). ISSN-2348-7968
4. Singh, P., Singh, K.: Image encryption and decryption using blowfish algorithm in matlab. Int. J. Sci. Eng. Res. 4(7), (2013). ISSN-2229-5518
5. Barhoom, T.S., Mousa, S.M.A.: A stegsnography LSB technique for hiding image within image using blowfish encryption algorithm. Int. J. Res. Eng. Sci. 3(3), (2015). ISSN-2320-9364
6. Ravali, S.V.K., Neelima, P., Sruthi, P., Sai Dileep, P., Manasa, B.: Implementation of blowfish algorithm for efficient data hiding in audio. Int. J. Comput. Sci. Inform. Technol. 5 (1), (2014). ISSN-0975-9646
7. Patel, K., Vishwakarma, S., Gupta, H.: Triple security of information using stegnography and cryptography. Int. J. Emerg. Technol. Adv. Eng. 3(10), (2013). ISSN-2250-2459
8. Gajjar, J.: Image stegnograpghy based on transform domain, blowfish and AES for second level security. Int. J. Adv. Comput. Technol. 4(6), (2015). ISSN-2320-0790
9. Suryakant, P.V., Bhosale, R.S., Panhalkar, A.R.: A novel security scheme for secret data using cryptography and stegnograpghy. Int. J. Comput. Netw. Inform. Secur. 2, 36–42 (2012)
10. Kulshreshta, A., Goyal, A.: Image stegnograpghy using dynamic LSB with blowfish algorithm. Int. J. Comput. Organ. Trends 3(7), (2013)
11. Wu, D.C., Tsai, W.H.: A steganography method for images by pixel value differencing. 24(9–10), 1613–1626 (2003)
12. Salunkhe, J., Sirsikar, S.: Pixel value differencing a stenographic method: a survey. Int. J. Comput. Appl. and International Conference on Recent Trends in Engineering & Technology (2013). ISSN-0975-8887
13. Thanikaiselvan, V., Subashanthini, S., Amirtharajan, R.: PVD based steganography on scrambled RGB cover images with pixel indicator. J. Artif. Intell. 7(2), (2014). ISSN-1994-5450
14. Thangadurai, K., Sudha Devi, G.: An analysis of LSB based image steganography techniques. In: International Conference on Computer Communication and Informatics (ICCCI), IEEE (2014)
15. Hanling, Z., Guangzhi, G., Caiqiong, X.: Image steganography using pixel-value differencing. IEEE (2009). doi:10.1109/ISECS.2009.139),109-112

Step-Stress Partially Accelerated Life Testing Plan for Rayleigh Distribution Using Adaptive Type-II Progressive Hybrid Censoring

Showkat Ahmad Lone, Ahmadur Rahman and Arif-Ul-Islam

Abstract This study deals with estimating data of failure times under step-stress partially accelerated life tests based on adaptive Type-II progressive hybrid censoring. The mathematical model related to the lifetime of the test units is assumed to follow Rayleigh distribution. The point and interval maximum likelihood estimations are obtained for distribution parameters and tampering coefficient. The Monte Carlo simulation algorithm along with R software is used to evaluate the performances of the estimators of the tempering coefficient and model parameters. The performances are carried out in terms of mean square errors and biases.

Keywords Life testing · Rayleigh distribution · Adaptive Type-II progressive hybrid censoring · Simulation study

1 Introduction

In the modern era, it is becoming more and harder to get failure information of items or systems with very high reliability under usual operating conditions. In such problems, an experimental process called accelerated life testing (ALT) is conducted, where items are experimented under higher stress than normal to find their failure information. Commonly used stress patterns are constant stress and step-stress (see, Nelson [1]). Thus, ALTs or partially accelerated life tests (PALTs) are conducted to minimize the lives of systems or items and to reduce the cost and the time incurred in the experiment. Using step-stress PALT (SSPALT), a product or system is first subjected to use (normal) conditions for the pre-specified duration of time, and if it survives then it is put into service at accelerated condition until the termination time is reached (see, Goel [2]).

S.A. Lone (✉) · A. Rahman · Arif-Ul-Islam
Department of Statistics & Operations Research,
Aligarh Muslim University, Aligarh 202002, India
e-mail: showkatmaths25@gmail.com

© Springer Nature Singapore Pte Ltd. 2018
M. Pant et al. (eds.), *Soft Computing: Theories and Applications*,
Advances in Intelligent Systems and Computing 584,
https://doi.org/10.1007/978-981-10-5699-4_8

Reliability practitioners have developed different censoring plans as a tool to minimize the total time and cost incurred in the experiment. The majority in use are Type-I and Type-II censoring plans (see, Balakrishnan and Ng [3]). Let us suppose that there are n items under particular experimental considerations. Under the traditional Type-I censoring plan, the specimens are tested up to a pre-fixed time point η, while the Type-II censoring plan needs the process to continue until we get a pre-specified number of failures $m \leq n$. The mixture of these two plans forms the hybrid censoring plan. Many of the researchers in the world have used this censoring scheme as a tool for their research (see e.g. Childs et al. [4], Gupta and Kundu [5], Kundu [6], Balakrishnan and Kundu [7]). Also, under SSPALT, Ismail [8] constructed the model for estimation purposes of Weibull failure data using hybrid censoring. One of the major drawbacks of this censoring scheme is the presence of inflexibility to remove some experimental units before the termination of the experimental process. Hence, to overcome this drawback, a new generalization of existing censoring plans called progressive Type-II censoring or progressively Type-II hybrid censoring plan is introduced.

Under progressive Type-II hybrid censoring plan, the experimenter prefixes the number of items or systems to be failed (say 'm') out of the n units placed under life testing experiment. At the time when the first failure occurs, R_1 units among the $n - 1$ remaining (surviving) units are randomly taken off from the experimental process. In the same way, at the time when the second unit fails, the experimenter randomly removes R_2 of the remaining $n - 2 - R_1$ units. The same procedure continues until the mth experimental unit gets failed. At this time point, all the surviving units $R_m = n - m - R_1 - R_2 - \cdots - R_{m-1}$ are withdrawn from the experimental process. Here, $R_1, R_2, \ldots, R_m$ are pre-fixed whole numbers. For an extensive review of the literature of progressive censoring, the readers may refer to Balakrishnan [9], Balakrishnan and Aggarwala [10].

Specifically, speaking of the progressively Type-II hybrid censoring, there is some work either on ordinary life testing or PALT, e.g. see Kundu and Joarder [11], Ng et al. [12], Lin et al. [13], Mokhtari et al. [14], Alma and Belaghi [15]. Recently, Ismail [16] introduced SSPALT for the first time under adaptive progressively Type-II hybrid censoring using Weibull failure data. Our study deals with estimating information about the failure times of test units under SSPALT using adaptive progressive Type-II hybrid censoring. The mathematical distribution of the failure times is assumed to follow Rayleigh distribution function. The following shall discuss the rest of the work.

In Sect. 2, we describe the method and material used in the test. The Rayleigh distribution as a lifetime model is also discussed. Section 3 describes the maximum likelihood method to find the point and interval estimates of parameters and tempering coefficient. Simulation studies for illustrating the theoretical results are presented in Sect. 4. Finally, conclusions are included in Sect. 5

2 The Material and Test Method

The Rayleigh distribution is an important lifetime distribution used to model the lifetime of the random process. It has many applications, including reliability, life testing and survival analysis. The probability density function (PDF) is given below as

$$f_Y(y; \theta) = \frac{y}{\theta^2} \exp\left(-\frac{y^2}{2\theta^2}\right), \quad y > 0, \ \theta > 0. \tag{1}$$

The survival function and hazard function are as follows:

$$S(y) = \exp\left(-\frac{y^2}{2\theta^2}\right), \tag{2}$$

$$h(y) = \frac{y}{\theta^2}, \tag{3}$$

Now, the probability density function (PDF) under SSPALT is given by

$$f(y) = \begin{cases} 0, & y \le 0 \\ f_1(y), & 0 < y \le \tau, \\ f_2(y), & y > \tau \end{cases} \tag{4}$$

where

$$f_2(y) = \beta \frac{[\tau + \beta(y - \tau)]}{\theta^2} \exp\left[-\frac{[\tau + \beta(y - \tau)]^2}{2\theta^2}\right], \quad \theta > 0, \ \beta > 1; \tag{5}$$

The above is obtained by applying the Variable-Transformation technique on (1) along with the given model, proposed by DeGroot and Goel [17].

$$Y = \begin{cases} T & \text{if } T \le \tau \\ \tau + \beta^{-1}(T - \tau) & \text{if } T > \tau \end{cases} \tag{6}$$

where T is the lifetime of an item under usual operating conditions, τ is the stress change time, and $\beta(>1)$ is the tempering coefficient.

It was Kundu and Joarder [11] who introduced a censoring scheme called progressive Type-II hybrid censoring (PHC) scheme. Under this scheme, the life test under progressive censoring scheme $(R_1, \ldots, R_m)$ is finished at a random time $\min(Y_{m:n:n}, \eta)$ where $0 < \eta < \infty$ and $1 \le m \le n$ are fixed in advance, and $Y_{1:m:n} \le Y_{2:m:n} \le \cdots \le Y_{m:m:n}$ are the ordered failure times resulting from the experiment. Specifically, the experiment terminates at time $Y_{m:m:n}$ if the mth progressively censored observed failure will occur before time η (i.e. $Y_{m:m:n} < \eta$). Otherwise, the experiment will be terminated at time η with $Y_{j:m:n} < \eta < Y_{j+1:m:n}$, and all the remaining $\left(n - \sum_{i=1}^{j} R_i - j\right)$ surviving items are censored at time η, where the symbol j denotes the number of failed units up to time η; hence, it is a random variable.

Using the censoring scheme discussed above, the experimenter may come out with a very small sample size (even equal to zero), and therefore, it is not possible to use the standard inference procedures for efficient results. To overcome this drawback, Ng et al. [12] proposed a more advanced censoring scheme called adaptive censoring scheme in which the observed number of failures (or effective sample size) m is fixed in advance and the experiment time is free to run over time η. Also, if $Y_{m:m:n} < \eta$, then the experiment continues along with the pre-fixed progressive censoring scheme $(R_1, \ldots, R_m)$; else, the surviving units following the $(j+1)$th, $\ldots$, $(m-1)$th observed failure are not eliminated from the experiment. Once the m observed numbers of failures are obtained, all the remaining units $R_m = n - m - \sum_{i=1}^{j} R_i$ are withdrawn from the experiment at the time point $Y_{m:m:n}$. That is, in this case, $R_{j+1} = \cdots = R_{m-1} = 0$. If $n \to \infty$, we will have a typical progressive Type-II censoring scheme. But if $n \to 0$, the scheme reduces to the case of the traditional Type-II censoring scheme.

Under the new censoring scheme called an adaptive progressive Type-II hybrid censoring scheme, the experimenter is free to change the value of η to adjust the optimum of shorter experimental time and a higher chance of observing many failures. According to Lin et al. [13], it guarantees us not only to acquire m observed failure times for efficiency of statistical inference but also to control the total time on the test to be not too far away from the ideal time η.

3   Maximum Likelihood Estimation

Let $Y_1, Y_2, \ldots, Y_n$ be the n independently and identically distributed lifetimes of units following the Rayleigh distribution. The m completely observed (ordered) lifetimes are denoted by

$$y_{1:m:n} < \cdots < y_{n_u:m:n} \leq \tau < y_{n_u+1:m:n} < \cdots < y_{J:m:n} \leq \eta < y_{J+1:m:n} < \cdots < y_{m:m:n}$$

where n_u is the number of failed units at use conditions.

3.1   Point Estimation

Using APHC scheme, the likelihood function under SSPALT for the m ordered lifetime data set is as follows:

$$L(\theta, \beta) \propto \prod_{i=1}^{m} f_1(y_{i:m:n}) f_2(y_{i:m:n}) \prod_{i=1}^{J} [s_1(\tau)]^{R_i} [s_2(y_{m:m:n})]^{\left(n-m-\sum_{i=1}^{J} R_i\right)}, \quad (7)$$

The natural logarithm of the above likelihood of the above function is given below

$$\ln L(\theta, \beta) = \sum_{i=1}^{m} \ln y_i + \sum_{i=1}^{m} \ln \phi_i - 4m \ln \theta + m \ln \beta - \frac{1}{2\theta^2} \left[\sum_{i=1}^{m} y_i^2 + \sum_{i=1}^{m} \phi_i^2 \right]$$
$$- \frac{\tau^2}{2\theta^2} \sum_{i=1}^{J} R_i - \frac{J}{2\theta^2} \phi_m^2 \left(n - m - \sum_{i=1}^{J} R_i \right), \quad (8)$$

where $\phi_i = \tau + \beta(y_i - \tau)$ and $\phi_m = \tau + \beta(y_{m:m:n} - \tau)$.

The partial derivatives of the above log-likelihood are equated to zero with respect to parameters, and the resulting equations are

$$\frac{\partial \ln L}{\partial \theta} = -\frac{4m}{\theta} + \frac{1}{\theta^3} \left[\sum_{i=1}^{m} y_i^2 + \sum_{i=1}^{m} \phi_i^2 \right] + \frac{\tau^2}{\theta^3} \sum_{i=1}^{J} R_i$$
$$+ \frac{J}{\theta^3} \phi_m^2 \left(n - m - \sum_{i=1}^{J} R_i \right) = 0, \quad (9)$$

$$\frac{\partial \ln L}{\partial \beta} = \frac{m}{\beta} + \sum_{i=1}^{m} \frac{(y_i - \tau)}{\phi_i} - \frac{1}{\theta^2} \sum_{i=1}^{m} \phi_i(y_i - \tau)$$
$$- \frac{J}{\theta^2} \phi_m(y_{m:m:n} - \tau) \left(n - m - \sum_{i=1}^{J} R_i \right) = 0, \quad (10)$$

From Eq. (9), we can obtain the point estimate of parameter θ as a function of $\hat{\beta}$ as

$$\hat{\theta} = \left[\frac{\sum_{i=1}^{m} y_i^2 + \sum_{i=1}^{m} \phi_i^2 + \tau^2 \sum_{i=1}^{J} R_i + J \phi_m^2 \left(n - m - \sum_{i=1}^{J} R_i \right)}{4m} \right]^{\frac{1}{2}}, \quad (11)$$

Since it is very hard to obtain the solution for the nonlinear equation in (10), the iterative technique, Newton–Raphson method, is used to find the ML estimate of β. And hence, the ML estimate of θ can be easily obtained from Eq. (11).

3.2   Interval Estimation

Here, the interval ML estimates of parameters θ and β are obtained using an obtained data from APHC. Miller [18] suggested that the asymptotic distribution of the ML estimates of θ and β is given by

$$\left(\left(\hat{\theta} - \theta\right), \left(\hat{\beta} - \beta\right)\right) \rightarrow N\left(0, I^{-1}(\theta, \beta)\right).$$

where $I^{-1}(\theta, \beta)$ is the variance–covariance matrix of the unknown model parameters. The elements of the 2×2 matrix I^{-1}, $I_{ij}^{-1}(\theta, \beta), i, = 1, 2$, approximated by $I_{ij}\left(\hat{\theta}, \hat{\beta}\right)$, under described censoring schemes are presented below.

$$
\begin{aligned}
I_{11} = -\frac{\partial^2 \ln L}{\partial \theta^2} = -\frac{4m}{\theta^2} + \frac{3}{\theta^4}\left[\sum_{i=1}^{m} y_i^2 + \sum_{i=1}^{m} \phi_i^2\right] \\
+ \frac{3\tau^2}{\theta^4}\sum_{i=1}^{J} R_i + \frac{3J}{\theta^4}\phi_m^2\left(n - m - \sum_{i=1}^{J} R_i\right),
\end{aligned}
\tag{12}
$$

$$
\begin{aligned}
I_{22} = -\frac{\partial^2 \ln L}{\partial \beta^2} = \frac{m}{\beta^2} + \sum_{i=1}^{m}\frac{(y_i - \tau)^2}{\phi_i^2} + \frac{1}{\theta^2}\sum_{i=1}^{m}(y_i - \tau)^2 \\
+ \frac{J}{\theta^2}(y_{m:m:n} - \tau)^2\left(n - m - \sum_{i=1}^{J} R_i\right),
\end{aligned}
\tag{13}
$$

$$
\begin{aligned}
I_{12} = I_{21} = -\frac{\partial^2 \ln L}{\partial \theta \partial \beta} = -\frac{\partial^2 \ln L}{\partial \beta \partial \theta} = \\
-\frac{2}{\theta^3}\sum_{i=1}^{J}\phi_i(y_i - \tau) - \frac{2J}{\theta^3}(y_{m:m:n})\left(n - m - \sum_{i=1}^{J} R_i\right),
\end{aligned}
\tag{14}
$$

The approximate $100(1 - \gamma)\%$ two-sided confidence limits for θ and β are, respectively, given by

$$\hat{\theta} \pm Z_{\gamma/2}\sqrt{I_{11}^{-1}\left(\hat{\theta}, \hat{\beta}\right)} \quad \text{and} \quad \hat{\theta} \pm Z_{\gamma/2}\sqrt{I_{22}^{-1}\left(\hat{\theta}, \hat{\beta}\right)}$$

4   Simulation Studies

The study is carried out to compare the performance of the MLEs in terms of their mean square errors and biases for different choices of $(n, m, \tau \text{ and } \eta)$ and parameter values. We consider the following three progressive censoring schemes, and for each setting, the bias and MSEs based on 10,000 simulations are estimated and reported in tabular form.

Scheme 1: $R_1 = R_2 = \cdots = R_{m-1}$ and $R_m = n - m$;
Scheme 2: $R_1 = n - m$ and $R_2 = R_3 = \cdots = R_m = 0$;
Scheme 3: $R_1 = R_2 = \cdots = R_{m-1} = 1$ and $R_m = n - 2m + 1$

For each, the simulation procedures are carried out according to the following steps.

Step 1. First, the values of (n, m, τ, η) and (θ, β) are specified.
Step 2. Generate a random sample of size n from Rayleigh distribution under both normal and accelerated conditions using inverse CDF method.
Step 3. For given values of $n, m, \tau, \eta (\eta > \tau), \theta$ and β, we generate the progressive hybrid censored sample using the model given by Eq. (4). The sample data sets under AHPC are given below (Tables 1, 2, 3, 4, 5 and 6).

$$y_{1:m:n} < \cdots < y_{n_u:m:n} \le \tau < y_{n_u+1:m:n} < \cdots < y_{J:m:n} \le \eta < y_{J+1:m:n} < \cdots < y_{m:m:n}$$

Table 1 Mean values of the bias and MSE when θ, β, τ and η are set at 0.5, 1.2, 2.5 and 6, respectively	(n, m)	Schemes	Values of θ		Values of β	
			Bias	MSE	Bias	MSE
	(20, 8)	1	0.472	0.531	0.518	0.588
		2	0.503	0.613	0.622	0.676
		3	0.485	0.576	0.546	0.624
	(30, 8)	1	0.353	0.461	0.446	0.516
		2	0.390	0.547	0.510	0.598
		3	0.379	0.512	0.485	0.576
	(50, 8)	1	0.252	0.403	0.311	0.344
		2	0.295	0.495	0.446	0.440
		3	0.267	0.423	0.395	0.393
	(20, 12)	1	0.209	0.311	0.238	0.300
		2	0.298	0.441	0.478	0.409
		3	0.226	0.309	0.404	0.365
	(30, 12)	1	0.153	0.215	0.209	0.204
		2	0.194	0.301	0.338	0.284
		3	0.167	0.248	0.230	0.277
	(50, 12)	1	0.090	0.103	0.101	0.107
		2	0.134	0.178	0.209	0.210
		3	0.112	0.139	0.137	0.121

Table 2 Mean values of the bias and MSE when θ, β, τ and η are set at 0.5, 1.2, 2.5 and 12, respectively

(n, m)	Schemes	Values of θ		Values of β	
		Bias	MSE	Bias	MSE
(20, 8)	1	0.192	0.266	0.267	0.267
	2	0.228	0.302	0.345	0.363
	3	0.206	0.280	0.318	0.302
(30, 8)	1	0.138	0.201	0.222	0.240
	2	0.205	0.254	0.298	0.309
	3	0.144	0.223	0.270	0.287
(50, 8)	1	0.112	0.145	0.170	0.209
	2	0.154	0.198	0.197	0.288
	3	0.128	0.167	0.183	0.244
(20, 12)	1	0.102	0.151	0.167	0.189
	2	0.198	0.202	0.224	0.301
	3	0.166	0.187	0.205	0.277
(30, 12)	1	0.091	0.108	0.119	0.141
	2	0.145	0.155	0.188	0.209
	3	0.120	0.126	0.167	0.178
(50, 12)	1	0.073	0.071	0.079	0.080
	2	0.103	0.120	0.119	0.167
	3	0.087	0.083	0.100	0.122

Table 3 Mean values of the bias and MSE when θ, β, τ and η are set at 1.5, 1.2, 2.5 and 6, respectively

(n, m)	Schemes	Values of θ		Values of β	
		Bias	MSE	Bias	MSE
(20, 8)	1	0.334	0.370	0.603	0.567
	2	0.379	0.423	0.698	0.637
	3	0.356	0.406	0.649	0.589
(30, 8)	1	0.278	0.311	0.517	0.488
	2	0.321	0.388	0.616	0.576
	3	0.294	0.358	0.568	0.512
(50, 8)	1	0.187	0.270	0.402	0.377
	2	0.250	0.306	0.485	0.450
	3	0.211	0.287	0.434	0.405
(20, 12)	1	0.186	0.208	0.366	0.300
	2	0.276	0.299	0.476	0.371
	3	0.223	0.287	0.424	0.341
(30, 12)	1	0.155	0.153	0.265	0.178
	2	0.199	0.223	0.349	0.259
	3	0.160	0.198	0.310	0.225
(50, 12)	1	0.101	0.067	0.111	0.112
	2	0.136	0.118	0.185	0.167
	3	0.113	0.091	0.131	0.132

Table 4 Mean values of the bias and MSE when θ, β, τ and η are set at 1.5, 1.2, 2.5 and 12, respectively

(n, m)	Schemes	Values of θ		Values of β	
		Bias	MSE	Bias	MSE
(20, 8)	1	0.217	0.250	0.405	0.441
	2	0.290	0.336	0.486	0.505
	3	0.257	0.290	0.433	0.468
(30, 8)	1	0.177	0.190	0.356	0.367
	2	0.226	0.287	0.407	0.419
	3	0.195	0.258	0.388	0.399
(50, 8)	1	0.127	0.155	0.267	0.302
	2	0.186	0.178	0.370	0.356
	3	0.150	0.186	0.301	0.327
(20, 12)	1	0.107	0.159	0.271	0.290
	2	0.190	0.166	0.357	0.361
	3	0.158	0.185	0.313	0.323
(30, 12)	1	0.077	0.087	0.166	0.178
	2	0.117	0.117	0.287	0.259
	3	0.097	0.103	0.217	0.225
(50, 12)	1	0.048	0.053	0.101	0.086
	2	0.087	0.087	0.137	0.167
	3	0.065	0.070	0.111	0.108

Table 5 Mean values of the bias and MSE when θ, β, τ and η are set at 1.5, 1.2, 4 and 6, respectively

(n, m)	Schemes	Values of θ		Values of β	
		Bias	MSE	Bias	MSE
(20, 8)	1	0.288	0.251	0.565	0.632
	2	0.334	0.341	0.706	0.734
	3	0.311	0.289	0.609	0.680
(30, 8)	1	0.223	0.192	0.491	0.503
	2	0.310	0.297	0.607	0.619
	3	0.276	0.268	0.565	0.565
(50, 8)	1	0.127	0.150	0.338	0.413
	2	0.301	0.183	0.450	0.496
	3	0.208	0.164	0.401	0.454
(20, 12)	1	0.205	0.159	0.370	0.395
	2	0.321	0.166	0.461	0.463
	3	0.267	0.185	0.411	0.421
(30, 12)	1	0.177	0.086	0.261	0.186
	2	0.237	0.127	0.277	0.268
	3	0.197	0.101	0.223	0.230
(50, 12)	1	0.048	0.043	0.131	0.085
	2	0.087	0.097	0.187	0.189
	3	0.075	0.067	0.165	0.123

Table 6 Mean values of the bias and MSE when θ, β, τ and η are set at 1.5, 1.2, 4 and 12, respectively	(n, m)	Schemes	Values of θ		Values of β	
			Bias	MSE	Bias	MSE
	(20, 8)	1	0.416	0.250	0.405	0.441
		2	0.596	0.336	0.486	0.505
		3	0.467	0.290	0.433	0.468
	(30, 8)	1	0.353	0.190	0.356	0.367
		2	0.426	0.287	0.407	0.419
		3	0.395	0.258	0.388	0.399
	(50, 8)	1	0.307	0.155	0.267	0.302
		2	0.386	0.178	0.370	0.356
		3	0.545	0.186	0.301	0.327
	(20, 12)	1	0.247	0.159	0.271	0.290
		2	0.299	0.166	0.357	0.361
		3	0.278	0.185	0.313	0.323
	(30, 12)	1	0.170	0.087	0.166	0.178
		2	0.217	0.117	0.287	0.259
		3	0.196	0.103	0.217	0.225
	(50, 12)	1	0.056	0.053	0.101	0.086
		2	0.107	0.087	0.137	0.167
		3	0.085	0.070	0.111	0.108

5 Results

As the sample size increases, the biases and MSEs of the estimated parameters decrease. This indicates that the maximum likelihood estimators are consistent and asymptotically normally distributed.

6 Conclusion

In this study, we have considered estimating and analysing data of failure times under SSPALT based on adaptive Type-II hybrid censoring using maximum likelihood approach. The mathematical model related to the lifetime of the test units is assumed to follow Rayleigh distribution. Using Newton–Raphson method, we obtain the numerical values of MLEs of model parameters. Their performances are analysed and discussed in terms of MSE and bias. It is seen that under adaptive progressive hybrid censoring scheme, a better efficiency in estimation of parameters is observed because of the large available sample size obtained under APHC. Therefore, APHC is better option to use in order to obtain a higher efficiency of estimates of model parameters.

References

1. Nelson, W.: Accelerated Life Testing: Statistical Models, Test Plans and Data Analysis. Wiley, New York (1990)
2. Goel, P.K.: Some estimation problems in the study of tampered random variables. Ph.D. thesis, Department of Statistics, Carnegie-Mellon University, Pittsburgh, Pennsylvania (1971)
3. Balakrishnan, N., Ng, H.K.T.: Precedence-Type Tests and Applications. Wiley, Hoboken (2006)
4. Childs, A., Chandrasekar, B., Balakrishnan, N., Kundu, D.: Exact likelihood inference based on Type-I and Type-II hybrid censored samples from the exponential distribution. Ann. Inst. Statist. Math. **55**, 319–333 (2003)
5. Gupta, R.D., Kundu, D.: Hybrid censoring schemes with exponential failure distribution. Comm. Statist. Theory Methods **27**, 3065–3083 (1988)
6. Kundu, D.: On hybrid censoring Weibull distribution. J. Statist. Plann. Inference **137**, 2127–2142 (2007)
7. Balakrishnan, N., Kundu, D.: Hybrid censoring: models, inferential results and applications. Comput. Stat. Data Anal. **57**, 166–209 (2013)
8. Ismail, A.A.: Estimating the parameters of Weibull distribution and the acceleration factor from hybrid partially accelerated life test. Appl. Math. Model. **36**(7), 2920–2925 (2012)
9. Balakrishnan, N.: Progressive censoring methodology: an appraisal. Test **16**, 211–296 (2007)
10. Balakrishnan, N., Aggarwala, R.: Progressive Censoring: Theory, Methods and Applications. Birkhäuser, Boston (2000)
11. Kundu, D., Joarder, A.: Analysis of Type-II progressively hybrid censored data. Comput. Statist. Data Anal. **50**, 2509–2528 (2006)
12. Ng, H.K.T., Kundu, D., Chan, P.S.: Statistical analysis of exponential lifetimes under an adaptive hybrid Type-II progressive censoring scheme. Nav. Res. Logist. **56**, 687–698 (2009)
13. Lin, C.T., Ng, H.K.T., Chan, P.S.: Statistical inference of Type-II progressively hybrid censored data with Weibull lifetimes. Commun. Statist. Theory Methods **38**, 1710–1729 (2009)
14. Mokhtari, E.B., Rad, A.H., Yousefzadeh, F.: Inference for Weibull distribution based on progressively Type-II hybrid censored data. J. Statist. Plann. Inference **141**, 2824–2838 (2011)
15. Alma, O.G., Belaghi, R.A.: (2015). On the estimation of the extreme value and normal distribution parameters based on progressive type-II hybrid censored data. J. Stat. Comput. Simul. 1–28
16. Ismail, A.A.: Inference for a step-stress partially accelerated life test model with an adaptive Type-II progressively hybrid censored data from Weibull distribution. J. Comput. Appl. Math. **260**, 533–542 (2014)
17. DeGroot, M.H., Goel, P.K.: Bayesian estimation and optimal design in partially accelerated life testing. Nav. Res. Logist. Q. **16**(2), 223–235 (1979)
18. Miller, R.C.: Survival analysis. Wiley, NewYork (1981)

A Three-Layer Approach for Overlay Text Extraction in Video Stream

Lalita Kumari, Vidyut Dey and J.L. Raheja

Abstract Overlaid texts are annotated text on video frames embedded externally for providing additional information to viewer of video sequences. The externally embedded texts can be used for auto-indexing and searching of video files in a video library using contextual contents inside video files. In this paper, we proposed a novel algorithm to detect and extract the overlaid text in digital video which allows users to get a much deeper understanding of video content. The proposed algorithm uses SVM as machine learning approach to filter/extract text more accurately. It uses multi-resolution processing algorithm due to which the proposed algorithm is able to extract embedded text of different font size from same video frame. Text detection from video sequences enables us to auto-indexing of video based on text embedded on video frames. Embedded texts enable deaf and hard-of-hearing users to watch videos. It is also useful for the people, who have hearing impairments from understanding the content of video. It also helps to those kinds of people who want to watch video in sound-sensitive environments.

Keywords Video stream · Overlaid text · Text identification · Local binary pattern · Video optical character recognition

L. Kumari (✉)
Department of Electronics and Communication Engineering, NIT Agartala, Jirania, India
e-mail: kumaril2003@yahoo.co.in

V. Dey
Department of Production Engineering, NIT Agartala, Jirania, India
e-mail: vidyut.pe@nita.ac.in

J.L. Raheja
Digital System Group, CSIR/CEERI, Pilani, Rajasthan, India
e-mail: jagdish.raheja.ceeri@gmail.com

© Springer Nature Singapore Pte Ltd. 2018
M. Pant et al. (eds.), *Soft Computing: Theories and Applications*,
Advances in Intelligent Systems and Computing 584,
https://doi.org/10.1007/978-981-10-5699-4_9

1 Introduction

Image processing includes task such as object classification, detection, and tracking in which many authors such as [1–3] has given remarkable contribution. Nowadays text detection extraction from natural scene becomes a popular, but it is still a challenging task due to different font style and font size. Recently on social network, unstructured video data populated rapidly. Text in a video frame can be categorized into two subparts: scene text and overlaid text. Scene text are those text in video frame which are captured naturally at the time of video recording [4]. Overlaid text are external text, embedded at latter stage for better understanding of the video [5–7]. Video-related acquisition faces many problem because of low quality input frame, which results long computation time in text detection [8–13]. Extracted text can be classified into text region or non-text region. For region growing, we have classification algorithm such as neural network, SVM k-mean clustering, in which we choose k-mean cluster.

Clustering forms a group of cluster by processing a set of data. k-mean clustering is used to select probable text blocks. We proposed k-mean clustering to identify text region and non-text region in video. In k-mean clustering we have used $k = 2$ and considered only high mean value in a cluster high mean value will be considered only. In result, k-mean clustering will give a set of blocks T. Shivakumara et al. [14] proposed a new text frame classification method in a combination of wavelet and median moment with k-means clustering to select probable text blocks. They used Max–Min Clustering approach to obtain dominant and high-contrast pixel. They divided whole frame of size 256 * 256 into 16 equally sized (64 * 64) to identify feature of the text.

Support vector machine is a supervised machine learning algorithm. SVM is placed as one of the most powerful families of machine learning approach. SVM is able to separate data in a higher dimension space. Moreover, SVM as a sword is capable of slicing and dicing complex dataset. SVM is useful to build a strong and powerful model. It works well when margin of separation is clear and when number of dimension is greater than the number of samples.

2 Related Work

SWT performed well for text detection from video [15–18] where as region growing is used in [19, 20]. Many approaches are being explored by researches for text detection such as Bayesian classifiers [19], Laplacian approach [8], MSER [21], Structure based partitioning [20], wavelet transform [22], etc. [23] proposed another approach to detect text from natural scene image with the help of local binarization followed by conditional random field (CRF). A two-stage scheme for text detection in video frames is proposed in [24]. In first stage, they used Canny edge detector to find the edge magnitude of image intensity, and after that, dilation is used to link the character edge of every text line. For second stage, they used SVM classifier and sliding window protocol to refine the bounding box which minimizes the false alarms. For text line detection, vertical and horizontal projection is used. Zhuge and Lu [25] proposed an algorithm to detect and locate text in videos. They explored MSER method for region detection. They developed GAM algorithm to solve the color bleeding problem. They also use top-/bottom-hat morphological operator to enhance the text contrast of complex background. Then, graph cuts method is applied for text segmentation. In this paper, they claim that GAM algorithm gives better edge detection compared to Sobel edge detection. The drawback of this paper is that it only works when text line is vertical or horizontal.

Motion perception field algorithm is proposed in [26] to detect both superimposed and scene text with multiple languages and multiple alignments in video. In this paper, MPF is applied to detect motion patterns. A mathematical morphology and region clustering-based text information extraction from Malayalam news videos is proposed in [27]. Maximum gradient (MGD) identifies text region and non-text region by assigning zero value for text region and nonzero value for non-text region. Thereafter Cross Correction Agglomerative clustering is used to detect video frame containing similar text. Lee et al. [28] proposed a method to detect the overlay name text in news program which works on rule-based characteristics in production of the TV news program. Canny edge detector is used to identify the overlay text beginning frame. This method is applied in such a way so density difference of beginning frame to current frame and previous frame is recognized. Yi in [29] proposed another method which work on color based clustering for text detection and Anthimopoulos in [30] used canny edge detector followed by connected components analysis multiple time with different resolutions.

3 Proposed Method

As discussed in Sect. 2, most of the algorithms are designed to be used in text detection from the scene image captured at specific external conditions such as text domain, light direction, image categories. Few algorithms are designed to detect overlaid text from video frame, but their efficiency/accuracy (as discussed in previous section) needs to be increased using machine learning approach for better result. Therefore, we proposed a three-layered algorithm for overlaid text detection

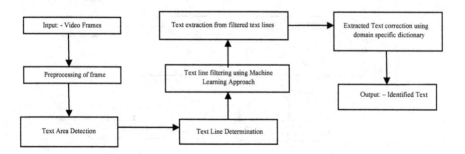

Fig. 1 Generalized flowchart of proposed algorithm

from large range of different video category using SVM and *k*-mean cluster. To
identify text having large range of font size simultaneously from a single video
frame, the proposed algorithm uses multi-resolution processing algorithm.
Overview flowchart of the proposed algorithm is shown in Fig. 1. This overview
flowchart describes the main work flow through which the proposed algorithm
moves forward in order to extract text from video frames. As Fig. 1 describes, our
proposed method constitutes following broad stages: image normalization and text
area detection through conventional method, text area refinement through machine
learning approach, text extraction from refined text area, and fourth one extracted
text validation/correction using domain-specific language dictionary. Each stage
shown in Fig. 1 contains multiple intermediate steps which is described in the detail
flowchart of the proposed algorithm is shown in Fig. 2.

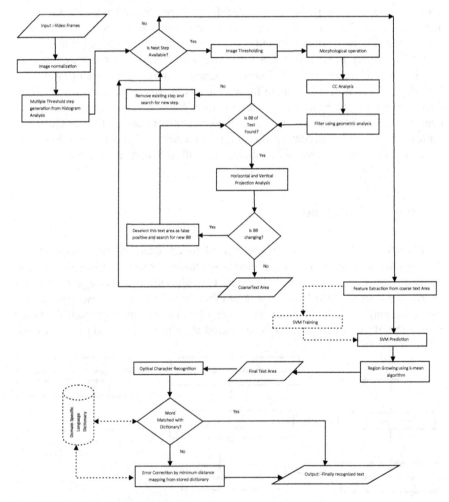

Fig. 2 Detailed flowchart of proposed algorithm

3.1 Text Area Detection Through Conventional Approach

This is the first layer of the text detection process. It consists of multiple subprocess within the first layer of the processing. Task of this layer is to normalize video frame, compute multiple thresholding step using histogram analysis, analyze connected components from the obtained different binary images, and determine text area.

First step is to normalize video frame image to achieve consistency in dynamic range as video frame may be of poor contrast level. After that, its histogram is analyzed and multiple threshold value is determined on the basis of found irregularity in histogram curve. Rest of the process is repeated for each level of this threshold value to identify/filter text areas. After image thresholding/edge detection, image dilation operation and morphological operation are applied to remove noise and to smooth up the text area candidates. Figure 3 shows a sample input video frame with its histogram and corresponding binary image having connected components as bounding box. Connected component analysis is performed to figure out geometric properties of each candidate text area in the form of bounding box. Geometric properties such as solidity, eccentricity, extent, Euler number, aspect ratio are analyzed to filter out candidate text area from non-text areas. After that, horizontal and vertical projection analysis is performed on each found candidate text area to get confidence on the selection. If projection of the text area suggests extension of bounding box area, it is rejected. After performing above-mentioned steps for all threshold level, we finally obtain coarse text area which is further refined through machine learning approach.

3.2 Text Area Refinement Through Machine Learning Approach

Text area detected through conventional approach produces false-positive areas along with the candidate text area. To implement method similar to optical human perception, so that false alarm would be avoided, machine learning approach is

Fig. 3 Text area detection through conventional approach. **a** Histogram of input video frame. **b** Thresholding from one of the multiple threshold steps. **c** Detected text areas

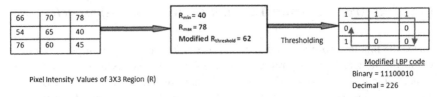

Pixel Intensity Values of 3X3 Region (R)

Modified LBP code
Binary = 11100010
Decimal = 226

Fig. 4 Modified LBP calculation for region R

required. For this purpose, our proposed algorithm uses support vector mechanism (SVM) for training and prediction followed by k-mean algorithm as region growing algorithm to refine the detected test area.

Texture segmentation is best suited for image analysis task which is captured by local binary pattern (LBP). We have adjusted LBP so that it will be best suited for our task. A 3X3 LBP kernel is used to measure local contrast in order to obtain efficient texture classification. Instead of general LBP, we used modified local binary pattern (mLBP), where neighboring pixels are compared with average of the intensity values in the 3X3 window. Modified LBP calculation for 3X3 region is shown in Fig. 4. On applying LBP on a 2-D image, we get another 2-D image containing texture pattern of corresponding pixel in original image. After SVM prediction, k-mean-algorithm-based region growing algorithm is applied to produce the final result. Insertion of machine learning approach into text detection algorithm is only for removing the false-positive result.

3.3 OCR and Word Validation/Auto-Correction

After final detection of text area, optical character recognition (OCR) process is performed in order to recognize characters from the obtained text area. Character-to-word formation is done using character orientation and spacing between characters. Standard space and standard deviation are computed and analyzed to group characters to form a word. These words are compared with the specific language dictionary of a particular domain and are validated. If it is not found in the dictionary, then corrected word is chosen from dictionary which gives minimum value of character conflict count.

4 Experimental Results and Conclusion

For testing of proposed algorithm, we have used both standard dataset from ICDAR 2013 briefly and non-standard dataset mainly developed locally. We have used ICDAR dataset for training as well testing purpose. To get confidence in overlaid text detection, we used several news videos and movies clip with embedded

(a)   (b)   (c)

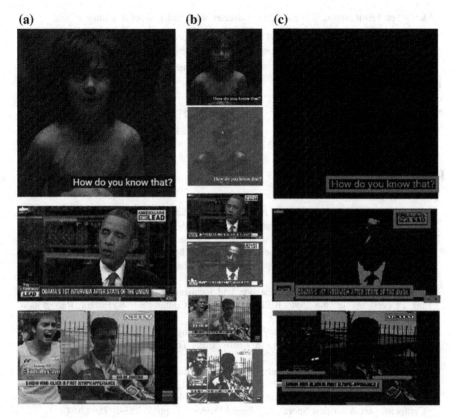

Fig. 5 Test result analysis

subtitles available at YouTube. This set has been generated by selecting mixture of both conventional dataset and non-conventional dataset. Text area selection, refinement, and text extraction analysis are shown in Fig. 5.

5 Conclusion

In this paper, we presented three-layered approach for overlaid text detection. This system consists of very efficient algorithm which combines conventional text area detection approach, machine-learning-based refinement approach, along with text extraction approach armed with domain-based auto-error detection and error correction algorithms. Comparative test result data is shown in Table 1. This unique three-layered approach improves the performance of the system. The main contribution of the system is texture/LBP intensity image mapping system and domain-based language dictionary implementation for error detection and error correction.

Table 1 Test result comparison on different dataset, with and without of dictionary use

Test input description	No. of tests performed	Accuracy without dictionary use		Accuracy with dictionary use	
		Precision	Recall	Precision	Recall
ICDAR 2013 scene text detection	10	0.82	0.84	0.82	0.85
Movies' scenes with subtitle	10	0.89	0.89	0.95	0.95
TV news video frames	10	0.84	0.86	0.84	0.89

References

1. Bosch, A., Zisserman, A., Munoz, X.: Image classification using random forests and ferns. In: 11th IEEE International Conference on Computer Vision, pp. 1–8 (2007)
2. Wolf, C., Jolion, J.: Object count/area graphs for the evaluation of object detection and segmentation algorithms. Int. J. Doc. Anal. Recognit. **8**(4), 280–296 (2006)
3. Li, H., Doermann, D., Kia, O.: Automatic text detection and tracking in digital video. IEEE Trans. IP **9**(1), 147–156 (2000)
4. Huang, X., Ma, H.: Automatic detection and localization of natural scene text in video. In: Proceedings of the 20th IEEE International Conference on Pattern Recognition, pp. 3216–3219, Aug 2010
5. Zhao, X., Lin, K.H., Fu, Y., Hu, Y., Liu, Y., Huang, T.S.: Text from corners: a novel approach to detect text and caption in videos. IEEE Trans. Image Process. **20**(3), 790–799 (2011)
6. Kim, W., Kim, C.: A new approach for overlay text detection and extraction from complex video scene. IEEE Trans. Image Process. **18**(2), 401–411 (2009)
7. Ekin, A.: Information based overlaid text detection by classifier fusion. In: IEEE International Conference on Acoustics, Speech and Signal Processing, pp. II-753–II-756 (2006)
8. Shivakumara, P., Phan, T.Q., Tan, C.L.: A laplacian approach to multi-oriented text detection in video. IEEE Trans. Pattern Anal. Mach. Intell. **33**(2), 412–419 (2011)
9. Li, Z., Liu, G., Qian, X., Guo, D., Jiang, H.: Effective and efficient video text extraction using key text points. IET Image Process. **5**(8), 671–683 (2011)
10. Ye, Q., Huang, Q.: A New text detection algorithm in images/video frames PCM. LNCS **3332**, 858–865 (2004)
11. Hua, X., Yin, P., Zhang, H.J.: Efficient video text recognition using multiple frame integration. IEEE Int. Conf. Image Process. (ICIP) **2**, 397–400 (2002)
12. Hua, X.-S., Chert, X.-R., Wenyin, L., Zhang, H.-J.: Automatic location of text in video frames. In: Proceedings of the 2001 ACM Workshops on Multimedia, 24–27 Sept 2001
13. Winger, L.L., Robinson, J.A., Jernigan, M.E.: Low-complexity character extraction in low-contrast scene images. Int. J. Pattern Recognit. Artif. Intell. **14**(2), 113–135 (2000)
14. Shivakumara, P., Dutta, A., Phan, T.Q., Tan, C.L., Pal, U.: A novel mutual nearest neighbor based symmetry for text frame classification in video. Pattern Recognit. **44**, 1671–1683 (2011)
15. Yang, H., Quehl, B., Sack, H.: A framework for improved video text detection and recognition. Multimed. Tools Appl. **69**(1), 217–245 (2014)
16. Liu, X., Wang, W.: Robustly extracting captions in videos based on stroke-like edges and spatio-temporal analysis. IEEE Trans. Multimed. **14**(2), 482–489 (2012)
17. Epshtein, B., Ofek, E., Wexler, Y.: Detecting text in natural scenes with stroke width transforms. In: IEEE Conference on Computer Vision and Pattern Recognition, San Francisco (2010)

18. Jung, C., Liu, Q., Kim, J.: A stroke filter and its application to text localization. Pattern Recogn. Lett. **30**(2), 114–122 (2009)
19. Shivakumara, P., Sreedhar, R.P., Phan, T.Q., Lu, S., Tan, C.L.: Multioriented video scene text detection through Bayesian classification and boundary growing. IEEE Trans. Circuits Syst. Video **22**(8), 1227–1235 (2012)
20. Yi, C., Tian, Y.: Text string detection from natural scenes by structure-based partition and grouping. IEEE Trans. Image Process. **20**(9), 2594–2605 (2011)
21. Chen, H., Tsai, S., Schroth, G., Chen, D., Grzeszczuk, R., Girod, B.: Robust text detection in natural images with edge-enhanced maximally stable extremal regions. In: Proceedings of the 18th IEEE International Conference on Image Processing, pp. 2609–2612, Sept 2011
22. Zhao, M., Li, S., Kwok, J.: Text detection in images using sparse representation with discriminative dictionaries. Image Vis. Comput. **28**(12), 1590–1599 (2010)
23. Pan, Y.F., Hou, X., Liu, C.L.: A hybrid approach to detect and localize texts in natural scene images. IEEE Trans. Image Process. **20**(3), 800–813 (2011)
24. Anthimopoulos, M., Gatos, B., Pratikakis, I.: A two-stage scheme for text detection in video images. Image Vis. Comput. **28**(9), 1413–1426 (2010)
25. Zhuge, Y.Z., Lu, H.C.: Robust video text detection with morphological filtering enhanced MSER. J. Comput. Sci. Technol. **30**(2), 353–363 (2015)
26. Huang, X., Ma, H., Ling, C.X., Gao, G.: Detecting both superimposed and scene text with multiple languages and multiple alignments in video. Springer Science + Business Media, LLC (2012)
27. Anoop, K., Gangan, M.P., Lajish, V.L.: Advances in Signal Processing and Intelligent Recognition Systems, Advances in Intelligent Systems and Computing. Springer International Publishing, Switzerland (2016)
28. Lee, S., Ahn, J., Lee, Y., Jo, K.: Beginning Frame and Edge Based Name Text Localization in News Interview Videos. ICIC 2016, Springer International Publishing, Switzerland, Part III, pp. 583–594 (2016)
29. Yi, J., Peng, Y., Xiao, J.: Color-based clustering for text detection and extraction in image. ACM MM 847–850 (2007)
30. Anthimopoulos, M., Gatos, B., Pratikakis, I.: Multiresolution text detection in video frames. In: International Conference on Computer Vision Theory and Applications, pp. 161–166 (2007)

A Comprehensive Review and Open Challenges of Stream Big Data

Bharat Tidke and Rupa Mehta

Abstract Research in big data becomes pioneer in the field of information system. Data stream is well-studied problem in traditional data mining environment, but still needs exploration while dealing with big data. This paper mainly reviewed different research activities, scientific practice, and methods which have been developed for streaming big data. In addition, examine well-known real-time platforms which are evolving to handle streaming problem and having existing similarity in terms of usage of main memory and distributed computing technologies for non-real-time data. Finally, summarize open issues and challenges faced by current technologies while acquisition and processing of big data in real time.

Keywords Big data · Stream data · Distributed mining

1 Introduction

Due to rapid growth of smart phones, access to Internet becomes quite easy, results in large amounts of unstructured data, which has been collected and stored from different areas of society [1]. In addition, real-time systems such as sensor-based technologies normally generate streams of data, which require quick storing as well as processing of incoming data. However, such data streams pose new features as compares to traditional stream data. Many applications such as traffic management, log data from Web search engines, Twitter, electronic mail also generates high volumes of stream data with velocity, which is difficult to handle with existing data streaming techniques [2, 3]. In past decade, big data comes into picture and can be

B. Tidke (✉) · R. Mehta
Department of Computer Engineering, SVNIT, Surat, India
e-mail: batidke@gmail.com1

R. Mehta
e-mail: rgm@coed.svnit.ac.in

© Springer Nature Singapore Pte Ltd. 2018
M. Pant et al. (eds.), *Soft Computing: Theories and Applications*,
Advances in Intelligent Systems and Computing 584,
https://doi.org/10.1007/978-981-10-5699-4_10

defined in terms of its characteristics volume, velocity, variety, value, and veracity [3–5]. Many researchers proposed different techniques, tools, and complex processes for getting insight into different characteristics of big data. Exploring stream data with velocity is key challenge in big data research, which has been focused by many researchers but still has potential to explore for many applications and domains. Also, processing of stream data in real time differs from non-real-time data processing, since data has to be analyzed based on historic stored data, before itself gets stored for further analysis and prediction of upcoming streams.

Motivation
Data has been generated and acquired at rapid speed which involves volume with it ultimately create challenge to develop methods which must be automated and can respond quickly to make decision in specified time. Since size of data is too big, such data needs to be moved and stored in distributed environment for further computation as traditional data warehouses are ill suited. Further analysis of such data using classic OLAP cube also does not work and replaced by distributed storage environment such as Hadoop which uses master–slave architecture for storing data, also map-reduced technique for processing data in batches and NoSQL databases which uses different storage techniques having columns, graphs, documents, key-value stores which can work on top of Hadoop to make it suited for streaming big data. Some assumption that has been followed by traditional system while dealing with stream data mining can be solved using distributed data processing frameworks or tools for handling the problem of big data.

- Possible to collect and store whole data stream, not the sample or summaries of data.
- Integration and indexing of data in real time irrespective of the format in which they came.
- Velocity with which data come can be processed using distributed streaming algorithm in real time and stored in distributed fashion for improving existing model for further analysis.
- Analyzing existing or past data is crucial while making targeted future prediction, but decision based on operational or transactional data needs real-time analysis has to be processed in parallel with low-latency time.

2 Related Work

Data stream can be conceived as a continuous and changing sequence of data that continuously arriving at a system to store and process. Stream data processing deals with some or all data input as one or more continuous data stream.

2.1 Data Stream Mining

They explained different factors which are necessary to mine information from streaming data including time window which further can be divided into landmark window basically takes whole new data stream as a window instead of sample and considered them equally important which may cause problem to build model with limited memory. Another one is sliding window, one of the most used windowing technique in the field of stream mining that only takes recent data stream and discarded the old ones, and also, it is flexible depending upon the accuracy needed by the model makes it popular. Next one is fading window usually assigned weight to the data according to its arrival time newer one having higher as compare to older one and finally Tilted time window lies between sliding and fading window in terms of its variance.

2.2 Big data Stream Mining

In past few years, mostly research is based on collecting and storing data due to growth of Web 2.0 technologies and increase in bandwidth for data transfer. Many sources of data have been evolved and are mostly need real-time analysis of such stream data and also information extracting algorithms to analyze it. Even Hadoop has been used by Yahoo in its earlier days to collect, store, and analyze large volume of click stream data, which later used by ecommerce enterprises to solve the problems of customers, while choosing product and other followed product normally user tend to purchase. In addition, recommending similar products they want to purchase in future is based on their purchased experience. Many data mining algorithms have been proposed to overcome the challenges of stream data as seen in above section, but to overcome volume, velocity, and volatility challenges [3], a standard framework based on Lambada [6] architecture having different phases ranging from stream data collection to data visualization has been overview.

Nowadays, concept of "**Smart Cities**" are on its evolving stages, and data gathered using different technologies such as sensors from different aspects such as users location, social gathering information, ITS, temperature changes produce data. Similarly social media sites such as Twitter Facebook, Linkedin also generates large amount of data are some of main sources of big stream real-world data [7]. After acquisition of data from different sources the most important phase in mining any unstructured data is preprocessing unfortunately it has not been yet explored fully in terms of big data. Since data in real world is dirty as well as noisy, same is applied for real streaming data which makes it worse to analyze data which has not been preprocessed.

Beginning of any streaming processing paradigm was based on hidden information that comes with incoming data that further can be used to obtain useful results to do analysis. In this process, since data is arriving continuously and in

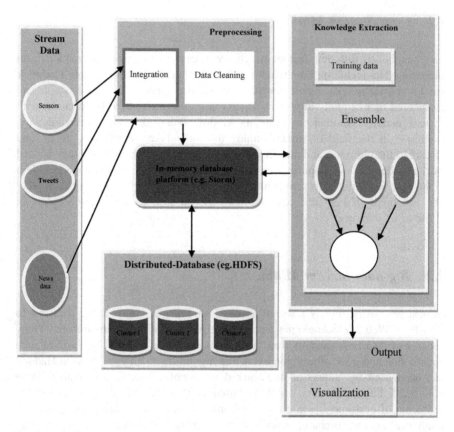

Fig. 1 Architecture for handling stream big data

huge amount, only a small fraction of stream data is stored in limited memory databases and process using stream processing system such as storm or kafka as shown in Fig. 1. Further, it can be stored in large and distributed databases such as HDFS for future use. Many machine learning algorithms have been used to extract hidden information from stream data. Different approaches present by authors has been discussed below and summarized in Table 1.

Rutkowski et al. [8] in one of his paper suggested that algorithms based on Hoeffding's bound which considered as one of the most used decision-tree technique in mining data stream needs to be revised and proposed a method using Mcdiarmid's inequality to split node in a tree by picking correct attribute, and they performed several experiments and evaluated their result using splitting measures Gini index and information gain.

Limitation of this method is that to split node among given n node, it needs to scan huge number of data elements before selecting right attribute, later this limitation has been overcome in [9] in that they used statistical method for selecting attribute to split node among given n node based on Taylor's theorem and

Table 1 Summary of various approaches proposed on stream big data

Authors	Handles big data	Platform	Technique	Model or algorithm	Evaluation measure	Dataset	Environment
Vu et al. [13]	Yes	SAMOA	Regression	VAMR (vertical adaptive model rules) and HAMR (hybrid adaptive model rules)	Mean absolute error (MAE) and root-mean-square error (RMSE)	UCI machine learning repository	Distributed
Bifet et al. [19]	Yes	SAMOA	Classification	Randomized ensembles	McNemar's test, sign test and Wilcoxon's signed-rank test. Statistic, kappa-temporal statistic ADWIN	UCI machine learning repository	Centralized
Fegaras et al. [14]	Yes	Hadoop and spark	Apache MRQL	Incremental	Group by, join-group by, and k-means clustering	Synthetic data	Distributed
Marron et al. [15]	No	GPU	Ensemble random forest	GVFDT (very fast decision trees on GPUs)	Classification accuracy and time	Synthetic data	Centralized
Khalilian et al. [27]	Yes	Vector space	Clustering	DCSTREAM	Mean difference, std. error, precision, recall, F-measure	KDDCUP 99 and synthetic datasets	Centralized
Fong et al. [28]	Yes	MOA	Accelerated particle swarm optimization (APSO)	Swarm search-feature selection (SS-FS)	Accuracy, kappa (kappa statistics), TP-FP rate, precision, recall, F-measure	Sensor data	Centralized

(continued)

Table 1 (continued)

Authors	Handles big data	Platform	Technique	Model or algorithm	Evaluation measure	Dataset	Environment
Yun et al. [16]	No	Sliding window	Frequent pattern mining	Weighted erasable pattern mining algorithm suitable for sliding window-based data stream (WEPS)	Runtime, memory usage, and scalability	FIMI repository, synthetic datasets r	Centralized
Agerri et al. [12]	Yes	STORM	Linguistic processors on virtual machines	NLP tools	Performance gain and time	Car dataset and wiki news dataset	Distributed
Duarte et al. [29]	No	MOA	Regression	Adaptive model rules	MAE and root-mean-squared error (RMSE)	UCI repository	Centralized

properties of the normal distribution to test evaluation of splitting criteria and proposed Gaussian decision-tree algorithm to improve the performance of mining streaming data. Again in [10, 11], they proposed firstly (mDT) algorithm based on splitting criteria called misclassification error combined with Gini index for creating tree node, which also decides the accurate attribute for existing and incoming stream data and secondly Decision Trees Based on the Hybrid Split Measure (hDT) which they tested on UCI repository dataset. Agerri et al. [12] presented new distributed and highly scalable architecture for analysis of stream textual news data using natural language processing (NLP). They performed experiment using different distributed pipeline modules on virtual machines and evaluated performance of the system using original incoming streaming news in which documents has been taken in reproducible manner. In addition, some limitation with proposed system still exists, and they suggested which can be solved by using distributed NoSQL databases like MongoDB.

Vu et al. [13] propose streaming algorithm based on AM rules Decision in distributed environment. This is first kind of experiment on adaptive rules in distributed platform for which they used SAMOA open-source software which basically built to deal large-scale data stream, and their main focus is to understand different decision rules in terms of regression. Fegaras [14] proposed framework based on incremental approach for distributed stream data, mainly focuses to improve traditional batch processing as used by map-reduced function in Hadoop, making it iterative incremental batch processing, enables it to store processing data in memory and tested their framework on dataset consist of complex arbitrary values, and also evaluated that their result are accurate instead of approximate. Marron [15] presented use of traditional classification algorithms such as random forest and VFDT for mining large amount of stream data using GPU, making these algorithm to run in parallel to deal with the volume of big data. They compare performance of Very Fast Decision Tree on GPU (GVFDT) and Random Forest algorithms with similar platform such as MOA and VFML having both algorithm, and they found that their results are better in terms of speed as well as accuracy. Yun et al. [16] worked on frequent pattern mining using sliding window technique and proposed algorithm WEPS (Weighted Erasable Pattern mining algorithm on sliding window-based data Streams) in that they assigned weight ti nodes of tree for creation and pruning purpose. The proposed architecture has been divided into two parts; first phase mainly concentrating on sliding window in that tree creation and recreation have been performed, and in second part, they prune the pattern based on weight assigned to it. Zliobaite and Gabrys [17] Proposed automated preprocessing technique based on adaptive technique for three different cases. In each one, different adaptive model has been used for preprocessing data as well for prediction using different techniques such as incremental approach ensemble classifier (Table 2).

Table 2 Summary of various processing models for streaming big data

System/tools	Processing model	Stream type	Operating system	Open source	Built-in language	Supportive languages	Current release/version	Developed at	Available on	Function	Features
Kafka	Batch and real time	Tuples	OS independent	Yes	Clojure	Any language	0.9.6	Back type	Apache software foundation	Distributed real-time computation	1. Secure 2. Multi-tenant deployment
Flink	Batch and real time	Strings	OS independent	Yes	Java and scala	Java, Scala, and Python	0.10.2	Data artisans	Apache software foundation	Distributed real-time computation	1. Low-latency stream processor 2. Flexible operator state and streaming windows
Spark	Batch and streaming	Discretized stream	Windows and Linux	Yes	Scala	Scala, Java, Python, R	1.6.0	UC, Berkeley	Apache software foundation	Large-scale data processing	1. Decentralized hides all cluster management tasks 2. Checkpointing and recovery minimize state loss
S4	Real time	Event	Windows and Linux	Yes	Java	Any language	0.6.0	Yahoo	Apache software foundation	Processing continuous Stream	Flexible deployment
Samza	Batch and real time	Messages	OS independent	Yes	Scala, Java	CQL, Pig	0.10.0	LinkedIn	Apache software foundation	Processing continuous stream	1. Simple API 2. Managed state 3. Fault tolerance 4. Durability 5. Pluggable processor isolation

3 Challenges in Various Domains

There are various challenges have been focused by different authors [18–21] and some of them are.

3.1 GIS

Existing technology available in geographic information system (GIS) is mainly concentrating on conventional databases, dealing normally with static data which limits it when it comes to analyzing big data. A new tools and techniques for GIS in terms of big data is require to meet new changing environment of spatial databases. Liu et al. [22] big data show that how big data can revolutionized the world of GIS system

3.2 Human Mobility Patterns

Due to large usage of smart phones, sensor-based mobiles are in the pockets of millions, so data of each individual and their traveling habits can be explore, but such data comes not only volume based, but with velocity and because of its spatial nature, in terms of variety as well. Gonzalez et al. [23] shows that individual human have high degree of temporal and spatial pattern regularity which can be used in epidemic prevention, emergency response urban planning as well as agent-based modeling [24].

3.3 Space Technology

The Sloan Digital Sky Survey has collected data which compromises of around 500 million photometric observations of objects from the sky which makes job of space scientist easy to get data without sending astronomer's into the sky. But still extracting knowledge from such vast collection of big data is tedious job. Business enterprises are also using big space data to carry their operation in different remote parts of the world [25, 26].

4 Conclusion and Unsolved Issues

Big Data means opportunities for different section of society to grow virtually, and most of the data are continuous in nature which creates research challenges to extract information from such huge volume of stream data. In this paper, we have presented concept of stream big data and highlighted the general architecture which can be for stream big data, also provided a literature survey on numerous techniques and mechanisms for getting information from stream big data. Still there are many issues need to be focused for usage of big data.

- *Decision science*

Building system for real-time analytics which can transfer data science into decision science becomes vital to cope up with enormous need of today's information system.

- *Distributed algorithms*

Many frameworks have been developed for distributed computing, but for analyzing and predicting accurate information for such application, there is a need to have distributed data mining algorithms.

References

1. Lohr, S.: The age of big data. New York Times 11 (2012)
2. Fan, W., Bifet, A.: Mining big data: current status, and forecast to the future. ACM SIGKDD Explor. Newsl. **14**(2), 1–5 (2013)
3. Labrinidis, Alexandros, Jagadish, H.V.: Challenges and opportunities with big data. Proc. VLDB Endow. **5**(12), 2032–2033 (2012)
4. Chen, C.P., Zhang, C.-Y.: Data-intensive applications, challenges, techniques and technologies: a survey on Big Data. Inf. Sci. **275**, 314–347 (2014)
5. Chen, M., Mao, S., Liu, Y.: Big data: a survey. Mob. Netw. Appl. **19**(2), 171–209 (2014)
6. Aggarwal, C.: Data streams: models and algorithms. Springer, Berlin (2007)
7. Nguyen, H.-L., Woon, Y.-K., Ng, W.-K.: A survey on data stream clustering and classification. Knowl. Inf. Syst. **45**(3), 535–569 (2015)
8. Rutkowski, L., Pietruczuk, L., Duda, P., Jaworski, M.: Decision trees for mining data streams based on the McDiarmid's bound. IEEE Trans. Knowl. Data Eng. **25**(6), 1272–1279 (2013)
9. Rutkowski, L., Jaworski, M., Pietruczuk, L., Duda, P.: Decision Trees for mining data streams based on the Gaussian approximation. IEEE Trans. Knowl. Data Eng. **26**(1), 108–119 (2014)
10. Rutkowski, L., Jaworski, M., Pietruczuk, L., Duda, P.: The CART decision tree for mining data streams. Inf. Sci. **266**, 1–15 (2014)
11. Rutkowski, L., Jaworski, M., Pietruczuk, L., Duda, P.: A new method for data stream mining based on the misclassification error. IEEE Trans. Neural Netw. Learn. Syst. **26**(5), 1048–1059 (2015)
12. Agerri, R., Artola, X., Beloki, Z., Rigau, G., Soroa, A.: Big data for natural language processing: a streaming approach. Knowl. Syst. **79**, 36–42 (2015)
13. Vu, A.T., De Francisci Morales, G., Gama, J., Bifet, A.: Distributed adaptive model rules for mining big data streams. In: IEEE International Conference on Big Data, pp. 345–353 (2014)

14. Fegaras, L.: Incremental query processing on big data streams. IEEE Trans. Knowl. Data Eng. **28**(11), 2998–3012 (2016). doi:10.1109/TKDE.2016.2601103
15. Marron, D., Bifet, A., De Francisci Morales, G.: Random forests of very fast decision trees on GPU for mining evolving big data streams. ECAI **14** (2014)
16. Yun, U., Lee, G.: Sliding window based weighted erasable stream pattern mining for stream data applications. Future Gener. Comput. Syst. (2016)
17. Zliobaite, I., Gabrys, B.: Adaptive preprocessing for streaming data. IEEE Trans. Knowl. Data Eng. **26**(2), 309–321 (2014)
18. Domingos, P., Hulten, G.: Mining high-speed data streams. In: Proceedings of Sixth ACM SIGKDD International Conference on Knowledge Discovery and Data Mining, pp. 71–80 (2000)
19. Bifet, A., de Francisci Morales, G., Read, J., Holmes, G., Pfahringer, B.: Efficient online evaluation of big data stream classifiers. In: Proceedings of the 21th ACM SIGKDD International Conference on Knowledge Discovery and Data Mining, pp. 59–68 (2015)
20. Krempl, G., Žliobaite, I., Brzeziński, D., Hüllermeier, E., Last, M., Lemaire, V., Noack, T., Shaker, A., Sievi, S., Spiliopoulou, M., Stefanowski, J.: Open challenges for data stream mining research. ACM SIGKDD Explor. Newsl. **16**(1), 1–10 (2014)
21. Gaber, M., Zaslavsky, A., Krishnaswamy, S.: Mining data streams: a review. SIGMOD Rec. **34**(2), 18–26 (2005)
22. Liu, J., Li, J., Li, W., Wu, J.: Rethinking big data: a review on the data quality and usage issues. ISPRS J. Photogramm. Remote Sens. (2015)
23. Yue, P., Jiang, L.: BigGIS: how big data can shape next-generation GIS. In: Third International Conference on Agro-geoinformatics (Agro-geoinformatics 2014), IEEE (2014)
24. Gonzalez, M.C., Hidalgo, C.A., Barabasi, A.-L.: Understanding individual human mobility patterns. Nature **453**(7196), 779–782 (2008)
25. SDSS-III: Massive Spectroscopic Surveys of the Distant Universe, the Milky Way Galaxy, and Extra-Solar Planetary Systems, Jan 2008, http://www.sdss3.org/collaboration/description.pdf/
26. Jagadish, H.V., Gehrke, J., Labrinidis, A., Papakonstantinou, Y., Patel, J.M., Ramakrishnan, R., Shahabi, C.: Big data and its technical challenges. Commun. ACM **57**(7), 86–94 (2014)
27. Madjid, K., Mustapha, N., Sulaiman, N.: Data stream clustering by divide and conquer approach based on vector model. J. Big Data **3**(1) (2016)
28. Fong, S., Wong, R., Vasilakos, A.V.: Accelerated PSO swarm search feature selection for data stream mining big data. IEEE Trans. Serv. Comput. **9**(1), 33–45 (2016)
29. Joao, D., Gama, J., Bifet, A.: Adaptive model rules from high-speed data streams. ACM Trans. Knowl. Discov. Data (TKDD) **10**(3), 30 (2016)
30. Beyer, M.A., Laney, D.: The importance of "Big Data: a definition. Gartner, Stamford (2012)

Biomass Estimation at ICESat/GLAS Footprints Using Support Vector Regression Algorithm for Optimization of Parameters

Sonika and Priya Rathi

Abstract Forest aboveground biomass is a great significance for a better understanding of global carbon cycle. Estimation of biomass at ICESat/GLAS footprint was done by integrating datasets from single sensors. The biomass estimation accuracy of support vector machine regression was studied. Multiple linear regression equations were established from some of the most important variables found using support vector machine algorithm. The study also introduced a more accurate method of data collection from ICESat footprints. The results of the study were very encouraging. SVM gives AGB prediction where the best 6 variables had an RMSE of 18.502 t/ha. The study conclusively established that SVM is capable of estimating the biomass single sensor approach in predicting AGB. The outcome of the study was the formulation of an approach that could result in predicted biomass estimation. The study finally analyzed its limitations and suggested improvements that would result in even better estimation accuracies.

Keywords ICESat · GLAS · Support vector machine · AGB · LiDAR

1 Introduction

Forestry plays an essential part in the worldwide carbon spending plan in the progression of the earthly carbon cycle [1]. Potential carbon discharges could be discharged to the environment because of deforestation, and precise biomass estimation is important to comprehend the effects of deforestation on worldwide

Sonika (✉)
Faculty of Computer Science & Engineering, Indraprastha Institute of Technology,
Amroha, India
e-mail: saxenasonalika12@gmail.com

P. Rathi
Banasthali Vidyapith, Jaipur, India
e-mail: priyarathii77@gmail.com

© Springer Nature Singapore Pte Ltd. 2018   101
M. Pant et al. (eds.), *Soft Computing: Theories and Applications*,
Advances in Intelligent Systems and Computing 584,
https://doi.org/10.1007/978-981-10-5699-4_11

change. In ranger service, biomass is characterized as the stove dry mass of the forest [2].

A worldwide evaluation of biomass and its flow is a key segment to climatic change gauging models and adjustment techniques and by assessing the basic qualities and attributes (AGB) of the environments. The measure of carbon isolated can be evaluated dry weight of biomass [3]. Generally, the AGB is measured by utilizing the width of a tree, yet this immediate strategy for biomass estimation is additional time taken, expensive, and now and then dangerous. There are two sorts of remote detecting; one is active remote sensing and another is passive remote sensing. Latent sensors must be utilized to identify vitality when they actually occur in energy is accessible. A dynamic sensor can be utilized for analyzing wavelengths that are not adequately given by the sun. LiDAR stands for light detection and ranging which is a dynamic sensor, and it specifically measures the vertical part of vegetation which has utilized the possibility to gauge the auxiliary vegetation qualities [4].

2 Literature Review

Carbon sequestration and biomass appraisal have been the most imperative things on the motivation of environmental change. Remote detecting consolidated with precise field information accumulation can possibly grow exceptionally dependable biomass models and maps. Woodland skeletal parameters contain overhang stature, tree volume, DBH (diameter of bosom tallness), and biomass.

2.1 Measurement of Biomass

Biomass measurement should be possibly utilizing immediate and backhanded strategies. The point direct techniques are to set up connections between the field measured biomass and parameters recorded by satellite. These strategies use systems, for example, different relapse approach; machine learning calculations; and factual troupe models for building up the relationship. Then, again roundabout strategies in light of stand qualities are obtained from the remote detecting information itself and after that used to foresee timber volume and biomass estimation. The evaluations acquired from both techniques are at last increased by a biomass development element (BEF) to arrive all out over the ground biomass (AGB).

2.2 Estimation of Biomass Using Different Datasets

There are numerous sorts of remote detecting methods which are generally utilized as a part of estimation of backwoods biomass. Prior to the rise of LiDAR, RADAR

(radio recognizing and extending) and passive optical remote detecting were utilized. The estimation of AGB utilizing remote detecting is typically accomplished
by setting up relapse conditions between field plot biomass and the remote detecting
information [3, 5]. There are three distinct sorts of LiDAR for the estimation of
biomass.

2.2.1 Airborne LiDAR

Laser vegetation imaging sensor (LVIS) is the sort of airborne system and echo
recovery (SLICER) of Canopies for Scanning LiDAR imager; it is helpful in
research applications. Airborne LiDAR which has sensor tallness under 2 km, it
conveying 20 or all the more than 20 focuses per m^2 and a mass number of reflec-
tions along the full waveform joined with various look points can recoup single tree
portrayal [6]. The provided details regarding straight biomass estimations in light of
full waveform created measurements accomplishing precision of around 72%.

2.2.2 Spaceborne LiDAR

Spaceborne LiDAR is the investigation of capacities of LiDAR on satellite stages.
This full-wave laser framework ICESat/GLAS has been use to find biomass
parameters to furthermore shade tallness for hugeland zones. The GLAS (geoscience laser altimeter framework) is the main LiDAR in working spaceborne
framework. Spaceborne information is basically used to demonstrate assessing
carbon spending plan of the worldwide covering stature [7].

2.2.3 Terrestrial LiDAR

The other one is area-based laser scanner that is terrestrial LiDAR, which is consolidated with a profoundly exact differential worldwide situating system. The
control of the utilization of earthly framework; there was issue on impact of item
shading and wind. Earthbound laser scanners give the rapidly and efficiently
detailed data and profoundly exact time-of-flight 3D information. Applications have
wide run including mining, topography, architecture, as-built surveying, monitoring, archaeology, civil engineering furthermore in city modeling.

3 Importance of Synergy of Sensors for AGB Estimation

The appropriateness of various remote detecting information sorts for certain
woodland parameters will remove the utilization of information sets to give the
asked for biomass mapping items [8]. In earlier year, study reported that models

that incorporate the mean Crown range, got from single tree delineation together with tree stature assess and sort extents, make accessible measurement, better wood quality and age evaluations where without the single tree data. In their study, the watched root mean square error (RMSE) for regard volume evaluation utilizing factors for tree tallness and vegetation structure was around 30% in profoundly organized, joined stands.

4 ICESat/GLAS Large Footprint Full-Waveform LiDAR in Forestry

ICESat was an exact satellite which is dispatched by the National Aeronautics and Space Administration (NASA) in January 2003. This satellite propelled with the geoscience laser altimeter system (GLAS), which contains three lasers (Laser 1, Laser 2, and Laser 3). NASA's GLAS remains for Ice, Cloud and Land Elevation Satellite (ICESat) that is intended to gauge the shelter statures, ice sheet height changes, and land surface geology plot [9]. In entire estimation, "Gaussian disintegration" is broadly used to recover data from GLAS waveform measurements.

5 Optimizing Techniques—Support Vector Machine Regression

Various philosophies have been worked upon the persistent vegetation parameters which are for the most part of three classes, viz. physically based models [10] and direct models. Direct models characterize straightforward straight relationship between ward factors (AGB, Canopy tallness, and so on.) and free factor (remotely detected information).

SVM is a directed grouping and relapse strategies in view of factual learning hypothesis. In the SVM, there are four bits, utilizing a piece capacity; support vector relapse maps the information space (e.g., free factors) to an extraordinary dimensional space where complex nonlinear choice limits between classes get to be straight. SVM regression applies the idea of a ε-harsh misfortune work that overlooks point mistakes inside a separation of ε from the genuine worth by weighting them with zero. The arrangement is gotten through a little subset of preparing focuses, and the bolsters (vectors from focuses closest to choice limit) contain all the obliged data to characterize the capacity and result in "to a great degree effective calculations." SVR minimizes grouping blunder on concealed information without earlier presumptions made on the likelihood appropriation of the information.

Fig. 1 Study area with overlaid ICESat footprints

6 Study Area

The critical standard for study site is the accessibility of the ICESat/GLAS waveform information in the study range. This is exceptionally essential keeping in mind the end goal to relate the biomass with the textural parameters, ghostly parameters, and LiDAR datasets.

The chosen region for the study is in Doon valley of western part (Timli range) (29.7–30.7 °N and 77.40–77.5 °E) (Fig. 1).

7 Methodology

See Fig. 2.

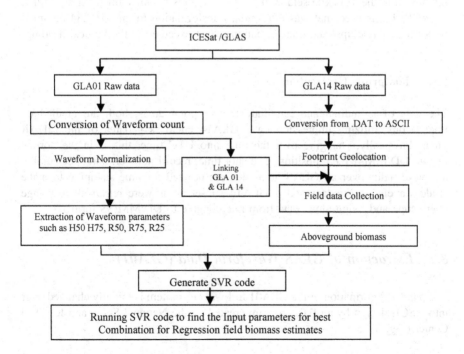

Fig. 2 Detailed methodology of the study

8 GLAS Data Processing

ICESat/GLAS is a spaceborne LiDAR system. GLAS has 15 information items, out of them two items (GLA14 and GLA01) are connected in our study. GLA01 remains for Level 1A Global Altimetry Data that encases data about file number, shot time, and shot number. GLA01 likewise gives the stature.

8.1 Extraction of GLAS Elevation Data (GLA14)

8.1.1 Conversion to ASCII File

The elevation product is assigned in paired organization (.DAT) that was changed over into ASCII group. For this renovation were utilized IDL, stands for interactive data dialect project, is created by NSIDC.

8.1.2 Ellipsoid Correction

For the extraction of the chosen factors from the ASCII documents, the ASCII record should be investigated to comprehend the information stockpiling. These records are expelled from the information set. The GLA14 information item is referenced to the Topex/Poseidon ellipsoid. An ellipsoid amendment from TOPEX to WGS84 was used that was ellipsoidal transformation to the WGS84 datum is performed. The ellipsoidal transformation is just executed in the vertical heading.

8.1.3 Footprints Geolocation

The data of latitudinal and longitudinal give point areas, both are cradled on impressions utilizing impression data. GLA14 encloses geoarea of the LiDAR pillar. Physically, changing over the time into UTC time, this is in the normal known DD/MM/YYYY position. The shot time, record number, and shot number are basic fields over all GLAS items that are utilized for time stamping to make crude waveform. Fundamentally, GLA14 subset items were removed to change over scope and longitudinal data from microdegree to DD:MM:SS format.

8.2 Extraction of GLAS Waveform Data (GLA01)

The parallel information of the GLA01 full-waveform item is initially changed over into ASCII design by an IDL program created by the National Snow and Ice Data Center (Fig. 3).

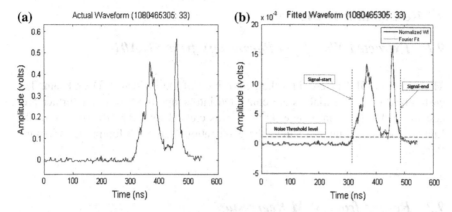

Fig. 3 **a** Voltage waveform before normalization and **b** gaussian fitted waveform

The changed over .dat records were the subset furthermore adjusted. Since the elevation from GLAS14 is for impression center and GLA01 for waveform, the following stride worried with normal field among GLA14 and GLA01.

8.3 Extraction of Waveform Variables

Finally, the code is able to generate all the 22 parameters out of which 18 relevant parameters excluding pBegin, pEnd, wCanopy, and wground were selected for biomass prediction. A detailed MATLAB code was generated to extract a total of 22. Some are detailed as shown in Table 1.

Table 1 Parameters of a single waveform and their physical explanation

	Waveform parameters	Physical significance
1	**Waveform signal begin** (wStart) and **waveform signal end** (wEnd)	**wStart**: highest capture point among surface and transmitted heartbeat **wEnd**: lowest rise reflected from earth surface
2	**Waveform centroid** (wCentroid)	**wCentroid** represents mean elevation within the footprint
3	**Waveform distance** (*wDistance*)	Represents top tree tallness and top shade stature
4	**Pick distance** (wpDistance)	**wpDistance** represents "Average Tree Height"
5	**Hx, e.g., H25, H50, H75 etc.**	The heights below which 25, 50, 75, and 100% of the waveform energy is reflected

9   Results and Discussion

9.1   Extracted Waveform Parameters from GLA01

The output of GLA14 provides the geolocation of the incident LiDAR beam. The customized GLA14 variables are converted into shape file format for further processing in ArcGIS from where .kml file was created for easy visualization of the LiDAR footprint. Figure 4 shows the geolocation of ICESat footprints in Tripura Region.

9.2   Results from SVM Regression

The SVM regression algorithm is heavily dependent on the choice of kernel function. Hence, SVM is first run separately on the whole dataset using all the 4 available kernel functions. The final choice of the kernel function is made on the basis of the best R^2 value obtained on predicted versus observed biomass. The loss parameter ε is fixed at 0.1. The choice of ε should be based on maximization of R^2 value for a given kernel function (Table 2).

The final choice thus lies between linear and Laplace Kernel. When compared with other Kernel, linear Kernel in SVM clearly has better prediction accuracy. The R^2 value obtained with linear kernel function is 0.763 as compared to 0.746 obtained from Laplace kernel. The root mean square error of linear, RBF, polynomial and Laplace Kernel at footprint level is 19.83, 25.00, 19.82, and 24.99,

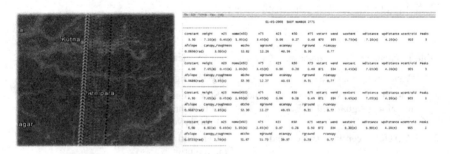

Fig. 4 Location of the 40 selected ICESat footprints in Patnipara area of Tripura state

	Kernel	R^2	Root mean square error
Table 2 Effect of kernel function on combined final prediction accuracies	Linear	0.763	19.834
	RBF	0.738	25.005
	Polynomial	0.728	19.824
	Laplace	0.746	24.998

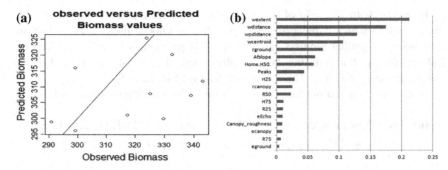

Fig. 5 Predicted versus observed values of biomass from SVM **a** using linear kernel of combined datasets and **b** predictor importance graph of ICESat parameters

respectively. The SVM predicted biomass versus observed biomass graphs on combined datasets (LiDAR, spectral, texture) is described in Fig. 5a.

Since SVM runs separately on the whole dataset using the entire 4 available kernel functions, the best choice of Kernel is linear Kernel which has maximum R^2 value. Hence, it was concluded that linear Kernel function is required for better biomass predictions. The other important comparison can be based on the R^2 values obtained on separate datasets. The SVM predicted biomass versus observed biomass graphs on LiDAR. The corresponding variable importance graph of LiDAR parameters graph using linear Kernel is shown in Fig. 5b.

The more important parameters can be used by multiple regression algorithms for modeling biomass equations. The biomass prediction R^2 values obtained for LiDAR datasets are shown in Table 3.

Table 3 SVM prediction accuracy on LiDAR datasets

Dataset	SVM R^2 value	Adj-R^2	Root mean square error
LiDAR	0.714	0.662	18.502

Table 4 Result table

Parameter	R^2
Wpdistance/wdistance	0.573
Wextent	0.545
H75	0.450
R50	0.585
Ecanopy	0.193
H25	0.089
Wpdistance + wextent + H75	0.633
Wpdistance + wextent + H75 + ecanopy	0.701
Wpdistance + wextent + H75 + ecanopy + R50	0.7140
Wpdistance + wextent + H75 + ecanopy + R50 + H25	0.7147

However, with the help of support vector machine regression algorithm, the number of variables is reduced. Now, we have to choose much lesser subset of parameters to predict biomass. These important parameters have to be taken up individually to find out the best combinations. The most important LiDAR parameters are combined. The results are displayed in Table 4.

Hence, the required biomass equation from SVM algorithm is given as follows:

$$\text{Biomass} = 3.542 * \text{wpdistance} + 3.971 * \text{wextent} + 4.592 * H75 - 2.540 * R50$$
$$+ 8.525 * \text{ecanopy} - 4.995 * H25 + 143.349$$

10 Conclusions

The present study was done for the estimation of the forest biomass using the waveform parameters from ICESat/GLAS. The individual parameters having R^2 values are shown in Table 4. The combination of the parameter predicts R^2 value. After the estimation of biomass, the R^2 value is increased as described in Table 4. All 4 and 5 parameters give the best R^2 value 0.7147, and it is approximately stable at 3 places. Hence, after taking all 5 parameters, we have predicted best biomass equation using SVM algorithm. The estimations were done using support vector machine regression algorithm. The most important variables were selected using SVM regression algorithm, and linear equations were developed using linear regression algorithm. The study was successful in achieving its objectives and in answering the research questions.

Acknowledgements We are grateful to Dr. Subrata Nandy and friends who have contributed toward the development of this research.

References

1. Dong, J., Kaufmann, R.K., Myneni, R.B., Tucker, C.J., Kauppi, P.E., Liski, J., Buermann, W., Alexeyev, V., Hughes, M.K.: Remote sensing estimates of boreal and temperate forest woody biomass: carbon pools, sources, and sinks. Remote Sens. Environ. **84**(3), 393–410 (2003)
2. Brown, S., Gaston, G.: Use of forest inventories and geographic information systems to estimate biomass density of tropical forests: application to tropical Africa. Environ. Monit. Assess. **38**, 157–168 (1997)
3. Drake, J.B., Dubayah, R.O., Clark, D.B., Knox, R.G., Blair, J.B., Hofton, M.A., Chazdon, R. L., Weishampel, J.F., Prince, S.: Estimation of tropical forest structural characteristics using large-footprint LiDAR. Remote Sens. Environ. **79**, 305–319 (2002)
4. Lefsky, M.A., Cohen, W.B., Harding, D., Parker, G., Acker, S.A., Gower, S.T.: Remote sensing of aboveground biomass in three biomes. International archives of the photogrammetry. Remote Sens. Spat. Inf. Sci. **34**(Part 3/W4), 155–160 (2001)

5. Austin, J.M., Mackkey, B.G., Van Niel, K.P: Estimation on forest biomass using satellite radar: an exploratory study in a temperate Australian Eucalyptus forest. For. Ecol. Manag. **176** (1–3), 575–583 (2003)

6. Wang, Z., Boesch, R., Ginzler, C.: Colour and data fusion: application to automatic forest boundary delineation in aerial images. Int. Arch. Photogramm. Remote Sens. Spat. Inf. Sci. **3** (B7), 1203–1207 (2007)

7. Xing, Y., Gier, A.D., Zhang, J., Wang, L.: An improved method for estimating forest canopy height using ICESat-GLAS full waveform data over sloping terrain: a case study in Changbai mountains, China. Int. Arch. Photogramm. Remote Sens. Spat. Inf. Sci. **3**(B7), 1203–1207 (2010)

8. Koch, B.: Status and future of laser scanning, synthetic aperture radar and hyperspectral remote sensing for forest biomass assessment. ISPRS J. Photogramm. Remote Sens. **65**, 581–590 (2010)

9. Zwally, H., Schutz, B., Abdalati, W., Abshire, J., Bentley, C., Brenner, A., Bufton, J., Dezio, J., Hancock, D., Harding, D.: ICESat's laser measurements of polar ice, atmosphere, ocean, and land. J. Geodyn. **34**, 405–445 (2002)

10. Kimes, D., Nelson, R., Manry, M., Fung, A.: Review article: attributes of neural networks for extracting continuous vegetation variables from optical and radar measurements. Int. J. Remote Sens. **19**, 2639–2663 (1998)

Novel Miniaturized Microstrip Patch Antenna for Body Centric Wireless Communication in ISM Band

Raghvendra Singh, Pinki Kumari, Pushpendra Singh, Sanyog Rawat and Kanad Ray

Abstract Body centric wireless communication is an emerging and active area of research for many applications such as identification, tracking, healthcare systems and radio frequency-linked telemetry. In this paper, on-body performance and analysis of various wearable square-shaped spiral cut MSPAs are investigated by measuring reflection coefficient, radiation pattern and impact of human body equivalent model. This antenna basically covers the industrial, scientific and medical band (ISM 2.45 GHz) which can be used for wearable applications in the field of health care and telemetry. In particular, designing and analysis are focused on the performance of a miniaturized square-shaped spiral cut MSPA for body centric wireless communication. The characterization of proposed antenna is analysed using numerical equivalent of three-layered canonical model of skin, fat and tissue. For all types of wearable square-shaped spiral cut MSPA, the main parameter under study is reflection coefficient and the effect of novel double C structured square-shaped spiral cut MSPA with miniaturized designs.

R. Singh
Department of Electronics and Communication Engineering, JK Lakshmipat University, Jaipur, India
e-mail: planetraghvendra@gmail.com

P. Kumari
Department of Electronics and Communication Engineering, PSIT, Kanpur, India
e-mail: pinki.kumari3007@gmail.com

P. Singh
Amity School of Engineering and Technology, Amity University Rajasthan, Jaipur, India
e-mail: singhpushpendra548@gmail.com

S. Rawat
Department of Electronics and Communication Engineering, Manipal University, Jaipur, Rajasthan, India
e-mail: sanyograwat@gmail.com

K. Ray (✉)
Amity School of Applied Sciences, Amity University Rajasthan, Jaipur, India
e-mail: kanadray00@gmail.com

© Springer Nature Singapore Pte Ltd. 2018
M. Pant et al. (eds.), *Soft Computing: Theories and Applications*,
Advances in Intelligent Systems and Computing 584,
https://doi.org/10.1007/978-981-10-5699-4_12

Keywords Body centric wireless communication · ISM band
On-body channel · Wearable antenna · MSPA (microstrip patch antenna)

1 Introduction

Antenna is the most important part in the field of wireless communication. With the increasing development in the field of wireless communication, antenna is being used in many applications. There are many areas where body centric communication systems can be used such as identification, tracking and healthcare systems. Wearable antennas can also be applied for youngsters, the aged and the athletes for the purpose of monitoring [1]. Due to growing technology, antenna requirement is also gaining heights. For this particular purpose, it has to be lightweight, low cost and small size [2].

Body centric wireless communication systems are becoming the important part for future communication. It consists of three basic communications as shown in Fig. 1, i.e. on-body, off-body and in-body communication [3]. On-body basically refers to the communication between on-body/wearable devices which is the centre of discussion. Off-body refers to the communication from off-body to on-body devices. In-body by the name itself suggests communication with the implantable device or sensor.

Body-worn antennas are much more suitable for biomedical telemetry than implantable antennas. Many challenges occur in case of medical implant devices (IMDs) such as miniaturization, compatibility with the human body and the patient safety [4]. So to overcome these, wearable antennas are being used, which can be easily worn. The main advantage of wearable antenna as compared to implantable is the SAR value. In case of on-body, SAR value is reduced and issues related to radiation also decreases.

Body-worn antenna so far is capable of size reduction. As in [5] due to body movements, a wearable antenna is constantly changing its direction of maximum radiation, which leads to significant changes in the radio link performance. This

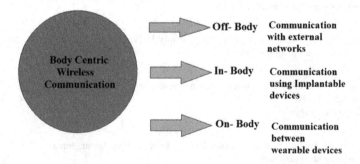

Fig. 1 Description of body centric wireless communication [3]

antenna helps in medical purpose. Defected ground can also be used for reducing the size of the microstrip components as in [6–8].

In the proposed work, antenna used is of comparatively small size and it is being placed over the skin layer. This particular antenna works in industrial, scientific and medical (ISM) band operating at 2.45 GHz.

2 On-body Communication

On-body communication is between on-body and wearable devices. The wearable antenna is worn on the body, and the communication can be done wirelessly.

In the proposed work, miniaturized antenna is used as a wearable device in medical application. Using this particular antenna, other devices can also communicate with it. In medical background, several devices are used which store data such as blood pressure, heartbeat. As shown in Fig. 2, wearable antenna not only stores the results but also communicates it to the concern physician about the data through the access point.

3 Design Strategy of Wearable Square-Shaped Spiral Cut MSPA

The flow chart of proposed strategy is given in Fig. 3. This flow chart clearly describes the steps followed to design the desired shape. First, the high permittivity substrate is selected, and then, the rectangular patch is designed upon its surface. This microstrip patch antenna is designed and simulated in computer simulation

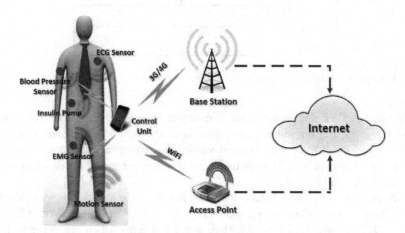

Fig. 2 On-body and off-body communication [9]

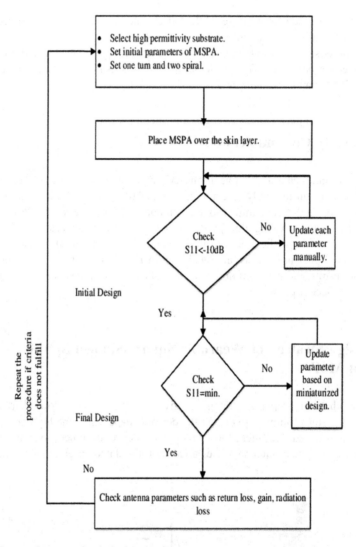

Fig. 3 Proposed strategy for designing the antenna

software, i.e. CST microwave suite. Simulation is done by using one-layer skin model as in [3, 10], but in this paper, three-layered model (skin, fat and muscle) is used for making it more effective and viable. This electrical equivalent model of three-layered human body is more realistic than one-layer model for simulations as shown in Fig. 4a [12, 13].

The proposed antenna is placed above the human tissue model. Figure 4b shows the layers, i.e. muscle ($\varepsilon_r = 54.417$, $\sigma = 1.882$ S/m), fat ($\varepsilon_r = 5.280$, $\sigma = 0.104$ S/m) and skin ($\varepsilon_r = 38.0066$, $\sigma = 1.464$ S/m) which are the electrical equivalent model of human tissue [11].

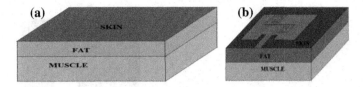

Fig. 4 **a** Layered structure of human tissue model. **b** Proposed antenna placed on human tissue model

After simulating the rectangular patch antenna in free space box, the strategy is to miniaturize and optimize this antenna for achieving minimum reflection coefficient ($|S1, 1|$). Many optimized designs are simulated for square-shaped spiral cut MSPA as double I, L and C structures. In this paper, initially square-shaped spiral cut and later on C-shaped mirror image are designed horizontally for getting more precise results in ISM band. The initial parameters of MSPA are set manually. Number of turns is taken according to the design as well as spirals were also drawn simultaneously. The design was placed over the skin layer as the study of interest is wearable antenna. After the simulation procedure in free space box, all depicted designs were simulated over the canonical tissue model and different results were obtained.

4 Design Configuration of Wearable Square-Shaped Spiral Cut MSPA

The proposed design parameters are being given in the tabular form. The material used in the patch region is lossy copper as given in Table 1. Substrate used in wearable square-shaped spiral cut MSPA is polytetrafluoroethylene (PTFE) with relative dielectric constant 10.2, conductivity 1 and tangent delta 0.0022. The dimensions are so chosen to be reasonable for body-worn devices. Volume of antenna is ($20 \times 17.75 \times 1.9 = 674.5 \text{ mm}^3$) which is miniaturized enough to be used on body applications in ISM band for body centric wireless communication. Front and back views of proposed antenna are shown in Fig. 5a, b. Here, in Fig. 6a–d, front view of the square-shaped spiral structure, double I, double L and double C cut is shown and it is simulated in CST microwave suite using FDTD (finite-difference time-domain).

Table 1 Parameter value of proposed antenna

Parameter	Value (mm)
Length of rectangular patch (L)	17.75
Width of rectangular patch (W)	20
Length of slot ($L1$)	9.5
Square slot on ground length (l)	3
Square slot on ground width (w)	3
Circle slot on ground r	2

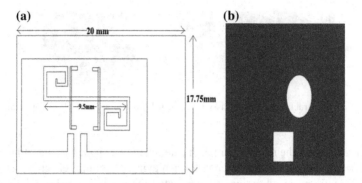

Fig. 5 **a** Front view of double C structure in proposed square-shaped spiral cut MSPA. **b** Back view of double C structure in proposed square-shaped spiral cut MSPA

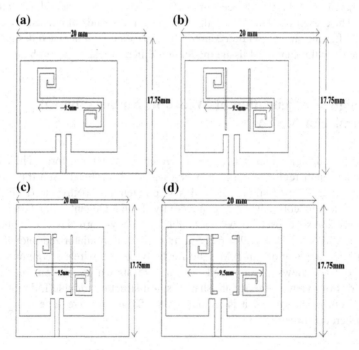

Fig. 6 Front view of **a** square-shaped spiral cut MSPA. **b** Double I structure. **c** Double L structure. **d** Double C structure

5 Results and Analysis

5.1 Reflect Ion Coefficient Versus Frequency Performance

In case of two square spiral shapes, resonant frequencies are 2.49, 2.45, 2.49 and 2.45 GHz for spiral cut MSPA, double I, double L and double C cut structures, respectively, as in Fig. 7a–d. All frequency bands are lying in ISM band. Impedance bandwidth of antenna is (40 MHz) in case of square-shaped spiral cut MSPA (Fig. 7a). The impedance bandwidth is now increased to 60 MHz (Fig. 7d). It is increased by 20 MHz compared with simple square-shaped spiral cut MSPA.

The reflection coefficient is −12.93, −22.20, −17.22, −24.8 dB, for square-shaped spiral structure, double I, double L and double C cut structures, respectively. It is observed that double C structure is showing the best results on comparison with respect to reflection coefficient (Fig. 8).

5.2 Radiation Patterns

The radiation patterns achieved at 2.4 GHz (Fig. 9a, b) are showing unidirectional pattern which are desired to provide wireless link for on-body and off-body communication and minimum radiation towards the user's body. It is highly desirable for wearable antenna.

5.3 Frequency Versus Gain Plot

Here, frequency versus gain plot has also been observed which is showing the gain (IEEE) at frequency variations from 1 to 3 GHz. The point of interest is ISM frequency that is in between 2.4 and 2.5 GHz. The gain values at 2.4, 2.45 and 2.5 GHz are 3.08, 3.17 and 3.26, respectively. The gain is positive in ISM frequency range, which is showing the appropriate radiation in bore sight of antenna (Fig. 10).

6 Conclusion

A complete strategy of ISM band antenna design for body centric wireless communication is developed. It Starts with a simple square-shaped spiral cut MSPA, after different modified designs the précised and miniaturized square-shaped spiral cut MSPA with double C structure is achieved. Theoretical results are showing the good agreement with simulated results. This antenna is showing −24 dB reflection coefficient at 2.45 GHz (ISM band) for on-body communication in stacked human

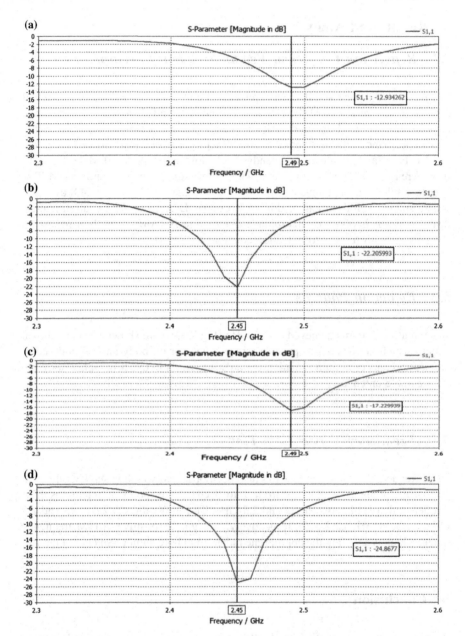

Fig. 7 **a** Reflection coefficient verus frequency performance of Fig. 6a–d

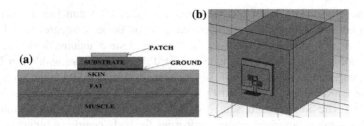

Fig. 8 **a** Placing of proposed antenna over layered tissue model. **b** Proposed antenna in CST microwave studio suite placed over the layered tissue model

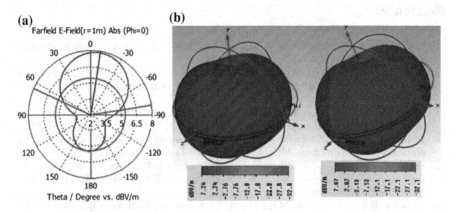

Fig. 9 **a** Two-dimensional radiation pattern of antenna double C structure and **b** three-dimensional radiation pattern at 2.4 GHz

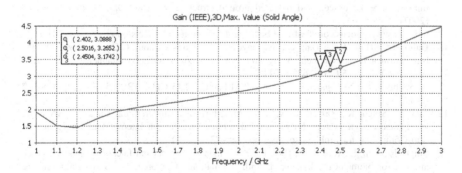

Fig. 10 Frequency versus gain plot of double C structure in CST microwave studio

tissue model. The volume of antenna is miniaturized (674.5 mm^3) so that it can be easily used in body centric wireless communication devices. Square-shaped spiral cut MSPA with double C structure is showing desired unidirectional radiation pattern, and the main lobe magnitude is 7.84 dbV/m with beam width 146.3°.

The impedance bandwidth of square-shaped spiral MSPA with double C structure is 60 MHz in ISM band (2.4–2.5 GHz), and the gain (IEEE) is 3.17 at 2.45 GHz. The miniaturized square-shaped spiral cut MSPA and its modified variations show appropriate antenna properties for body centric communication in wearable applications.

References

1. Conway, G.A., Scanlon, W.G.: Antennas for Over-Body-Surface Communication at 2.45 GHz, vol. 57, pp. 844–855. IEEE, April 2009
2. Mackowiak, M., Correia, L.M.: A Statistical Model for the Influence of Body Dynamics on the Gain Pattern of Wearable Antennas in Off-Body Radio Channels, pp. 381–399. Springer, Berlin (2013)
3. Liu, C., Guo, Y.-X., Xiao, S.: A review of implantable antennas for wireless biomedical devices. Forum for Electromagnetic Research Methods and Application Technologies (2016)
4. Wong, K.L., Lin, C.I.: Characteristics of a 2.4-GHz compact shorted patch antenna in close proximity to a lossy medium. Microw. Opt. Technol. Lett. **45**, 480–483 (2005)
5. Roblin, C., Laheurte, J.-M., D'Errico, R., Gati, A., Lautru, D., Alvès, T., Terchoune, H., Bouttout, F.: Antenna Design and Channel Modeling in the BAN Context—Part I: Antennas (pp. 139–155). IEEE, Jan 2011
6. Kanaujia, B.K., Khandelwal, M.K., Dwari, S., Kumar, S., Gautam, A.K.: Analysis and Design of Compact High Gain Microstrip Patch Antenna with Defected Ground Structure for Wireless Applications. Springer, Berlin (2016)
7. Dwari, S., Sanyal, S.: Compact sharp cutoff wide stopband low-pass filter using defected ground structure and spurline. Microw. Opt. Technol. Lett. (2006)
8. Dwari, S., Sanyal, S.: Compact wide stopband low-pass filter using rectangular patch compact microstrip resonator cell and defected ground structure. Microw. Opt. Technol. Lett. **49** (2007)
9. https://www.elprocus.com/best-ieee-based-real-time-m-tech-projects-on-electronics-2014/
10. Gallo, M., Hall, P.S., Bai, Q., Nechayev, Y.I., Constantinou, C.C., & Bozzettil, M.: Simulation and measurement of dynamic on-body communication channels. IEEE Trans. Antennas Propag. **59**, 623–630 (2011)
11. Al-Shaheen, A.: New patch antenna for ISM band at 2.45 GHz. ARPN J. Eng. Appl. Sci. **7** (2012)
12. Jain, L., Singh, R., Rawat, S., Ray, K.: Miniaturized meandered and stacked MSA using accelerated design strategy for biomedical applications. In: 5th International Conference of Soft Computing for problem solving, pp. 725–732 (2016)
13. Singh, R., Jain, L., Rawat, S., Ray, K.: Performance of wideband falcate implantable patch antenna for biomedical telemetry. In: 5th International Conference of Soft Computing for problem solving, pp. 757–765 (2016)

Recognition of Noisy Numbers Using Neural Network

Chanda Thapliyal Nautiyal, Sunita Singh and U.S. Rana

Abstract In this paper, noisy numbers have been recognized with the help of artificial neural network. Literatures divulge that recently there are many character recognition algorithms for recognition of handwriting, numbers, and alphabets using artificial neural network. These algorithms used multilayer perceptron neural network and large number of input neurons for the recognition. Here, we propose an efficient supervised single-layer perceptron learning with very few input neurons for training. The proposed recognition algorithm is tested on 8 noisy numbers and yields good level of recognition accuracy. Up to 49.28% noise, this algorithm successfully recognizes 87.5% of the given numbers.

Keywords Character recognition · Artificial neural network · Single-layer perceptron · Multilayer perceptron neural network

1 Introduction

In recent times, character recognition has become one of the most alluring and challenging research areas of pattern recognition. The scope of character recognition systems is very wide. Some of its diverse applications include data entry for business documents, reading assistance for blind, bank checks, passport, and conversion of any handwritten document into structural text form [1, 2]. An artificial neural network provides credible and expeditious calculations in performing

C.T. Nautiyal (✉)
Uttarakhand Technical University, Dehradun, Uttarakhand, India
e-mail: chanda.nautiyal@gmail.com

S. Singh
IRDE (DRDO), Ministry of Defence, Dehradun, Uttarakhand, India
e-mail: sunnidma@gmail.com

U.S. Rana
Department of Mathematics, DAV (PG) College, Dehradun, Uttarakhand, India
e-mail: drusrana@yahoo.co.in

© Springer Nature Singapore Pte Ltd. 2018
M. Pant et al. (eds.), *Soft Computing: Theories and Applications*,
Advances in Intelligent Systems and Computing 584,
https://doi.org/10.1007/978-981-10-5699-4_13

classification and recognition task [3]. Literatures divulge that recently there are many character recognition algorithms for recognition of handwriting, numbers, and alphabets using artificial neural network. These algorithms used multilayer perceptron neural network and large number of input neurons for the recognition. But reduction in processing time and simultaneously yielding desired level of recognition accuracy is still main focus of research.

Perwej and Chaturvedi [4] proposed handwritten English alphabets recognition technique using artificial neural network architecture with 25 input layers. Every letter was trained with 20 samples and tested for 5 samples. Proposed scheme recognized the alphabets with an average accuracy of 82.5%. Hirwani and Gonnade [5] used multilayer perceptrons for the recognition of handwritten English characters with 200 training samples and 200 testing samples. The training was done through backpropagation algorithm. Mehta and Kaur [6] proposed a neural network classifier for isolated character recognition. Backpropagation model of neural network has been proposed for recognizing numbers 0–9. In the present paper, an efficient supervised perceptron learning algorithm has been developed with very few numbers of input neurons, only 15. To reduce processing time and faster convergence, it is desirable to use minimum number of input neurons. Use of minimum number of input neurons increases the learning rate. The proposed algorithm is tested on 8 noisy numerals. Up to 49.28% noise, this algorithm recognizes 87.5% characters successfully. This paper has been organized as follows. Section 2 briefs about perceptron learning algorithm. Section 3 describes methodology and Sect. 4 and 5 simulation results and conclusion, respectively.

2 Perceptron Learning

In living organisms, neuron is the fundamental information processing cell. A large number of neurons are connected with each other inside the brain. A single-layer perceptron is mathematical modeling of biological processing unit, neuron. McCulloch and Pitts developed the first mathematical model of neuron. Typical single-layer perceptron consists of an input and an output layer with adjustable weights and bias. It calculates weighted sum of numerous neurons in input layer [7, 8]. In perceptron learning, weights are upgraded in each iteration and are more powerful than the Hebb rule [9]. The single-layer feedforward network is used for the classification and prediction among the classes that are linearly separable. Linear separability refers to the case when a linear hyperplane exists to place the instances of one class on one side and those of the other class on the other side of the plane. In this way, perceptron learning algorithm is a binary classifier: a function that maps its input x (a real-valued vector) to an output value $f(x)$ (a single binary value) [10]. If the sum of weighted inputs is greater than its threshold, then output unit will take value 1. In terms of classification, an object will be classified by unit j into class 'A' if

$$\sum W_{ji}X_i > \theta_j$$

where w_{ji} is the weight from unit i to unit j, X_i is the input from unit i, and θ_j is the threshold on unit j. Otherwise, the object will be classified as class '*B*.' Suppose there are n inputs. The equation

$$\sum W_{ji}X_i = \theta_j$$

forms a hyperplane in the n-dimensional space, dividing the space into two halves. When $n = 2$, it becomes a line. The architecture for single-layer perceptron is shown in Fig. 1.

Network training begins with the inputs set consisting of neurons $x_1, x_2 \dots x_{15}$ and target y. It has three layers, input, weight, and output. The middle layer weights are adjustable; they are modified up to desired error. Error is the difference between desired and actual value. Common bias is 1 and learning rate, a number between 0 and 1.

The training algorithm is as follows [9, 11].

Step 1   Set learning rate $\alpha > 0$ and threshold θ. Initialize weights and bias to zero (or some random values).

Step 2   Repeat the following steps, while cycling through the training set, until all training patterns are classified correctly.

(a)   Set activations for input vector. Here, we have set identity function as activation function

$$x_i = s_i \quad \text{for } i = 1 \text{ to } 15$$

(b)   Compute total input for the output neuron:

$$y_{\text{in}} = b_j + \sum_{i=1}^{15} x_i w_i$$

Fig. 1 Architecture of single-layer perceptron

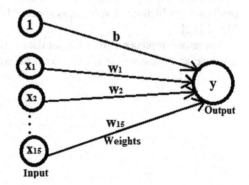

the activation output (Binary Step Function) used is

$$y = f(y_{in}) = \begin{cases} 1 & y_{in} \geq \theta \\ 0 & -\theta \leq y_{in} \leq \theta \\ 1 & y_{in} \leq \theta \end{cases}$$

(c) Update the weights and bias only if that pattern is misclassified, i.e., if the target is not equal to the output response. If $y_j \neq t_j$ and the value of $x_i \neq 0$, then

$$w_{ij(new)} = w_{ij(old)} + \alpha t_j x_i$$

$$b_{j(new)} = b_{j(old)} + \alpha t_j$$

Else

$$w_{ij(new)} = w_{ij(old)}$$

$$b_{j(new)} = b_{j(old)}$$

(d) Test for stopping condition: The stopping conditions may be the weight changes.

3 Methodology for the Recognition of Numbers

Network is trained for learning pattern of each of the numbers. The objective is to train the network to recognize numbers 1, 2, 3, 4, 5, 6, 7, 8. In order to keep number of inputs to a minimum, they are represented by a 5×3 array of dots, hence generating 15 inputs. For example, in Fig. 2, graphical representation of the number '1' is shown. Highlighted position is represented by the pound sign (#) in dot array, input value corresponding to it is 1, and dot (.) represents the non-highlighted position for which input value is assigned −1. Character inputs for '1' are shown in Fig. 2 [12].

Therefore, representation of input data x_i ($i = 1$ to 15), for each of the numbers rowwise and columnwise starting from the top, is in Table 1.

Fig. 2 Example of number
'1'

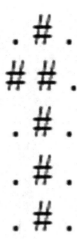

Table 1 Representation of input data

Number	Representation of input data x_i				
1	−1 1 −1	1 1 −1	−1 1 −1	−1 1 −1	−1 1 −1
2	1 1 −1	−1 −1 1	−1 −1 1	−1 1 −1	1 1 1
3	1 1 −1	−1 −1 1	1 1 −1	−1 −1 1	1 1 −1
4	1 −1 1	1 −1 1	1 1 1	−1 −1 1	−1 −1 1
5	1 1 1	1 −1 −1	1 1 −1	−1 −1 1	1 1 −1
6	−1 1 1	1 −1 −1	1 −1 −1	1 1 1	−1 1 1
7	1 1 1	−1 −1 1	−1 −1 1	−1 −1 1	−1 −1 1
8	1 1 1	1 −1 1	1 1 1	1 −1 1	1 1 1

Initial weight matrices are set to zero. After the training, weights corresponding to the correct output are noted. Net is trained for the recognition of these numbers for given test data as described in previous sections. The training takes very few number of epochs.

4 Simulation Results and Discussion

Here, algorithm is simulated in MATLAB to train a neural network for classifying the 8 noisy numbers. The algorithm successfully classifies 87.5% noisy numbers with various percentages of noise (13.04, 23.19, 31.88, 42.03, and 49.28%). MATLAB program outputs are shown below.

Input Pattern for training

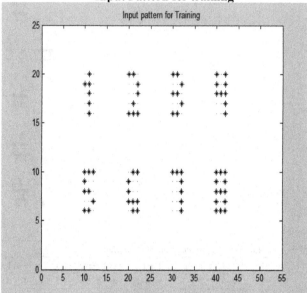

13.04% Noisy Input Pattern

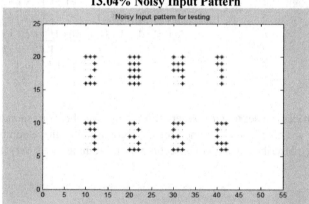

Output for13.04% Noisy Input

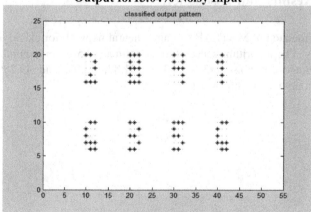

23.19% Noisy Input Pattern

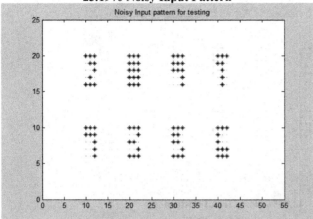

Output for 23.19% Noisy Input

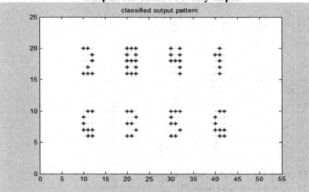

31.88% Noisy Input Pattern

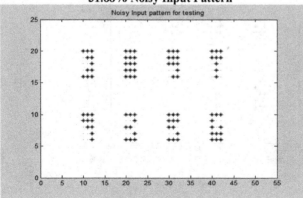

Output for 31.88% Noisy Input

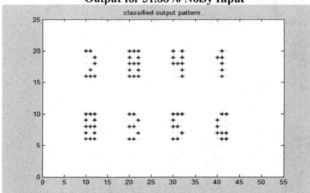

42.03% Noisy Input Pattern

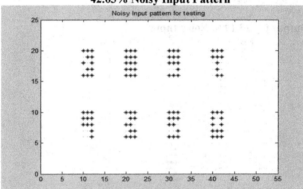

Output for 42.03% Noisy Input

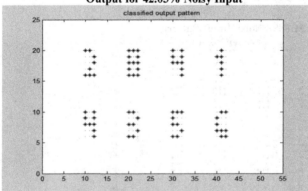

49.28% Noisy Input Pattern

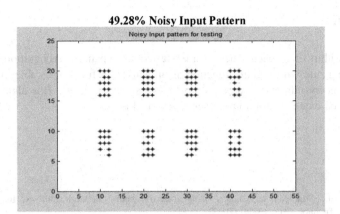

Output for 49.28% Noisy Input

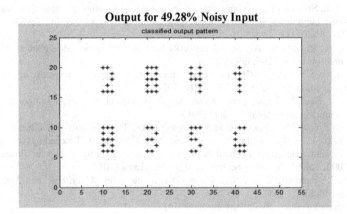

5 Comparison with Existing Techniques

Table 2 shows that present algorithm uses very few numbers of neurons with higher level of accuracy.

Table 2 Comparison table

Model	Neurons in input layer	Recognition accuracy
Perwej et al. [4]	25	82.5%
Zebardast et al. [13]	77	81.2677 and 85.1535%
Proposed model	15	87.5%

6   Conclusion

The above algorithm has been successfully tested for pattern recognition. Perceptron learning rule with identity function as an activation function is able to recognize 87.5% noisy numbers accurately up to 49.28% noise level. This algorithm can further be extended for some other types of objects.

References

1. Prasad, K., et al.: Character recognition using matlab's neural network toolbox. Int. J. u-e-Serv. Sci. Technol. **6**(1), 13–20 (2013)
2. https://en.wikipedia.org/wiki/Opticalcharacterrecognition. (2016)
3. Pradeep, J., Srinivasan, E., Himavathi, S.: Diagonal based feature extraction for handwritten alphabets recognition system using neural network. Int. J. Comput. Sci. Inf. Technol. (IJCSIT) **3**(1), 27–38 (2011)
4. Perwej, Y., Chaturvedi, A.: Neural networks for handwritten english alphabet recognition. arXiv preprint arXiv:1205.3966 (2012)
5. Hirwani, A., Gonnade, S.: Character recognition using multilayer perceptron. Int. J. Comput. Sci. Inf. Technol. (IJCSIT) **5**(1), 558–661 (2014)
6. Mehta, R., Kaur, R.: Neural network classifier for isolated character recognition. Int. J. Appl. Innov. Eng. Manage. (IJAIEM) **2**(1) (2013)
7. Fu, L.-M.: Neural networks in Computer Intelligence. Tata McGraw-Hill Education (2003)
8. Rutkowski, L., (ed.): Neural Networks and Soft Computing: Proceedings of the Sixth International Conference on Neural Network and Soft Computing, Zakopane, Poland, 11–15 June 2002, vol. 19. Springer Science & Business Media (2013)
9. Fausett, L.: Fundamentals of Neural Networks Architectures, Algorithms, and Applications. Pearson (2006)
10. https://en.wikipedia.org/wiki/Perceptron (2016)
11. Maren, A.J., Harston, C.T., Pap, R.M.: Computing Applications. Academic Press (2014)
12. Silver, H.: Neural networks in electrical engineering. In: Proceedings of ASEE New England Section Annual Conference (2006)
13. Zebardast, B., Maleki, I., Maroufi, A.: A novel multilayer perceptron artificial neural network based recognition for kurdish manuscript. Indian J. Sci. Technol. **7**(3), 343 (2014)

Pollution Check Control Using License Plate Extraction via Image Processing

Shivani Garg and Nidhi Mishra

Abstract The main objective of this paper is pollution check control using license plate detection. In this paper, we extract the information about the vehicle owner by using registration number. So initially, we extract the registration number from the license plate after that compare the registration number to the local database. If registration number is stored in the database, then we retrieve the detail of the vehicle owner. The first key step of this work is to extract the registration number from the license plate for the retrieval of information about the vehicle owner. The extraction of license plate is divided into four parts: first one is preprocessing of image, second is localization of license plate, next step is segmentation of the characters in the license plate, and final step is recognition of characters from the segmented license plate. Apart from that we have also extended our research work to pollution fine implementation and vehicle entry restriction for particular area, vehicle which are more than 10 years old.

Keywords Number plate recognition · License plate · Pollution fine control

1 Introduction

License plate detection plays a major role in the world because vehicles are increasing day-by-day and pollution is also increasing due to increase of vehicle number. In this paper, we introduce the pollution fine control; in this, we impose fine on the vehicle owner whose pollution check date is expired. In commercial cities, the commercial places or near the buildings, parking areas are built, but the problem with those areas is that it does not have enough space. So to manage space

S. Garg (✉) · N. Mishra
Department of Computer Science & Engineering,
Poornima University, Jaipur, India
e-mail: garg.shivani14@gmail.com

N. Mishra
e-mail: nidhi.mishra@poorinma.edu.in

© Springer Nature Singapore Pte Ltd. 2018
M. Pant et al. (eds.), *Soft Computing: Theories and Applications*,
Advances in Intelligent Systems and Computing 584,
https://doi.org/10.1007/978-981-10-5699-4_14

of these parking lots, proper management and proper use of parking space are necessary, as this will improve the efficiency and reduces management cost. A single person in the parking areas is not enough to manage all kind of activities such as charging persons to place vehicle, how to control vehicle's in and out space. This has made necessary to use the system of automatic plate recognition which in turn provide benefit like reduces management cost required to control and manage parking areas. Also to reduce road accidents and stolen vehicles cases, the number plate recognition system is highly effective which can not only help police (by reducing their man power), but also can track vehicles whenever needed.

The images of various vehicles have been acquired, and thereafter submitted to the MATLAB software where they are first resize the image and then convert it into gray scale images. Then, we use median filter to remove the noise in the image. Brightness, contrast, and intensity adjustments are made to optimum values to enhance the number plate and its digits. The output is saved in a notepad. Objective of proposed work is to create the modules for traffic regulation using a system of number plate detection. The input of the module is a part of the captured image, and output is a number which is a editable form of license plate. Basically, the proposed system recognizes all kinds of number plates of different varieties.

The rest of the paper is organized as the follows: Sect. 2 presents the background and related research of the proposed work. Section 3 describes the architectural design of proposed work. Section 4 elaborates the methodologies of proposed system. Section 5 highlights the experimental set up and results, and finally, Sect. 6 concludes the proposed work.

2 Backgrounds and Related Work

Previously, different work had already been carried out in the area of license plate recognition. There are five divisions in this system, which are input image, pre-processing of image, localization of the license plate in vehicle image, character segmentation from the license plate, and character recognition from the segmented license plate. The work of first two divisions should be done in an accurate way so that license plate can be easily searched and localized. Thus detection will greatly affect the efficiency and the speed of the whole license plate recognition system.

In the earlier works, the edge detection techniques was found to be an important technique that works by changing brightness of the plates, but this technique could not be used for images. But they can be applied to complex images due to their sensitivity to the unwanted edges.

Seyed Hamidreza Mohades Kasaei et al. (2011) have presented a real-time and robust technique of license plate detection and recognition; it uses the morphology and template matching. The performance of corrected plate's identification and localization was 98.2%, and correctly recognized characters were 92% [1].

Abdul Mutholib et al. (2013) have presented the optimization of ANPR algorithm on limited hardware of Android mobile phone. Proposed algorithm was based on Tesseract library. The template matching-based OCR will be compared to artificial neural network (ANN)-based OCR. Results on 30 images showed that the recognition rate of license plate was 97.46%, and the processing time in recognition of license plate was 1.13 [2].

Farajian et al. [3] studied different techniques and compared them. Author finds out that every technique has own limitations, and some method gives better results under some conditions. This paper will lead to select the best technique for detection just according to circumstance. After this survey, author concludes that there was no ideal technique which is suitable for all circumstances [3].

Anagnostopoulos et al. (2008) have applied a method to locate the license plates, which consist of three stages: initially, used the Sobel operator that will fetch the vertical edges of the image of vehicle. After that applied the integral image and HSV color space [4].

Pratik Madhukar Manwatkar et al. (2014) In this paper, author has reviewed and analyzed different methods to find out the text characters from the images. Author has discussed some application of text recognition system [5].

3   Architectural Design of Proposed Work

Figure 1 presents the architectural design of our proposed system of traffic regulation using number plate detection. In this work, we have performed data preprocessing as the first step of number plate detection.

The overall description of the design specification as shown in Fig. 1 is explained below.

3.1   Input Image

The initial step of license plate extraction system is to select an input vehicle image. This image contains license plate of vehicle. After the selection of the image, preprocessing of the image is performed.

3.2   Preprocessing of Image

This is the second step of license plate recognition system. Preprocessing is very essential for the good quality performance of character segmentation.

Preprocessing steps are:

1. Resizing image
2. Rgb-to-gray conversion

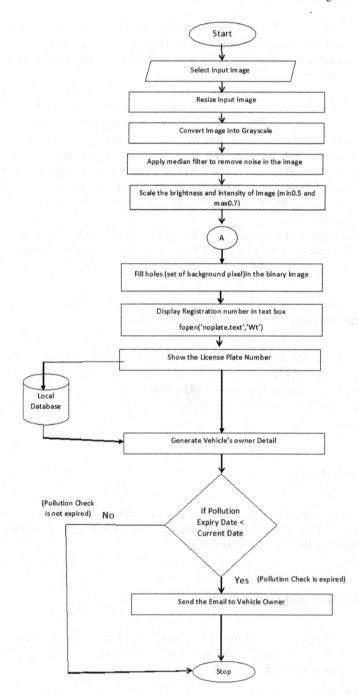

Fig. 1 Process flow diagram

3. Noise removal (median filter)

Preprocessing of an image means "preparation" of the sample/image to introduce it to an algorithm for specified task such as recognition of image, feature extraction.

3.3 Brightness and Intensity of Image

After preprocessing of the image, we adjust the brightness and intensity of the image. To scale the intensity in MATLAB, we use imadjust() function. Intensity of the image is the average of the intensities of all pixels in image. Brightness can be increased or decreased by simple addition or subtraction to the image matrix.

3.4 Fill Holes in the Binary Image

A general use of the flood-fill process is to fill "holes" in image. Holes fill in the binary image are performed because of representation of foreground objects in the image. In our system, we use imfill() function to fill the holes in the binary image.

3.5 Display Registration Number in the Text Box

After filling the holes in the image, we obtain a number from the image. This number is displayed in the text box. This is the registration number of the vehicle.

3.6 Local Database

We create a local database which contains the information of vehicle owner. When registration number is obtain, then our system compared this registration number to the database. When match is found in the database, then all the details of the vehicle owner are obtained.

3.7 Pollution Fine

This system mainly works on the pollution fine control. In this, if pollution check date is expired, then fine is imposed on the vehicle's owner. This fine is impose by sending the mail to the owner of the vehicle, and when pollution check is not expired, then no mail is sent to the vehicle owner and no fine is imposed on the owner.

4   Methodologies

The main goal of this system is to focus on the localization and recognition of the numbers present in license plate of a vehicle. We locate the license plate part of the vehicle from every images first and then register them to a reference image by template matching in same precision. We use a simple templates matching to recognize the letters in the estimated trained images. Before the processing of two proposed phase for the solution, input images need to preprocess of accurate output. The following sections describe the process of these methods step-by-step. This paper, based on the image processing methods, gives the image pretreatment, for the position of license plate on the whole image of vehicle.

To make the vehicle license plate recognition, multiple small phases must implement such as image acquisition, preprocessing of image, localization of license plate, segmentation of alphanumeric code, recognition of license plate code.

4.1   Image Preprocessing

Image preprocessing considerably increase the consistency of an optical examination for the specific feature detection. In the license plate, we generally use resizing of image and then convert into gray scale image. To resize the image, we are using imresize(); in this, width of the image is 400 and height of the image is adjusted accordingly to maintain the aspect ratio. After that we use rgb2gray() to convert the resized image into gray scale image.

4.2   License Plate Localization

In license plate localization, first we must have to find the location of the number plate in whole vehicle image for identification of registration number. To execute the system from input image to the output characters as in editable form, localization process takes important role. For the localization, basically, we use median filter to filter out the noise in the image; in this, noise type is pepper and salt. Dilation is performed after the noise removal from the image; then, we locate the license plate in the input image.

4.3   Segmentation of Alphanumeric Code

Segmentation of alphanumeric code is a key step in license plate extraction system. Many difficulties are in this step, which are noise presence in the plate area, alphanumeric codes are not properly aligned. If the alphanumeric codes in license plate are in ideal condition, that is, codes are sufficiently separated, code segmentation may be accomplished directly from vertical and horizontal segmentation. In this, it detects the horizontal lines in the image with a pixel value zero. After that, the image is converted into gray scale image, and then, we simply use 'for' loop to detect the portion of the image that had connected objects with pixel value of zero.

4.4 Recognition of Alphanumeric Code

Recognition of alphanumeric code is a process for the automatic conversion of alphanumeric code into editable text. After segmentation, alphanumeric codes of the license plate system need to carry on recognition for all isolated alphanumeric code. The template matching technique is an appropriate technique for the recognition of particular alphanumeric code. This technique is used in binary images, properly built templates, and it also obtained fine results for gray scale images. In this character recognition, we adjust the brightness and intensity of the image. After that, we fill the holes present in the binary image, and in the end, we use template matching technique to recognize the characters from the image. Finally, the output is displayed in the text box that is registration number of the vehicle.

4.5 Pollution Fine Controller and Driver Information Fetcher

After the character recognition, we checked the owner name, city, address, registration number of vehicle, and contact number. After recognition of the vehicle registration number, we compare the registration number to the local database. If the match is found, then we get the detail of the vehicle owner. Then, we compare the registration number to pollution control table, then we get the expiry date of the pollution check of vehicle owner. If current date is in the valid range, then all are OK, otherwise the fine is to be calculated and the fine can be send on manual basis or automatically on e-mail, and the details will be send by fetching the details required from the local database record using the registration number. Structure of the database of the local database will look like (Fig. 2).

Structure of the database of the **pollution control** will look like (Fig. 3).

Vehicle plate number	Name	City	E-mail Id	Date of registration
RJ14CJ5252	Krishan Pal	Jaipur	garg.shivani41@gmail.com	12/16/2013

Fig. 2 Structure of the database of the local database

Fig. 3 Structure of the
database of the pollution
control

Registration number	Date of pollution check	Expiry date
RJ14CJ5252	6/18/2015	12/18/2015

The main objective of this work is to present more proficient way of pollution control testing for vehicle by using alphanumeric code segmentation and recognition. In proposed system, we manually created the database of vehicle license plate image. The license plate image is converted into text image by using alphanumeric code recognition.

5 Experiment Results

The proposed system was evaluated by taking vehicle input image manually, and the result obtained is shown in the following steps

- Input image
- Resize the input image
- Convert image into gray scale image
- Apply median filter to remove the 'pepper and salt' noise
- Scale brightness and intensity of the image
- Fill holes in the binary image
- Display the license plate number
- Compare with database
- Display the Vehicle owner detail
- If pollution check is expired, then mail is send to vehicle owner (Figs. 4, 5, 6, 7, 8, 9 and 10).

Fig. 4 Input image

Fig. 5 Resize input image

Fig. 6 Gray scale conversion

Fig. 7 Scale brightness and intensity

The experiments have been performed to test the performance of proposed system and to measure the processing time and recognition rate. Totally, 22 images were taken under various time, conditions, and angle as shown in Tables 1 and 2.

Fig. 8 Fill holes in binary
image

Fig. 9 Final image

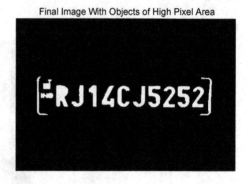

Fig. 10 Registration number

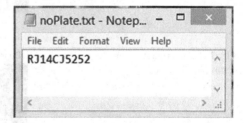

After the character recognition, registration number of the license plate is recognized; then, we compare the registration number to the manually created local database of vehicle owner detail. If vehicle registration number is found in the local database, then detail of the vehicle owner is displayed as shown in Fig. 11.

The registration number will be matched with the local database record of pollution check control. If the previous registration number is found in with the pollution control database, we get the expiry date and the date of pollution check shown in Fig. 12.

Table 1 Results of license plate detection

S. No.	Number plate	Recognized output	Processing time (s)	Recognition rate (%)
1	RJ14CJ5252	RJ14CJ5252	7.17	100
2	AFR420	AFR4Z0	3.90	83.33
3	AKH343	AKH343	4.11	100
4	AED632	AEQ632	3.99	83.33
5	AWR310	AWR310	3.25	100
6	AXZ016	AXZ016	4.10	100
7	RJ14CH7058	RJ14CH7058	6.89	100
8	RJ07CA6393	RJ07CA6393	8.01	100
9	RJ14CJ0724	QJ14CJ0724	8.45	90
10	RJ142C2901	RJ142C2009	6.81	80
11	RJ14CU5794	RJ14CU5794	6.96	100
12	RJ14CK8175	RJ14CK8175	6.76	100
13	RJ26CA2638	RJZ6CAZ638	7.64	80
14	MH12DE1433	RH1ZDE1433	6.10	80
15	RJ14CU5794	RJ14CU5794	7.05	100
16	RJ07CA6393	RJ07CA6393	8.59	100
17	RJ14TD5156	RJ14TD5956	8.42	90
18	RJ14LC7419	RJ14LC7489	8.54	90
19	RJ26CA2638	RJZ6CAZ638	8.26	80
20	RJ27CD4244	RJZ7CD4244	8.37	90
21	RJ26TA1260	RJZ6TA1260	8.22	90
22	RJ26CA3767	RJ26CA3767	6.80	100

Table 2 Performance of proposed license plate detection

	Total accuracy	In percentage (%)	Recognition rate (in %)
License plate detection	(11/22) * 100	50	92.57

Fig. 11 Display vehicle owner detail

Personal Details:
Name: Krishan Pal
Email-id – garg.shivani41@gmail.com
Address- Jaipur
Polution fine on car RJ14CJ5252 of Rs 1000
Confirm email id: garg.shivani41@gmail.com
Vehicle wit License Plate Number RJ14CJ5252 is allowed to enter

Fig. 12 Pollution check
control database

vechplatenc ▾	checkupdate ▾	validdate ▾	(
⌿ RJ14CJ5252	6/18/2015	12/18/2015	
RJ14CU5794	6/18/2016	12/18/2016	
✱			

Fig. 13 E-mail sent to the
vehicle owner regarding
expiry date of pollution check

helpdesk@ranthamboretigervilla.com
to me
Vechicle Details for number AKH343
 Personal Details:
Email-id – garg.shivani41@gmail.com
Name: Rashmi Rathi
Address- Jaipur
Polution valid upto 2016-10-31

Fig. 14 E-mail sent to the
vehicle owner regarding fine

helpdesk@ranthamboretigervilla.com
to me
Vechicle Details for number RJ14CJ5252
 Personal Details:
Email-id – garg.shivani41@gmail.com
Name: Krishan Pal
Address- Jaipur
Fine imposed of Rs 1000

Case 1: If pollution check expiry date is valid, then an e-mail is sent to vehicle owner, which contains the information of pollution check expiry date as shown in Fig. 13.

Case 2: If pollution check date is not valid, then e-mail is sent to the vehicle owner regarding fine imposed,. as shown in Fig. 14.

6 Conclusion

The main aim of this work is to improve security of registered number plate in India. The proposed approach finds pollution check control of vehicle and extracts the driver information using license plate detection method. In proposed work, if pollution check date is expired, then mail is sent to the vehicle owner regarding fine imposed, and if pollution check date is not expired, then mail is sent to the vehicle owner regarding pollution check expiry date. We can also restrict the vehicles which are 10-year old or more than 10-year old in our work. The experimental results show that this approach can achieve outstanding and robust detection performance in pollution fine controller and vehicle restriction area. For the experimentation, totally 22 images of license plate were tested. The overall system performance of the system is measured on the basis of accuracy and recognition rate; recognition rate is 92.57% and accuracy is 50%. The limitation of the proposed system is that it fails to recognize some characters, which are '1', '2', and 'R'. Future work has also been discussed to recognize the double line license plate and multiline license plate. We will also increase the template matching database.

References

1. Liu, C. C., Luo, Z.: Extraction of vehicle license plate number using license plate calibration. In: IET International Conference on Frontier Computing. Theory, Technologies and Applications, Taichung, 2010, pp. 187–192 (2010)
2. Mesleh, A., Sharadqh, A., Al-Azzeh, J.: An optical character recognition. Contemp. Eng. Sci. 5(11), 521–529 (2012)
3. Farajian, N., Rahimi, M.: Algorithms for license plate detection: a survey. In: First International Congress on Technology, Communication and Knowledge (ICTCK 2014) 26–27 November 2014
4. Babu, C.N.K., Subramanian, T.S., Kumar, P.: A feature based approach for license plate-recognition of Indian number plates. In: IEEE International Conference on Computational Intelligence and Computing Research (ICCIC), Coimbatore, pp. 1–4 (2010)
5. Madhukar, P., Singh, K.R.: Text recognition from images. Int. J. Adv. Res. Comput. Sci. Softw. Eng. 4(11) (2015)
6. Rabee, A., Barhumi, I.: License plate detection and recognition in complex scenes using mathematical morphology and support vector machines. In: International Conference on Systems, Signals and Image Processing, Dubrovnik, Croatia, 12–15 May 2014
7. Chen, B., Cao, W., Zhang, H.: An efficient algorithm on vehicle license plate location. In: Proceedings of the IEEE International Conference on Automation and Logistics Qingdao, China September 2008
8. Aytekin, B., Altuğ, E.: Increasing driving safety with a multiple vehicle detection and tracking system using ongoing vehicle shadow information. In: IEEE International Conference on Systems Man and Cybernetics (SMC), Istanbul, pp. 3650–3656 (2010)
9. Wazalwar, D., Oruklu, E., Saniie, J.: Design flow for robust license plate localization. In: IEEE International Conference on Electro/Information Technology (EIT), Mankato, MN, pp. 1–5 (2011)
10. Sulaiman, N., Jalani, M., Mustafa, M.: Development of automatic vehicle plate detection system. In: IEEE 3rd International Conference on System Engineering and Technology, 19–20 Aug 2013
11. Bhatia, N.: Optical character recognition techniques: a review. 4(5) May (2014)
12. Aghdasi, F., Ndungo, H.: Automatic licence plate recognition system. In: 7th AFRICON Conference in Africa, Gaborone, vol. 1, pp. 45–50 (2004)
13. Vamvakas, G., Gatos, B., Stamatopoulos, N., Perantonis, S.J.: A complete optical character recognition methodology for historical documents, DAS'08. In: The Eighth IAPR International Workshop on Document Analysis Systems, Nara, pp. 525–532 (2008)
14. Haneda, K., Hanaizumi, H.: A flexible method for recognizing four-digit numbers on a license-plate in a video scene. In: IEEE International Conference on Industrial Technology (ICIT), Athens, pp. 112–116 (2012)

15. Zheng, L., Samali, L., Yang, L.T.: Accuracy enhancement for license plate recognition. In: 10th International Conference on Computer and Information Technology (CIT), Bradford, pp. 511–516 (2010)
16. Kulkarni, P., Khatri, A., Banga, P., Shah, K.: A feature based approach for localization of Indian number plates. In: IEEE International Conference on Electro/Information Technology, Windsor, ON, pp. 157–162 (2009)

Watermarking Technology in QR Code with Various Attacks

Azeem Mohammed Abdul, Srikanth Cherukuvada, Annaram Soujanya, G. Sridevi and Syed Umar

Abstract Devour preservation and substantiation is becoming increasingly important on day-to-day life. Digital watermarking is the best technique developed to solve this problem. This is built using a digital watermark_wavelet_turn invisible for a QR code from the image. Converted into the investment process, a binary image, Logo, watermark adequate water, then put them in sub-selected band. Experimental results show that in all cases, this resistance to attack, and as such, can serve the protection of copyright and a viable authentication tool.

Keywords QR code · Salt–pepper attacks · Gaussian attack

1 Introduction

A digital watermarking is a pattern contains of bits in a digital image, audio or video will be inserted in recognized copyright and authentication mode. The purpose of image is to add watermarks perfect hidden secret information of the original

A.M. Abdul (✉)
Department of Electronics and Communication Engineering, KL University,
Vaddeswaram, India
e-mail: mohammedazeem123@gmail.com

S. Cherukuvada · A. Soujanya
Department of Computer Science Engineering, Institute of Aeronautical Engineering,
Hyderabad, India
e-mail: srikanth.cherukuvada@gmail.com

A. Soujanya
e-mail: annaramsoujanya@gmail.com

G. Sridevi
Department of Computer Science Engineering, KL University, Vaddeswaram, India

S. Umar
Department of Computer Science Engineering, GIST, Jaggayyapet, India
e-mail: umar332@gmail.com

© Springer Nature Singapore Pte Ltd. 2018   147
M. Pant et al. (eds.), *Soft Computing: Theories and Applications*,
Advances in Intelligent Systems and Computing 584,
https://doi.org/10.1007/978-981-10-5699-4_15

message, which is a powerful attack. In recent years, some scientists have proposed the adoption of techniques for the watermark. In [1], the main drawback is the space, and it is easy to cut and attacked. In [2], the method proposed in the original image with the image rights (n, n) embedded secret sharing system. This process could withstand pollution, such as JPEG, and the size of the additional noise and [3] multi spectral imaging technique by wavelet watermark. In [4], the authors developed a system with wavelet transform color images based on texture features and secret sharing. In this paper, we are presenting blind watermarking with discrete wavelet algorithm with two levels of transformation (DWT) have at a picture of the QR code.

1.1 QR Code [5]

QR code (quick response code) is the trademark of bar code in 2D. A bar code optically machine readable label assigned to an object, and it records the relevant information. The coded information of QR can make four types of standards of data (numeric, text, byte/binary, Kanji) or supported by extension, almost all types Facts. QR code system became popular outside. Thank you to the automotive industry for its excellent readability and fast storage capacity compared to UPC bar codes default. Applications include monitoring products, product identification, time tracking, document management, marketing, and generally more (Fig. 1).

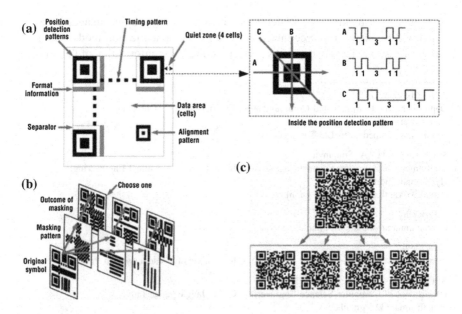

Fig. 1 QR code analysis and structure

1.2 Watermark

The watermark is the process of hiding the digital information the carrier signal, such as voice, images, video, etc. In addition, found buried digital data to identify a property Copyright. Watermarking methods cannot have divided into two categories Areas, which are:

1. The integration process spatial domain watermark is directly modifying the pixel values.
2. In the watermark embedding, the frequency domain the process is the inclusion of information.

For example, changing the space transformed frequency [6] coefficients. However, most of the signal processing paradigms in the recent literature can also be a frequency be characterized field of surgery. In addition, there are many good models of perception frequency domain, which has developed a great success reported.

2 Processing Methodology of Watermarking

Watermark remained within their frequency. The image of the QR code is broken first_wavelet_converted into two levels in two dimensions, as shown in Fig. 2 [7]. The extraction then binder of watermark that does not require an original image of the QR code to retrieve the incorporation of the watermark. There were two steps of the algorithm, watermark, mining our water watermark register.

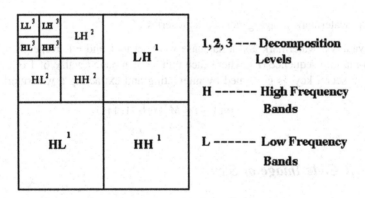

Fig. 2 Two-dimensional wavelet of QR code with various levels

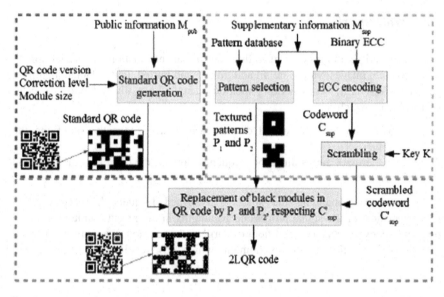

Fig. 3 Merging of watermarking in QRCODE

2.1 Registration Watermark

I. The stage of the merger process described in (Fig. 3): Step watermark image with a secret key.

$$S = \{s_i, 1 \leq i \leq N\}, s_i\{-1, 1\} \tag{1}$$

II. The watermark image generated as a series of

Bit values S watermark and a database were set to 1 and -1, respectively. The pseudo-random sequence (P), where each number can have a value of 1 or -1 with a random secret key, is generated by integrating and extracting a watermark

$$P = \{p_i, 1 \leq i \leq M\}, p_i\{-1, 1\} \tag{2}$$

2.2 QR Code Image of Step

The image required coding stages shown in Fig. 4. In which, encoding and decoding analysis is clearly explained.

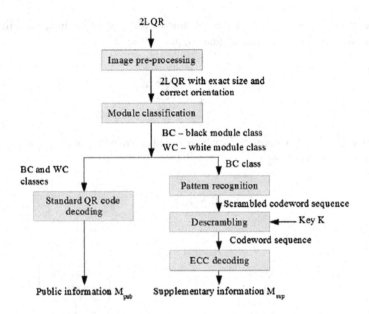

Fig. 4 Flow analogy of the coding process

1. Two levels DWT MXM Image (your) was computed QR code in the image.
2. Watermark subsequently incorporated into the sub-band or HL, LH, or HH_2. Under Rule:

$$t_i = t_i + ap_i.s_i, \quad i = 1, 2, 3, \ldots, N \tag{3}$$

where the t_i is the image input and t_i' is the output of the image with water mark embedded in the QR code.
3. Then, the inverse DWT (IDWT) was then applied to obtain the image with a watermark.
4. Calculate PSNR.

3   Watermark Extraction

Image watermark original water of a QR code the mining algorithm [8–10]. The provision of the original amount the pixel with everything necessary. Thus, the prediction of the original pixel value is performed using noise canceling technique. In this article, we use a mask averaging 3×3 whose entries were 1/9. The extraction procedure is described as follows (Fig. 4).

i. The imaginary image is $\widehat{t}_i$ will be obtained with input of the image t_i^* will be of convolution mask then the original signal is (Fig. 5).

$$\widehat{t}_i = \frac{1}{c \times c} \sum_{i}^{c \times c} t_i^* \qquad (4)$$

ii. Total Estimated watermark $\widehat{s}_i$ which is indicated with the difference of input and imagination image

$$\delta = t_i^* - \widehat{t}_i = \alpha P_i \cdot \widehat{s}_i \qquad (5)$$

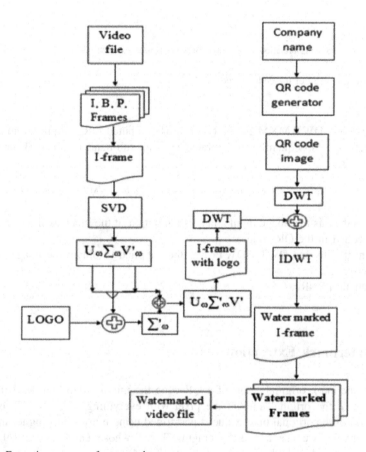

Fig. 5 Extraction process of watermark

iii. So, the difference can be easily being predicted with the below formula

$$\text{sgn}(\delta_i) = P_i \cdot \hat{s}_i \tag{6}$$

iv. Compute NC analysis.

4 Simulation and Experimental Analysis

4.1 Evaluation of Performance

Based on the NC count and PSNR ratio, we can clearly expect the measurements of the original watermark (S) and the evaluated watermark (S'). Formula used for given in Eqs. 7 and 8.

$$\text{NC} = \frac{\sum_{i=1}^{M} S_i S'_i}{\sum_{i=1}^{M} S_i^2} \tag{7}$$

$$\text{PSNR} = 10 \log \frac{(2^b - 1)}{\left(\frac{1}{m \times n} \sum_{i=0}^{m-1} \sum_{j=0}^{m-1} (O(i,j) - R(i,j))^2\right)} \tag{8}$$

4.2 Text Watermark with the Following Parameters

Coefficient magnitude (of α) 5, 10, 15, ..., 50 and the key 100. The QR image and Table 1 and the CN code [11–13] PSNR Big difference in a factor of 2 sub-band HH. All cases QR code correctly. On the other value of the factor of the Big Hand What to lower PSNR therefore the possibility of failure largely.

α	PSNR	NC	Decode QR code
5	51.2128	0.8224	✓
10	43.5675	0.8451	✓
15	39.808	0.8552	✓
20	37.6548	0.8825	✓
25	**35.7849**	**0.899**	✓
30	34.1767	0.9116	✓
35	32.6423	0.9161	✓
40	31.5694	0.9181	✓
45	30.6368	0.9201	✓
50	29.7186	0.9214	✓

Table 1 PSNR and NC of QR code

Watermark image (see Fig. 6). On the other hand, The NC value to increase water extraction (see point Fig. 7) and reduces noise. In view of the fact, this compromise Between PSNR and NC, empirical measure Factor 25. In addition, the

Fig. 6 QR code with extraction of watermark

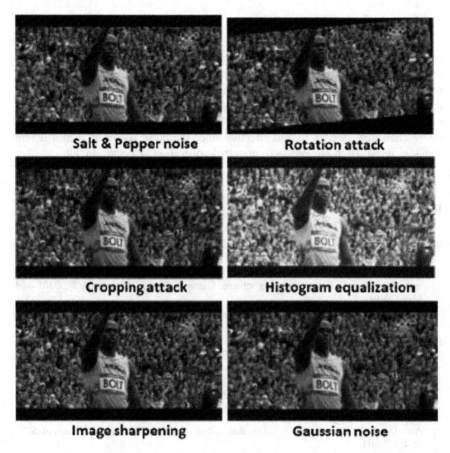

Fig. 7 Image with watermark extraction with various attacks

sub-band watermark 2. Converting the comparison February LH HL HH in Table 2 we see that it's different PSNR negligible, while HH 2 NC baseline Watermarks are most like the original image. We have the following experiment embedded watermark in the HH 2, after he applied DWT, and important information is collected at low frequency Subgroup. This allows a subsequent watermark is embedded, Loss of a variety of information that is not, in the frequency band Then, instead of the high NC resulting in the other direction. Figure 7 shows that the watermark extracted difference the magnitude of the coefficients. All images are extracted contains watermarks some visual noise for the watermark extraction method [14]. Picture QR natives, not a service. In fact, the transmission, you might want to harm the image contaminated unexpected noise in the network Communication. It is necessary to test the robustness of the several attacks like the noise of the salt and pepper, the algorithm.

5 Conclusion

In this article, the digital watermarking technology we announced if the watermark of the binary image is embedded the QR Picture Code. LH, and the presentation process in HL Based on the wavelet transform or sub-band HH. Experience the result shows that it is possible that the algorithm for recovering acceptable image quality and watermarks. Goal Measures, such as NC PSNR and size Factor. As future challenges, we try to find effective since the noise power of such a severe attack to resist, High compression, geometric distortion or occlusion, etc.

References

1. Cheung, W.N.: Digital image watermarking in spatial and transform domains. In: Proceedings of the TENCON, pp. 374–378 (2000)
2. Yalamanchili, S., Rao, M.K.: Copyright protection of gray scale images by watermarking technique using (N, N) secret sharing scheme. J. Emerg. Technol. Web Intell. 2(2), 101–105 May (2010)
3. Rangsanseri, Y., Panyaveraporn, J., Thitimajshima, P.: PCA/wavelet based watermarking of multispectral images. In: 2005 International Symposium on Remote Sensing (ISRS2005), Korea, 12–14 Oct 2005
4. Dharwadkar, N.V., Amberker, B.B.: Watermarking scheme for color images using wavelet transform based texture properties and secret sharing. Int. J. Inf. Commun. Eng. 6(2), 94–101 (2010)
5. QRCode. http://en.wikipedia.org/wiki/QR_code
6. Podilchuk, C.I., Delp, E.J.: Digital watermarking algorithms and applications. Proc. IEEE Signal Process. Mag. 18(4), 33–46 (2001)
7. Artz, D.: Digital steganography, hiding data within data. Proc. IEEE Internet Comput. 5(3), 75–80 (2001)
8. Wolfgang, R.B., Podilchuk, C.I., Delp, E.J.: Perceptual watermarks for digital images and video. Proc. IEEE 87(7), 1108–1126 (1999)
9. Kundur, D., Hatzinakos, D.: Toward robust logo watermarking using multiresolution image fusion. IEEE Trans. Multimedia 6(1), 185–198 (2004)
10. Hsieh, M.-S., Tseng, D.-C., Huang, Y.-H.: Hiding digital watermarks using multiresolution wavelet transform. IEEE Trans. Ind. Electron. 48(5), 875–882 (2001)
11. Cheung, W.N.: Digital image watermarking in spatial and transform domains. Proc. TENCON 3, 374–378 (2000)
12. Barni, M., Bartolini, F., Capellini, V., Magli, E., Olmo, G.: Near-lossless digital watermarking for copyright protection of remote sensing images. Proc. IGARSS 3, 1447–1449 (2002)
13. Kaarna, A., Toivanen, P.: Digital watermarking of spectral images in PCA/wavelet transform domain. Proc. IGARSS 6, 3564–3567 (2003)
14. Ho, A., Shen, J., Hie Tan, S., Kot, A.: Digital image-in-image watermarking for copyright protection of satellite images using the fast hadamard transform. Proc. IGARSS 6, 3311–3313 (2002)

Cognitive Networked Redemption Operational Conception and Devising (CNROCD)

Azeem Mohammed Abdul, Srikanth Cherukuvada, Annaram Soujanya, G. Sridevi and Syed Umar

Abstract Improving infrastructure, information and communication technologies to significantly increase the use of the Internet is still in use in cyber risks and network threats. Complex movements and new risks and threats systematically paved the way for the language has become a new field of land, air, sea, and space. It is quite clear that the cyber threat is likely to continue to intervene in theaters in the world of cyber year. But sometimes it is difficult to determine, first, column's Cyber-specific threats—remotely simple computer code into cash Cyber sure that the cyber attacks, cyber terrorism, and industrial espionage. Because, combined with the price of "exponential language and complex cyber attacks, multivariate caused, it becomes a requirement for designers, how to deal with cyber activities and problems in this area in the operational planning process. In this article, we tried to achieve a continuous production flow Cyber in order to understand and visualize complex question of the complexity of cyber crime, and the scope and systematic prevention provide a standard Internet operating model. Disguised as standard, we aim to help designers understand the complex task Cyber demonstrates, the advantage of using the force of gravity analysis of the factors (COG), which is usually dealt with military decision-making (MDMP) at the end of using

A.M. Abdul (✉)
Department of Electronics and Communication Engineering, KL University,
Vaddeswaram, India
e-mail: mohammedazeem123@gmail.com

S. Cherukuvada · A. Soujanya
Department of Computer Science Engineering, Institute of Aeronautical Engineering,
Hyderabad, India

G. Sridevi
Department of Computer Science Engineering, KL University, Vaddeswaram, India

S. Umar
Department of Computer Science Engineering, GIST, Jaggayyapet, India
e-mail: umar332@gmail.com

© Springer Nature Singapore Pte Ltd. 2018   157
M. Pant et al. (eds.), *Soft Computing: Theories and Applications*,
Advances in Intelligent Systems and Computing 584,
https://doi.org/10.1007/978-981-10-5699-4_16

technicians have an understanding of business planning. Operational cyber proposal will be presented as an example in this study; there are plans to provide a comprehensive Internet-based leader.

Keywords Cognitive mapping · Cyber informatics · Cyberspace · CCOD

1 Introduction

Throughout, history has been around it has always been a struggle between the forces of the nation. The fighting, clashes or fighting to get into the modern era and the different types of equipment, tactics and strategy [1]. Techniques have been developed for command and control of subtle already hostile forces. These techniques generally depend on the intention commander "to the desired end state of the enemy, we can make various differences to see. conditions, such as Sun Tzu usually means that the power in serious condition and use. If an enemy without a fight, which is better, or as recommended and intact condition [2] on the one hand, Clausewitz emphasized the theoretical importance of "total war" or "absolute" war. As we know, were called to fight the enemy with all the resources and momentum is destroyed the enemy [2]. Complex and multidimensional security environment today examine leadership and new strategies as appropriate cyberspace, which is becoming a new area for air, land, sea, and determine space [3] into account. Operational environment consists of friends, enemies, and neutral system is faced with a new supporting factor cyberspace and cooperates with the business change as a political, military and economic, social, information, infrastructure, etc. [4, 5]. Work Environment (œ) is separated from the products of the information system for many network data [3]. Technology development and increased use of the Internet and called on governments and institutions, the first situation, and more.

2 Literature Survey

In particular, the increased use of information and communication technologies (ICT), engineering equipment (phones and tablets), and more than three billion people who have access to the Internet, people use blogs and social networks [6, 7]. And with these facts, cyberspace has become a good place for criminal and terrorist organizations [8]. For example, Isis used the media to spread their ideology and message after the capture of Mosul, the second largest city in Iraq [9]. Especially, Isis can eat in creating a climate of fear in Iraq releasing videos and photos on social networks like Twitter and YouTube [10]. The power of social networks during elections, the events of the street to oppressive regimes or natural disasters, demonstrates the ability to change the traditional way of understanding the media. With this change in the environment, the major news agencies benefits of user-generated movies [11]. Thus, certain public services considered interference and restrictions on access to information, particularly in the network social, which provides a direct source.

In this new environment, it is easy to place a long time, information warfare in a manner similar to Example [12]. It is true that physical appearance, cyber wars begin in cyberspace, and impacts the of events in real life [13]. An increasing number of cyber attacks and different look for people, institutions, and countries to take drastic measures against them. These measures ranging from personal acts, because he is aware of the risks of cyber to reflect awareness events of the strategic position such as documents of the cyber security of soil and computer response teams arrived (Cirta). Surgical methods online ahead with the civilian government and cyber military force to protect property, security, and cooperation. In these groups, depending on the degree of vulnerability of cyber security, incident is treated cold practice of cyber education. As an institution, the military organization is to ensure that cyber protection of property, and they must be willing to procedures, planning, and scientific work in this government and the ambiguous land, site of ships, which is full of people and groups (3 passes are colored in); the target of cyber attacks vary depending on the motives and the desired end state that try hard to reach the planners and perpetrators of men. Qiao and Wang's China strategy, which means the war. "The arena is close to you and the enemy is the only network, there is no smell of gunpowder and the smell of blood" [14]. The army decided to treat (MDMP), designs, and "thing. We also organize product data illustrate zoom Cyber purposes discussed in the next section: for maintaining the relationship between Germany and the thoughts and actions of cyber only these two. In this section, we have a map of Cyber support system "prepared by a heavy factor computer networks. In addition, we have the right to protect cyber; we know it is in the case of cyber expert services well organized army.

3   The Proposed Approach for Cyber Security

Go to the complex as cyber warfare is a combination of "art and science, and you can see Blend, experience, observation and memories that will be made to the methods and techniques and tools, and support of the community [15]. In this world, we believe that the Cyber one (peace, tranquility, and kill), a change in the strategy of the group work with your wisdom and righteousness found the strategic objectives. That day we will come; variables edition instructions cyber experts call a "cyber-war", one of the great edition program stunt "computer work contrary conference facilities in the country, and was widely adopted e timber nation and the people. War and the Rest cyber probably less absolute, Duqûn Regina Red October and others. Some thought to remain in the organization's program, attention, and some have been supported by the state, because of the complexity of the language and wants a lawyer is still not enough evidence to show you the source code. In our discussions, we want the illegal activities of cyber war plan and examine the philosophy itself. We are interested in the idea of the program cyberspace (CO) are missing in the program. They also report a clear description of the work programs on cyber warriors met in detail, and lawyers "in the necessary national team, but managers and planners working within the system. In this article, we will organize a very special Cyber Coalition before expected. We try to adapt; the work have cyber expectations, business strategy and the military process (MDMP), hereinafter referred to as "elements malls necessary websites (CO) to build line as Cyber Cooperation (ClOO) focuses Cyber (CCOG) evaluation Cyber Police (CDC) and Cyber would last. Although there are elements of the design work, I will do all in our results require more work and more strategic phase without help or advice Court orders. In the morning, we saw "complex, currently living in "interoperability, combined with the latest technology, which plays a role much.

4 Military Decision Making Process with Effective Planning and (MDMPEP)

4.1 Internet Planning

Plan an activity to prevent the leader kaua'ōnohi work "to reach work on military targets [16]. Given the ambiguous nature of the war, many temporary". A teacher and some areas of the ho "need ohana"ana plan a work in progress as Field Manual 5–0 pale tongue, making partner, first, and then think one of the ways to achieve the goal [17]. m-level scheduled for any reason whatsoever, "secret of all.

4.2 Factor Analysis

Factor analysis or three-column format, staff officers of all factors, and what you get is a kind of test. It is often used MDMP method and can be used at all levels of the company. This is an area of functionality, the proposed plan, and work to achieve the conditions of management personnel, as well as the environmental impact assessment submitted target mode. The commander and his staff, "how they think requires Table 1. Factor cyber analysis and factor analysis results due creates discipline. In general, the shape of the three columns (the vulnerability of critical capture and output), and he is ready for it. Action critically critical vulnerabilities in their functional areas and financial plan represents CO helping to clear the focus of both enemies and friends CO., and plans to teach to determine.

Table 1 Cyber factor analysis for a sample

Mission/critical activity/functional area/critical vulnerability	Deduction	Output
Lack of talented cyber specialists in military organizations	Being unable to envision the cyber risks	To plan professional cyber security trainings on theory and hands on
	Being exposed to cyber incidents and unaware of them for a long time	To plan cybersecurity lessons in military high schools and academies
		Station some personnel on job training to scientific organizations and institutes dealing with cybersecurity
	Being unable to sustain situational awareness among commanders and staff	To plan cyber threat situational awareness training for commanders and staff to remind that cybersecurity is the commander responsibility
To defend army critical information systems against cyber attacks	Enough workforce assignment of cyber professional and a clear definition of "defense, active defense and offense' in procedures	To defend information systems 24/7
		To hire part lime or full time civilian contractors, engineers, hackers and malware analysts
	Building strong coordination with intelligence organizations	To assign liaison personnel mutually between cyber command and intelligence units
The risks of open source intelligence (OSINT) and social networks	Being an easy target to fishing attacks	To plan cyber exercises to draw attention of leading cyber attacks (fishing, waterhole attacks, etc.)
	Gathering OSINT via social networks with masked and fake social network accounts	To limit the use of social networks in military organizations
	Using metadata of uploaded contents and EXIF information of uploaded photos [21]	To assign a content operator to control, erase and change the metadata and other information of contents

4.2.1 The Center of Gravity (COG) Analysis

Among the design elements for computer work, Cog is a key to reach the desired destination: the center of the analysis of gravity, a number of questions: What is finally announce that the enemy is there? What measures will the enemies of the end of the finale? The conditions to support the activities of the enemy? What works to prevent friends to reach a final state? CCOG Analysis of the benefits of computers factors, as well as a friend and faithful partner opportunities to address the vulnerability.

4.3   The Work of Art and Design

Joint Pub 1-02, define graphics to integrate military force military targets (technical work) and the planning, organization and technology. The work has an important role in the intuitive operation and technical design, and a list of war-related events.

4.4   Enthusiasm of Creative Thinking

Faced with a do, need designers with resolution of difficult and complex issues to apply less than [15] memory. Positive thinking is a method to conceptualize, analyze, synthesize, and/or analyze or how they are collected, production, experienced, or observed. The emphasis on creative thinking and careful planning of the taxonomy of Bloom's learning shed light on this subject. As shown in Fig. 2, 1956, the old model, which was developed for educational purposes, a model that can help you connect with your creative thinking and creative thinking, and link to model elements [15]. It is a recognized leader in the brainchild less optimal experience and training, as well as the law (Table 2).

5   Computer Cyberspace to Do the Design Work

5.1   Cyberspace Operations (CO)

Held in Cyberspace (CO) for access to the Internet, where the main goal is political or in cyberspace [29]. The computer-theater lead the world and management system, to take steps to protect the information and communication technologies, and take total. The soldiers and the management of the design of e-support troops should be included, depending on the activity level. These are military integration CO detailed indene. Taking into account, the designers covered criteria detained at the indene. And indene, CO designers develop soft enemies and the environment. Depending on the lens, designers commander or CO designers must prepare computerized design CO, indane and even the first stage targetwhen the computer is in active measures to promote the objectives of the integrated cooperation of designers and the biggest military operation (Fig. 1).

5.2   Working of Cyber Operation Design (COD)

Construction of computer functions, the joint efforts of the picture or Indian, will affect the poor and most vulnerable. The creation of the arts and stimulate creative

Table 2 CCOG and its critical capabilities

Cyber center of gravity (CCOG)	Critical capabilities (CC)
SCADA systems and military information systems infrastructure	To implement cyber attacks against SCADA Systems by manual means (hiring a person/supporter to use a malware/worm infected hard drive on these systems)
	To infect enemy's military information systems with a computer virus, worm, or malware to steal or gather information (screen shots, key strokes, and files) by infiltrating into the systems or spear fishing by using social networks, open source intelligence (OSINT), and social engineering
	To implement a zero day exploit to database, e-mail, and file servers that are connected on internet
	To implement DDOS attacks
	To implement cyber-electronic warfare activities in order to get electronic intelligence from frequency logs of command and control systems by using airborne warning/intelligence systems or drones
Critical vulnerabilities (CV)	Critical requirements (CR)
Limited use of cyber intelligence activities	To gather OSINT about SCADA/Cyber physical systems and their system requirements by using TOR networks or spoofed IP addresses
National and international legal challenges and NATO perspective on cyber issues (accepting it as an act of war or not, ambiguity of cyber rules of engagements etc.)	Preparing a legal frame document that clearly defines the activities, tasks, and duties about cyber
Lack of talented cyber specialists and cyber manpower planning in military organizations	Reverse engineering and multi criteria analysis of some well known malwares directed to gather information from the systems they infected
Lack of a roadmap and strategy designed to reach the national cyber security policy, lack of information sharing with institutions, universities, and defense firms	To initiate a collaboration and among technical universities, civil institutions and defense firms about research and development (R&D) in cyber
Lack of a clear task definition of cyber activities among institutions	Forming a social network team, working 24/7 and solely focusing on Facebook, Twitter, LinkedIn, Vine, Instagram and the others also
The lack of integration of cyber and intelligence units and the dilemma of whose job that is, limited cooperation between these two functional areas	To have a national vulnerability database and enough cyber experts and contractors
The legal challenges	A great number of zombie computers, bodies
The difficulty in integrating and implementing cyber and electromagnetic activities in tactical level under a command	Strong cooperation between intelligence cyber units

Fig. 1 Design flow of
operation cognitive mapping

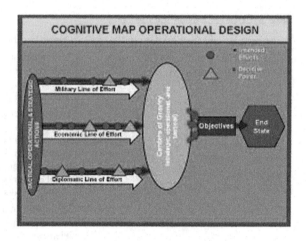

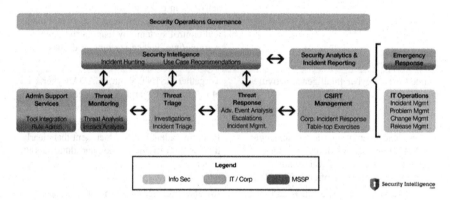

Fig. 2 Cyber security operational model sam

thinking and rational management of these measures gives a complete picture of the
work, called cognitive maps of the war. To identify the elements of the intellectual
computer, online business planning is used. In accordance with the behavior Cyber
works of the computer (the government, the people and the armed forces), cycling,
the center of gravity (dial) announced the end (John finally). All design elements IT
resources Fig. 2 are used to the four stages of a disease that starts ending:
"Introduction confusion" and ends with "masters." can protect your computer may
be adapted on military expansion and Government of the teeth (dental strategic
work), or the final state (the last military government of the state).

6 Conclusion

A multidimensional and multivariable complex security environment has become to prepare for a must-have for all types of threats. What happens to the development and transfer of advanced technology systems or information is clearly the best way to threats. National Infrastructure comprehensive system administrators create their own data centers, government and military organizations and networks, which operate to manage the functions of the software and hardware. Smart devices, mobile phones, home appliances, and more human interaction cars on a daily basis increase the use of technology. A logical consequence of the evolution of social networks and other technological advances that make it easier to increase the use of criminal informants. Open source, and to collect information on social networking sites like Face book, Twitter, Instagram and other Internet users, the standard has been an easy task even more. What the government and military organizations in a changing security situation, we may sooner or later adapt an unexpected support for peacekeeping operations on the threats and dangers, is a revision of the plans and procedures.

Cyber events and organizations should play a major role in the theater of global networks, databases, and command and control systems play must be designed according to assume that we protect critical assets. In this regard, it is difficult to cyber threats and targeted/attacked by complex problems is to detect a bad analogy. So, with the emphasis on the importance of planning and the elements, and a first and a comprehensive cyber threats, cyber attacks, cyber espionage operation is designed to provide a design we can use cyberspace activities offer. If they cannot design a cyber activity, plans to cyber experts believe that the common point and understand each other, and strengthen cooperation, the institutional and national cyber security closer.

References

1. Trade, M.I.: Master of war: classic strategic thinking. Routledge (2005)
2. Trade, M.I., Clausewitz, S.T.: The Art of War and the War College, the US Army Institute for Strategic Studies in Carlisle Barracks, Pennsylvania War (1991)
3. The brochure, A. T. R. C. (2010). TRADOC Cyberspace Operations Concept Plan 2016–2028 capabilities brochure. Washington, DC: DOD
4. Göztepe, K., Kilic, R., Food, A.: Depth cyber defense: design cyber security agency in Turkey. Mar. Sci. Technol. J. **10**(1), 1–24 (2014)
5. Göztepe, K.: Based on fuzzy rules based system design expert in computer security. Int. J. Inf. Secur. Sci. **1**(1), 13–19 (2012)
6. Singh, A.K., Sahu, R.: A better quality of internet, telephone and call center integration of e-government for all citizens. Gov. Inf. Q. **25**(3), 477–490 (2008)
7. Gungor, V.C., Falcon, D., Kocak, T., Ergüte, S., Buccella, C., Cecati, C., Hancke, G.P.: Smart Grid and communications technologies and standards. IEEE Trans. Comput. Ind. **7**(4), 529–539 (2011)

8. Choo, K.K.R.: Organized criminal groups in cyberspace: the typology. Organized Crime Trends **11**(3), 270–295 (2008)
9. Kay M.J.: The ISIS tactic Social Media a new place in modern warfare [online]. http://techcrunch.com/2014/10/15/isis-tactics-illustrate-socialmediatonewlocation,modernwar(2014)
10. Gray, G.: (2014) Isis and social media [online], Ankara Strategy Institute. http://www.ankarastrateji.org/haber/isis-and-social-media1399/
11. Newman, N., Dutton, W.H., Blank, G.: Social media is changing the ecology of Britain in the fourth and fifth block. Internet Sci. Int. J. **7**(1), 6–22 (2012)
12. Héjas, A.J., Héjas, H., Héjas, J., JA (2015) Cyber war in Lebanon awareness: intelligence research. Int. J. Cyber Secur. Justice Digit. (IJCSDF) **4**(4):482–497
13. Al-Ahmad, W.: Cyber security corporation has developed a strategy for the management of the war. Int. J. Cyber Secur. Justice Digit. (IJCSDF) **2**(4), 1–9 (2013)
14. Liang, G., Xiangsui, W.: Free-of-war, pp. 551–563. PLA Literature and Art Publishing House, Beijing (1999)
15. Joint Staff, the operational model of the J-7. Planner Manual, joint and coalition warfare in Suffolk, Virginia (2011)
16. Lussier, J.W., Shadrick, S.B., Prevost, M.I.: Think of it as the commander of a prototype: an adaptive approach (no ARI-RP-2003–02), instruction manual. Army Research Inst. Behavioural and Social Sciences, Alexandria, VA (2003)
17. FM 5–0. (2010) Process Operations Defence Staff of the Army
18. FM 101-5. (1997) the organization and operation of the Army, Washington, DC, Headquarters Division staff
19. Kober, (2010), the AA Design I remove: Army military planning decision-making, coordination, the Advanced Military Studies, United States Army Command and General Staff College at Fort Leavenworth School
20. Grigsby Jr., W.W., Gorman, S., Marr, J., McLamb, J., Stewart, M., Schiffer, P.: Integrated planning, design, military operations in decision making. Mil. Rev. **92**(4), 15 (2012)

Secure Group Authentication Scheme for LTE-Advanced

M. Prasad and R. Manoharan

Abstract Due to the increased popularity of group-oriented applications and protocols, group communication occurs from network layer multicasting to application layer telephonic and video conferencing. In this scenario, security services are necessary to provide authenticity, integrity, and communication privacy. Even though one-to-one communication is well established with good standard and security, still the group communication remains unexplored, since the group communication is not as simple as the one-to-one communication extension. In group Diffie–Hellman (GDH) key agreement protocol, all the group members combined to generate a group key, since many GDH key agreement protocols utilize the original Diffie–Hellman (DH) key agreement protocol to arrange all the group members in a logical ring or a binary tree to exchange DH public key. From these, there are two significant aspects which influence the efficiency of the GDH when the member of the groups gets increased. Here, we propose GDH key agreement protocol constructed on secret key sharing and forward secrecy with single-mode key confirmation and digital certificate of the DH public key to afford authentication of group keys. By using this algorithm, messages and key updates will be restricted to the group; hence, the computation load is distributed among the user equipment. BAN logic analysis and the experimental results show that our proposed protocol performs better for the key establishment in terms of computational cost and memory.

Keywords Diffie–Hellman · GDH · LTE-A · Authentication · BAN logic

M. Prasad (✉) · R. Manoharan
Pondicherry Engineering College, Puducherry, India
e-mail: Prasad.psd@gmail.com

© Springer Nature Singapore Pte Ltd. 2018   167
M. Pant et al. (eds.), *Soft Computing: Theories and Applications*,
Advances in Intelligent Systems and Computing 584,
https://doi.org/10.1007/978-981-10-5699-4_17

1   Introduction

1.1   Background

Cryptography is not about simply hiding the message from others, so that no one can find the message. But rather to provide the message in public in such a way that no one except the original receiver recognizes the message. Group Diffie–Hellman schemes for authenticated key exchange are designed to deliver a group of users communicating over an open network and holding the secrets with a session key to be used to accomplish multicast with data integrity and confidentiality. In the open network environment, applications such as multiparty computing, collaboration workspaces, video conferencing are very popular these days. In these criteria, there is a need for secured services to afford privacy of group communication and reliability of communication data. To attain the required level of security, it is significant to establish a mutual secret key for encrypting or signing the group communication data. The communication channel is vulnerable to eavesdropping, and each group member subsidizes their consistent secret information shared which suits a part of overall group key. In this paper, we consider the scenario in which the group members never communicate before. The group is dynamic, where the users may join and leave the multicast group at any time.

1.2   Related Works

The issue of authentication is addressed in [1] by Hailong et al. They had proposed a scheme for authenticated group key agreement among the members of a group. They have used the ideas of 1D-based encryption scheme and threshold cryptography to split the cryptographic operation among multiple users. However, they have not concentrated on automatic authentication of the group key in a distributed environment.

In 2001, Boneh and Franklin [2] followed Shamir's idea to propose a real-world ID-based encryption scheme based on the Weil pairing. From this, the ID-based cryptography received noteworthy.

The mobile devices are generally resource-constrained because they possess only low power in terms of energy and limited computing capability. In this scenario, cryptographic operations with exclusive computations would become heavy load for mobile devices. Hence, it is a critical issue to reduce the computational load of mobile devices in AKE or ID-based AKE protocols. To overcome the resource-constrained situation on the client side, several AKE protocols for mobile devices have been proposed for conventional public key systems. Also, based on Boneh and Franklin's ID-based public key setting, a number of ID-based AKE protocols for mobile devices have been proposed to focus on the computation issue for mobile devices. These protocols above adopted an imbalanced computation technique to reduce the client's computational cost by shifting computational burden to a powerful server.

The offline precomputation technique is engaged to reduce the online computational load of mobile devices. In this phase, ephemeral secrets (or random values) are required to generate some values in advance. The ephemeral secrets and these precomputed values are stored in the memory of mobile devices for the usage in the online phase. As an outcome, a new type of attacks would occur, called ephemeral-secret-leakage (ESL) attacks, in the sense that an adversary can reveal the private keys of clients from those precomputed values or the corresponding exchange messages if the ephemeral secrets are compromised. To our knowledge, the existing ID-based AKE protocols did not address the ESL attacks at all.

Chuang and Tseng's protocol is secure against all known attacks and delivers forward secrecy. However, all the ID-based AKE protocols implemented the precomputation technique to reduce the computational load of the mobile client. In this case, those ID-based AKE protocols would be vulnerable to ESL attacks under mobile client–server environments.

In a set of n group members to compute a group DH key is an exceptional case of secure multiparty computation in which a group of n members when each retains a private key k_i and implement a function securely as $f(k_1, k_2, ..., k_n)$. Tzeng [1] proposed a protocol using random oracle model which is secure against both active and passive attacks with $f(k_1 + k_2, ..., + k_n) = gk_1, k_2, ..., k_n$ (modp). It never provides any useful information to passive attackers and accomplishes fault tolerance against any alliance of malicious attackers.

A practical conference key distribution systems [2] based on public keys, which authenticate the users and provide the Diffie–Hellman problem, is inflexible. Only few interactions are needed, but the overall cost for computation is low. Even though it is simple in several means, it has its own drawbacks such as constant communication and number of rounds which depends on the network used. It is secure against any type of attack, and it ensures that the discrete logarithm problem is inflexible. Burmester et al. [2], Ingemarsson et al. [3], Steiner et al. [4] in their proposal utilized this methodology to organize group members in a logic ring and to interchange DH public keys.

1.3   Contribution and Organization

The main factors that affect the efficiency of a GDH protocol are the computational cost and the communication rounds, and this is happening when the group size increases to be huge. Here, we propose a GDH protocol using discrete logarithm and Diffie–Hellman key exchange algorithm.

Our proposed GDH protocol is secure, tough, and efficient. In our proposed GDH protocol, each member employs a one-way hash function to generate the DH key validation and uses it to afford the authentication of group keys. The proposed GDH protocol can provide key secrecy, forward secrecy, and key independence of group keys. The proposed GDH protocol can counterattack against unknown key-share attack and key compromise impersonation attack. The contributions of this paper are listed below.

- It is efficient in terms of computational cost and the communication rounds.
- The security of the GDH protocol is proven to be secure with the DH.
- Our GDH protocol is robust to accommodate dynamic change of group members.
- The GDH protocol can provide key secrecy, forward secrecy, and key independence and can counterattack unknown key-share attack and key compromise impersonation attack.

The organization of this paper is as follows. In Sect. 1, we present the model for GDH systems, including security, attacks, and adversaries. In Sect. 2, we present a basic GDH protocol, which is based on a secure broadcast encryption scheme using the secret sharing. In Sect. 3, we present an authenticated GDH protocol; in Sect. 4, we discuss the security; and in Sect. 5, we compare the performance of proposed protocol with other protocols. We conclude in Sect. 6.

2 Group Diffie–Hellman Key Agreement Protocol

In this section, we define our GDH protocol which includes the overall architecture, the challenges, security goals, and possible attacks of group key.

2.1 Protocol Description

There has no key generation server in our proposed GDH protocol. All the group members collaboratively determine the group key. The one-time DH secret key k_i is contributed by each group members. Let us assume a group of five members with individual secret keys k_1, k_2, k_3, k_4, k_5, respectively such that our proposed GDH enables the members of the group to compute the group key $g^{2(k_1 k_2 + k_1 k_3 + k_1 k_4 + k_1 k_5 + k_2 k_3 + k_2 k_4 + k_2 k_5 + k_3 k_4 + k_3 k_5 + k_4 k_5)} (\bmod p)$. For every communication session, a group key is generated. Our protocol is capable of accommodating the dynamic change in group size. When a new member entered into the group, a new key is generated.

The public keys with digital certificate will be used to authenticate group keys. In our protocol, each member uses a one-way function to generate a key confirmation of the group key. Since the computation of a one-way function is faster than generation and verification of a digital signature, our proposed GDH protocol is more efficient than authenticated GDH protocols using the digital signatures. The communication rounds is another important factor affecting the efficiency of a GDH protocol. The basic GDH protocol needs two rounds to authenticate, and the authenticated GDH protocol needs three rounds.

2.2 Security Parameters in Group Key

Let us assume that a sequence of group keys is represented as $k = \{k_1, k_2, \ldots, k_n\}$. The security parameters are listed below:

(a) Privacy: No one except the group member can obtain the established key k_i. It is computationally infeasible.
(b) Authentication: No one can able to run the terminating protocol as normal; in particular, no one can make a group member believe there is an authentic member to talk with.
(c) Forward secrecy: It is a property of key agreement protocols certifying that a session key derived from a set of long-term keys cannot be compromised if one of the long-term keys is compromised in the future. This must not be used to derive any other keys.
(d) Key independence: If a session key is compromised, the adversary is not able to identify the group key or any long-term keys.

2.3 Conceivable Attacks

(a) Unknown key-share attack: An entity A ends up believing that he/she shares a key with B, and although this is in fact the case, where B mistakenly believes that the key is instead shared with an entity, $E \neq A$.
(b) Key compromise impersonation attack: An attack in which a user from a group masquerades as a trusted user. Suppose A's long-term private key is compromised. The adversary who knows A's long-term private key can imitate A, since this value identifies A. However, this attack enables the adversary to imitate other entities to A.
(c) Ephemeral-secret-leakage (ESL) attack: In the sense that an adversary can reveal the private keys of clients from those precomputed values or the corresponding exchange messages if the ephemeral secrets are compromised.

3 Authenticated Group Diffie–Hellman Protocol

In this section, we propose a GDH protocol to allow a group of n members secretly decide a group key collaboratively. This GDH protocol especially provides secrecy of the group key to all group members. In the GDH protocol, each member in the group selects a one-time DH secret and broadcasts the one-time DH public key initially. A session is established between two members after receiving each DH public key from one of the other group members. Likewise, all member can establish $n - 1$ DH secrets with other members in the group. Hence, all the members use an absolutely secure encryption scheme to distribute these $n - 1$-shared DH secrets to other members. The absolutely secure encryption scheme is based on the secret sharing scheme.

3.1 Robust Broadcasting (RB) Scheme with Absolute Security

Let us assume a member UE_1 likes to transmit the message m to a group of members (UE_2, UE_3, ..., UE_n) securely in a broadcasting channel. Some of the following parameters are listed as follows:

S_x: a preshared key between the members UE_1 and members UE_x, where $x \neq 1$.
p: large prime number.
Z_x: be the public identity of each member UE_x with $Z_x \in [1, n - 1]$.

1. UE_1 constructs an interpolating polynomial $P(x)$ of degree $\leq (n - 1)$ which passes through n points $(x_1, y_1 = f(x_1)), (x_2, y_2 = f(x2)), \ldots, (x_n, y_n = f(x_n))$ and is given by $P(x) = \sum_{j=1}^{n} P_j(x)$

where

$$P_j(x) = y_j \prod_{\substack{k=1 \\ k \neq j}}^{n} \frac{x - x_k}{x_j - x_k} \tag{1}$$

Then, UE_1 computes the publicly shareable key as $P_j(x)$, for $x = 1, 2, \ldots, n - 1$, and broadcasts these values.

2. For each member UE_x where $x \in [2, n]$ which includes the shared secret, S_x and $n - 1$ publicly shareable keys, $P_1(x)$ for $x = 1, 2, \ldots, n - 1$.

Using Lagrange interpolating polynomial gives

$$S_x \prod_{j=1}^{n-1} \frac{-j}{Z_x - j} + \sum_{x=1}^{n-1} P_1(x) \frac{-Z_x}{x - Z_x} \prod_{j=1, j \neq x}^{n-1} \frac{-j}{x - j} (\bmod p) = m.$$

3.2 Diffie–Hellman Key Agreement Protocol

Assume that n members in the group, $UE_1, UE_2, \ldots, UE_n$, need to set up a group key collaboratively in a broadcast channel. The main objective of the GDH protocol is to afford the secrecy of the group key to all group members. The following parameters are utilized:

- p: a large prime number such that is $2q + 1$, where q is also a large prime number.
- g: a generator for the subgroup Gq.

Our proposed GDH protocol consists of three phases: group authentication key exchange, authenticate and forward, and get_key. In the group authentication phase, all entities form a target group and cooperatively generate the group authentication message with their secret keys. Then, they transmit messages within the group. The algorithm is illustrated below:

Algorithm GroupAuth_KeyExch (Group, 1, n)

Let us assume Group = $\{UE_1, UE_2, ..., UE_n\}$, and n represents the number of groups

The aim of this algorithm is to authenticate all the devices and establish a session key for the defined group.

```
{
        fori = 1 to n do
        {
                Recive_Msgi1 = null;
                Recive_Msgi2 = null;
                Obtained_Keyi = False;
        }
// Key Forwarding Phase
AuthFwd(Un, AS);
// User equipment authentication and key processing phase
// Server authenticates all UE in the group and generate a group session key
// UE Authentication
AS receives {IDi || Ni} by sorting Xn1 for i = 1, 2, ..., n;
 AS process {IDi ||
Ni} and Ki  to validate the reliability of Xn2 for i = 1, 2, ..., n;
If Xn2 validation fails then rejected; //
if UE fails then eNb terminates the protocol
// Key processing phase
Choose a session key KG randomly;
Identity of the groups is taken as IDG;
fori = 1 to n do
{
        EKEYi = HashKi (IDG|| Ni)  ⊕ KG;
        Mi = {IDi, IDG, EKEYi, HashKi, (IDG || Ni || KG)};
}
```

```
AS  →  Un : {M1, M2, ... , Mn};
GetKey(Group, 1, n);
}
```

In this phase, the group is defined and the key is generated with randomly generated session key. It incorporates the key forwarding phase, which forwards the key to authentication server for verification. Next is the authentication forward phase, where the sender generates its own authentication message and forwards to the receiver in the group with the help of hash function.

Algorithm AuthFwd (UE_i, UE_j)

UE_i is the sender, and UE_j is the receiver. In this algorithm, the sender generates its own authentication message and forwards it to the receiver.

```
{
        Xi1 = (Idi ‖ Ni ‖ Recive_Msgi1);
        ω = (Idi ‖ Ni ‖ Recive_Msgi2);
        Xi2 = HashKi(ω);
        Xi = {Xi1 , Xi2};
        Ui ⇒ Uj : Xi ;
        if (Ui ≠ AS) then
                {
                        Recive_Msgi1 = Recive_Msgi1 ‖ Xi1;
                        Recive_Msgi2 = Recive_Msgi2 ‖ Xi2;
                }
}
```

The session key is verified by the authentication server and forwarded to all the members in the group. This session is valid for only one communication session.

Algorithm GetKey(Group, 1, h)

//From this algorithm, all the UE receives the group session key.

```
{
        if (1 ≥ h) then returns;
        m = ⌈(1 + h)/2⌉ ;
        if (!Obtained_Keyh) then
                {
                        //UEh obtains the group session key
                        DKEYh = HashKh (IDG‖ Nh);
                        K'G = EKEYh ⊕ DKEYh;

if (HashKh (IDG ‖ Nh ‖ K'G) = HashKh (IDG ‖ Nh ‖ KG)) then
                        K'G is the group session key;
```

```
                        else
                                returns;  //UEh ends
                        Obtained_Keyh = True;
                }
    UEh  →  UEm :  {M1, M1 + 1, ... , Mm};
    if (1 < m) then
                {
                        Simultaneous begin
                        GetKey(Group, 1, m);
                        GetKey(Group, m + 1, h);
                        Simultaneous ends;
                }
        else
                {
                        // U1 receives the group session key
                        DKEY1 = HashK1 (IDG ‖ N1);
                        K'G = EKEY1 ⊕ DKEY1;

if (HashK1 (IDG ‖ N1 ‖ K'G) = HashK1 (IDG ‖ N1 ‖ KG)) then
                                K'G is the group session key;
                        else
                                returns;  //UE1 ends
                        Obtained_Key1 = True;
                }
}
```

After the completion of the three phases, the secret session key is distributed among the group with authentication server's authentication. When the group size changes, the session key needs to be generated again with the same process as above illustrated.

4 Security Analysis

Initially, we analyses the security concerns of the proposed protocol and further comparison with the existing protocols.

To analyze the proposed protocol, we define some reasonable assumptions in advance:

(1) Each *UE* shares a secret key with the authentication server AS. Each *UE* protects its shared secret key.

(2) All the secret keys are protected by AS which is a trusted third party, authenticating individualities of *UE* in the group, and choosing a group session key randomly in an unpredicted way so that no *UE* can predict any generated group session key.

(3) The *UE* never exposes the key intentionally to other groups.

4.1 BAN Logic

In BAN logic, we distinguish these objects: agents (*A*, *B*), cryptographic keys (*k*), identifiers (*n*), and statement (*X*, *Y*). Some formulas represent Boolean expressions and idealized messages. With these specifications, we can define the following paradigms:

- $n| \equiv X$: (*n* believes *X*) The identifier *n* is allowed to accept though *X* is true.
- $A \triangleleft X$: (*A* sees *X*) Someone has sent a message to *A* containing *X* so that he can read *X* and repeat it.
- $A| \sim k$: (*A* once said *k*) At some time, *A* used key *K*.
- $A| \sim X$: (*A* once said *X*) At some time, *A* said a message containing *X*.
- $A \Rightarrow X$: (*A* has influence over *X*) *A* is an authority on *X* and can be trusted on *X*.
- $A \overset{k}{\leftrightarrow} B$: (*A* and *B* share key *k*) *A* and *B* can use key *K* to communicate. The key is unknown to others.
- $\#X$: (*X* is fresh) It specifies that *X* is never revealed before.
- $\longrightarrow kB$: (*B* has public key *k*) and the corresponding private key $K - 1$.
- $A \overset{X}{\rightleftharpoons} B$: (*A* and *B* share secret *X*) and possibly some trusted comrades.
- $\{X\}_k$: The statement *X* is encrypted by a secret key *k*.
- $\langle X \rangle_k$: The statement *X* is connected with a secret key *k*, and *k* may be used to prove the source of *X*.

From the above-said paradigms, we inferred with various rules for manipulating our proposed protocol which are listed below:

1. Message-meaning rule: It defines that if *A* believes (*A* share (*k*) *B*) and *A* sees $\{X\}k$, then *A* believes (*B* said *X*). Such that

$$\frac{A| \equiv \left(A \overset{k}{\leftrightarrow} B \right), A \triangleleft \{X\}_k}{A| \equiv (B| \sim X)} \tag{4.1}$$

2. Nonce verification: In this verification process, if *A* believes *X* is fresh and *A* believes *B* once said *X*, then *A* believes *B* believes *X*. such that

$$\frac{A| \equiv (\#(X)), A| \equiv (B| \sim X)}{A| \equiv \sim (B| \equiv \sim X)} \tag{4.2}$$

3. Jurisdiction: If A believes B has influence over X and A believes B believes X, then A believes X.

$$\frac{A| \equiv (B \Rightarrow X), A| \equiv (B| \equiv X)}{A| \equiv X} \tag{4.3}$$

4. Freshness: If A believes that the statement X is fresh, then A believes the entire statement (X, Y) is fresh.

$$\frac{A| \equiv \text{fresh}(X)}{A| \equiv \text{fresh}(X, Y)} \tag{4.4}$$

5. Message decryption: If A believes that he shares a secret key k with B and sees the statement X is encrypted by K, then A sees X.

$$\frac{A| \equiv A \overset{k}{\leftrightarrow} B, A \vartriangleleft \{X\}_k}{A \vartriangleleft X} \tag{4.5}$$

Our proposed protocol is illustrated with BAN logic as follows:
Key forward phase:

1.
$$UE_1 \rightarrow UE_2 : \left(X_1, \langle X_1 \rangle_{k_1} \right) \tag{4.6}$$

2.
$$UE_2 \rightarrow AS : \left(X_1, X_2, \left\langle X_2, \langle X_1 \rangle_{k_1} \right\rangle_{k_2} \right) \tag{4.7}$$

Key verification phase:

1.
$$AS \rightarrow UE_2 : \left(\left\{ UE_1 \overset{K}{\leftrightarrow} UE_2 \right\}_{k_1}, \left\langle X_1, UE_1 \overset{k}{\leftrightarrow} UE_2 \right\rangle_{k_1}, \left\{ UE_1 \overset{K}{\leftrightarrow} UE_2 \right\}_{k_1}, \left\langle X_2, UE_1 \overset{k}{\leftrightarrow} UE_2 \right\rangle_{k_1} \right) \tag{4.8}$$

2.
$$UE_2 \rightarrow UE_1 : \left(\left\{ UE_1 \overset{K}{\leftrightarrow} UE_2 \right\}_{k_1}, \left\langle X_1, UE_1 \overset{k}{\leftrightarrow} UE_2 \right\rangle_{k_1} \right) \tag{4.9}$$

In our proposed protocol key forwarding phase is between UE_1 and UE_2 the statement X_1 is encrypted with key k_1 is given in Eq. (4.6) is proved by BAN logic by the Eq. (4.1). Then, the UE_2 forwards the key to authentication server with both the statements from UE_1 and UE_2, which includes both the UE's keys given in Eq. (4.7). In the key verification phase, authentication server forwards the encrypted message from both UE which is specified in Eq. (4.8). In Eq. (4.9), UE_2 returns acknowledgment UE_1 with the encrypted message.

4.2 Principles of Diffie–Hellman

Here, we analyze the robust broadcast scheme, GDH protocol, and examine the security properties of group key.

Lemma 1: The RB scheme is absolutely secure.

Proof: UE_1 constructs the polynomial $f(x)$, which includes $n-1$ degree and broadcasts $f(x)$, where $x = 1, 2, \ldots, n-1$. For every envisioned group member, UE_i, where $i \in [2, n]$, knows a shared secret S_x. Both the points (Z_x, S_x) are on the polynomial $f(x)$.

DH scheme security relies on Computational Diffie–Hellman assumption (CDH assumption) with two values ga and gb, where a computationally bounded opponent cannot mend the DH secret gab.

Lemma 2: The GDH protocol is safe, delivers the CDH assumption, and is rigid.

Proof: The shared secret S_x is broadcast to other users using RB scheme which is absolutely secure. In this case, we have to verify whether the opponents can able to reveal the secret from the broadcast information.

5 Performance Evaluation

Let us assume to be the time unit for each round and $n = 2m$ (m is an integer) be the number of the UE's, and also we assume $|f|$ which is the sizes of all attributes of the transmitted message are the same. In that, the attributes specified are the identity, the nonce, and the secret key is $|f|$. The performance evaluation of the proposed protocol in terms of the completion time complexity, the total rounds of transmitted messages, the communication overhead, the computational complexity, and the execution time is illustrated below.

The time complexity of our protocol is calculated as $((m+1) \times t + (m+1) \times t) + 1 = (2(\log n + 1) \times t) + 1$.

To establish a session key for the group, we require $2 \times \left(\left(\sum_{i=1}^{m} \frac{n}{2^i} \right) + 2 \right)$ rounds of transmitted messages.

The complexity of the total rounds of the transmitted message is $O(n + 1)$ and depends upon the numbers of the *UE*s participated in the group. In the implementation, it requires $n|f|$ communication overhead to synchronize and proceed with the communication at the same time.

6 Conclusion

We have proposed a protocol for group-oriented authentication with key exchange, in which a group of n *UE*s can generate and authenticate with each other. They can able to share a group session key among them with the authenticated authentication server. In performance, it requires only $O(n + 1)$ rounds of messages, $O(\log n)$ completion time, and $O(\log n)$ waiting time. In addition, the proposed scheme has the following advantages:

1. We use BAN logic to prove the modal logic of belief with *UE*s involved in communication.
2. Our proposed protocol is more efficient than the existing protocols in terms of completion time, waiting time, and rounds of transmitted messages.
3. Our protocol is efficient, even though the group is dynamic.

References

1. Tzeng, W.-G., Tzeng, Z.-J.: Round-efficient conference key agreement protocols with provable security. In: Okamoto, T. (ed.) Advances in cryptology—ASIACRYPT 2000, pp. 614–627. Springer, Berlin Heidelberg (2000)
2. Burmester, M., Desmedt, Y.: A secure and efficient conference key distribution system. In: De Santis, A. (ed.) Advances in cryptology—EUROCRYPT'94, vol. 950, pp. 275–286. Springer, Berlin Heidelberg (1995)
3. Ingemarsson, I., Tang, D., Wong, C.K.: A conference key distribution system. Inf. Theory IEEE Trans. **28**(5), 714–720 (1982)
4. Steiner, M., Tsudik, G., Waidner, M.: Diffie-Hellman key distribution extended to group communication. In: Proceedings of the 3rd ACM Conference on Computer and Communications Security, New York, NY, USA, 1996, pp. 31–37

Modeling the Alterations in Calcium Homeostasis in the Presence of Protein and VGCC for Alzheimeric Cell

Devanshi D. Dave and Brajesh Kumar Jha

Abstract Alzheimer's disease is one of the neurodegenerative diseases which cannot be cured completely, but only the progression of it can be prevented. The causes of the Alzheimer's are many, but the present piece of work focuses on the alterations taking place in the calcium homeostasis in the presence of protein, i.e., buffer and voltage gated calcium channel (VGCC). The fundamental nature of the buffer is to decrease the cytosolic calcium level, whereas the role of VGCC is to increase the same. Hence, the combination of both the parameters helps in maintaining the calcium concentration, thus preventing the cell loss in Alzheimer's disease. On the basis of this, a two-dimensional mathematical model is developed using these parameters and is solved analytically using Laplace transform which is further simulated in MATLAB.

Keywords Alzheimer's disease · Calcium concentration · VGCC
Protein · Laplace transform

1 Introduction

Millions of people around the world are suffering from dementia [1]. Alzheimer's disease falls under this widespread umbrella [2]. It was first coined by Alois Alzheimer who studied the physiology of Alzheimeric brain of patient named Auguste D [3]. The striking changes in the human behavior which can be noticed straightforwardly by the surrounding people are the loss in the thinking and the planning capacity and the memory loss [3]. The person faces difficulty in performing all the tasks which have its coordination administered by the brain. Furthermore, the physiological changes which cannot be seen but is the reason for

D.D. Dave (✉) · B.K. Jha
Department of Mathematics, School of Technology, Pandit Deendayal
Petroleum University, Raisan, Gandhinagar 382007, Gujarat, India
e-mail: ddave1822@gmail.com

B.K. Jha
e-mail: brajeshjha2881@gmail.com

© Springer Nature Singapore Pte Ltd. 2018
M. Pant et al. (eds.), *Soft Computing: Theories and Applications*,
Advances in Intelligent Systems and Computing 584,
https://doi.org/10.1007/978-981-10-5699-4_18

the changed behavior is noteworthy. As Alzheimer's is sporadic in nature, the actual reason or the cause for the same is not known [1]. The two main reasons which may be the cause of Alzheimer's disease are senile plaques and neurofibrillary tangles. But Small D. mentioned that the twisted tangles may be present in almost all the neurodegenerative diseases but senile plaque is the only hallmark of Alzheimer's disease [4]. Plaques take place extracellularly, whereas the tangles are inside the cells. Physiologically plaques are the excessive accumulation of the sticky protein named amyloid beta peptide which is obtained by the splitting of amyloid precursor protein (APP), whereas neurofibrillary tangles are the results of the twisted strands of another protein named tau. Both plaques and tangles play crucial role in progression of Alzheimer's disease [4, 5]. Due to these, the areas involved in regulating the thinking, planning, and remembering the activities are damaged. As the disease progresses, the whole of the cortex gets impaired which finally results in shrinking of the brain. Therefore, the development of the new memories or other routine activities cannot be performed with an ease. These plaques and tangles disturb the transport and signaling system of the healthy mammalian brain. The disproportionate accumulation of amyloid beta hinders the signaling pathways between different nerve cells and synapses, whereas the twisted tau hinders the passing of necessitates intracellularly which further results in the cell loss and cell death, thus resulting in the brain contraction. The perfect result for these cell deaths is not known, but increased level of calcium concentration may be one of them [6, 7]. The alteration and deregulation in the calcium signaling with increasing age affects various cellular processes which lead to the neurodegenerative diseases [6, 8]. Calcium being the second messenger plays pivotal role in regulating various functions of the human brain such as synaptogenesis, excitotoxicity, neurotransmitter release, enzyme activation, gene expression, cell differentiation [1, 4, 6, 9, 10], dysregulation of which leads to the diseases like Alzheimer's or Parkinson's [11]. Talking about tau, it fails in controlling the calcium concentration, as a result of which the cytosolic calcium concentration increases. Thus, to control the increasing amount of calcium in nerve cells, buffers come into action [12, 13]. Also, the role of voltage-gated calcium channel is to be noted in the same context. Buffers on the one hand decrease the amount of calcium, and the voltage-gated calcium channel on the other hand increases the calcium in the nerve cells, thus harmonizing the amount of calcium. So we can conclude that the calcium channel (L-type) has its role in pathogenesis of AD [4].

Keeping in mind the above physiology, a two-dimensional mathematical model, which is taken into consideration over here, is as follows:

$$\frac{\partial C}{\partial t} = D_{Ca_x} \frac{\partial^2 C}{\partial x^2} + D_{Ca_y} \frac{\partial^2 C}{\partial y^2} - u(t) \frac{\partial C}{\partial x} - v(t) \frac{\partial C}{\partial y} + f(x, y, C) \qquad (1)$$

where $[Ca^{2+}] = C$ is the calcium profile, u and v are the velocity components along x- and y-axes, respectively, and $f(x, y, C)$ denotes internal processes due to reaction taking place, source, and sink; i.e., in our case, it will be calcium buffering (reaction term) and the voltage-gated calcium channel (source).

Retrospective survey of the literature supported the fact that the impairment in calcium homeostasis leads to the neurodegeneration. The alteration in the accumulation and production of amyloid beta protein and swirling tangles leads to the pathogenesis of AD. In 2003, Mattson and Chan studied the neuronal and glial calcium signaling [10]. Their study supported the fact that the alteration in brain physiology due to the increased intracellular calcium levels leads to excitotoxic and apoptotic cell death. Several authors have shown that the alteration in the calcium homeostasis leads to Alzheimer's disease [1, 6, 7, 14]. Although it is not necessary that it is the only cause of AD, it surely has its hand in the progression of AD. Also, it is found that the calcium signaling is not only altered in the neurons, but also the changes in astroglial calcium signaling, i.e., astrocytes and microglia, have its effect on Alzheimer's [8, 14–18]. Calcium signaling is a phenomenon in which several parameters take place actively. The role of ER, mitochondria, VGCC, and the proteins or buffers is different but remarkable which further results in deciding the progressiveness of the neurodegeneration. Lafrela in 2002 discussed the role of calcium dyshomeostasis and intracellular signaling in Alzheimer's and concluded that the increase in calcium elicits the two primary suspects of AD and vice versa [7].

A bird's view on the literature so far suggests that the mathematical models have not been developed on large scale on the physiology taking place in the Alzheimeric brain, by keeping in mind the calcium homeostasis. Mathematically, the modeling of the physiology of calcium diffusion in mammalian brain is done on a large scale. Various analytical and numerical techniques have been employed by several authors depicting the phenomenon of calcium diffusion in different cells [19–29].

Thus, in the view of above, one can conclude that the physiology of Alzheimer's disease with respect to calcium signaling is not studied extensively using the mathematical models. Hence, the mathematical formulation of the above-discussed phenomenon in the presence of VGCC and buffer, i.e., protein, is shown in the next section.

2 Mathematical Formulation

The geometry of the cell varies according to the geometry and region of the cell. Here, we have taken square region of the cell having calcium buffering process in the presence of VGCC which is shown in Fig. 1.

The mathematical formulation of calcium buffering and entry of free calcium ions Ca^{2+} through VGCC is given in the next sub-sections.

2.1 Calcium Buffering

The physiology of the calcium buffering and the mathematical formulation for the same is shown as [30]:

Fig. 1 Calcium buffering in the presence of VGCC in cells [22]

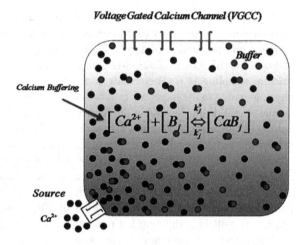

$$[Ca^{2+}] + [B] \Leftrightarrow [CaB] \tag{2}$$

where $[Ca^{2+}]$ and $[B]$ are the calcium and buffer concentrations, respectively, and $[CaB]$ is the concentration of the calcium-bound buffers.

2.2 Voltage-Gated Calcium Channel

For mathematical formulation of VGCC, the Goldman–Hodgkin–Katz (GHK) current equation is used and given as [22]:

$$I_{Ca} = P_{Ca} z_{Ca}^2 \frac{F^2 V_m}{RT} \frac{[Ca^{2+}]_i - [Ca^{2+}]_o \exp\left(-z_{Ca} \frac{FV_m}{RT}\right)}{1 - \exp\left(-z_{Ca} \frac{FV_m}{RT}\right)} \tag{3}$$

where $[Ca^{2+}]_i$ and $[Ca^{2+}]_o$ are the intracellular and extracellular concentrations, respectively; P_{Ca} is the permeability of calcium ion; z_{Ca} is the valency of calcium ion; F is Faraday's constant; V_m is membrane potential; R is real gas constant; and T is the absolute temperature.

Equation (3) is converted into moles/second by using the following equation:

$$\sigma_{Ca} = \frac{-I_{Ca}}{z_{Ca} F V_{nervecells}} \tag{4}$$

The negative sign in Eq. (4) is due to the inward current which is negative.

Including the calcium buffering and VGCC as internal processes, the term $f(x, y, C)$ is calculated as the sum of the free calcium ion entering the cytosol through VGCC and the amount of the calcium being buffered.

Further, the non-dimensionalization of the dependent variables C and B by scaling the dissociation constant of the buffer (K) and the total buffer concentration $[B]_T$, respectively, is done [27]. Thus, we have

$$c = \frac{C}{K}, \quad c_\infty = \frac{C_\infty}{K}, \quad b = \frac{[B]}{[B]_T} \tag{5}$$

Now incorporating the obtained term of $f(x, y, C)$ in Eq. (1), we obtain the final model for the physiology of calcium diffusion as

$$\frac{\partial c}{\partial t} = D_x \frac{\partial^2 c}{\partial x^2} + D_y \frac{\partial^2 c}{\partial y^2} - u \frac{\partial c}{\partial x} - v \frac{\partial c}{\partial y} - k_m^+ [B]_\infty (c - c_\infty) + \sigma_{VGCC} \tag{6}$$

where k_m^+ is buffer association rate constant, $[B]_\infty$ is the amount of buffer used in computation, C_∞ is the steady-state calcium concentration, and σ_{VGCC} is the calcium influx through VGCC.

The appropriate initial and boundary conditions which have been employed are:

$$c = 0, \quad t = 0, \quad x \geq 0, \quad y \geq 0 \tag{7}$$

$$c = c_0, \quad t > 0, \quad x = 0, \quad y = 0 \tag{8}$$

$$\frac{\partial c}{\partial x} = 0, \quad \frac{\partial c}{\partial y} = 0, \quad t \geq 0, \quad x \to l_1, \quad y \to l_2 \tag{9}$$

The solution to Eq. (6) is found using the technique adopted by Jha et al. [19]. Equation (6) is transformed into one-dimensional advection reaction equation and given as

$$\frac{\partial c}{\partial T} = D \frac{\partial^2 c}{\partial X^2} - U \frac{\partial c}{\partial X} - pc \tag{10}$$

where p is the obtained $f(x, y, C)$ term. Thus, the analytic solution of above equation is found using similarity [31] and Laplace transforms as follows:

$$c(X, T) = \frac{c_0}{2} e^{-\sqrt{\frac{p}{D}}(X - UT)} \left[2 - \text{Erfc} \left(\frac{2\sqrt{DpT} - (X - UT)}{2\sqrt{DT}} \right) + e^{2\sqrt{\frac{p}{D}}} \text{Erfc} \left(\frac{2\sqrt{DpT} + (X - UT)}{2\sqrt{DT}} \right) \right] \tag{11}$$

3 Results and Discussion

Figure 2 depicts the phenomenon of calcium diffusion taking place in two dimensions having protein of a different amount and voltage-gated calcium channel as the parameters affecting $[Ca^{2+}]_i$. From the figure, it is shown that as the amount

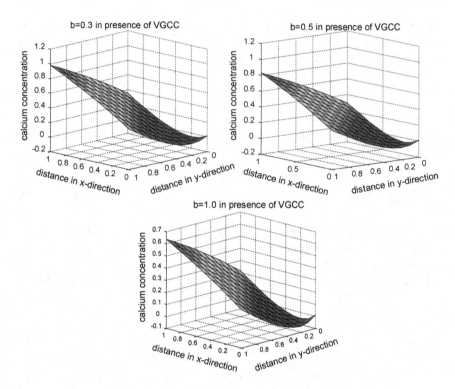

Fig. 2 Calcium advection diffusion in the presence of different amount of protein and VGCC

of the protein increases, the level of the calcium concentration decreases. From Alzheimeric view, we can conclude that the increase in the amount of the protein (i.e., tau) is directly related to the decrease in the calcium concentration of the cell. Over and above this, the presence of the voltage-gated calcium channel maintains the calcium concentration of the cell, because the proteins bind with the calcium and decrease the level, whereas the voltage-gated calcium channel increases it. Thus, it is observed that VGCC and protein maintain the cell concentration of the cell followed by cell protection against excitotoxicity and apoptotic cell death. This would prevent the cell loss and hence the brain shrinkage.

Figure 3 shows the calcium diffusion taking place in two dimensions having protein of different amount and absence of voltage-gated calcium channel. Having discussed in Fig. 1, it can be easily concluded that the absence of VGCC results in comparatively lower levels of calcium concentration in the cell. The increasing amount of protein results in decreasing amount of calcium concentration which finally converges to the background concentration of the cell. As we know that the role of the L-type calcium channel is there in the pathogenesis of Alzheimer's disease, the effect of it on the cytosolic calcium concentration is important. Thus, the alteration in calcium homeostasis and the role of VGCC in maintaining the cell concentration of the calcium are clearly seen from both of the figures (Table 1).

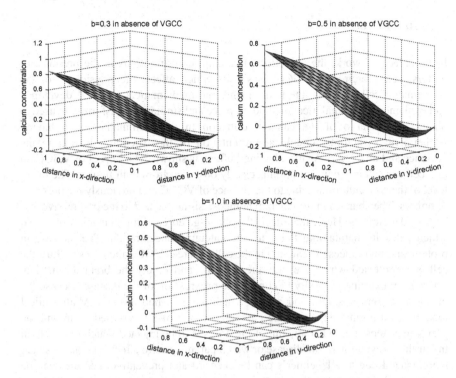

Fig. 3 Calcium advection diffusion in the presence of a different amount of protein and absence of VGCC

Table 1 Value of physiological parameters [19, 22]

Symbol	Parameter	Values
D	Diffusion coefficient	200–300 μ m^2 s^{-1}
k^+	Buffer association rate	1.5–600 μ m^{-1} s^{-1}
$[B]_\infty$	Buffer concentration	50–150 μM
$[Ca^{2+}]_\infty$	Background calcium conc.	0.1 μM
u, v	Velocity	10–100 μm s^{-1}
σ	Source amplitude	1 pA
V	Volume of cytosol	5.233 * 1013 1
F	Faraday's constant	96,485 C/mol
R	Ideal gas constant	8.31 J/(mol k)
T	Temperature	300 K
P_{out}	Rate of calcium efflux from cytosol	0.5 s^{-1}
Z_{ca}	Valency of Ca^{2+} ion	2

4 Conclusion

In this piece of work, the alteration in calcium concentrations in the presence of buffers, i.e., protein and VGCC, is shown which affects the physiology of the Alzheimeric cell. For this, the two-dimensional mathematical model is developed and explained using analytical method. Laplace and similarity transforms are used to obtain the scenario of the calcium diffusion in the mammalian brain having Alzheimeric symptoms. Graphical presentation of the calcium diffusion in the presence and absence of VGCC, having buffers, is shown. The impact of VGCC on intracellular calcium concentration is clearly observed from both figures. The lower level of the concentration is due to the absence of VGCC and obviously the presence of buffers. The changes in the calcium homeostasis are related to the progressiveness of the Alzheimer's. Here, the effect of voltage-gated calcium channel is observed which helps in maintaining the cytosolic calcium concentration. The increase in protein amount reduces the calcium in the cell, whereas VGCC increases it; thus, the cell gets protected which prevents the cell loss and finally the brain declination. There is an extreme necessity to control the progression of the disease because the patient experiences stupor and apathy which leads to helplessness. Mathematical models for the same depicting the pathophysiology of the diseased brain and the preventive steps (i.e., what amount of the drugs is to be injected which may help in the further prevention of the disease or which parameters are affected at the most and are damaged due to Alzheimer's can be checked and preventive measures can be incorporated computationally which can be further experimented) are the works of this study which needs to be unraveled in future. Thus, it can work as a barrier to the progressing Alzheimer's disease. As Alzheimer's cannot be cured completely, the protective measures can surely be found which helps in preventing the further progression.

References

1. Brawek, B., Garaschuk, O.: Network-wide dysregulation of calcium homeostasis in Alzheimer's disease. Cell Tissue Res. (2014). doi:10.1007/s00441-041-1798-8
2. Abramov, A., Canevari, L., Duchen, M.: Calcium signals induced by amyloid beta peptide and their consequences in neurons and astrocytes in culture. Biochem. Biophys. Acta. **1742**, 81–87 (2004)
3. Maurer, K., et al.: Auguste D and Alzheimer's disease. Lancet **349**, 1546–1549 (1997)
4. Small, D.: Dysregulation of calcium homeostasis in Alzheimer's disease. Neurochem. Res. **34**, 1824–1829 (2009)
5. Green, K., LaFrela, F.: Linking calcium to Aβ and Alzheimer's disease. Neuron **59**, 190–194 (2008)
6. Disterhoft, J., et al.: The calcium rationale in aging and Alzheimer's disease-evidence from an animal model of normal aging. Ann. N. Y. Acad. Sci. **747**, 382–406 (1994)
7. Lafrela, F.: Calcium dyshomeostasis and intracellular signalling in Alzheimer's disease. Nat. Rev. Neurosci. **3**, 862–872 (2002)

8. Verkhratsky, A., et al.: Astroglial calcium signaling in Alzheimer's disease. Biochem. Biophys. Res. Commun. (2016). doi:10.1016/j.bbrc.2016.8.088
9. Korol, T., et al.: Disruption of calcium homeostasis in Alzheimer's disease. Neurophysiology **40**, 457–464 (2008)
10. Mattson, M., Chan, S.: Dysregulation of cellular calcium homeostasis in Alzheimer's disease: bad genes and bad habits. J. Mol. Neurosci. **17**, 205–224 (2001)
11. Rappold, P., Tieu, K.: Atrocytes and therapeutics for Parkinson's disease. Neurother. J. Am. Soc. Exp. Neuro. Ther. **7**, 413–423 (2010)
12. Schwaller, B.: Cytosolic Ca^{2+} Buffers. Cold Spring Harb. Perspect. Biol. **2**, a004051 (2015)
13. Wang, Z., Tymianski, M., Jones, O.T., Nedergaard, M.: Impact of calcium buffering on the spatial and temporal characteristics of intercellular calcium signals in astrocytes. J. Neurosci. 7359–7371 (1997)
14. Mattson, M., Chan, S.: Neuronal and glial calcium signaling in Alzheimer's disease. Cell Calcium **34**, 385–397 (2003)
15. Barreto, G., et al.: Role of Astrocytes in Neurodegenerative diseases. In: Raymond, Chang C.-C. (eds.) Neurodegenerative diseases-processes, prevention, protection and monitoring, pp. 257–272. InTech (2011)
16. Zatta, P., Nicolini, M.: Non-neuronal cells in Alzheimer's disease. World Scientific (1995)
17. Verkhratsky, A., Butt, A.: Glial neurobiology: a textbook. Wiley, New York (2007)
18. Verkhratsky, A., Olabarria, M., Noristani, H., Yeh, C., Rodriguez, J.: Astrocytes in Alzheimer's disease. Neurother. J. Am. Soc. Exp. Neuro. Ther. **7**, 399–412 (2010)
19. Jha, B.K., Adlakha, N., Mehta, M.N.: Analytic solution of two dimensional advection diffusion equation arising in cytosolic calcium concentration distribution. Int. Math. Forum. **7** (3), 135–144 (2012)
20. Jha, B.K., Adlakha, N., Mehta, M.N.: Two dimensional finite element model to study calcium distribution in astrocytes in presence of excess buffer. Int. J. Biomath. **7**(3), 1–11 (2014)
21. Jha, A., Adlakha, N.: Analytical solution of two dimensional unsteady state problem of calcium diffusion in a neuron cell. J. Med. Imaging Health Inf. **4**, 1–7 (2014)
22. Jha, B.K., et al.: Two dimensional finite element model to study calcium distribution in astrocytes in presence of VGCC and excess buffer. Int. J. Model Simul. Sci. Comput. (2013). doi:10.1142/S1793962312500304
23. Kotwani, M., Adlakha, N., Mehta, M.N.: Finite element model to study calcium diffusion with excess buffer approximation in fibroblast cell. Int. J. Comput. Appl. Math. **7**(4), 503–514 (2012)
24. Kotwani, M., Adlakha, N., Mehta, M.N.: Finite element model to study effect of buffers, source amplitude and source geometry on spatio-temporal calcium distribution in fibroblast cell. J. Med. Imaging Health Inf. **4**, 1–8 (2014)
25. Naik, P., Pardasani, K.: One dimensional finite element model to study calcium distribution in oocytes in presence of VGCC, RyR and Buffers. J. Med. Imaging Health Inf. **5**(3), 471–476 (2015)
26. Panday, S., Pardasani, K.R.: Finite element model to study effect of advection diffusion and Na^+/Ca^{2+} exchanger on Ca^{2+} distribution in oocytes. J Med Imaging Health Inf **3**(3), 374–379 (2013)
27. Smith, G.D.: Analytical steady state solution to rapid buffering approximation near an open Ca^{2+} channel. Biophys. J. **71**, 3064–3072 (1996)
28. Tewari, S.: A variational-ritz approach to study cytosolic calcium diffusion in neuron cells for a one-dimensional unsteady state case. GAMS J Math Math Biosci **2**, 1–10 (2009)
29. Tripathi, A., Adlakha, N.: Finite volume model to study calcium diffusion in neuron cell under excess buffer approximation. Int. J. Math. Sci. Eng. Appl. **5**(3), 437–447 (2011)
30. Keener, J., Sneyed, J.: Mathematical physiology, second edn. Springer (1998)
31. Crank, J.: The mathematics of diffusion. Oxford University Press, London (1975)

Solution of Multi-objective Portfolio Optimization Problem Using Multi-objective Synergetic Differential Evolution (MO-SDE)

Hira Zaheer and Millie Pant

Abstract Portfolio optimization plays an important role in managing the financial assets of an individual. The investments are made such that an individual attains the maximum benefit out of it. In this paper, a bi-objective portfolio optimization model is considered, where the objectives are to maximize the return and minimize the risk, and is solved using multi-objective synergetic differential evolution (MO-SDE).

Keywords Multi-objective synergetic differential evolution (MO-SDE)
Portfolio optimization · Return and risk

1 Introduction

The investor's expectations for financial output are very uncertain in nature; that is, some would expect a return greater than 40%, and some expect risk not higher than 15%. Constructing appropriate portfolio by taking care of such uncertain expressions is quite interesting and challenging because investor's expectations cannot be easily fulfilled. Mathematically, such a problem can be modeled in terms of optimization and can be solved by applying a suitable technique. In the present study, a multi-objective model is considered, where the objectives are to maximize the return and to minimize the profit. The Pareto front obtained provides all possible solutions for the user. In the literature, we can find several instances, where multi-objective portfolio problem is solved by applying different approaches. Some

H. Zaheer (✉) · M. Pant
Department of Applied Sciences and Engineering,
Indian Institute of Technology (IIT) Roorkee, Roorkee, India
e-mail: hirazaheeriitr@gmail.com

M. Pant
e-mail: millifpt@iitr.ernet.in

© Springer Nature Singapore Pte Ltd. 2018
M. Pant et al. (eds.), *Soft Computing: Theories and Applications*,
Advances in Intelligent Systems and Computing 584,
https://doi.org/10.1007/978-981-10-5699-4_19

examples are as follows: Gupta et al. [1] presented a fuzzy approach in which multi-objective portfolio optimization problem is solved by converting it into single objective. Saborido et al. [2] solved a mean–downside risk–skewness (MDRS) model which optimizes the skewness, expected return, and downside risk by considering cardinality constraints and budget bounds. The solution method used is evolutionary multi-objective with a novel mutation, crossover, and reparation operators. Pouya et al. [3] added some P/E criteria to Markowitz mean–variance model and solved using invasive weed optimization (IWO). P/E ratio is an important criterion for investment in stock market. They solved multi-objective problem by converting it into single-objective problem using IWO algorithm and compared the results with particle swarm optimization (PSO) and reduced gradient method (RGM). Yoshimoto [4] incorporated transaction cost into Mean Markowitz basic model and solved it using nonlinear programming. Konno and Suzuki [5] added skewness into Mean Markowitz model. Chang et al. [6] introduced cardinality constraints to portfolio optimization problem. Soleimani et al. [7] induced market value constraints to Markowitz model first time and consider cardinality constraints and minimization of transaction cost. Golmakani and Fazel [8] also worked on market sectors in portfolio and applied PSO for obtaining the solution. Sadjadi et al. [9] also worked on cardinality constraints in Markowitz model using genetic algorithm. Ghahtarani and Najafi [10] used goal programming and robust optimization approach for solving portfolio optimization model. In Table 1, important development of some authors is summarized.

Table 1 Summary of work done by different authors in portfolio optimization

Author(s)	Development	Year
Konno and Suzuki	Assets skewness	1995
Yoshimoto	Transaction costs	1996
Katagiri and Ishii [11]	Using fuzzy method in Markowitz model	1999
Chang et al.	Cardinality constraints	2000
Goldfarb and Lyengar [12]	Using robust optimization approach in Markowitz model	2003
Oh et al. [13]	Beta portfolio selection	2006
Soleimani et al.	Sector constraint	2009
Golmakani and Fazel	Market sector in portfolio	2011
Sadjadi et al.	Cardinality constraints in Markowitz model	2012
Ghahtarani and Najafi	Goal programming and robust optimization	2013
Saborido	Mean–downside risk–skewness (MDRS)	2016
Pouya et al.	P/E criteria to Markowitz mean–variance model	2016

2 Multi-objective Synergetic Differential Evolution (MO-SDE)

Differential Evolution is robust stochastic technique for solving unconstrained and constrained problems with single-objective function. However, with some modifications, it can be used for solving multi-objective problems as well. In the present study, the authors have considered MO-SDE proposed by Ali et al. [14] for solving a bi-objective portfolio optimization problem. It starts in the same manner as that of a basic DE algorithm.

After initialization and mutation, MO-SDE undergoes crossover as shown in Eq. 1, where $X_{i,G} = \{x_{1,i,G}, x_{2,i,G}, x_{3,i,G}, \ldots, x_{D,i,G}\}$ be the target vector and $V_{i,G} = \{v_{1,i,G}, v_{2,i,G}, v_{3,i,G}, \ldots, v_{D,i,G}\}$ be the mutant vector and $U_{i,G} = (u_{1,i,G}, \ldots, u_{D,i,G})$ be the trial vector:

$$u_{j,i.G} = \begin{cases} v_{j,i.G} & \text{if } \text{rand}_{i,j}\,[0,1] \leq C_r \vee j = j_{\text{rand}} \\ x_{j,i.G} & \text{otherwise} \end{cases} \tag{1}$$

Then comes selection procedure, the most crucial step in MOPs because selection of good solutions results in obtaining a good Pareto optimal front. Selection procedure is defined as follows.

The idea of MO-SDE is taken from the concept of PDEA Madavan [15] and DEMO Robic and Filipic [16]. In PDEA [15] approach, NP solution obtained after mutation and crossover is combined with the original parent population, and the size of new population became 2NP. After that, non-dominated sorting takes place and each solution (parent and candidate) is given a non-dominated rank. An explicit preservation strategy (used in NSGA) is used for giving non-dominated ranking, and NP solutions are selected on the basis of non-dominating rank and the rank given by crowding distance. In DEMO [16], the target or the trial solution is discarded if it gets dominated by one another. Therefore, the size of population will be in between NP and 2NP. In the next step, population is shrunk on the basis of non-dominated sorting and by calculating the solution of same front with crowding distance metric. The shrunk population contains only the best NP solutions.

In MO-SDE, both trial and target solution vectors are compared, and if trial solution dominates the target solution, then trial solution will come in current population and target solution will be added to the advanced population metrics; otherwise, target solution will be added to current and trial solution will be added to advanced population metrics. At each generation, both populations are combined and the total size of population will be 2NP. After that, population is truncated with the help of elite and explicit preservation strategy borrowed from NSGA-II, Deb [17, 18]. Pseudo-code of MO-SDE is shown in Table 2.

Opposition-based initial population: A population of size NP is generated using Eq. 2, and its corresponding opposite population of same size is generated using Eq. 3 and combined together, so the new population size became 2NP.

Table 2 Pseudo-code of MO-SDE

	Pseudo-code of MO-SDE illustrating how the procedure acts on a population of individuals, repeating mutation, crossover, and selection until the convergence criteria is met
Step 1	Generate randomly *NP* individuals, using Eq. (2) and *NP* opposite individuals using Eq. (3). Merge these two and select *NP* fittest solutions as initial population with the help of non-dominated sorting and crowding distance metric. Set the values of control parameters *F* and *Cr*
Step 2	Set $i = 0$
Step 3	$i = i + 1$
Step 4	Corresponding to target individual X_i, select three distinct individuals from population say X_{r1}, X_{r2}, and X_{r3} such that $i \neq r1 \neq r2 \neq r3$ and generate mutant vector V_i using strategy DE/rand/1 (also explained in [19]). For this operation, tournament best is selected on the basis of non-domination
Step 5	Shuffling of target vector X_i with mutant individual V_i generated in step 4 to generate trial vector U_i using Eq. (1)
Step 6	If all components of the trial vector are lying in the given range, then go to step 7 else uniformly generate that parameter within given range using Eq. (2) and go to step 7
Step 7	Obtain fitness values for vector U_i. If trial vector U_i dominates target vector X_i, then it immediately replace target vector in current population and target vector is added to auxiliary population otherwise trial vector is added to auxiliary population
Step 8	If $i < NP$, move to step 3 otherwise switch to step 9
Step 9	Merge these two population (current and auxiliary) and select *NP* fittest solutions for next generation using non-dominated sorting and crowding distance metric
Step 10	If stopping criteria met, then stop otherwise move to step 2

$$x_{j,i,0} = l_j + \text{rand}_{i,j}[0,1] \times (u_j - l_j) \tag{2}$$

$$ox_{j,i,0} = l_j + u_j - x_{j,i,0} \tag{3}$$

are generated and merged together to form a population of size 2NP, where $x_{j,i,0}$ and $ox_{j,i,0}$ are the *j*th component of *i*th population vector and its opposite population vector, respectively. Among these 2NP solutions, NP best solutions are consider as the initial population of the SDE algorithm.

3 Numerical Illustration

The numerical problem is borrowed from Gupta et al. [1]. Ten different companies are randomly selected from NSE, which are taken as 10 assets for the problem. Yearly return is calculated with the help of monthly return of all assets. Calculation of yearly return, variance, covariance is explained by Zaheer et al. in [20]. With the

Table 3 Yearly return of 10 assets

Company	Return
A B B Ltd. (ABL)	0.19278
Alfa Laval (India) Ltd. (ALL)	0.13587
Bajaj Hindustan Ltd. (BHL)	0.40086
Crompton Greaves Ltd. (CGL)	0.29892
Hero Honda Motors Ltd. (HHM)	0.14921
Hindustan Construction Co. Ltd. (HCC)	0.30107
Kotak Mahindra Bank (KMB)	0.23818
Mahindra & Mahindra Ltd. (MML)	0.23114
Siemens Ltd. (SIL)	0.26122
Unitech Ltd. (UNL)	0.56246

help of yearly return (Table 3), variance, and covariance matrix (Table 4), the mathematical model is formulated.

By using Tables 3 and 4, the following mathematical model is formulated:

$$\max f_1(x) = 0.19278x_1 + 0.13587x_2 + 0.40086x_3 + 0.29892x_4$$
$$+ 0.14921x_5 + 0.30107x_6 + 0.23818x_7 + 0.23114x_8 + 0.26122x_9 + 0.56246x_{10}$$
$$\min f_2(x) = 0.1401x_1x_1 + 0.0868x_2x_2 + 0.4232x_3x_3 + 023027x_4x_4 + 0.07587x_5x_5$$
$$+ 0.18957x_6x_6 + 0.16280x_7 + 0.11891x_8x_8 + 0.18988x_9x_9 + 0.58191x_{10}x_{10}$$
$$+ 0.10078x_1x_2 + 187110x_1x_2 + 0.19250x_1x_4 + 0.08633x_1x_5 + 0.12689x_1x_6$$
$$+ 0.13209x_1x_7 + 0.06919x_1x_8 + 0.21966x_1x_9 + 0.02598x_1x_{10} + 0.20718x_2x_3$$
$$+ 0.15458x_2x_4 + 0.05175x_2x_5 + 0.14749x_2x_6 + 0.08768x_2x_7 + 0.03790x_2x_8$$
$$+ 0.14741x_2x_9 + 0.10203x_2x_{10} + 0.35904x_3x_4 + 0.09484x_3x_5 + 0.2647x_3x_6$$
$$+ 0.0592x_3x_7 + 0.1196x_3x_8 + 0.308x_3x_9 + 0.0895x_3x_{10} + 0.0523x_4x_5$$
$$+ 0.1874x_4x_6 + 0.0249x_4x_7 + 0.007x_4x_8 + 0.2351x_4x_9 + 0.0945x_4x_{10}$$
$$+ 0.0602x_5x_6 + 0.1165x_5x_7 + 0.1237x_5x_8 + 0.11404x_5x_9 - 0.03522x_5x_{10}$$
$$+ 0.09074x_6x_7 + 0.0536x_6x_8 + 0.2089x_6x_9 - 0.01913x_6x_{10} + 0.1863x_7x_8$$
$$+ 0.1549x_7x_9 + 0.1405x_7x_{10} + 0.1066x_8x_9 - 0.0291x_8x_{10} + 0.1312x_9x_{10}$$

subjected to

$$x_1 + x_2 + x_3 + x_4 + x_5 + x_6 + x_7 + x_8 + x_9 + x_{10} = 1 \quad x_i \geq 0 \quad i = 1, 2. \dots 10$$

Table 4 Variance–covariance matrix

Company	ABL	ALL	BHL	CGL	HHM	HCC	KMB	MML	SIL	UNL
ABL	0.1401	0.0503	0.0935	0.0962	0.0431	0.0634	0.0660	0.0346	0.1098	0.0129
ALL	0.0503	0.0868	0.0935	0.0962	0.0431	0.0634	0.0660	0.0346	0.1098	0.0129
BHL	0.0935	0.0935	0.4232	0.1795	0.0474	0.1323	0.0296	0.0598	0.1541	0.0447
CGL	0.0962	0.0962	0.1795	0.2302	0.0261	0.0937	0.0125	0.0035	0.1175	0.0472
HHM	0.0431	0.0431	0.0474	0.0261	0.0758	0.0301	0.0582	0.0619	0.0570	-0.0176
HCC	0.0634	0.0634	0.1323	0.0937	0.0301	0.1895	0.0453	0.0268	0.1044	-0.0095
KMB	0.0660	0.0660	0.0296	0.0125	0.0582	0.0453	0.1628	0.0931	0.0774	0.0702
MML	0.0346	0.0346	0.0598	0.0035	0.0619	0.0268	0.0931	0.1189	0.0533	-0.0145
SIL	0.1098	0.1098	0.1541	0.1175	0.0570	0.1044	0.0774	0.0533	0.1898	0.0656
UNL	0.0129	0.0129	0.0447	0.0472	-0.0176	-0.0095	0.0702	-0.0145	0.0656	0.5819

4 Results and Discussion

The above model is solved using MO-SDE algorithm, and working of the algorithm is explained in Sect. 2. Since the constraint is linear in nature, the problem can be converted into an unconstrained one by omitting one variable in the form of other.

Gupta et al. [1] modified multi-objective problem into single-objective problem and solved using Lingo software. Their result is shown in Table 5, and the optimal point is also marked in Fig. 1. The results obtained from MO-SDE are reported in Tables 5 and 6, and the obtained optimal Pareto front is shown in Fig. 1.

Table 5 Results obtained from Gupta et al. using Lingo software and result obtained by MO-SDE

Algorithm	Return	Risk
Lingo	0.4546	0.2055
MO-SDE	0.3037	0.0886

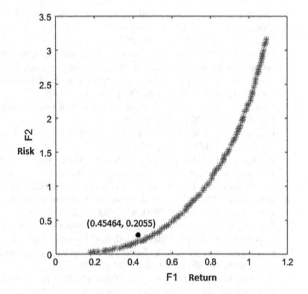

Fig. 1 Pareto optimal front obtained by MO-SDE

Table 6 Solution of multi-objective portfolio optimization in terms of asset value

Asset	Allocation
ABL	0.12014
ALL	0.19278
BHL	0.00257
CGL	0.01885
HHM	0.13457
HCC	0.01100
KMB	0.29878
MML	0.08350
SIL	0.14314
UNL	0.04700

5 Conclusion

This paper showed the implementation of MO-SDE for solving a bi-objective portfolio optimization problem. An advantage of MO-SDE is that we can solve both the objectives (maximize returns and minimize risk), which are conflicting in nature simultaneously. The numerical results and graphs indicate that MO-SDE is an attractive alternative for dealing with multi-objective portfolio optimization problems. Further, other soft computing techniques [21, 22] and formal method [23] will also be investigated to improve the performance.

Acknowledgements The reported study was partially supported by DST, research project No. INT/RFBR/P-164.

References

1. Gupta, P., et al.: Fuzzy portfolio optimization. Springer (2014)
2. Saborido, R., et al.: Evolutionary multi-objective optimization algorithms for fuzzy portfolio selection. Appl. Soft. Comput. **39**, 48–63 (2016)
3. Pouya, A.R., Maghsud, S., Mustafa, J.R.: Solving multi-objective portfolio optimization problem using invasive weed optimization. Swarm Evol. Comput. **28**, 42–57 (2016)
4. Yoshimoto, A.: The mean-variance approach to portfolio optimization subject to transaction costs. J. Oper. Res. Soc. Jpn. **39**(1), 99–117 (1996)
5. Konno, H., Suzuki, K.-I.: A mean-variance-skewness portfolio optimization model. J. Oper. Res. Soc. Jpn. **38**(2), 173–187 (1995)
6. Chang, T.-J., et al.: Heuristics for cardinality constrained portfolio optimization. Comput. Oper. Res. **27**(13), 1271–1302 (2000)
7. Soleimani, H., Golmakani, H.R., Salimi, M.H.: Markowitz-based portfolio selection with minimum transaction lots, cardinality constraints and regarding sector capitalization using geneticalgorithm. Expert Syst. Appl. **36**(3), 5058–5063 (2009)
8. Golmakani, H.R., Fazel, M.: Constrained portfolio selection using particle swarm optimization. Expert Syst. Appl. **38**(7), 8327–8335 (2011)
9. Sadjadi, S.J., Gharakhani, M., Safari, E.: Robust optimization framework for cardinality constrained portfolio problem. Appl. Soft Comput. **12**(1), 91–99 (2012)
10. Ghahtarani, A., Najafi, A.A.: Robust goal programming for multi-objective portfolio selection problem. Econ. Model. **33**, 588–592 (2013)
11. Katagiri, H., Ishii, H.: Fuzzy portfolio selection problem. In: IEEE SMC'99 Conference Proceedings, International Conference on Systems, Man, and Cybernetics. IEEE (1999)
12. Goldfarb, D., Iyengar, G.: Robust portfolio selection problems. Math. Oper. Res. **28**(1), 1–38 (2003)
13. Oh, K.J., et al.: Portfolio algorithm based on portfolio beta using genetic algorithm. Expert Syst. Appl. **30**(3), 527–534 (2006)
14. Ali, M., Millie, P., Ajith, A.: Improving differential evolution algorithm by synergizing different improvement mechanisms. ACM Trans. Auton. Adapt. Syst. (TAAS) **7.2** (2012)
15. Madavan, N.K.: Multiobjective optimization using a Pareto differential evolution approach. In: Proceedings of the Congress on Evolutionary Computation 2002, pp. 1145–1150 (2002)
16. Robic, T., Filipic, B.: DEMO: differential evolution for multiobjective optimization. In: Proceedings of the 3rd International Conference on Evolutionary Multi Criterion Optimization, LNCS 3410, pp. 520–533 (2005)

17. Deb, K.: Multi-objective optimization using evolutionary algorithms. Wiley, Chichester (2001)
18. Deb, K., Pratap, A., Agarwal, S., Meyarivan, T.: A fast and elitist multiobjective genetic algorithm: NSGA-II. IEEE Trans. Evol. Comput. **6**, 182–197 (2002)
19. Zaheer, H., Pant, M., Kumar, S., Monakhov, O., Monakhova, E., Deep, K.: A new guiding force strategy for differential evolution. Int. J. Syst. Assur. Eng. Manag. 1–14 (2014). doi:10. 1007/s13198-014-0322-6
20. Zaheer, H., et al.: A portfolio analysis of ten national banks through differential evolution. In: Proceedings of Fifth International Conference on Soft Computing for Problem Solving, pp. 851–862. Springer, Singapore (2016)
21. Ansari, I.A., Pant, M., Ahn, C.W., Jeong, J.: PSO Optimized multipurpose image watermarking using SVD and chaotic sequence. In: Proceedings of Bio-Inspired Computing–Theories and Applications: 10th International Conference, BIC-TA 2015 Hefei, China, 25–28 Sept 2015, vol. 562. Springer (2016)
22. Jauhar, S., Pant, M., Deep, A.: Differential evolution for supplier selection problem: a DEA based approach. In: Proceedings of the Third International Conference on Soft Computing for Problem Solving, pp. 343–353. Springer, India (2014)
23. Singh, N., Chandra, M., Yadav, D.: Formal specification of asynchronous checkpointing using Event-B. In: 2015 International Conference on Advances in Computer Engineering and Applications (ICACEA), pp. 659–664. IEEE (2015)

Frequency Fractal Behavior in the Retina Nano-Center-Fed Dipole Antenna Network of a Human Eye

P. Singh, R. Doti, J.E. Lugo, J. Faubert, S. Rawat, S. Ghosh, K. Ray and A. Bandyopadhyay

Abstract The retina nano-antenna shows a major characteristic of the center-fed dipole antenna's working in the visible region. The cellular assembly that might work as a network of antennas is analyzed here. The collective response of various cone cells holds the geometric features of the antenna network. The fractal arrangement of the antenna lattice holds various symmetries during electromagnetic signal processing, and each symmetry generates a peak in the resonance band. Using true biological structural data, we have identified the resonance frequency spectrum of entire nano-network of cone and rod cells in a human eye.`

Keywords Bio-inspired antenna · Result analysis

P. Singh · K. Ray (✉)
Amity School of Applied Science, Amity University Rajasthan, Jaipur, India
e-mail: kanadray00@gmail.com

P. Singh
e-mail: singhpushpendra548@gmail.com

R. Doti · J.E. Lugo · J. Faubert
Visual Perception and Psychophysics Laboratory, School of Optometry,
Universite de Montreal, Montreal H3T 1P1, Canada

S. Rawat
Manipal University, Jaipur, India
e-mail: sanyograwat@gmail.com

S. Ghosh
Natural Products Chemistry Division, CSIR-North East Institute
of Science & Technology, Jorhat 785006, Assam, India

A. Bandyopadhyay
Advanced Key Technologies Division, Advanced Nano Characterization Center,
National Institute for Materials Science, 1-2-1 Sengen, 3050047 Tsukuba
Ibaraki, Japan

© Springer Nature Singapore Pte Ltd. 2018
M. Pant et al. (eds.), *Soft Computing: Theories and Applications*,
Advances in Intelligent Systems and Computing 584,
https://doi.org/10.1007/978-981-10-5699-4_20

1 Introduction

Thus far, protein was believed to vibrate mechanically and ionic transmission governed by electric potential was believed to be the key feature of a protein's operation. The concept has recently been challenged, and it has been argued that the proteins vibrate in the presence of electromagnetic signal like a cavity resonator [1–3]. The interaction between the living cells and the electromagnetic signals is not new. Protein synthesis is stimulated by electromagnetic fields of the specific frequencies in the RF range [4–6]. Electromagnetic field can damage proteins [7]. There have been plenty of works done analyzing the electromagnetic interactions of the living cells. In most of these cases, the studies are biological. Since we have now evidence that proteins vibrate electromagnetically, we can revisit the electromagnetic interactions of the cells, considering the whole cell as a cavity resonator [8, 9]. Here, we describe the biological aspects of the retina nano-center and justify the reasoning for considering it as a dipole antenna network.

The retina is the innermost of the three layers in our eye; it resembles a thin fragile meshwork, which is no thicker than a postal stamp, and it lies below the vitreous chamber. The thickness of the retina is not homogeneous, and there is a region at the center of the macula where the thickness is minimum (known as the fovea); consequently, it reduces the light absorption and scattering before reaching the photoreceptors. Moreover, the retina is a direct extension of the central nervous system [10] [secular]. The retina contains collector cells and photoreceptors. The photoreceptors are neurons specialized in capturing light. The collector cells are other neurons specialized on gathering information from photoreceptors and then transfer it along to the brain. The retina has two categories of photoreceptors, namely rods and cones.

Herein, the cylindrical rods and cones located at the back of the retina are modeled as a uniform nano-antenna. By avoiding atomic-scale structural changes, we considered that all cylindrical antennas are identical and simulated their operation in the visible region which is consistent with the rod's dimensional and geometrical structure. Each nano-antenna (or each light detecting site on the retina) absorbs the electromagnetic energy and translates that into a quantized electronic charge that is subsequently used (electrically) in the nerve spikes of vision. The results of the computer simulation study showed that if the simulated antennas have the original photoreceptor cell dimensions, the frequency responses for both the approaches would be very close to each other. An array of the novel-modeled antennas is also discussed. Those are used in biomedical applications of artificial retinal photoreceptors in medicine, although the main scope is not designing artificial retinal photoreceptor prosthesis. The cones and the cylindrical rod cell are shown in Fig. 1 [11, 12].

Fig. 1 Mechanism figure, the
light interaction with the rods
and cones [11, 12]

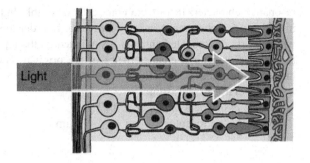

2 Model of Antenna Array

An apparent similarity exists between the octagonal symmetry of solid-state
nano-pillars grown on silicon and the octagonal motif seen in the biological retina
of the eye. A perfect octagonal symmetry is present on the retina of the eye at 7°–8°
of eccentricity where statistically the density of rods is sufficient to completely
surround each of the diminishing number of cones. That is illustrated in Fig. 2i
(according to UC Berkeley group). The ratio of the dimensions of cones and rods
is ∼1.8:1. As nano-antenna interacts with light, these individual octagonal sites
uniquely provide the geometrical definition of the precise middle of the visible
band. An array of model antennas for the cone cells is developed to analyze the

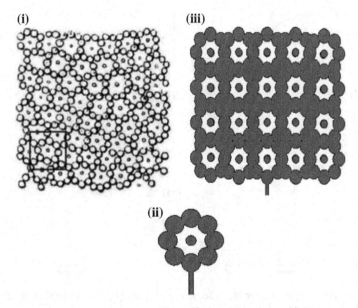

Fig. 2 (i) 2D view of the receptor system, (ii) modeled single-unit cell antenna, (iii) 20-unit cell
modeled antenna

resonance characteristics of the cone cells at visible light frequencies and also to investigate the photoreceptor array structure. In order to see how such array model would work, the modeled antenna is simulated on the top of a silicon substrate with the same dimensional characteristics of the original nano-antenna network in the human retina. That is:

Cones diameter = 1.4 μm.

Middle cylindrical diameter = 0.77 μm,

Height of the receptor system = 35–75 μm.

3   Results and Discussion

3.1   Resonating Frequency Curve

Case 1   By taking the one unit of receptor system (cones and rods) on the retina, then modeled (shown in Fig. 2ii) antenna resonates at the visible band (0.3–3 PHz) from 0.85 to 1.15 PHz. The antenna parameter reflection coefficient (S_{11}), gain (dBi), and VSWR are shown in Fig. 3i. The gain of the single unit of our modeled antenna runs toward the positive values, while the value of the VSWR is always positive.

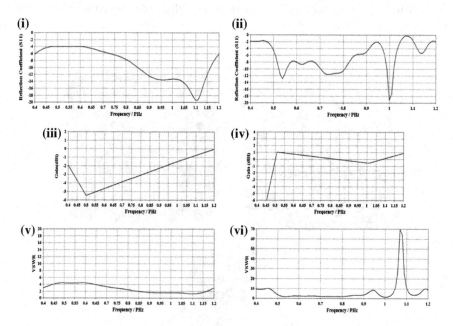

Fig. 3 Modeled antenna characteristic: reflection coefficient S11 (dB), gain (dBi), and VSWR: (i), (iii), (v) for a single unit and (ii), (iv), (vi) for multiple units

Case 2   In this case, modeled antenna is simulated in the CST microwave studio by using 20 unit cells of the receptor system as shown in Fig. 2iii. A high improvement in the result of the modeled antenna is shown in Fig. 3ii. Now, the modeled antenna shows a triple band resonating effect in between 0.52–0.55, 0.70–0.80, and 0.98–1.01 PHz, respectively, in the visible region. Gain curve also shows a better response compared to the Case 1. Simulated gain value of the modeled antenna is nearly +1 dBi which holds a good agreement with the center half-wave dipole antenna's (length is slightly less than half of the wavelength) gain 1.76 dBi. VSWR swims exactly as Case 1 from 0.5 to 1.0 PHz.

To increase the cell unit of the modeled antenna by taking the original dimensional retina receptor system cell (rods and cones), we may follow the original view of the retina receptor system.

The trichromatic theory of the retina receptors suggests that a given response rate by a given cone is ambiguous. For example, a low response rate by a middle wavelength cone might indicate low intensity 540 nm, or brighter 500 nm or still brighter 460 nm [10]. The nervous system can determine the color and the brightness (incidence of light of different wavelengths) of the light only by comparing the response of the different types of the cones. The different response rates of the cone as a function of light intensity can be explained by Lambert's illumination law. According to Lambert's illumination law, the intensity I is proportional to radiated power $I \propto P$. The cone response at different wavelengths is explained by the simulated radiation curve of the single-unit modeled antenna as depicted in Fig. 4.

Both the rods and the cones contain photo-pigments that release energy when struck by light. Photo-pigments consist of 11-cis retinal bound by proteins (opsins). When a single photon of light is incident on the 11-cis retinal, it converts into another form 11-trans retinal form as shown in Fig. 5 [13, 14].

Generally, opsins change its nature of polarity by a single incident photon, similar to a center-fed dipole with sinusoidal current distribution as shown in

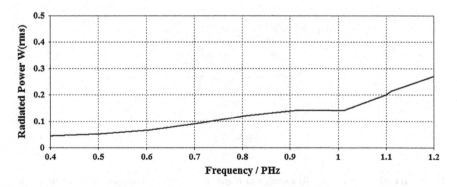

Fig. 4 Radiated power W (rms)/frequency (PHz) of single modeled antenna

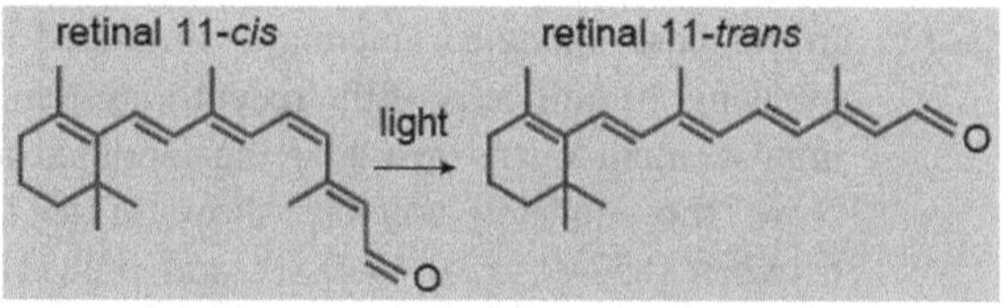

Fig. 5 Conversion shape of proteins [14]

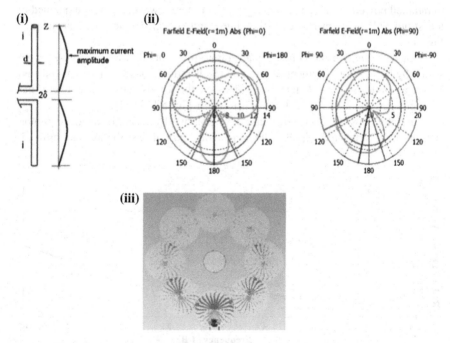

Fig. 6iii, because here the cones are treated as filamentary conductors and they are replaced by a current element. The directivity of the half-wavelength dipole (D) is 1.64. The current amplitude is maximum at the center of the dipole. For such center-fed dipole, the total radiated power and the electric field equation are given by [15]:

$$P_{rad} = 0.609\eta I_{max}^2/2\pi,\tag{1}$$

$$E\theta = [j60I_m e^{j(\omega t - \vec{k}\cdot\vec{r})}\{\cos(\pi/2\cos\theta)/\sin(\theta)\}]/r.\tag{2}$$

In the retinal structure, the rod (center-fed dipole) occurs in the form of tabular, with a negligible thin wall that is composed of a perfect conductor, and the current density is linear, Z-directed which flow on the surface S (S with a cylinder of

Fig. 6 (i) Center- fed dipole, (ii) variation of E and H field, (iii) current distribution on the patch geometry

counter C extending from $-l$ to $+l$). $E\theta(s,z)$ is equal to 0 within the interval $-\delta \le z \le \delta$. The voltage at the dipole terminal shows a sharp peak; thus,

$$V = - \int_{-\delta}^{\delta} E\theta(s,z)dz. \tag{3}$$

Thus, $E\theta = f(z)$ and $\ne f(s)$ within $-\delta \le z \le \delta$. This is appropriate to express as a Dirac delta function. This implies that a positive unit voltage is applied to the center-fed dipole and the amount of current (Current expression) can be obtained with the help of using Hallen's integral equation and the method of moments [15]. Here, the center-fed dipole (receptor system) is slender as depicted in Fig. 6i. The Dirac Delta function represents a periodic oscillation in the cone cell, and due to cyclicity, we consider the signal as sinusoidal and therefore is given by $I(z) = I_m \sin(k(l - z))$. Then, recall Eq. (1), $(P_{rad})_{max}/(P_{rad})_{min} = (I_{max}/I_{min})^2$, and I_{max}/I_{min} (VSWR) is equal to 1.45 which matches exactly with Fig. 3ii, vi in the visible region range and approximately with the Dirac delta function. The directivity value of the average wavelength in the visible region is 2.05, which is shown in Fig. 7ii. The surface current density is maximum at the center of the cones as shown in Fig. 6iii. Both the results are a good match with the theoretical results.

It is observed that the polar plot of the center-fed dipole appears as doughnut-shaped, with the null along the θ equal to 0° and 180° [12]. The pattern variation of E and H of the modeled antenna is depicted in Fig. 6ii, which also shows that the major lobe direction highly matches with the center-fed dipole (0° and 180°). The momentum (pl) and the energy (Ul) of the center-fed dipole in the electric field are shown in Fig. 6i.

$$pl = 2lq, \tag{3}$$

$$Ul = plE\theta \cos(\theta), \tag{4}$$

where $2l$ is the total length (height 35 μm) of the cylinder and q is the electric charge (1.6×10^{-19} C). After solving Eq. (3), we get $pl = 0.112 \times 10^{-22}$ kg m/s². Maximum electric field radiation pattern (doughnut-shaped) arises almost in the

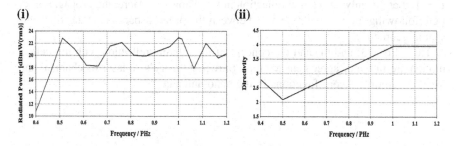

Fig. 7 (i) Radiation power [dBmv (rms)]/frequency (PHz), (ii) directivity/frequency (PHz)

perpendicular direction and a null along the 0° and the 180° directions. Recall Eq. (2), put $\theta = 90°$, we get the magnitude of the maximum value of the electric field.

$\text{Mod}(E\theta_{\max}) = [60 \bmod (I_m)]/r_{\min}$; here, $r_{\min}$ (cones radius 0.77 μm) and mod (I_m) is 1.45; and after solving Eq. (4), we get numeric value of Ul as 0.126×10^{-14} J. If the entering electron in the cone generates some angle with the respective field, then the resulting path of the electron will be helical.

If a rotation of the light wave underlies the laser emission, then the possibility of helical electron transmission increases, and we concentrate on this particular feature largely, because the produced standing wave sets the primary criterion in the non-conducting and extremely insulating environment of the biological structure. Since the propagation of the electrons essentially involves a helical *"internal reflection"* around, or within space between pillar and pillar (octagonal order appears in rods around cones), the network of cells acts as an array of helical antennas. To be specific, this occurs within the 300 nm dimension of a single nano-pillar or it is below as within the near field of light wavelength. The light travels outside this space (Mazur group at Harvard) by reducing the dimensional of the existence space. There is a possibility of a similar rotational motion occurring in the biological nano-antenna structures of the retina. The angular momentum (p_2) and energy U_2 of the rotating object (for single unit) is

$$p_2 = I\omega, \tag{5}$$

$$U_2 = (I\omega^2)/2 = (mv^2/4), \tag{6}$$

where I is the moment of the inertia of rod [cylinder $(mr^2/2)$ and ω is the angular velocity (v/r)]. Putting the standard values in Eqs. (5) and (6), we obtained $p_2 = 0.475 \times 10^{-26}$ kg m/s^2 and $U_2 = 0.204 \times 10^{-14}$ J.

Now, recall the VSWR curve of Fig. 3(v) the value of VSWR from the range 0.5–1.0 PHz was nearly close to VSWR curve of Fig. 3(vi). Beyond this region, VSWR rapidly increases until reaching its maximum value of 69.47 at the 1.06 PHz frequency (280 nm approximately). We presume that, if we check the value of the reflection coefficient by using the formula VSWR $= (1 + \Gamma)/(1 - \Gamma)$, then it provides 0.97 which is very close to 1. We are also familiar with the statement that at $\Gamma = 1$, there is only an internal reflection not transmission. Here, the propagation of light following the internal reflection between the pillar's spaces is reduced by this spatial geometry.

Here, Eqs. (3) and (4) and the concept of the internal reflection with the helical motion are in major agreement with the Berkeley theory's "a laser mechanism and Mazur's group's theory functions in the nano-antenna."

3.2 Angle with Retinal Eccentricity or Nature of the Color Vision

According to Gerald C. Huth's theory of "A Modern Explanation for Light Interaction with the Retina of the Eye Based on Nano structural Geometry," a complete octagonal order appears in the rods around cones at around 7°–8° angle due to the hexagonally arrayed cones in the fovea to a large angle and the continuing introduction of statically distributed rods. The length between adjacent cones and rods constitutes an antenna that defines the precise middle of the visual band 550 nm at 7°–8° of the retinal eccentricity, short wavelength interacting beyond 20°, while the long wavelengths are sensitive in the central the fovea. The wavelength of the visible light interaction is controlled by the geometric dimensionality of the pillar-to-pillar spacing exactly replicating the retinal antenna model proposed here. The distance and the phase difference between the receptor system (rods and cones) are 0.58 micrometers and 0° [16, 17], respectively. Here, there is no variation allowed in the length (amplitude) and in the spacing between the adjusted dipole in the modeled antenna taken as an array of the individual dipoles. It means that all individual dipoles have equal length and spacing in the modeled antenna network (octagonal arrays). It is the case of arrays of n-isotropic sources with equal magnitude and spacing. Then, the required angle expression is [16]:

$$\theta = \cos^{-1}[\pm\{(2N+1)/2n\}\lambda/a], \tag{7}$$

where n is the source number, a is the space between the point sources, and N is an integer number (1, 2, 3, 4, ...). Here, n is 9 and a is 0.58 Å µm (standard view).

Case 1   Middle wavelength (M cones sensitive)
         The value of the λ is 550 nm, for balancing mode situation N is 8, and then $\theta = 18°$ (approximately).
Case 2   Short wavelength (S cones sensitive)
         The value of the λ is 400 nm, again N is 8, and then $\theta = 49°$ (approx.). This value is beyond the 20°.
Case 3   Long wavelength (L cones sensitive)
         The value of the λ is 700 nm, again N is 8, and then θ has no value.

According to the trichromatic theory, these three types of cones are sensitive to absorb different wavelengths of light which are responsible for producing the trichromatic nature of color vision shown in Fig. 8. It gives the correct explanations of the plotted curve relative proportion of incident light absorption versus wavelength. The simulated and theoretical results of the model antenna study along with the retinal angle are very close to each other.

Fig. 8 Color vision diagram
with cones and bipolar cell.
[10]

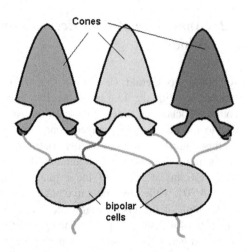

4   Conclusion

The antenna network simulation of the proposed retinal center-fed dipole antenna
shows a good matching with the theoretical and experimental observations of
Berkeley groups and Gerald C. Huth's theory. The proposed modeled retinal
antenna shows a complex triplet of electromagnetic resonance band in the visible
region. The self-similarity in the nature of the distribution of the electromagnetic
field reflects the rod and cone arrangement symmetry.

Acknowledgement J.E. Lugo thanks the magnetophotonics material SEP-PRODEP grant.

References

1. Sahu, S., Ghosh, S., Fujita, D., Bandyopadhyay, A.: Live visualizations of single isolated
   tubulin protein self-assembly via tunneling current: effect of electromagnetic pumping during
   spontaneous growth of microtubule. Sci. Rep. **4**, 7303 (2014)
2. Sahu, S., Ghosh, S., Hirata, K., Fujita, D., Bandyopadhyay, A.: Multi-level
   memory-switching properties of a single brain microtubule. Appl. Physicslett. **102**, 123701
   (2013)
3. Sahu, S., Ghosh, S., Ghosh, B., Aswani, K., Hirata, K., Fujita, D., Bandyopadhyay, A.:
   Atomic water channel controlling remarkable properties of a single brain microtubule:
   correlating single protein to its supramolecular assembly. Biosens. Bioelectron. **47**, 141–148
   (2013)
4. de Pomerai, D.I., Daniells, C., David, H., Allan, J., Duce, I., Mutwakil, M., Thomas, D.,
   Sewell, P., Tattersall, J., Jones, D.: Non-thermal heat-shock response to microwaves. Nature
   6785, 417–418 (2000)
5. Goodman, R., Blank, M., Lin, H., Khorkova, O., Soo, L., Weisbrot, D., Henderson, A.S.:
   Increased levels of hsp70 transcripts are induced when cells are exposed to low frequency
   electromagnetic fields. Bioelectrochem. Bioenerg. **33**, 115–120 (1994)

6. Goodman, R., Henderson, A.S.: Exposure of salivary gland cells to low frequency electromagnetic fields alters polypeptide synthesis. Proc. Nat. Acad. Sci. (US) **85**, 3928–3932 (1998)
7. O'Carroll, M.J., Henshaw, D.L.: Aggregating epidemiological evidence: comparing two seminal EMF reviews. Risk Anal. **28**, 225–234 (2008)
8. Newman, E. A.: Electrophysiology of retinal glial cells. Eye Research Institute of Retina Foundation, 20 Staniford Street, Boston, MA 02114, USA
9. Baylor, J.D.A., Fuortes, M.G.F., O'bryan, P.M.: Receptive fields of cones in the retina of the turtle. J. Physiol. **214**, 265–294 (1971)
10. Sekuler, R., Blake, R.: Perception, Second Edition. (1990). ISBN 0-07-056065-X McGraw-Hill
11. The Interaction between Light and Matter. www.springer.com/cda/content/…/cda…/9783642322600-c1.pdf
12. Rotanowska, M., Sarna, Y.: Light-induced damage to the retina: role of rhodopsin chromophore revisited. Photochem. Photobiol. **81**, 1305–1330 (2005). doi:10.1562/2004-11-13-1R3-371
13. Saari, J.C., Garwin, G.G., Van Hooser, J.P., Palczewski, K.: Reduction of all-trans-retinal limits regeneration of visual pigment in mice. Vis. Res. **38**, 1325–1333 (1998)
14. Marc, T. L.: Visual Processing by the Retina. www.weizmann.ac.il/neurobiology/labs/…/kandel_ch26_retina.pdf
15. Kalal, J. W.: Biological Psychology, Fifth Edition. North Carolina State University (1995). ISBN 0-534-21108-9
16. Elliott, R.S.: Antenna Theory and Design, Revised Edition. IEEE Antenna & Propagation Society, Sponsor (2006)
17. Canbay, C., Unal, I.: Electromagnetic modeling of retinal photoreceptors. Prog. Electromagn. Res. PIER **83**, 353–374 (2008)

DNA as an Electromagnetic Fractal Cavity Resonator: Its Universal Sensing and Fractal Antenna Behavior

P. Singh, R. Doti, J.E. Lugo, J. Faubert, S. Rawat, S. Ghosh, K. Ray and A. Bandyopadhyay

Abstract We report that 3D-A-DNA structure behaves as a fractal antenna, which can interact with the electromagnetic fields over a wide range of frequencies. Using the lattice details of human DNA, we have modeled radiation of DNA as a helical antenna. The DNA structure resonates with the electromagnetic waves at 34 GHz, with a positive gain of 1.7 dBi. We have also analyzed the role of three different lattice symmetries of DNA and the possibility of soliton-based energy transmission along the structure.

Keywords Biological living system · Antenna · DNA vibration

P. Singh (✉) · K. Ray
Amity School of Applied Science, Amity University Rajasthan, Jaipur, Rajasthan, India
e-mail: singhpushpendra548@gmail.com

K. Ray
e-mail: kanadray00@gmail.com

R. Doti · J.E. Lugo · J. Faubert
Visual Perception and Psychophysics Laboratory, School of Optometry,
Universite de Montreal, Montreal H3T 1P1, Canada

S. Rawat
Manipal University, Jaipur, India
e-mail: sanyograwat@gmail.com

S. Ghosh
Natural Products Chemistry Division, CSIR-North East Institute of Science & Technology,
Jorhat 785006, Assam, India

A. Bandyopadhyay
Advanced Key Technologies Division, Advanced Nano Characterization Center,
National Institute for Materials Science, 1-2-1 Sengen, Tsukuba 3050047, Japan

© Springer Nature Singapore Pte Ltd. 2018
M. Pant et al. (eds.), *Soft Computing: Theories and Applications*,
Advances in Intelligent Systems and Computing 584,
https://doi.org/10.1007/978-981-10-5699-4_21

213

1 Introduction

An antenna is capable of transmitting and receiving electromagnetic waves (EMs). Biomaterials are insulators, whether insulators have any potential to radiate like a metallic antenna or not depends on multiple factors. An antenna emits electromagnetic signals only when the standing wave formed inside absorbs integral multiple of its energy in the cavity and only if the radiation resistance is very low. Certain fields could literally break apart DNA (fields as low as 0.18 µT) as suggested recently from studies showing associations with damage to DNA repair genes [1, 2]. Insulators like biomaterials, do not have highly reflecting metallic outer boundary that reflects and develops electromagnetic signals multiple times, and develops high-quality factor or produces high-quality standing waves. For example, DNA literally needs to breathe, else it would break apart [3]. Our attempt to theorize that biomaterials could also act as an antenna includes the factor that even such insulated biomaterials would somehow produce standing waves, which is fundamentally against the concept of physics. Such problems are not addressed seriously, and the predictions for the electromagnetic interactions continue [4, 5]. Even though claims are already been made that DNA is a fractal antenna [6], there are several questions need to be answered before we could confirm whether such a possibility could really exist. There are different kinds of antennas, but recently fractal antennas have received a lot of attention. Fractal antenna is an antenna design that uses the most important fractal characteristic known as self-similarity. The self-similarity in the DNA structure does exist; however, that alone cannot promise electromagnetic resonance in the extremely low frequency (ELF) and radio frequency (RF) ranges. The electromagnetic field interacts with DNA in the ELF range and during DNA strand break, such as described in [7]. EMF can break apart DNA [2]; that is, by maximizing its length and perimeter, they can receive or transmit electromagnetic radiation at many different frequencies simultaneously thus suitable for multiband or wideband applications. Since fractal antenna designs are tightly packed, they are very useful in telecommunication applications such as microwave communication and cellular phones.

DNA is a molecule that contains genetic information utilized by all life-forms and numerous viruses. The human DNA in form of the double helix and various models of conductions has been proposed thus far. The DNAs' interaction with the electromagnetic fields (EMFs) has a wide frequency bandwidth, which does not limit to an optimal frequency. DNA within cells has a compact structural property similar to a fractal antenna with a bandwidth lying at the RF range. By fractal behavior in the electromagnetic resonance spectrum, it is meant that the arrangement of the frequencies has a similar distribution at different scales. In this particular case, we cover a specific domain; however, in future, we would provide detailed frequency database [8–10]. Herein, we developed a new electromagnetic interaction model for the double-helix-shaped human DNA (see Fig. 1). Moreover, we have performed computer simulations in the RF range, with the help of antenna

Fig. 1 DNA structure [11]

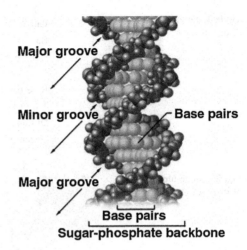

Major groove

Minor groove — Base pairs

Major groove

Base pairs

Sugar-phosphate backbone

theory, and compared the results with the biological data. The human DNA is modeled as a uniform medium with a helical shape. The proposed antenna has the physical dimensions of the same order of magnitude that the original human DNA dimensions [11]. The simulated results were very close to the biological results.

The present work is divided into three parts: In Sect. 2, we present the antenna designing characteristic. Section 3 describes the analysis of the simulation and comparison with biological data. Finally, we wrap up the work with the conclusions.

2 Modeled Antenna Design

DNA is a double helix in which molecules contain two polynucleotides (nucleotide is the basic unit of the DNA structure) wrapped around one another arranged in such a way that their base pair molecules are inside and sugar phosphate molecules "backbone" are outside to form the famous double-helix structure. There are three kinds of weak interactions run in parallel in a DNA, along the linear chain; then, there are two spirals, one spiral if we consider the DNA as a normal cylinder and then if the other spiral if we consider a twisted cylinder. Therefore, we would always get three distinct kinds of transmissions along the DNA molecule simultaneously. Thus, helical structure is not absolutely regular. Superposition of three distinct dynamics would change the structure in homogeneously, and we can distinguish minor and major grooves which are shown in Fig. 1. At room temperature, the conductivity of DNA is 2.4 mho/cm. [12] and the value of the dielectric constant is 4.7 [13]. Here, the 2D view of modeled antenna is designed to take the original view of the A-DNA structure, which has the following physical structural data as shown in Table 1. The 3D view of the modeled antenna based on A-DNA structure is shown in Fig. 2.

Table 1 Dimensional scale of parameter

S. No.	Parameter	Dimension
1	Base pair diameter	0.23 nm
2	Helix packing diameter	2.55 nm
3	Base pair/turn	1
4	Base pairs per turn of helix	11
5	Tilt of base normal to the helix axis	19°
6	Distance per complete term	3.2 nm
7	Distance between base pairs	0.34 nm

Fig. 2 Single-turn 3D view of modeled antenna

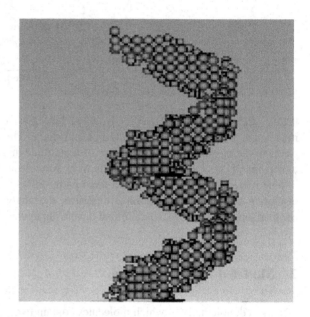

3 Analysis of Simulated and Biological Result

3.1 Radiation Analysis

The reflection coefficient S_{11} (reflection coefficient is an antenna parameter that shows the amount of lost power by the load and does not return as a reflection) of the modeled antenna is shown in Fig. 3a. The reflection coefficient indicates that modeled antenna resonates at the 34 GHz frequency and the corresponding gain is positive (1.7 dBi). The gain value of the single turn of the 3D modeled antenna swims toward the positive value as depicted in Fig. 3b.

The antenna theory suggests that if the dimension of the helix antenna is small compared to their wavelength, then the maximum radiation pattern would always be in the normal direction [14]. Here, the DNA fractal antenna consisting of a small loop where the helix packing diameter is the same as the small loop diameter and

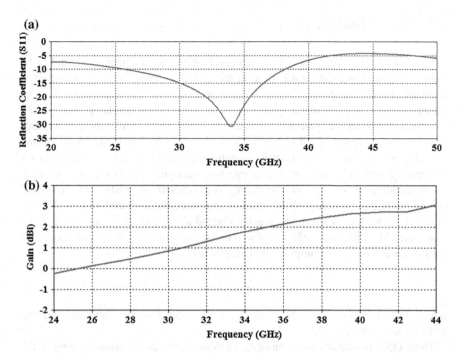

Fig. 3 a Variation of the reflection coefficient (dBi) versus to frequency (GHz), **b** variations of the efficiency versus to frequency (GHz)

the length of the loop is almost the same as one-turn distance $(n \cdot L < \lambda)$. The modeled DNA antenna radiates with the maximum field in the normal direction to the helical axis. The simulated E and H radiation pattern plane of DNA one-turn double-helix modeled antenna is depicted in Fig. 4 which is significantly close to the theoretical assumption made by direct transmission calculation.

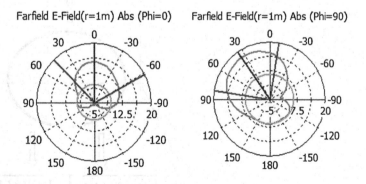

Fig. 4 E and H plane radiation patterns

3.2 Power Analysis

Experimentally, the actual current distribution on the helix antenna is found to be more complex. Since the total current at the end of a DNAs cylindrical surface is constant, the traveling wave's asymmetric or nonlinear terms generated by the superposition of three different kinds of transmissions would get damped due to the structural stretch in the molecular structure. Longer stretches of DNA are entropically elastic under tension; it undergoes a continuous structural variation due to the available energy, continual collisions with the water molecules. The modeled DNA antenna behaves in a way that during structural damping, the DNA molecules are cyclically stressed; as a result, the energy is dissipated internally. The dissipated energy holds the relation with the current amplitude as described in Eq. (1). The shape of the hysteresis curve remains unchanged along with the amplitude and the strain rate [14]. The loss coefficient also remains constant. The dissipated energy by the structural vibration (damping) is then given by

$$P_{st} = \alpha I^2, \tag{1}$$

where α is the loss coefficient and I is the variation in the amplitude of the steady-state current.

Here, DNA helical fractal antenna may be considered as a dynamic system of the DNA molecules which undergo collisions due to structural or thermal vibration. Now, we consider the following case of energy dissipation as shown in Fig. 5, and the variation is demonstrated with the viscous damping. Let I_n be the small amount of the current displacement from the steady-state current (I_1) and the damping voltage in this case such that:

Fig. 5 Dynamic system

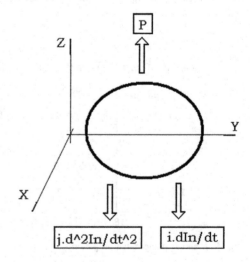

$$I_n = I_1 \sin(\omega t - \varphi), \tag{2}$$

$$V_{\text{damp}}(dV/dt) = jdI_n/dt \tag{3}$$

The power loss per cycle $(2\pi/\omega)$ of the current flow due to the applied potential say damped potential (voltage) following the interaction with the neighbor's molecule is given by:

$$P = \int V_{\text{damp}} dI_n \tag{4}$$

Considering Eqs. (2) and (3), we get

$$P = j\pi\omega^2 I_1^2. \tag{5}$$

where j is a constant, and then, Eq. (5) is close to Eq. (1). This means that the vibration amplitude of the current at the steady state holds the proportional relationship with the power loss. The simulated result also well followed the theoretical assumption related to the vibrational study of the DNA molecules with EM wave interaction. The VSWR curve is depicted in Fig. 6a, and we know that VSWR is equal to the ratio of the maximum and minimum current or the voltage value and its value is close to 1 at the modeled antenna's resonance frequency.

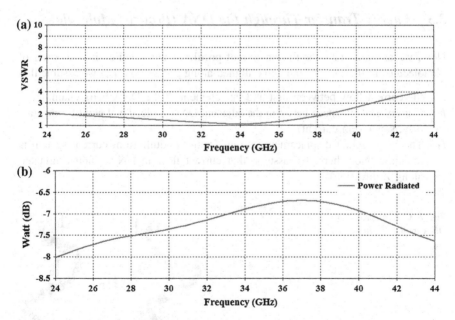

Fig. 6 a Variation of the VSWR versus frequency (GHz), b radiate power (dB) versus frequency (GHz)

Suppose the wavelength range of these structural vibration $= \lambda_{min} - \lambda_{max}$ mm. Let $P_{(vis)max}$ and $P_{(vis)min}$ correspond to the maximum and minimum power for $I_{p(max)}$ and $I_{p(min)}$ current amplitudes. The required value can be obtained from the depicted Fig. 6b, by using the following parameter values $P_{(vis)max} = -6.6$ dB, $P_{(vis)min} = -7.9$ dB, $VSWR_{max} = 4$, and $VSWR_{min} = 1$, and considering Eq. (5), we get

$$P_{(vis)max}/P_{(vis)min} = (\lambda_{min}/\lambda_{max})/(I_{\rho(max)}/I_{\rho(min)})^2 = (\lambda_{min}/\lambda_{max})/(VSWR)^2$$

After putting the values, we get

$$\lambda_{min}/\lambda_{max} = 0.05 - 0.83$$

The theoretical models of DNA vibration and the model of related twisting vibration suggest that the frequency of such vibrations was estimated within the range $\lambda = 1 - 0.1$ mm (approximately). A model of conformational mobility of DNA, where the motion of nucleotides was considered like pendulum, has been used to obtain the vibration frequency range of $\lambda = 1 - 0.6$ mm (approximately) [15]. Following the theoretical model $\lambda_{min}/\lambda_{max} = 0.1$, the simulated result accepts this ratio for the wide frequency range. Here, simulated results of the modeled antenna are well matched with the theoretical results (Figure 7).

3.3 Energy Transfer Through the DNA (Protein) Molecule

Now, we assume that a nonlinear current profile of DNA lattice is composed of N molecules in the form of the helix aligned along Z direction. Let we assume that

I the current flow between first and last element of the DNA helix chain.

I_0 Current exits in space between the adjacent DNA molecule at equilibrium state (say, separation current)

I_d The longitudinal displacement current from its equilibrium current system is dynamic. Now here, we assume that current flow in DNA chain originated along Z direction)

Fig. 7 Current profile of DNA molecules in helix form

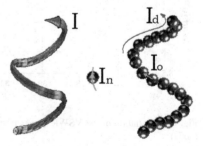

I_{rd} Relative displacement current (we may define the relative displacement current between the adjacent molecules from displacement current I_d)

We may define the relative displacement current between adjacent molecules in the form of I_d and then

$$I_{rd} = I_d(n+1) - I_d(n) \qquad (6)$$

The total interaction or displacement power between two adjacent DNA molecules in dynamic states can be written in the form:

$$P_{rd} = j(I_{rd} + I_0)^2 - P_0 \qquad (7)$$

From Eq. (7), three different cases can be discussed which is consistent with the energy or power profile of the DNA helix shape

Case 1: By using Eq. (7), the relative displacement power between the two adjacent molecules will be minimum, if the neighboring molecules tried to maintain the equilibrium states:

$$\lim_{I_{rd} \to 0} P_{rd} = 0 \qquad (8)$$

Then, $dP_{rd}/dI_{rd} = 0, d^2P_{rd}/dI_{rd} = 0$

Case 2: Equation (8) is found to be in the increasing order if $I \to I_D$, so

$$\lim_{I \to I_D} P_{rd}f = +\text{value}$$

Case 3: If $I_{rd} < 0$, then $d^3P_{rd}/dI_{rd}^3 < 0$.

By the power or energy profile, Friesecke and Pego proved that the continuous limit for relative displacement profile is:

$I(t) = [P'_{rd}/P'''_{rd}] \cdot [\text{amplitude. Sech (angle)}]^2$ [15, 16].

Thus, obtained equation of the current or power shows a solitonic waveform which provides the comprehensive explanation of the theoretical statement "A process of vibrational energy transfer along the protein molecule is considered on the basis of a hypothesis of the soliton" [17].

3.4 Dynamic Power Spectrum

The power spectrum of the dynamic system (in case of cherry flow) is shown in [18] which approximates very well with the simulated power curve as shown in Fig. 8. In case of absolute continuous spectrum, the finite length coverage indicates

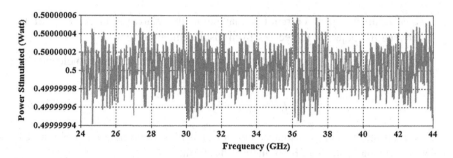

Fig. 8 Power stimulated profile of modeled fractal antenna

a bounded spectral curve, whereas the absence of such convergence indicates the presence of singularities in spectrum measure.

4 Conclusions

In this work, we report that a 3D-A-DNA structure of the human body behaves as a fractal antenna which may interact with electromagnetic fields over a wide range of frequencies. The radiation analysis showed that the modeled antenna resonated at a frequency of 34 GHz with a corresponding positive gain of 1.7 dBi. Additionally, the maximum radiation pattern is mainly in the normal direction just similar to that antenna theory predicts. From our power analysis, we found that the power loss is proportional to the vibration amplitude squared of the current at steady state, a result found before [3]. The wavelength range of these structural vibrations $\lambda_{min} - \lambda_{max}$ was investigated. We found that the ratio value between $\lambda_{min}/\lambda_{max}$ ranged from 0.05 up to 0.83. Other theoretical DNA vibrational models predicted a value of 0.1. We also found that the energy transfer through the DNA molecule can be calculated by the displacement power between two adjacent DNA molecules whose displacement current is described by a soliton waveform at equilibrium. This result was hypothesized before in [17]. Finally, we found that the simulated dynamic power spectrum of the antenna is similar to the one presented in [18].

Acknowledgement J.E. Lugo thanks the magnetophotonics material SEP-PRODEP grant.

References

1. Xing, H., Wilkerson, D.C., Mayhew, C.N., Lubert, E.J., Skaggs, H.S., Goodson, M.L., Hong, Y., Park-Sarge, O.K., Sarge, K.D.: Mechanism of HSP 70i gene bookmarking. Science **307**, 421–423 (2005)
2. O'Carroll, M.J., Henshaw, D.L.: Aggregating epidemiological evidence: comparing two seminal EMF reviews. Risk Anal. **28**, 225–234 (2008)

3. Alexandrov, B.S., Gelev, V., Bishop, A.R., Usheva, A., Rasmussen, K.O.: DNA breathing dynamics in the presence of a terahertz field. 29 October 2009. http://arxiv.org/abs/0910.5294
4. Adair, R.K.: Vibration resonances in biological systems at microwave frequencies. Biophys. J. **82**(3), 1147–1152 (2002)
5. Meyl, K.: DNA and cell resonance: magnetic waves enable cell communication. DNA Cell Biol. **31**(4), 422–426 (2012)
6. Blank, M., Goodman, R.: DNA is a fractal antenna in electromagnetic fields. Int. J. Radiat. Biol. **87**(4), 409–415 (2011)
7. Sage C, Carpenter D (Eds.) A scientific perspective on health risk of electromagnetic fields. Published online 31 August 2007 at: http://www.bioinitiative.org/report/index.htm
8. Sahu, S., Ghosh, S., Fujita, D., Bandyopadhyay, A.: Live visualizations of single isolated tubulin protein self-assembly via tunneling current: effect of electromagnetic pumping during spontaneous growth of microtubule. Sci. Rep. **4**, 7303 (2014)
9. Sahu, S., Ghosh, S., Hirata, K., Fujita, D., Bandyopadhyay, A.: Multi-level memory-switching properties of a single brain microtubule. Appl. Phys. Lett. **102**, 123701 (2013)
10. Sahu, S., Ghosh, S., Ghosh, B., Aswani, K., Hirata, K., Fujita, D., Bandyopadhyay, A.: Atomic water channel controlling remarkable properties of a single brain microtubule: correlating single protein to its supramolecular assembly. Biosens. Bioelectron. **47**, 141–148 (2013)
11. Watson, J.D., Crick, F.C.H.: Molecular structure of nucleic acids, a structure for deoxyribonucleic acids. Nature **171**, 737–738 (1993)
12. Dewarrat, F. C.: Electric characterization of DNA thesis (2002). https://nanoelectronics.unibas.ch/.../theses/Dewarrat-PhD-Thesis.pdf
13. Flock, S., Labarbe, R., Houssier, C.: Dielectric constant and ionic strength effects on DNA precipitation. Biophys. J. **70**, 1456–1465 (1996)
14. Thomson William, T., Dahleh Dillon, M.: Theory of vibration with applications. 5th edition (1998)
15. Moleron, M., Leonard, A., Daraio, C.: Solitary waves in a chain of repelling magnets. J. Appl. Phys. **115**, 184901 (2014)
16. Friesecke, G., Pego, R.: Nonlinearity **12**, 1601 (1999)
17. Reshetnyak, S.A., Shcheglov, V.A., Blagodatskikh, V.I., Gariaev, P.P., Maslov, MYu.: Mechanisms of interaction of electromagnetic radiation with a biosystem. Laser Phys. **6**(4), 621–653 (1996)
18. Zaks, M.: Fractal fourier spectra in dynamic system. Instituted fur Physik (2001). https://publishup.uni-potsdam.de/opus4-ubp/files/145/zaks.pd

Compact Half-Hexagonal Monopole Planar Antenna for UWB Applications

Ushaben Keshwala, Sanyog Rawat and Kanad Ray

Abstract In this paper, half-hexagonal monopole planar antenna designed using FR-4 substrate is presented for UWB applications. The antenna structure consists of half-hexagonal monopole with microstrip line feed. The antenna is designed using FR-4 substrate. The proposed antenna has compact size and offers impedance bandwidth of 6.12 GHz (3.03–9.15 GHz). The proposed antenna provides wide bandwidth in the UWB range which can be used for UWB applications. The gain characteristic is obtained to be stable in the passband with average gain of 2.86 dBi.

Keywords Half-hexagonal monopole · Monopole antenna · Ultra-wideband antenna

1 Introduction

With the declaration of 3.1–10.6 GHz frequency band for UWB applications by Federal Communication Commission (FCC) [1] for commercial use, large amount of research is being carried out in UWB technology. UWB system has many plusses such as ultra-wide uninhibited bandwidth, high data rate, compact system. Microstrip antenna, in general, comprises of a radiating conducting patch of desired

U. Keshwala (✉)
Electronics and Communication Engineering Department, Amity School of Engineering, Amity University, Noida, Uttar Pradesh, India
e-mail: Usha_keshwala30@yahoo.com

S. Rawat
Electronics and Communication Engineering Department, Manipal University Jaipur, Jaipur, India
e-mail: sanyog.rawat@gmail.com

K. Ray
Amity School of Applied Science, Amity University Rajasthan, Jaipur, India
e-mail: kanadray00@gmail.com

© Springer Nature Singapore Pte Ltd. 2018
M. Pant et al. (eds.), *Soft Computing: Theories and Applications*,
Advances in Intelligent Systems and Computing 584,
https://doi.org/10.1007/978-981-10-5699-4_22

225

shape and a conducting infinite ground plane parted by a squeaky dielectric sub-strate [2].

UWB antennas have many applications such as radar, imaging remote sensing, and medical applications. As per FCC, the bandwidth for UWB applications encompasses from 3.1 to 10.6 [1], and techniques have been investigated and reported to achieve specifiable ultra-wide bandwidth. One of the techniques is to use differently shaped monopoles. The differently shaped monopoles such as elliptical, annular, hexagonal, rectangular, square, triangular, circular are used to obtain ultra-wide bandwidth [3–6]. The mentioned monopole shapes with some modification have been analyzed for many UWB applications [7, 8].

There are numerous wideband antennas types which are reported; however, the printed monopole wideband antennas have gained a lot of attention due to its compact size, ease of integration with another circuit, and flat structure [9, 10]. By keeping these advantages of printed monopole antennas in view, half-hexagonal antenna is designed for UWB applications. The proposed configuration is the simplest compact structure of monopole, which is relaxed to fabricate.

The antenna proposed in this paper is simple planar half-hexagonal monopole with microstrip line feed. The antenna proposed is comprehended in two stages. In the first stage, a hexagonal monopole is shown in Fig. 1. In the second stage, a hexagonal monopole is modified to half-hexagonal monopole which acquires ultra-wide bandwidth of 6.12 GHz. The simulation results obtained on CST microwave studio for the same are mentioned and deliberated.

Fig. 1 Hexagonal monopole

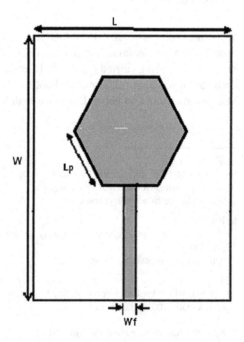

2   Antenna Scheme and Structure

The design initially begins with geometry of the planar hexagonal monopole as shown in Fig. 1. The proposed structure is designed using FR-4 substrate with $\varepsilon_r = 4.3$, tan $\delta = 0.025$, and thickness $(h) = 1.59$ mm. The configuration of patch and ground is made up of perfect electrical conductor (PEC). The antenna has a compact size of 28 mm × 43 mm × 1.59 mm. The length l_p of the hexagon is taken 13 mm, and it is fed by 50 Ω microstrip line of width $W_f = 2.54$ mm. The return loss and VSWR characteristic for the antenna are shown in Fig. 2a, b, respectively. The return loss graph shows the two resonances at 2.8 and 7.6 GHz, which falls in the UWB range, but the antenna presented acts as a dual band; to enhance the bandwidth, the modifications need to be done in the present structure.

The bandwidth in the UWB range can be achieved by amending the hexagonal shape and optimizing the dimensions of the ground plane as depicted in Fig. 3. The hexagonal shape is modified to half hexagon. The length of the initial ground is reduced and converted to partial ground to attain ultra-wide bandwidth The final design of the antenna consists of half-hexagonal monopole with $L = 28$ mm, $W = 43$ mm, $W_g = 12.5$ mm, $P = 1.5$ mm, $W_f = 2.54$ mm, $L_p = 13$ mm. The proposed printed half-hexagonal monopole acquires ultra-wide bandwidth of 6.12 GHz in the UWB range. The proposed half-hexagonal antenna is shown in Fig. 3a, b. The obtained wide bandwidth in UWB range makes the proposed antenna a worthy contender for ultra-wideband uses.

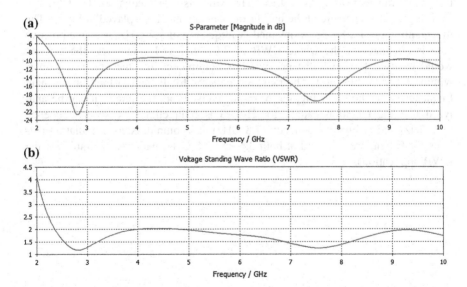

Fig. 2 **a** Variation of reflection coefficient (S_{11}) as a function of frequency, **b** variation of VSWR as a function of frequency

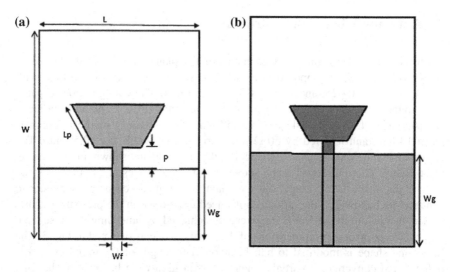

Fig. 3 Half-hexagonal monopole **a** front view, **b** back view

3 Results and Discussion

The simulated results for return loss characteristics and VSWR are shown in Fig. 4a, b, respectively. From the return loss characteristic, it can be observed that antenna resonates at 3.6 and 7.8 GHz frequencies. The simulated impedance bandwidth ranges from 3.03 to 9.15 GHz, which is nearly equal to 107.02%.

The gain characteristic of the half-hexagonal antenna is displayed in Fig. 5. The maximum gain of 2.89 dB is obtained at frequency of 7.2 GHz. A stable positive gain is obtained for the whole UWB of 3.01–9.15 GHz range. As the antenna would be used for small-distance wireless communication, the obtained gain is within an acceptable limit. The radiation patterns for E- and H-Field at resonant frequencies are shown in Fig. 6a, b, respectively. From the graph, it can be analyzed that the radiation pattern for H-filed is bidirectional at lower frequency (3.6 GHz) and at higher frequency (7.8 GHz). The omnidirectional radiation patterns for E-field are achieved at both 3.6 and 7.8 GHz, which is desirable for the UWB applications.

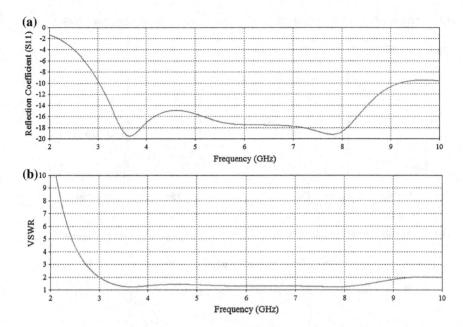

Fig. 4 **a** Variation of reflection coefficient (S_{11}) as a function of frequency, **b** variation of VSWR as a function of frequency

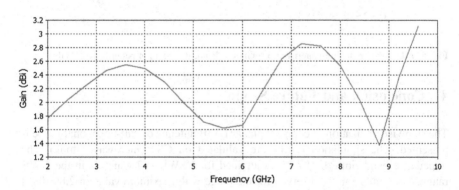

Fig. 5 Variation of gain as a function of frequency

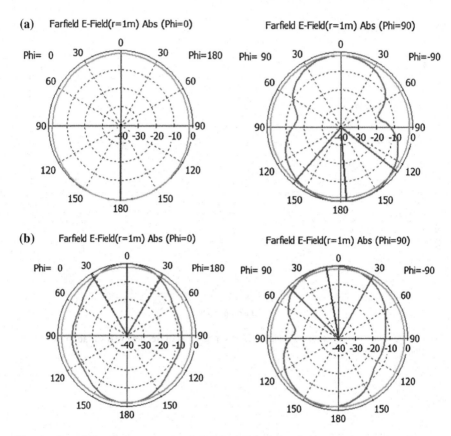

Fig. 6 **a** E and H radiation patterns at 3.6 GHz, **b** E and H radiation patterns at 7.8 GHz

4 Conclusion and Future Scope

The stepwise realization of compact planar UWB antenna of half-hexagonal shape is presented. The simulation results are validated on CST microwave studio. The impedance bandwidth of 107.2% is attained for VSWR < 2 which is in the UWB range. The stable gain is observed for the range with maximum value of 2.89 dB at 7.2 GHz. The antenna characteristic can be improved further by introducing multiple notches in the obtained ultra-wide band to avoid the interference to the existing communication systems.

References

1. Federal Communications Commission (FCC). Revision of part 15 of the commission's rules regarding ultra-wide band transmission systems. First Report and Order. ET Docket 98–153, FCC 02-48, 2002

2. Rawat, S., Sharma, K.K.: Annular ring microstrip patch antenna with finite ground plane for ultra-wideband applications. Int. J. Microw Wirel. Technol. **7**, 179–184 (2015)

3. Agrawall, N.P., Kumar, G., Ray, K.P.: Wideband planar monopole antennas. IEEE Trans. Antenna Propag. **46**, 294–295 (1998)

4. Lee, E., Hall, P.S., Gardner, P.: Compact wideband planar monopole antenna. Electron. Lett. **35**, 2157–2158 (1999)

5. Roh, Y., Chung, K., Choi, J.: Design of a microstrip-fed ultra wideband monopole antenna having band rejection characteristics. IEEE Antenna. Propag. Soc. Int. Symp., 556–559 (2005)

6. Su, S.W., Wong, K.L., Tang, C.L.: Ultra wideband square monopole antenna for IEEE 802.16a operation in the 2–11 GHz band. Micro. Opt. Tech. Lett., 42, 463–466 (2004)

7. Rawat, S., Sharma, K.K.: Stacked configuration of rectangular and hexagonal patches with shorting pin for circularly polarized wideband performance. Cent. Eur. J. Eng. **4**, 20–26 (2013)

8. Reddy, G.S., Mishra, S.K., Mukherjee, J.: Compact Bluetooth/UWB Dual-Band planar antenna with quadruple Band-Notch characteristics. IEEE Antenna. Wirel. Propag. lett. 13, 872–875 (2014)

9. Ellis, M.S.: Miniature staircase profile printed monopole antenna for ultra wide band applications. Proceedings of Cross Strait Quad-Regional Radio Science and Wireless Technology Conference, 237–240 (2013)

10. John, M., Ammann, M.J.: Optimization of impedance for printed rectangular monopole antenna. Microw. Opt. Technol. Lett. **47**, 153–154 (2005)

Effective Data Acquisition for Machine Learning Algorithm in EEG Signal Processing

James Bonello, Lalit Garg, Gaurav Garg and Eliazar Elisha Audu

Abstract The aim of this paper is to demonstrate that small dataset can be used in machine learning for seizure monitoring and detection using smart organization of multichannel EEG sensor data. This reduces training time and improves computational performance in terms of space and time complexities on hardware implementations. The proposed approach has been tested and validated using CHB-MIT dataset containing EEG recordings of 24 clinically verified seizure and non-seizure pediatric patients. The predictability is discussed in terms of the latency and the required length of data for the proposed approach over the state-of-the-art method in the field of EEG-based seizure prediction.

Keywords EEG · Multichannel data · Sensor data · Machine learning
Automated seizure detection

1 Introduction

The traditional process of manually deciphering information and analysis of electroencephalography (EEG) data for medical diagnosis is practically challenging and technically demanding to experts. With the emergence of machine learning and their applications in classification, nonlinear approximation and pattern recognition,

J. Bonello (✉) · L. Garg
University of Malta, Msida, Malta
e-mail: jamesbonello9@gmail.com

L. Garg
e-mail: lalit.garg@um.edu.mt

L. Garg · E.E. Audu
University of Liverpool, Liverpool, UK
e-mail: eliazar.audu@online.liverpool.ac.uk

G. Garg
University of Ulster, Magee Campus, Londonderry, UK
e-mail: garg-g@email.ulster.ac.uk

© Springer Nature Singapore Pte Ltd. 2018
M. Pant et al. (eds.), *Soft Computing: Theories and Applications*,
Advances in Intelligent Systems and Computing 584,
https://doi.org/10.1007/978-981-10-5699-4_23

a new frontier in diagnosing neurological and physiological conditions has risen to prominence. Today, machine learning techniques are visible in virtually all areas of biomedical data analysis; thanks to their resilience and ability to learn from past information (experience) and produce a global generalization. Various feature selection methods have been implemented to enhance correct identification of pattern of interest [1, 2]. However, there is no "silver bullet" in terms of which algorithm is suitable for all "weather" especially in EEG analysis of epileptic seizures. This has led to the development of different machine learning techniques tailored to meet the requirement of specific applications [2]. Artificial neural network (ANN), consisting of network of simple but highly interconnected elements (neurons), has been popularly used in EEG feature classification for epileptic seizures [3–7]. Other techniques used in EEG features classifications include support vector machine (SVM) [3, 8–11], extreme learning machine (ELM) [3, 8, 12, 13], and genetic algorithm with ANN [3, 14]. The SVM being the most popular classifier, the proposed analysis is considered using it. While, another advantage with SVM classifier is its requirement to set relatively less number of parameter which makes it more suitable to automate the feature classification process [3, 9–11]. The main novel focus of this paper revolves around using the minimum amount of data possible while still achieving desirable results.

The remaining of the manuscript is organized in mainly three further sections, where the next section would describe about the datasets and methods used in this analysis. Later in the results section, reliability of the proposed method is presented in terms of quantitative details. Finally, a conclusion has been drawn based on the outcome of this analysis. References for the supporting background work are provided in the end of this manuscript.

2 Materials and Methods

2.1 EEG Datasets

The datasets used for this study are taken from publicly (online) available data collected at the Children's Hospital Boston (CHB)[1] [6, 7, 15, 16]. EEG data is recorded from pediatric patients suffering from intractable seizures. The 24 EEG recordings are taken from 23 subjects (note that dataset chb21 was taken 1.5 years after chb01 and both EEG datasets belong to the same patient). More specifically, 5 male subjects (ages ranging from 3 to 22 years) and 17 female subjects (ages ranging from 1.5 to 19) characterize the database. A text file corresponding to each subject contains the number of hours of both seizure and non-seizure EEG, including a summary containing the number of seizures and the seizure start/end times of each hour of EEG. This text file of information proved key to the method

---

[1]https://physionet.org/pn6/chbmit/.

in this analysis. The EEG data has been sampled at the sampling rate of 256 Hz with a 16-bit resolution. For this project, the feature space was built solely from EEG data and any non-EEG; therefore, secondary signals found in the datasets were regarded as artifacts.

2.2 EEG Signal Processing

In this paper, we explore the use of non-invasive data-acquisition method for EEG analysis of epileptic seizures using CHB-MIT Scalp EEG Database [15]. It contains EEG signals of confirmed cases of seizure and non-seizure from 24 pediatric patients [8, 17]. For this implementation, EEGLab (v. 9.8.6b) [18] was used for visualization and manipulation of the signals. From a high-level perspective, this tool allows for visual analysis of both the time and frequency domains of EEG signals. It also allows for the processing of continuous EEG in both an interactive and also automated mode. While such a tool is indeed useful for performing EEG preprocessing tasks such as artefact removal and segmenting data into epochs, spectral feature extraction is more efficient when performed directly on MATLAB [19], and hence, EEGLab was only used to easily visualize EEG and automate signal preprocessing operations.

2.2.1 EEG Preprocessing

The datasets obtained from CHB-MIT database were in European data format (EDF) format.[2] All raw data were preprocessed prior to their training and classi-fication by retaining all the seizure hours in the dataset. Indeed, prior to focusing on how this will be achieved, the datasets are first segmented into 2 s epochs so that a suitable unit element for feature extraction and selection can be used. Moreover, for some of the patients, EEG was recorded with included "dummy" channels that did not read any signals. These vacant channels were treated as artefacts and were appropriately removed from the final compilation of data.

The main novel focus of this paper revolves around using the minimum amount of data possible while still achieving desirable results. As all non-seizure hours are not needed, the aim is to see if the classifier might effectively be trained using only a small subset of non-seizure EEG. The novelty here is that the size of the datasets produced by this data-acquisition technique is significantly smaller in comparison with other implementations [7]. Consequently, this would significantly reduce the computation time required for learning while still achieving desirable results. Hence, the number of non-seizure hours was kept to a necessary minimum. With this in mind, 10-min' subsets of EEG were extracted appropriately for each seizure

---

[2]https://physionet.org/pn6/chbmit/.

Fig. 1 Illustration of the processing of epochs in the EEG data

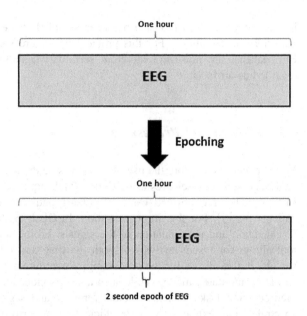

of each patient. The 10-min duration was randomly decided based on the duration of each seizure as it is sufficiently small than the complete one hour dataset, yet it has sufficient number of seizure and non-seizure epochs to effectively train the classifier for both classes. However, for some specific cases, 15-, 5-, and even 2-min subsets had to be taken instead. The reason being is that some seizures were longer than 10-min in the case of the third seizure of chb11. Another reason was that for some patients' EEG (chb12, chb13, chb16), a 10-min set contained more than 1 seizure. The SVM used in this project is constructed in such a way that the subset making up the final input set must either contain no seizures or exactly 1. When these relevant segments are extracted automatically using a program, the chosen data is compiled into one single patient dataset on which future processing and training will be done. The advantage of acquiring data in this way is that while the relevant data required will be in the final dataset, this dataset would be very small in size compared to other methods [6, 7]. Furthermore, the seizure hours need to be handled in a different way than the non-seizure hours. Figure 1 shows the "epoching" process from an "un-epoched" EEG dataset.

This epoch-based annotation was used to locate the start and end points of the seizure in the EEG. The database that contains the patient summary has information regarding the number of channels used to record the patient EEG, the number of seizures (if any) in each hour, and their respective start and end times. During the extraction of seizures, there are two cases to consider: hours containing one seizure and hours containing multiple seizures. Similarly, seizure details must be inferred to make sure that every set extracted contains, if any, exactly one seizure. Moreover, seizure sets are stored in appropriate database of each patient, which will also contain the chosen non-seizure sets. The process of creating patient database by

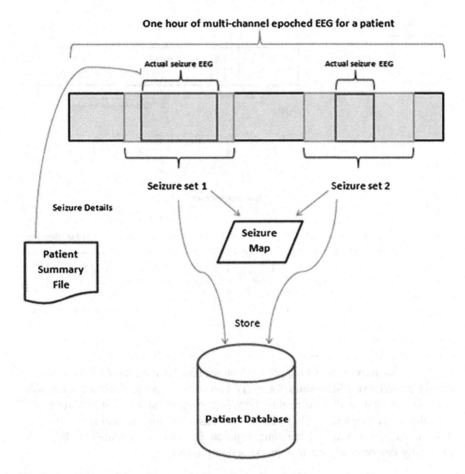

Fig. 2 Acquisition of the seizure sets to form the patient database

acquisition of the seizure sets and non-seizure sets is illustrated in Figs. 2 and 3, respectively. Only this patient database will then be used for future processing in the system. For both of these cases, the start and end epoch details of every seizure (for each hour) are added in a reference map (the seizure map) which proved its importance when creating the target vector.

2.2.2 Spectral Feature

It is important to note the simplicity of the feature extraction process. The only feature that was considered for machine learning was the power spectral density (PSD) of each EEG epoch. In simple terms, the PSD gives a map of how the power is distributed across the different frequencies of the signal. MATLAB signal

Fig. 3 Acquisition of the non-seizure sets

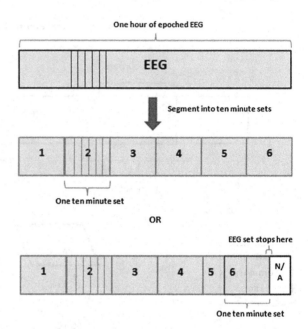

processing toolbox is used to apply the Fourier transform on the EEG and also to extract the relative PSD values for every two-second epoch. Such a feature was simple to calculate and as the results show, is also sufficient for learning. Moreover, only the lower frequencies (1–10 Hz) were considered for calculating the feature. This, in turn, also reduced the time required for such a computation. Figure 4 pictorially demonstrates the feature extraction process.

3 Support Vector Machine (SVM)

Support vector machine learning technique has emerged as one of the most powerful algorithm in solving classification problems [7, 9–11, 20]. It overcomes the problem associated with neural networks and provides good generalization performance and efficient computation by maximizing performance while minimizing the complexity of the learned model [20–22]. In this paper, we explore the potentials of SVM in supervised learning mode to train feature space for classification and seizure detection. Seizure classification is a binary classification problem, that is, the SVM will classify a feature vector into one of two classes [7, 21]. From a geometrical perspective, the SVM's goal is to construct an optimal hyperplane such that the distance between seizure points and non-seizure points is as large as possible [23]. In seizure prediction, the classifier parameters used may be exploited such that the boundary created between seizure and non-seizure points is

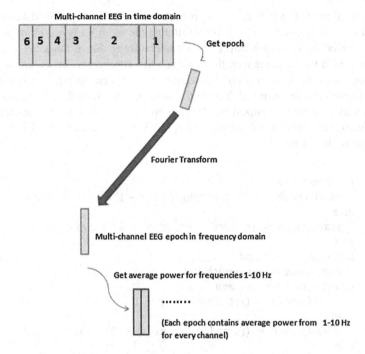

Fig. 4 Steps involved in the feature extraction process

extended to include the earliest possible seizure points. This may result in the false inclusion of non-seizure points and essentially increase the rate of false positives [7].

3.1 Label Vector Creation

To create the label vectors for every patient, first the respective seizure map is loaded and the number of patient-specific data sets in the input set is calculated. The set length (2, 5, 10, or 15 min) in terms of epochs is set accordingly depending on the current patient. The MATLAB script then iterates over the number of patient-specific data sets and performs the appropriate operations. If the first patient-specific data set is currently loaded, the target vector is created as a vector of -1's, and its dimensions will be set to amount $\times 1$. First, the current patient-specific data set is parsed for the hour number and also checked if it carries the seizure annotation. If the annotation exists, further processing will follow. If not, the set can be ignored as it is non-seizure and its respective labels would only consist of -1's (which are preset). It is important to note that an offset variable where offset $=$ (set length $\times (i - 1)$) is also created to keep track of the current patient-specific data set position in the label vector. i is the index of the patient-specific data set. The order of the patient-specific data sets constituting the input set is according to their order as sorted by corresponding dataset name/number. Since the label vector is being created in the same manner, the order remains consistent. For seizure sets, the hour number is searched within the seizure map. When found, the set/seizure start/end epoch details are noted. Taking the difference between the seizure start and set start will give the normalized seizure start epoch (a value from 1 to set length), and similarly, the normalized seizure end epoch is also obtained. The label vector algorithm is shown below.

$$
\begin{aligned}
&if\ s\_start\ <\ set\_start\ +\ 3 \\
&\quad start\_epoch\ =\ 1\ +\ (set\_length\ \times\ (i-1)) \\
&else \\
&\quad start\_epoch\ =\ (s\_start\ -\ set\_start\ +\ 1)\ +\ offset\ -\ 2 \\
&end \\
&if\ s\_end\ ==\ set\_end \\
&\quad end\_epoch\ =\ set\_length\ \times\ i \\
&else\ if\ s\_end\ ==\ set\_end\ -\ 1 \\
&\quad\quad end\_epoch\ =\ (set\_length\ \times\ i)\ -\ 1 \\
&\quad else \\
&\quad\quad end\_epoch\ =\ ((s\_end\ -\ set\_start)\ +\ 1)\ +\ offset\ -\ 2 \\
&end \\
&end \\
&target\_vector(start\_epoch{:}\ end\_epoch)\ =\ 1;
\end{aligned}
$$

Once the label vector is fully created, its MATLAB structure is renamed Seizure Class so as to be recognized by the SVM. Execution times for these steps are very fast compared to the signal processing functions provided in EEGLab. This difference may be attributed to the overheads generated by the large amount of data being handled by EEGLab.

3.2 SVM Training

Together with Seizure Data and Seizure Class structure, another structure needs to be created that represents the training structure. In the training, cross-validation is applied to the dataset. The training structure is divided into three parts, and the fifth, sixth, and seventh columns of this structure are filled out with these details accordingly as shown in the code below:

% divide input set in three parts for cross validation and populate
cross_valid_size = floor(struct_size/3);
epoch_part = floor(size(SeizureClass, 1)/3);

for i = 1: struct_size
  if i <= cross_valid_size
      train_structure (i, 5) = 1;
      train_structure (i, 6) = 1;
      train_structure (i, 7) = cross_valid_size × set_length;

   else if i > cross_valid_size && i <= (cross_valid_size × 2)
        train_structure (i, 5) = 2;
end
end
end
train_structure(i, 6) = (cross_valid_size × set_length) + 1;
train_structure(i, 7) = 2 × cross_valid_size × set_length;
train_structure(i, 5) = 3;
train_structure(i, 6) = (2 × cross_valid_size × set_length) + 1;
train_structure(i, 7) = size(SeizureClass, 1);

This training structure is then renamed as HourlySeizureData so as to be recognized by the SVM. Finally, when all three structures (SeizureData, SeizureClass, HourlySeizureData) are created, these are saved into a separate MATLAB workspace for the SVM. The feature space was cross-validated as described in the above algorithm. The SVM classifier was set to run for 25 iterations. In each iteration, it trains and tests the respective cross-validated sets, with an incrementing σ divisor in the Gaussian kernel, where $\sigma = \{1, 2, 3, 4, \ldots, 25\}$.

4 Results

The simulation was performed with EEGLab on patient-specific dataset (data format. edf file) from CHB-MIT database containing recorded cases from 24 pediatric patients. EEGLab provides the eeg_regepochs() function which takes in the EEG dataset and the epoch length as parameters. This function will then output the new epoch EEG as a dataset on EEGLab. EEGLab can be exploited by using eegh() to select multiple datasets for processing. Table 1 shows patient-specific results, and Table 2 shows a summary of the overall results obtained comparable to those of Shoeb and Guttag [7]. Referring to Table 2, it is clear that the results are very desirable. The first two results demonstrate that the sensitivity is marginally lower and the latency is improved compared to Shoeb and Guttag [7]. The most significant improvement is emphasized in the total number of hours of EEG data used, where Shoeb and Guttag [7] required 916 h EEG data for learning compared to the

Table 1 Results recorded for each patient for a small dataset

Patient	Sigma divisor	Average latency	No. of seizure epoch	No. of non-seizure epoch	No. of seizures	No. of seizures correctly detected	Number of false positives	Sensitivity (%)
chb01	2	3.714285714	232	3344	7	7	3	100
chb02	9	4	98	3478	3	3	153	100
chb03	7	2.857142857	213	3363	7	7	255	100
chb04	14	11.5	195	3381	4	4	94	100
chb05	12	2.4	291	3285	5	5	228	100
chb06	14	1.5	88	3488	10	10	988	100
chb07	12	2.33333333	175	3401	3	3	77	100
chb08	3	7	472	3104	5	5	0	100
chb09	15	1.333333333	153	3423	4	3	62	75
chb10	9	3.285714286	236	3340	7	7	7	100
chb11	15	1	412	3620	3	3	1	100
chb12	20	2.852941176	824	3352	40	34	719	85
chb13	19	3.916666667	274	2390	12	12	429	100
chb14	11	2.125	96	3480	8	8	767	100
chb15	23	6.764705882	918	5042	20	17	231	85
chb16	16	1.4	56	3520	10	10	1193	100
chb17	24	4	158	3418	3	3	20	100
chb18	19	2.5	171	3405	6	6	93	100
chb19	25	3.33333333	130	3446	3	3	30	100
chb20	5	9.5	133	3145	7	6	149	85.71
chb21	20	3.25	111	3465	4	4	175	100
chb22	18	2.666666667	114	3462	3	3	700	100
chb23	21	3	229	3347	7	6	404	85.71
chb24	18	3.923076923	284	2380	16	13	70	81.25

Table 2 Summary of the dataset performance based on sensitivity, latency, and specificity compared Shoeb and Guttag [7]

Performance	Our result	Shoeb and Guttag [7]
Sensitivity	92.39%	96%
Latency	3.72 s	4.6 s
Specificity	91.55%	N/A
Total hours used	49.48 h	916 h

50 h of EEG data required in this study. This ∼95% reduction in data space size greatly relaxes the processing power required for both time and memory.

5 Conclusion

In the presented approach, both the overall result and the majority of the patient results are very positive and found comparable to Shoeb and Guttag [7]. It has been shown that good performance can be obtained by using small data sets with efficient extraction of features from EEG data in machine learning training. The most important inference made from the proposed method is the requirement of reduced dataset size by ∼95% and computational latency while still achieving sufficient results for future practical implementations.

Compliance with Ethical Standards This article does not contain any studies with human participants or animals performed by any of the authors.

References

1. Martis, R.J., Acharya, U.R., Min, L.C., Mandana, K.M., Ray, A.K., Chakraborty, C.: Application of higher order cumulant features for cardiac health diagnosis using ECG signals. Int. J. Neural Syst. **23**(4), 1350014 (2013). doi:10.1142/S0129065713500147
2. Orosco, L., Correa, A.G., Laciar, E.: Review: a survey of performance and techniques for automatic epilepsy detection. J. Med. Biol. Eng. **33**(6), 526–537 (2013)
3. Bugeja, S., Garg, L.: Applications of machine learning techniques for the modelling of EEG data for diagnosis of epileptic seizures. In: The 3rd Workshop on Recognition and Action for Scene Understanding (REACTS 2015) Valletta, Malta, 5 Sept. (2015)
4. Hindarto, Haridi, M., Purnomo, M.H.: EEG signal identification based on root mean square and average power spectrum by using backpropagation. J. Theor. Appl. Inform. Technol. **66**(3) (2014)
5. Husain, S.J., Rao, K.S.: A neural network model for predicting epileptic seizures based on fourier-bessel functions. Int. J. Signal Proc., Image Process. Pattern Recognit. **7**(5), 299–308 (2014)
6. Shoeb, A.H.: Application of machine learning to epileptic seizure onset detection and treatment. PhD thesis, Massachusetts Institute of Technology, Cambridge (2009)
7. Shoeb, A.H., Guttag, J.V.: Application of machine learning to epileptic seizure detection. In: Proceedings of the 27th International Conference on Machine Learning (ICML-10), pp. 975–982 (2010)

8. Bugeja, S., Garg, L., Audu, E.E.: A novel method of EEG data acquisition, feature extraction and feature space creation for early detection of epileptic seizures. In: Conference: 38th Annual International Conference of the IEEE Engineering in Medicine and Biology Society, Orlando, Florida, August 16–20, At Orlando, Florida (2016)

9. Li, X., Chen, X., Yan, Y., Wei, W., Wang, Z.J.: Classification of EEG signals using a multiple kernel learning support vector machine. Sensors **14**(7), 12784–12802 (2014)

10. Nanthini, B.S., Santhi, B.: Seizure detection using SVM classifier on EEG signal. J. Appl. Sci. **14**(14), 1658 (2014)

11. Shen, C.-P., Chan, C.-M., Lin, F.-S., Chiu, M.-J., Lin, J.-W., Kao, J.-H., … Lai, F.: Epileptic seizure detection for multichannel EEG signals with support vector machines. In: 2011 IEEE 11th International Conference on Bioinformatics and Bioengineering (BIBE), pp. 39–43 (2011)

12. Chen, L.-L., Zhang, J., Zou, J.-Z., Zhao, C.-J., Wang, G.-S.: A framework on wavelet-based nonlinear features and extreme learning machine for epileptic seizure detection. Biomed. Sig. Process. Control **10**, 1–10 (2014)

13. Song, Y., Crowcroft, J., Zhang, J.: Automatic epileptic seizure detection in EEGs based on optimized sample entropy and extreme learning machine. J. Neurosci. Meth **210**(2), 132–146 (2012)

14. Palaniappan, R., Raveendran, P., Omatu, S.: VEP optimal channel selection using genetic algorithm for neural network classification of alcoholics. IEEE Trans. Neural Netw **13**(2), 486–491 (2002)

15. Goldberger, A.L., Amaral, L.A.N., Glass, L., Hausdorff, J.M., Ivanov, P.C., Mark, R.G., Mietus, J.E., Moody, G.B., Peng, C.-K., Stanley, H.E.: Physiobank, physiotoolkit, and physionet components of a new research resource for complex physiologic signals. Circulation **101**(23), e215–e220 (2000)

16. Shoeb, A., Edwards, H., Connolly, J., Bourgeois, B., Treves, S., Guttag, J.: Patient-specific seizure onset detection. Epilepsy Behav. **5**, 483–498 (2004)

17. Bugeja, S.: Applications of machine learning techniques for the modelling of EEG data for diagnosis of epileptic seizures (M.Sc dissertation). University of Malta (2015)

18. Delorme, A., Makeig, S.: EEGLAB: an open source toolbox for analysis of single-trial EEG dynamics including independent component analysis. J. Neurosci. Meth **134**(1), 9–21 (2004). doi:10.1016/j.jneumeth.2003.10.009

19. MATLAB (R2015a).: Computer Program, Natick. The Math Works Inc, Massachusetts. Retrieved from www.mathworks.com (2015)

20. Schuldt, C., Laptev, I., Caputo, B.: Recognizing human actions: a local SVM approach. In: ICPR 2004. Proceedings of the 17th International Conference on Pattern Recognition, vol. 3, pp. 32–36 (2004)

21. Cherkassky, V., Ma, Y.: Practical selection of SVM parameters and noise estimation for SVM regression. Neural Netw **17**(1), 113–126 (2004)

22. Garrett, D., Peterson, D.A., Anderson, C.W., Thaut, M.H.: Comparison of linear, nonlinear, and feature selection methods for EEG signal classification. IEEE Trans. Neural Syst. Rehabil. Eng. **11**(2), 141–144 (2003)

23. Balakrishnan, G., Syed, Z.: Scalable Personalization of long-term physiological monitoring: active learning methodologies for epileptic seizure onset detection. In: AISTATS. pp. 73–81 (2012)

Solving Nonlinear Optimization Problems Using IUMDE Algorithm

Pravesh Kumar, Millie Pant and H.P. Singh

Abstract A lot of science and engineering problems are nonlinear in nature. Due to various limitations of the traditional methods, we need some robust and efficient non-traditional techniques to solve these nonlinear and complex problems. In the current study, a new variant of differential evolution (DE) algorithm named information utilization-based modified differential evolution (IUMDE) algorithm is proposed. Further the proposed variant is implemented on six benchmark problems which are taken from literature. Experiments, results, and comparison confirm the competence and effectiveness of the proposed variant algorithm over the others.

Keywords Differential evolution algorithm · Selection · Mutation operation Nonlinear optimization

1 Introduction

Optimizations point out to find the feasible solution corresponding to the extreme point of the objective function. The purpose to find these optimal solutions comes generally either designing a problem to minimize total cost or to maximize probable reliability, or any others. For a reason of such great characteristics of optimal solutions, we have a big importance of optimization methods to perform, mainly in scientific, engineering, and business decision-making problems. We can categorize the optimization methods in traditional and non-traditional methods. Due to some

P. Kumar (✉)
Jaypee Institute of Information Technology, Noida, India
e-mail: praveshtomariitr@gmail.com

M. Pant
IIT Roorkee, Roorkee, India
e-mail: millidma@gmail.com

H.P. Singh
Cluster Innovation Centre, University of Delhi, New Delhi, India
e-mail: harendramaths@gmail.com

© Springer Nature Singapore Pte Ltd. 2018   245
M. Pant et al. (eds.), *Soft Computing: Theories and Applications*,
Advances in Intelligent Systems and Computing 584,
https://doi.org/10.1007/978-981-10-5699-4_24

limitation, traditional methods do not make sure the optimum solution and also have limited uses.

In some recent years, some non-traditional methods are very popular for solving optimization problems. Some of the famous algorithms are genetic algorithm (GA), ant colony algorithm (ACO), particle swarm optimization (PSO), differential evolution algorithm (DE), artificial bee colony (ABC), firefly algorithm, firework algorithm, teaching–learning-based optimization, (TLBO) and so on. These algorithms are population-based nature-inspired algorithm with probabilistic transition rule.

Storn and Price suggested the differential evolution (DE) algorithm in 1997. DE algorithm is a well-liked stochastic, dynamic, and shortest exploring-based algorithm to work out global optimization problems. The strength of DE algorithm over other evolutionary algorithms (EAs) has been proved in many studies in solving many real-world problems. A few of the most important and relevant advantages of DE include small formation, effortlessness employing, and a small number of control parameters are necessitate to be set. Furthermore, some advance advantages of DE include capability of handling nonlinear, discontinuous, non-differentiable, and multi-objective functions. A broad range of applications of DE have been successfully established in many real-life applications of engineering and science field such as chemical engineering, mechanical engineering design problems, pattern identification, power engineering, image processing, and so on [3–5].

Various researches have been proposed to improve the performance of DE in recent years. A few of the modified variants of DE during recent years are LeDE [6], TDE [7], DEahcSPX [8], Cauchy mutation-based DE (CDE) [9], mixed mutation strategy-based DE [10], jDE [11], ODE [12], SaDE [13], JADE [14], MDE [15], DERL [16], MRLDE [17], IUDE [18], cultivated differential evolution algorithm [19], and so on. A detailed literature survey of DE variants is given in [20, 21].

In this paper, two new modifications for DE are proposed. The first modification is proposed for mutation, and another is proposed for selection operation. The corresponding DE variant is named as information utilization-based modified DE (IUMDE). The significance of proposed modifications is discussed later in the paper.

The paper is organized as follows: in Sect. 2, the introduction of differential evolution algorithm is given. Proposed modified DE variant is described in Sect. 3. In Sect. 4, benchmark functions and experimental settings are given. Numerical results and comparison are discussed in Sect. 5, and finally, the conclusion of the study is presented in Sect. 6.

2 Differential Evolution (DE) Algorithm

DE algorithm starts with a set of random solution which is called population, and the solution is called individual or vector. Let $P = \{X_i(G), i = 1, 2, \ldots NP\}$ be the population of size NP of generation G and each vector is D-dimensional vector, i.e. $X_i(G) = (x_{1,i}(G), x_{2,i}(G) \ldots, x_{D,i}(G))$. The other operations of DE are described as below:

(i) **Mutation**: In this operation, three dissimilar vectors say $X_a(G)$, $X_b(G)$, and $X_c(G)$ randomly select from the population corresponding to a target vector $X_i(G)$ such that $a \neq b \neq c \neq i$. Now a vector called mutant vector $V_i(G) = (v_{1,i}(G), v_{2,i}(G)..., v_{D,i}(G))$ is generated by Eq. 2:

$$V_i(G) = X_a(G) + \text{SF}(X_b(G) - X_c(G)) \tag{1}$$

Here, $X_a(G)$ is base vector, SF is a real and constant value called scaling factor and used to control the amplification of differential variation $(X_b(G) - X_c(G))$.

(ii) **Crossover**: The second operation is crossover operation in which a trial vector $Y_i(G) = (y_{1,i}(G), y_{2,i}(G)...y_{D,i}(G))$ is generated as defined in Eq. 2:

$$y_{j,i}(G) = \begin{cases} v_{j,i}(G), & \text{if } \text{Cr} < \text{rand}_j \forall \, j = k \\ x_{j,i}(G) & \text{otherwise} \end{cases} \tag{2}$$

Here, Cr is the crossover factor having value between 0 and 1, rand_j is uniformly distributed random number between 0 and 1, and $k \in 1, 2, ..., d$; is the randomly chosen index. Here, k is used to ensure that at least one component of trial vector will be selected from mutant vector.

(iii) **Selection**: selection operation is the final operation of DE for any generation. In this operation, select the best vector between trial vector and target vector, and update the population $Q = \{X_i(G+1), i = 1, 2, ...NP\}$ of the next generation.

$$X_i(G+1) = \begin{cases} Y_i(G), & \text{if } f(Y_i(G)) < f(X_i(G)) \\ X_i(G) & \text{otherwise} \end{cases} \tag{3}$$

3 Information Utilization-Based Modified DE (IUMDE)

Information utilization-based selection operation for DE (IUDE) was proposed in our previous research [18]. In IUDE, a new selection operation was introduced for DE algorithm. The new selection operation uses two-way array to store the trial vector and reuse them by comparing the worst vector in the population. Here, worst vector means the individual which gives the maximum fitness value in the current population. This approach reduces the searching domain part in every generation, and therefore, gets better convergence speed as well as reduces the redundant computer memory.

In order to accelerate the convergent speed, a modified mutation strategy is also combined with IUDE algorithm and named it information utilization-based modified DE (IUMDE) algorithm. In IUMDE, the base vector is randomly selected from the subpopulation of best individuals. By taking base vectors from the best individuals, It will boost the convergence speed of the DE algorithm as suggested by Kaelo and Ali [16] in DERL algorithm, and Kumar and Pant [17] in MRLDE. In IUMDE, the size of subpopulation of the best individuals plays a very important role in boosting up the convergence speed. In this paper, we have taken the size of the subpopulation as 25% of the total population in all generation.

The pseudo-code of IUMDE is given as below;

Working of IUMDE Algorithm

```
Start
Produce a random uniform distributed population P=
{X_i(G), i=1,2,...NP}.
      X_i(0)    =    X_lower    +(X_upper    -X_lower)*rand(0,1);
rand(0,1)is uniform random number.
Estimate fitness of X_i(G): f{X_i(G)}
while (Stopping condition is not met )
{
  For(i=0;i<=NP;i++)
   {
    Make a Subpopulation P'={X_i(G), i'⊂ (1,2,..N)}of
    best Individuals
    Select a vector randomly from P'
    execute mutation operation

    execute crossover operation
    Estimate fitness of Y_i(G): f{Y_i(G)}

//**        Information        Preserve        Selection
Operation***////

    If f{Y_i(G)}< f{X_i(G)}
      {
        X_i'G+1)= Y_i(G)
      }
      Else
    {
       If f{Y_i(G)}< f{X_worst(G)}
        X_worst(G)=Y_i(G)
      }
  }/* END*/
} /* END*/
Terminate
```

4 Benchmark Problems and Experimental Settings

4.1 Benchmark Problems

Six benchmark problems are taken from the literature to evaluate the performance of the newly proposed variant IUMDE. The details of these problems are given in Table 1.

4.2 Performance Criterion

Following performance criteria have been considered in the paper:

NFEs: NFEs are defined as the number of fitness function evaluations to reach a fixed error; that is, we set the termination criteria as $|f_{opt} - f_{global}| \leq$ error.

Acceleration rate (AR): AR is employed to make a comparison between the convergence rate of algorithms A and B:

$$AR = 1 - \frac{NFE_B}{NFE_A} \%$$

Convergence graphs: The convergence graphs explain the mean fitness performance with respect to the NFE's of the total runs.

4.3 Parameter Settings

The experiments are executed on a computer with 2.00 GHz Intel (R) core, (TM) 2 duo CPU, and 2-GB of RAM. DEV C++ is used to execute the algorithms. The parameters are taken as follows;

- Population size (NP) is taken as 100.
- Dimension of each problems are taken as 30 and 50.
- The value of F and Cr are taken as 0.5 and 0.9, respectively.
- Value to Reach (VTR) and MaxNFE are fixed at 10^{-06} and 500,000, respectively.

5 Numerical Results

Numerical results and comparison for DE and IUMDE are presented in Table 2 in term of average NFE's of 50 runs.

Table 1 Benchmark functions with their global value

Function	Test function	Initial range	Min f_i
F1: Sphere functions	$\sum_{i=1}^{D} x_i^2$	$[-100, 100]^D$	0
F2: Schefel function	$\sum_{i=1}^{D} |x_i| + \prod_{i=1}^{D} |x_i|$	$[-10, 10]^D$	0
F3: Rosenbrok function	$\sum_{i=1}^{D-1} \left[100(x_{i+1} - x_i^2)^2 + (x_i - 1)^2 \right]$	$[-30, 30]^D$	0
F4: Ackley function	$-20 \exp\left(-0.2 \sqrt{1/D \sum_{i=0}^{D} x_i^2} \right) - \exp\left(\sqrt{1/D \sum_{i=0}^{D} \cos 2\pi x_i} \right) + 20 + e$	$[-32, 32]^D$	0
F5: Grienwenk function	$\frac{1}{4000} \sum_{i=1}^{D} x_i^2 - \prod_{i=1}^{D} \cos\left(\frac{x_i}{\sqrt{i}}\right) + 1$	$[-600, 600]^D$	0
F6: Penalty function	$\frac{\pi}{D} \left\{ 10\sin^2(\pi y_1) + \sum_{i=1}^{D-1} (y_i - 1)^2 [1 + 10\sin^2(\pi y_{i+1})] + (y_D - 1)^2 \right\}$ $+ \sum_{i=1}^{D} u(x_i, 10, 100, 4)$ where $y_i = 1 + \frac{1}{4}(x_i + 1)$ and $u(x_i, a, k, m) = \begin{cases} k(x_i - a)^m & \text{if } x_i > a \\ 0 & \text{if } -a \leq x_i \leq a \\ k(-x_i - a)^m & \text{if } x_i < -a \end{cases}$	$[-50, 50]^D$	0

Table 2 Numerical results and comparison of DE and IUMDE (VTR = 10^{-06})

Fun	$D = 30$			$D = 50$		
	DE	IUMDE	AR (%)	DE	IUMDE	AR (%)
F1	79,300	**19,700**	75.16	124,700	**32,800**	73.70
F2	124,900	**28,200**	77.42	184,200	**46,300**	74.86
F3	394,800	**124,630**	68.43	$2.3e+01^a$	$8.0e-02^a$	–
F4	115,900	**27,700**	76.10	175,900	**45,300**	74.25
F5	81,300	**19,400**	76.14	125,000	**32,700**	73.84
F6	69,500	**16,960**	75.60	107,600	**43,000**	60.04

[a]VTR is not reached successfully. Results in error obtained in MaxNFE.

Table 3 Comparison with other enhanced variants of DE algorithms ($D = 30$ and VTR = 10^{-09})

Fun	ODE [12]	DERL	LeDE [3]	JADE [14]	IUMDE
F1	67,524	54,880	49,494	29,000	**25,400**
F2	140,170	92,210	77,464	52,000	**39,600**
F3	NA	263,050	282,972	**150,000**	153,000
F4	106,694	103,360	76,111	45,000	**40,400**
F5	79,888	70,210	50,579	33,000	**26,200**
F6	63,710	64,110	41,384	27,000	**23,800**

It is clear from Table 2 that IUMDE obtains fewer NFE's than DE algorithm for all benchmark problems. We have also calculated the acceleration rate of IUMDE over DE which shows the convergence speed of IUMDE over DE. For example, from the table it can be seen that for function F1, IUMDE is 75.16% faster than DE algorithm. Similarly, we can see the same results for other functions. Hence, it can be concluded that the proposed IUMDE makes help to boost the convergence speed of DE.

In Table 3, the results of IUMDE are compared with some other popular enhanced variants LeDE [3], ODE [12], JaDE [14], and DERL [16] of DE. From the table, it is clear that for each function, IUMDE is obtaining the required accuracy in less NFE's. In case of function F3, JADE is taking less NFE than IUMDE, yet IUMDE is performing better than other remaining algorithms.

The convergence graphs of function F1 and F5 obtained by DE and IUMDE are shown in Fig. 1.

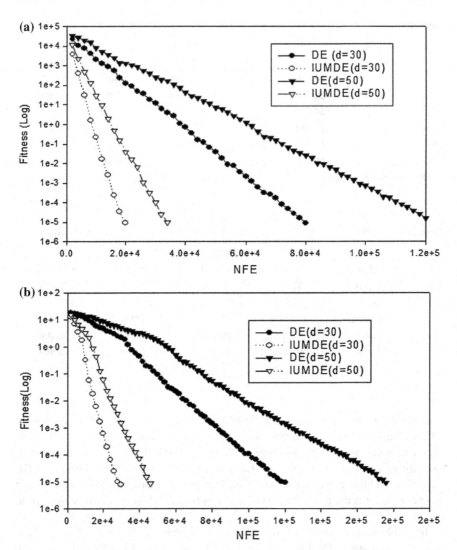

Fig. 1 Convergence graphs of function $F1$ and $F2$

6 Conclusions

In the present study, two modifications are suggested for DE algorithm. The modifications are made in mutation and selection operation, respectively. The proposed modification in mutation helps to choose a better base vector, and then modified selection utilized this mutant vector in the population of next generation. The corresponding variant is named as information utilization-based modified DE (IUMDE). From the numerical results, it is observed that the proposed modified variant IUMDE facilitate in improving the capability of DE algorithm besides maintaining the solution quality.

References

1. Storn, R., Price, K.: Differential evolution—a simple and efficient adaptive scheme for global optimization over continuous spaces. Technical Report, Berkeley, CA, TR-95-012 (1995)
2. Vesterstrom, J., Thomsen, R.: A comparative study of differential evolution, particle swarm optimization and evolutionary algorithms on numerical benchmark problems. In: Congress on Evolutionary Computation, pp. 980–987 (2004)
3. Kumar, P., Pant, M., Singh, V.P.: Modified random localization based DE for static economic power dispatch with generator constraints. Int. J. Bio-Inspir. Comput. 6(4), 250–261 (2014)
4. Kumar, P., Singh, D., Kumar, S.: MRLDE for solving engineering optimization problems. In: Proceedings of IEEE conference ICCCA-2015, pp. 760–764 (2015). doi:10.1109/CCAA. 2015.7148512
5. Kumar, S., Kumar, P., Sharma, T.K., Pant, M.: Bi-level thresholding using PSO, artificial bee colony and MRLDE embedded with Otsu method. Memet. Comput. 5(4), 323–334 (2013)
6. Cai, Y., Wang, J., Yin, J.: Learning enhanced differential evolution for numerical optimization. Soft. Comput. (2011). doi:10.1007/s00500-011-0744-x
7. Fan, H., Lampinen, J.: A trigonometric mutation operation to differentia evolution. J. Global Optim. 27, 105–129 (2003)
8. Noman, N., Iba, H.: Accelerating differential evolution using an adaptive local search. IEEE Trans. Evol. Comput. 12(1), 107–125 (2008)
9. Ali, M., Pant, M.: Improving the performance of differential evolution algorithm using cauchy mutation. Soft. Comput. (2010). doi:10.1007/s00500-010-0655-2
10. Pant, M., Ali, M., Abraham, A.: Mixed mutation strategy embedded differential evolution. In: IEEE Congress on Evolutionary Computation, pp. 1240–1246 (2009)
11. Brest, J., Greiner, S., Boskovic, B., Mernik, M., Zumer, V.: Self adapting control parameters in differential evolution: a comparative study on numerical benchmark problems. IEEE Trans. Evol. Comput. 10(6), 646–657 (2006)
12. Rahnamayan, S., Tizhoosh, H., Salama, M.: Opposition based differential evolution. IEEE Trans. Evol. Comput. 12(1), 64–79 (2008)
13. Qin, A.K., Huang, V.L., Suganthan, P.N.: Differential evolution algorithm with strategy adaptation for global numerical optimization. IEEE Trans. Evol. Comput. 13(2), 398–417 (2009)
14. Zhang, J., Sanderson, A.: JADE: adaptive differential evolution with optional external archive. IEEE Trans. Evol. Comput. 13(5), 945–958 (2009)
15. Babu, B.V., Angira, R.: Modified differential evolution (MDE) for optimization of non-linear chemical processes. Comput. Chem. Eng. 30, 989–1002 (2006)
16. Kaelo, P., Ali, M.M.: A numerical study of some modified differential evolution algorithms. Eur. J. Oper. Res. 169, 1176–1184 (2006)
17. Kumar, P., Pant, M.: Enhanced mutation strategy for differential evolution. In: Proceeding of IEEE Congress on Evolutionary Computation (CEC-12), pp. 1–6 (2012)
18. Kumar, P., Pant, M.: Modified single array selection operation for DE algorithm. In: Proceedings of Fifth International Conference on Soft Computing for Problem Solving. Advances in Intelligent Systems and Computing, vol. 437, pp. 795–803 (2016)
19. Pooja, Chaturvedi, P., Kumar, P.: A cultivated differential evolution algorithm using modified mutation and selection strategy. Innov. Syst. Des. Eng. 6(2), 67–74 (2015)

20. Neri, F., Tirronen, V.: Recent advances in differential evolution: a survey and experimental analysis. Artif. Intell. Rev. **33**(1–2), 61–106 (2010)
21. Das, S., Suganthan, P.N.: Differential evolution: a survey of the state-of-the-art. IEEE Trans. Evol. Comput. **15**(1), 4–13 (2011)

Health Recommender System and Its Applicability with MapReduce Framework

Ritika Bateja, Sanjay Kumar Dubey and Ashutosh Bhatt

Abstract There has been high influx of data over the Web in the past few decades. In order to convert this data into meaningful information, data analytics platforms such as big data analytics have emerged. Recommendation system (RS) is one such analytic system which uses predictive analysis to present and target the useful information to the users in order to help them arriving at desired conclusions. RS saves the time of most of the Internet users by reducing the number of actions needed to get to the required information which helps in effective decision making. RS has been widely used in multiple domains such as e-commerce, social networks to present recommended options to users in their current context. Recommended options thereby presented are based on the action taken by other users in similar context. RS is equally useful in healthcare services. In fact, they are a complementary tool to healthcare system which allows for effective decision making in healthcare services. RS uses either collaborative filtering approach, content filtering approach, or a blend of them (hybrid approach). This paper discusses the overview of what recommender systems are, how they are built, and its classifications. It also elaborates health recommender system (HRS) and gives a clear picture of how MapReduce Framework and Hadoop technology will help in improving the scalability and efficiency of HRS by stating illustrations.

Keywords Big data · Hadoop · MapReduce Framework · Recommender system
Social media

R. Bateja (✉) · S.K. Dubey
Department of Computer Science and Engineering, A.S.E.T,
Amity University, Sec.-125, Noida UP, India
e-mail: ritika.fet@mriu.edu.in

S.K. Dubey
e-mail: skdubey1@amity.edu

A. Bhatt
Department of Computer Science and Engineering, Birla Institute
of Applied Sciences, Bhimtal, Uttrakhand, India
e-mail: ashutoshbhatt123@gmail.com

© Springer Nature Singapore Pte Ltd. 2018
M. Pant et al. (eds.), *Soft Computing: Theories and Applications*,
Advances in Intelligent Systems and Computing 584,
https://doi.org/10.1007/978-981-10-5699-4_25

1 Introduction

Recommender systems are quite useful in data-intensive application as they analyze huge amount of data to provide better recommendations [1]. Application of RS is not only limited to the e-commerce/social networks, but they are equally important in healthcare services. In healthcare services, RS can be used by healthcare service providers to prescribe treatment to the patients based on diagnoses of the disease and patient's profile and as well as by the patients while doing self-study of the disease and choosing the efficient treatment. It is also used in the recommendation of physicians, hospitals and even in the selection of medicines.

Health care in the past has been heavily focused on disease-centric model where diagnosis is done using historical medical tests, trending diseases, healthcare provider experience, or medical laboratory results. However, the element of personalized treatment was missing in such treatments. Arisen of RS in health care is causing a shift from disease-centric care to patient-centric healthcare. In the patient-centric health care, profile of the patient—genetics, lifestyle, demographics, medical history, age, gender—is taken into account and patient themselves is actively participating in their own care, following instructions and prescriptions properly as advised by the physicians. Personalized suggestion on treatments, medicines, hospitals is helping patients in early detection, cure, and even prevention of disease, thus improving the quality of life.

RS is implemented using frameworks such as Hive, Mahout, and Hadoop, where Hive acts as a query processing and managing software [2]. Mahout provides us the important prediction models and Hadoop forms the base for these two frameworks as well as it uses MapReduce to write applications to process huge amounts of data, in parallel, on large clusters of commodity hardware in a reliable manner [3]. MapReduce Framework also forms the distributed solution for health care and reduces the information overload by using it with RS. The objective of this paper is to analyze various existing techniques by using RS in healthcare services and also to improve the efficiency and scalability of RS by implementing it using MapReduce Framework.

The remaining of this paper is organized as follows. Section 2 presents the overview of recommender systems. Section 3 discusses its classifications. Section 4 comprises of discussion on health recommender system (HRS). In Sect. 5, the detailed survey is presented. Section 6 comprises of research questions and analysis as well as provides justification of using MapReduce Framework and Hadoop technology with recommendation followed by conclusion in Sect. 7.

2 Recommender Systems

Recommender systems are defined as the software tools and techniques that explore and filter the relevant data in the context of the users for effective decision making. RS classifies the well-matched pairs by analyzing different data sources to develop concept of similarity between users and items [4]. It also helps in transforming information consumers into active contributors. There are four modules in recommender systems.

- User,
- Web server,
- Dynamic HTML generator,
- Recommender system.

The whole process of how RS works is explained in Fig. 1.
The architecture has the four functional modules explained as follows:

- User—The customer interacts with the Web pages over the Web site(s) which sends a request to the Web server.
- Web server—The Web server software further communicates with the recommender system to choose information/pages to be targeted to the user along with requested information/pages.
- Dynamic HTML generator—The HTML generator stores the requested pages in the repository.
- Recommender system—The recommender system uses its database to find the information associated with users' request and make recommendations.

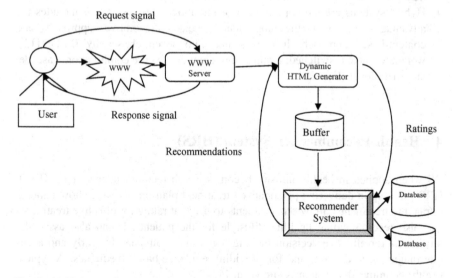

Fig. 1 Architecture of recommender system

The correlation database helps the recommender systems to generate recommendations and responses back to the user. The Web server software displays the recommended products to the user.

3 Classification of Recommender Systems

RS is classified into three major groups [5].

- Collaborative filtering systems—In this system, user gets recommendation for the particular item based on ratings given by other users for the same item. This is basically based on collecting users past behavior, reviews, or preferences [6]. Collaborative filtering (CF) methods are divided into two categories: neighborhood-based and model-based approaches. Neighborhood-based methods are also known as memory-based methods [7]. Memory-based methods are further subdivided into user-based and item-based recommendations, which are taken as input item, user, and rating to build recommendations of this form.
- Content-based systems—It recommends items to the user, based on the previous buying history of the user [8]. This methodology compares the representations of content that describes an item to representations of contents that interests the user. It is also known as an information retrieval (IR) task, where the query is referred to as the content related to the preferences of users, and the unrated documents are scored with relevance/similarity to this query [9]. It is mostly used where semantic meaning of content is important [10]. Some of the classification algorithms associated with such recommendations include k-nearest neighbor, decision trees, and neural networks [11].
- Hybrid systems are the combination of the above two systems. It includes the advantages of both the approaches, i.e., collaborative approach and content-based approach. It is classified into seven classes by Burke [12]: weighted, mixed, switched, feature augmentation, feature combination, cascade, and meta level (Fig. 2).

4 Health Recommender System (HRS)

RS when applied in health industry becomes health recommender system (HRS). HRS can help in delivering personalized treatment plan for patients, allow patients to rate the treatments, allow other patients to use that rating for effective treatments and also recommending healthier lifestyle for the patients. It can also assist the physicians for effective decision making regarding treatments, for early and accurate diagnosis of disease, and for providing evidence-based medicines. A typical health recommender system is shown in Fig. 3.

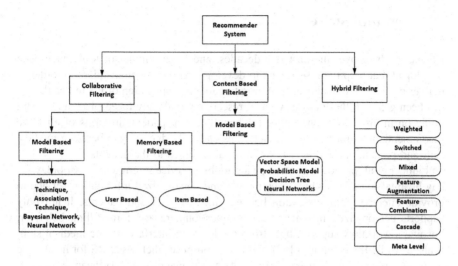

Fig. 2 Types of recommender system

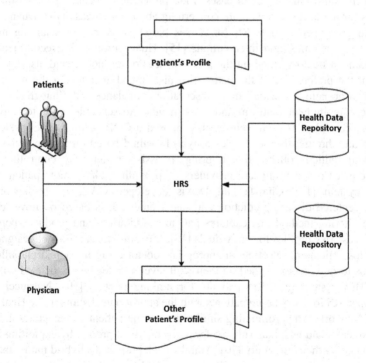

Fig. 3 Health recommender system (HRS)

5 Literature Survey

RS has evolved over the past few decades, and large number of tools have been developed that help the researchers to develop effective recommenders. In order to deliver personalized healthcare solution to patients, application of RS in health care has been growing these days. A lot of research had already taken place and is still going on in order to provide effective health care to patients in terms of effective recommendations for health and diet plans and to bridge the gap between medical staff and patients, providing 24 * 7 medical facilities to the patients.

Lamber et al. proposed context-aware mobile service referred to as Mobiday for day hospital workflow which uses mobile and personalized information service to provide effective communication between medical staff and patients. It not only supports personalized instant messages, questionnaire and form filling behavior, hospital's workflow support, but also provides Web interface for the medical staff. It also uses RFID techniques [13]. Kim et al. proposes diet based RS for healthcare by using information of users. It prevents and manages coronary heart disease. It provides the customized diet to customers by considering the basic information, vital sign, family history of diseases, food preferences according to seasons, and intakes for the customers who are concerning about the disease [14]. Duan, Street, and Xu developed clinical RS for nurses based on correlations among nursing diagnoses, outcomes, and interventions [15]. Huang proposed a recommendation algorithm for doctors based on the performances of the doctor and also takes into account the preferences of the patients. It also aimed at relieving the problem of doctor information overload and "reservation imbalance" of Shanghai Medical League Appointment Platform and helps patients to schedule a medical appointment successfully [16]. Hussein et al. proposed a CDD recommender system for diagnosing chronic disease. It uses a hybrid method based on multiple classifications and unified collaborative filtering. It assists patients by controlling their chronic disease and healthcare providers by providing 24/7 remote patient monitoring system [17]. Chawla and Dawis developed CARE, a framework for patient-centered disease prediction and management. It is based on novel collaborative filtering method that captures patient's similarities and produces personalized disease risk profiles for individuals [18]. Mohammadi and Babaei designed RS that helps patients in selecting an appropriate doctor using fuzzy text classification methods to rank doctors based on their skill level and degrees associated with that skill. This system relies on patients for ranking doctors [19]. Narducci et al. developed RS for associating patients with the appropriate doctors in the Health Net Social Network. After computing similarities among patients, it generates a ranked list of doctors and hospitals suitable for a given patient profile, by exploiting health data shared by the community [20]. Ambika et al. proposed a hybrid model that will provide personalized recommendations by analyzing the patient health record (PHR) of the patients for the prediction and diagnosis of non-communicable disease [21]. The summary of literature survey is given in Table 1.

Table 1 The summary of literature survey

Year	Ref.	Tool/technique/model/dataset	Issues/challenges	Analysis
2009	[13]	Mobile user interface, RFID (radio frequency identification) techniques and push information delivery techniques	Problem with user interface. Disturbances and interruptions making interaction complex while accessing the system on move. Providing effective communication between patients and clinician	This paper aims to increase the effectiveness of communication between patients and the medical staff by using mobile user interface, personalized information service, and RFID techniques to provide valuable information to the patients about disease
2009	[14]	BMR (base metabolic rate) based on individual's height, weight, and other existing data (age, sex, etc.) and vital signed standard agent to transform the vital signs into XML format	Providing customized service to user by considering their basic information such as vital signs, family history, family preferences	This paper aims to provide a customized diet recommender system to prevent and manage coronary heart diseases in healthcare services
2011	[15]	Prefix tree data structure, nursing diagnosis datasets extracted from community hospital, and evaluation measures like random selection, greedy selection	Creating effective comprehensive care plans for the patients by the nurses. Testing effectiveness of different recommendation strategies by using a new evaluation method based on average ranking positions	In this paper, RS for nursing care plan was developed, which uses correlations among nursing diagnosis, outcomes, and interventions
2012	[16]	Large amount of multidimensional medical data	Information overloading and reservation imbalance problem. Providing accurate medical recommendation while handling complexity of medical data and to	In this paper, a recommendation algorithm for doctors is proposed based on their performance and patient's preferences, thereby helping patients to schedule

(continued)

Table 1 (continued)

Year	Ref.	Tool/technique/model/dataset	Issues/challenges	Analysis
			improve the performance of Shanghai Medical League Appointment Platform	a medical appointment successfully
2012	[17]	Training data consist of diabetic patient's records, demographic data of active patient consisting of name, age, and medical datasets consisting of two types of data, home tests that include blood pressure, blood sugar level, weight, and the other including tests from laboratory	Improving the accuracy of RS while handling nature and complexity of medical and 2D data. Providing more accurate and efficient recommendations to patients in controlling their chronic disease and 24-h medical care and also assisting healthcare providers to have 24/7 remote patient monitoring system	In this paper, a chronic disease diagnosis RS was developed based on hybrid method using multiple classifications based on decision tree algorithms and unified collaborative filtering
2013	[18]	Medicare database of 13 million patients with 32 million visits was analyzed using CARE. Each record also includes diagnosis codes, i.e., ICD-9-CM (International classification of diseases, ninth revision, clinical modification)	The disease does not have a rating system or a preferential scoring system. Challenge is to develop a patient-centered model for personalized health care using big data analytics by producing personalized disease risk profiles for each individual	In this paper, CARE (Collaborative Assessment and Recommendation Engine) was introduced for patient-centered disease prediction and management
2014	[19]	Fuzzy linguistic approach and fuzzy text classification	Accuracy in patient's profile	In this paper, RS was developed which helps the patients in the selection of an appropriate doctor according to disease and medical

(continued)

Table 1 (continued)

Year	Ref.	Tool/technique/model/dataset	Issues/challenges	Analysis
				requirements using fuzzy linguistic approach
2015	[20]	Patient's personal health data including conditions, treatments like drugs, dosages, and health indicators	Selection of a doctor can be viewed as a problem of information asymmetry. Searching a doctor or hospital with best expertise for solving health problems of the patients	In this paper, RS is embedded in Health Net Social Network, which generates a ranked list of doctors as well as hospitals suitable for a given patient profile by exploiting health data shared by the community, after computing similarities among patients
2016	[21]	Electronic patient health records (PHR) are generated using patient's support, which includes demographic, psychological, cognitive, social, behavioral, and functional measurements	Challenge in developing a supervised learning model, which can deal with multiclass problems and also in generating user profiles	In this paper, PHRs are analyzed to provide personalized recommendation using hybrid filtering approach and it also introduces an approach to predict and diagnose non-communicable disease (NCD)

6 Research Questions and Analysis

6.1 Research Questions

The goal of our review is to find the issues and challenges involved in HRS while using it with MapReduce Framework and Hadoop. During the study, following research questions evolved:

- RQ1: Does Effective recommendations are possible using MapReduce Framework?
- RQ2: How often Hadoop technology can be used in it?

These research questions will help in identifying gaps in the related area.

6.2   Analysis

Justification for using MapReduce Framework and Hadoop with HRS is explained below.

HRS and MapReduce Framework

Traditional HRS suffers from scalability and efficiency problems as they are not able to handle large volume of health data with speed and efficiency. To overcome this problem, MapReduce Framework is used, which distributes the load of large health-related datasets over multiple nodes—running on commodity hardware—in order to support the parallel processing. Other features supported by using MapReduce paradigm in HRS include its simplicity of usage, fault tolerance, and flexibility [22].

In MapReduce, the data are broken down to smaller datasets, which is processed separately and the results of these smaller datasets are sum-reduced and mapped back to provide final outcome of processed data. It scales well too across many thousands of nodes and can handle petabytes of data. MapReduce Framework helps in the implementation of HRS, thereby improving the scalability, flexibility by managing, and analyzing large volume of patients' records. It helps in calculating vector of different items and uses their correlation while presenting the recommendations.

In a patient-centric healthcare model, vectors of different patient profiles based on their medical history can be created and correlated to present representation based on symptoms and condition of disease and to learn from successful treatment received by other patients.

HRS and Hadoop

Apache Hadoop is an open-source implementation of MapReduce Framework. The two core components of Hadoop are as follows:

- HDFS (Hadoop distributed file system) for data storage
- MapReduce Framework for providing parallel and distributed computing environment

HDFS supports implementation of MapReduce algorithm, which comprises of Map () and Reduce () functions for processing data by splitting the large volume of data using map function and then combining these data after processing takes place using reduce function. Moreover, Hadoop also supports various features such as robustness, scalability, simplicity, accessibility, and cost-effectiveness, which are not present in other big data analytics tools [23].

Figure 4 above depicts the graphical view of the data analyzed from the year 1997–2016. It clearly indicates that the research on using RS for health care is growing at a faster rate.

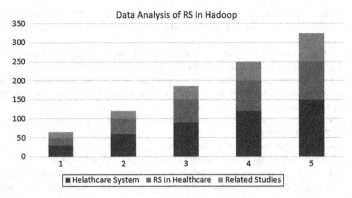

Fig. 4 Graphical representation of analyzed data from year (1997–2016)

7 Conclusion

As research in the HRS is growing at a pace and a huge volume of health-related data is increasing on the Web, RS is looking at possible real-time analytics of these data to deliver the personalized healthcare solution based on patient's profile. Attention should be given to patient-centric approach for this purpose. The main contribution of our work is to present a detailed overview of working of RS, provide a detailed description of HRS, and also show the usage of MapReduce methodology for aggregating the necessary information on health care from the large amount of ubiquitous data available online. It also elaborates the usage of Hadoop, an open-source-based platform, which consists of HDFS as file system (for storing unstructured data), MapReduce as a framework for parallel processing and can use Hive, Pig or Mahout for efficient retrieval of processed data. The future work will be based on developing efficient algorithm for using recommender system in developing a patient-centric healthcare model to deliver personalized solution for patient's healthcare needs.

References

1. Saravanan, S.: Design of large-scale content-based recommender system using Hadoop MapReduce framework. In: IC3, IEEE Computer Society, pp. 302–307 (2015). doi:10.1109/IC3.2015.7346697
2. Rakesh, et al.: Comparison of SQL with Hive QL. Int. J. Res. Technol. Stud. **1**(9), 2348–1439 (2014)
3. Chandan, V., et al.: Increasing efficiency of recommendation system using big data analytics. Int. J. Recent Innov. Trends Comput. Commun. **3**(10), 5742–5744 (2014)
4. Prem, M., Vikas, S.: Recommender system. In: Encyclopedia of Machine Learning, pp. 829–838. Springer, US (2011)

5. Adomavicius, G., Tuzhilin, A.: Towards the next generation of recommender systems: a survey of the state-of-the-art and possible extensions. IEEE Trans. Data Knowl. Eng. **17**(6), 734–749 (2005)

6. Ghodake, S.B., Paswan, R.S.: Survey on recommender system using distributed frame work. Int. J. Sci. Res. (IJSR) **5**(1) (2016). ISSN 2319–7064

7. Breese, J.S., Heckerman, D., Kadie, C.: Empirical analysis of predictive algorithms for collaborative filtering. In: Proceedings of the Fourteenth Conference on Uncertainty in Artificial Intelligence, Madison, WI (1998)

8. Pereira, N., et al.: Survey on content based recommendation system. Int. J. Comput. Sci. Inf. Technol. **7**(1), 281–284 (2016)

9. Balabanovic, M., Shoham, Y.: Content-based, collaborative recommendation. Commun. Assoc. Comput. Mach. **40**(3), 66–72(1997). doi:10.1145/245108.245124

10. Zhang, Q., et al.: A framework of hybrid recommender system for personalized clinical prescription. In: International Conference on Intelligent Systems and Knowledge Engineering, pp. 189–195. IEEE (2015)

11. Pazzani, M.J., Billsus, D.: Learning and revising user profiles: the identification of interesting web sites. Mach. Learn. **27**(3), 313–331 (1997). doi:10.1023/A:1007369909943

12. Burke, R.: Hybrid recommender systems: survey and experiments. User Model. User-Adap. Inter. **12**(4), 331–370 (2002). doi:10.1023/A:1021240730564

13. Lamber, P., Girardello, A., et al.: MobiDay: a personalized context-aware mobile service for day hospital workflow support. In: Proceedings of the AIME09 International Workshop on Personalization for e-Health, pp. 15–19 (2009)

14. Kim, J.H., Lee, J.H., et al.: Design of diet recommendation system for healthcare service based on user information. In: Fourth International Conference on Computer Sciences and Convergence Information Technology, pp. 516–518. ICCIT (2009)

15. Duan, L., Street, W.N., Xu, E.: Healthcare information systems: data mining methods in the creation of a clinical recommender system. EIS **5**(2), 169–181 (2011)

16. Huang, Y.F.: A doctor recommendation algorithm based on doctor performances and patient preferences. In: International Conference on Wavelet Active Media Technology and Information Processing (ICWAMTIP), pp. 92–95 (2012)

17. Hussein, A.S., Omar, W.M., Li, X., Ati, M.: Efficient chronic disease diagnosis prediction and recommendation system. In: Biomedical Engineering and Sciences (IECBES), IEEE EMBS Conference, pp. 209–214 (2012). doi:10.1109/IECBES.2012.6498117

18. Chawla, N.V., Dawis, D.: Bringing big data to personalized healthcare: a patient centered framework. J. Gen. Intern. Med. **28**, 660–665 (2013)

19. Mohammadi, N., Babaei, M.H.: Recommending an appropriate doctor to a patient based on fuzzy logic. Int. J. Curr. Life Sci. **4**(2), 403–407 (2014)

20. Narducci, F., Musto, C., et al.: A recommender system for connecting patients to the right doctors in the HealthNet Social Network. WWW (Companion Volume), pp. 81–82. ACM (2015)

21. Ambika, M., et al.: Hybrid model for behaviour-based recommender system: a healthcare perspective. Int. J. Adv. Eng. Technol. **7**(3), 88–91 (2016)

22. Dean, J., Ghemawat, S.: MapReduce: simplified data processing on large clusters. Commun. ACM **51**(1), 107–113 (2008)

23. Vinodhini, S., Rajalakshmi, V., Govindarajalu, B.: Building personalised recommendation system with big data and hadoop MapReduce. Int. J. Eng. Res. Technol. (IJERT) **3**(4) (2014). ISSN 2278–0181

Trigonometric Probability Tuning in Asynchronous Differential Evolution

Vaishali, Tarun Kumar Sharma, Ajith Abraham
and Jitendra Rajpurohit

Abstract Asynchronous differential evolution (ADE) has been derived from differential evolution (DE) with some variations. In ADE, the population is updated as soon as a vector with better fitness is found hence the algorithm works asynchronously. ADE leads to stronger exploration and supports parallel optimization. In this paper, ADE is incorporated with the trigonometric mutation operator to enhance the convergence rate, and the performance of the algorithm is tested for various values of trigonometric mutation probability; that is, the tuning of trigonometric mutation probability has been done to obtain its optimum setting. The proposed work is termed as ADE–trigonometric probability tuning (ADE-TPT). For tuning, the tests have been done over widely used benchmark functions referred from the literature, and the results obtained using different probabilities are compared.

Keywords Asynchronous differential evolution · Trigonometric mutation
ADE-TMO · ADE-TPT · Convergence

1 Introduction

Storn and Price [1] proposed differential evolution (DE) algorithm in 1995. It emerged as an attractive and powerful tool for optimization under the umbrella term evolutionary algorithms. It is a simple and efficient global optimizer especially for continuous optimization. This standard differential evolution employs a synchronous generation-based evolution strategy [2]. The efficacy of DE has been proven in various fields such as mechanical engineering [3], pattern recognition [4] and communication [5].

Vaishali (✉) · T.K. Sharma · J. Rajpurohit
Amity University Rajasthan, Jaipur 303007, India
e-mail: vaishaliyadav26@gmail.com

A. Abraham
Machine Intelligence Research Lab, Auburn, WA, USA

© Springer Nature Singapore Pte Ltd. 2018
M. Pant et al. (eds.), *Soft Computing: Theories and Applications*,
Advances in Intelligent Systems and Computing 584,
https://doi.org/10.1007/978-981-10-5699-4_26

In this paper, an efficient metaheuristic algorithm ADE is discussed which is a variation of DE and works asynchronously. ADE performs efficiently for possible nonlinear and non-differentiable global optimization problems [6]. The basic idea of ADE is based on DE, and the difference is there is no concept of generation increments in ADE; the evolution strategy in ADE is asynchronous in nature and based on individual target vector. Once a vector with better fitness than the target vector is found, it is immediately replaced with the target vector, and it participates in the evolution. Hence, considering the fitness value of the trial and target vector, the population is updated without any delay, whereas in DE, the vector with the better fitness has to wait until next generation to participate in evolution. In this paper, ADE is hybridized with trigonometric mutation operator (ADE-TMO) instead of the basic mutation operator to fasten the convergence speed of the algorithm, and the tuning of trigonometric probability (P_{TMO}) is done for the ADE. The performance of ADE is tested for various values of P_{TMO}, and a rigorous analysis is carried out on the basis of the results.

The paper is designed as follows: in Sect. 2, the working of ADE algorithm and a brief survey of ADE are discussed. Trigonometric mutation operation and ADE-TMO and the proposed work ADE-TPT, i.e. tuning of P_{TMO}, is discussed in Sect. 3. Section 4 details the experimental setup, performance metrics and results. Results analyses are summarized in Sect. 5. Conclusions and future scopes are discussed in Sect. 6.

2 Asynchronous Differential Evolution

ADE outperforms DE and many of the DE variants in most of the cases. Over the past years, some modifications have also been proposed to ADE. The results have shown that these modified algorithms perform better than ADE itself and various other algorithms. Milani and Santucci [7] used a parameter called synchronization degree (SD) to regulate the synchrony of the current population evolution, i.e. how fast a better candidate vector becomes the population member. In the proposed strategy, SD synchronizes the number of donor vectors produced by mutation and given in input to the crossover-selection phase. In [6], Zhabitskaya and Zhabitsky used a modified way of choosing the target vector for enhancing the convergence rate of the algorithm. In the proposed work, the target vector chosen was either random or worst. Ntipteni et al. [8] used asynchronous approach with master–slave architecture for parallel optimization. In [9], Zhabitskaya discussed the constraints on control parameters for one strategy of DE and four strategies of ADE. In [10], Zhabitskaya and Zhabitsky proposed ADE with Restart (ADE-R). This approach automatically restarts calculations with an increased population size while stagnation or population degeneration encounters. In [11], a crossover operator based on adaptive correlation matrix was used instead of Cr adaptation. In [12], Zhabitskaya et al. showed that ADE outperforms Simplex and Migrad method for solving the optimization problem. In [13], ADE was incorporated with convex mutation

(ADE-CM) in which the parent's information is used and hence it accelerates the convergence rate of the algorithm.

Basic ADE

Population Initialization
Population of NP candidate solutions is initialized:

$$X_i = \{x_{1,i}, x_{2,i}, x_{3,i}, \ldots, x_{D,i}\}$$

In the population, the ith individual is represented by index i and $i = 1, 2,\ldots$, NP; D is dimension of the problem.

In the search space, the individuals have constraints over the lower (l) and upper (u) limits: $l = \{l_1, l_2, \ldots, l_D\}$ and $u = \{u_1, u_2,\ldots, u_D\}$.

jth component of the ith vector is initialized as follows:

$$x_{j,i} = l_j + \text{rand}_{i,j}[0, 1] \times (u_j - l_j)$$

where $\text{rand}_{i,j}[0,1]$ is a random number and its value lies between 0 and 1.

Mutation Operation

Once the population is initialized, a target vector (X_i) is chosen randomly, and then it is mutated to form a mutant vector (V_i). This operation of mutation is performed using three randomly chosen mutually exclusive numbers r_1, r_2, r_3 and a scaling factor F.

$$V_i = X_{r1} + F \times (X_{r2} - X_{r3})$$

r_1, r_2 and r_3 are also different from the base index i. There are many other mutation schemes also other than the one specified above. The value of F lies between 0 and 1.

Crossover Operation

In crossover operation, the mutant vector (V_i) exchanges its components with the target vector (X_i) and creates a trial vector (U_i). This operation is performed to increase the diversity of the population.

$$u_{j,i} = \begin{cases} v_{j,i} & \text{if } \text{rand}_{i,j}\,[0, 1] \leq \text{Cr} \vee j = j_{\text{rand}} \\ x_{j,i} & \text{otherwise} \end{cases}$$

where $j = 1, \ldots, D$ $j_{\text{rand}} \in \{1,\ldots,D\}$ and chosen once for each i.

Cr is crossover probability set by the user generally ranges between 0 and 1.

If $\text{rand}_{i,j}[0,1] \leq \text{Cr}$, then the parameter in the trial vector comes from the mutant vector, otherwise from the target vector.

Selection Operation

This operation will decide whether the target vector or the trial vector will be a part of the population. Out of two, the vector having lower objective function value will survive. If target vector is having better fitness, the population will not be updated and if trial vector is having better fitness, then the population is updated and the target vector is replaced by trial vector without any delay. The selection operation is performed as follows:

$$X_i = \begin{cases} U_i & \text{if } f(U_i) \leq f(X_i) \\ X_i & \text{Otherwise} \end{cases}$$

3   Trigonometric Mutation Operation

Trigonometric mutation was proposed by Fan and Lampinen [14] with DE to enhance its performance. The inclusion of TMO enables the algorithm to converge faster without affecting its robustness. So the algorithm optimizes the problem with less number of function evaluations. But to avoid the premature convergence, trigonometric mutation probability is used.

In the basic mutation operation, three random individuals r_1, r_2 and r_3 are taken which are different from the target vector to be mutated. These three individuals form a hypergeometric triangle in the search space in which these individuals are the vertices. By incorporating the value of objective function in the mutation, the process can be biased towards the vertices having lowest function value.

In trigonometric mutation, the vector to be mutated is taken to be the centre of the hypergeometric triangle, and the sum of three weighted difference vectors is done for mutation. Hence, all the three individuals along with the target vector have an equivalent role in the mutation operation. The weight terms $(p_2 - p_1)$, $(p_3 - p_2)$ and $(p_1 - p_3)$ are used to enable the mutation to produce a better vector.

The trigonometric mutation operation is performed as follows:

$$\begin{aligned} V_{i,G+1} = & \frac{(X_{r1,G} + X_{r2,G} + X_{r3,G})}{3} \\ & + (p_2 - p_1)(X_{r1,G} - X_{r2,G}) \\ & + (p_3 - p_2)(X_{r2,G} - X_{r3,G}) + (p_1 - p_3)(X_{r3,G} - X_{r1,G}) \end{aligned}$$

where

$$p_1 = |f(X_{r1,G})|/p'$$
$$p_2 = |f(X_{r2,G})|/p'$$
$$p_3 = |f(X_{r3,G})|/p'$$
$$p' = |f(X_{r1,G})| + |f(X_{r2,G})| + |f(X_{r3,G})|$$

ADE-TMO Algorithm:

// Population Initialization

$$X_i = \{x_{1,i}, x_{2,i}, x_{3,i}, ..., x_{D,i}\}$$

do {

i=select_target_vector();

// Mutation Operation

if rand[0,1]< P_{TMO} then

$$V_i = \frac{(X_{r1} + X_{r2} + X_{r3})}{3} + (p_2 - p_1)(X_{r1} - X_{r2}) + (p_3 - p_2)(X_{r2} - X_{r3})$$
$$+ (p_1 - p_3)(X_{r3} - X_{r1})$$

else

$$V_i = X_{r1} + F \times (X_{r2} - X_{r3})$$

// Crossover Operation

for (j=0; j<D; j=j+1)

$$u_{j,i} = \begin{cases} v_{j,i} & if\ rand_{i,j}[0,1] \le Cr \vee j = j_{rand} \\ x_{j,i} & otherwise \end{cases}$$

// Selection Operation (for next iteration)

$$X_i = \begin{cases} U_i & if\ f(U_i) \le f(X_i) \\ X_i & Otherwise \end{cases}$$

} while the termination criteria is met.

Trigonometric probability tuning (ADE-TPT) has been done using the above algorithm by testing ADE with different values for P_{TMO}.

4 Experimental Setup and Results

The proposed work is tested over five benchmark functions having different characteristics. The functions considered for minimization are taken from the literature [15]. Each function is described in Table 1 with search space's lower and

Table 1 Benchmark
functions

Benchmark function	Search space	VTR
Sphere	−100 to 100	10^{-6}
Hyper ellipsoid	−50 to 50	10^{-6}
Ackley	−30 to 30	10^{-6}
Griewank	−30 to 30	10^{-6}
Rastrigin	−5.12 to 5.12	10^{-3}

upper bounds. The function is tested in MATLAB environment. Every function is optimized over 25 runs for different values of dimensions and P_{TMO} given below, and rest of the parameters are kept same.

4.1 Experimental Settings

Population size (NP): 100;
Dimension (D): 10, 20;
Scaling factor (F): 0.5;
Trigonometric mutation probability (P_{TMO}):
$P = 0.05$; $P_1 = 0.1$; $P_2 = 0.2$; $P_3 = 0.3$; $P_4 = 0.4$; $P_5 = 0.5$; $P_6 = 0.6$;
Crossover rate (Cr): 0.9;
Value to reach (VTR): 10^{-3} (for Rastrigin function) and 10^{-6} (for others);
Maximum number of function evaluations: 10^6.

4.2 Performance Evaluation Metrics

For tuning of P_{TMO}, ADE is tested for seven values of P_{TMO}. The metrics considered from [15] for the performance evaluation of the algorithm are convergence graph, average number of function evaluations (NFEs), standard deviation (SD) and success rate (SR), where SR is determined by division of number of successful runs and the total number of runs.

Also a statistical analysis [16, 17] is done for testing the algorithms' efficiency for each value of P_{TMO}. Bonferroni–Dunn [18] test is also done to calculate the significant difference. Critical difference (CD) for Bonferroni–Dunn's graph is calculated as:

$$CD = Q_\alpha \sqrt{\frac{k(k+1)}{6N}}$$

Q_α is critical value for a multiple nonparametric comparison with control [19], and k and N represent the total number of algorithms and number of problems considered for comparison, respectively. To show two levels of significance at $\alpha = 0.05$ and 0.10, horizontal lines are drawn.

4.3 Results

In Tables 2 and 3, the average NFE and standard deviation (SD) are summarized when the function reaches the VTR with different values of P_{TMO}. The value of D is 10 and 20 for Tables 2 and 3, respectively.

The convergence graphs in Figs. 1, 2, 3, 4 and 5 elaborate the performance of ADE-TMO for benchmark functions with $D = 10$ and different values of P_{TMO} (from P_1 to P_6).

Table 4 (for $D = 10$) summarizes the ranking of different values of P_{TMO} for the ADE-TMO algorithm and the CD for Bonferroni–Dunn's test. Table 5 shows the results of statistical analysis (for $D = 10$) done using Friedman's test.

Figure 6 represents the Bonferroni–Dunn's graph which notifies the significant difference between probabilities.

Table 2 Results for benchmark functions for $P - P_6$ at $D = 10$

P_{TMO}	$D = 10$				
	Sphere Avg. NFE (SD)	Hyper-ellipsoid Avg. NFE (SD)	Ackley Avg. NFE (SD)	Griewank Avg. NFE (SD)	Rastrigin Avg. NFE (SD)
P	2.11e+04 (7.00e−08)	2.23e+04 (5.33e−08)	2.95e+04 (1.33e−07)	1.50e+05 (6.09e−08)	3.25e+05 (5.86e−05)
P_1	1.83e+04 (3.49e−08)	1.97e+04 (3.57e−08)	2.52e+04 (1.36e−07)	1.27e+05 (3.23e−08)	2.84e+05 (5.39e−05)
P_2	1.42e+04 (3.76e−08)	1.52e+04 (3.49e−08)	1.84e+04 (9.49e−08)	4.13e+04 (4.30e−03)	1.69e+05 (3.62e−05)
P_3	1.16e+04 (4.56e−08)	1.26e+04 (3.17e−08)	1.42e+04 (8.89e−08)	1.62e+04 (2.30e−03)	6.89e+04 (3.14e−01)
P_4	9.81e+03 (2.10e−08)	1.05e+04 (2.96e−08)	1.15e+04 (5.17e−08)	9.19e+03 (2.81e−08)	2.75e+04 (3.14e−01)
P_5	8.38e+03 (2.15e−08)	9.12e+03 (2.24e−08)	9.99e+03 (1.36e−07)	7.74e+03 (2.30e−03)	1.38e+04 (6.71e−01)
P_6	7.50e+03 (1.41e−08)	8.13e+03 (9.31e−09)	1.81e+04 (3.89e−02)	6.58e+03 (5.00e−03)	1.03e+04 (1.02e+00)

Table 3 Results for benchmark functions for $P - P_6$ at $D = 20$

P_{TMO}	$D = 20$				
	Sphere Avg. NFE (SD)	Hyper-ellipsoid Avg. NFE (SD)	Ackley Avg. NFE (SD)	Griewank Avg. NFE (SD)	Rastrigin Avg. NFE (SD)
P	4.47e+04 (8.06e−08)	4.93e+04 (6.54e−08)	6.21e+04 (2.03e−07)	4.47e+04 (7.15e−08)	7.60e+05 (4.60e−03)
P_1	3.62e+04 (6.83e−08)	4.06e+04 (8.21e−08)	4.96e+04 (1.63e−07)	3.05e+04 (5.26e−08)	7.81e+05 (1.30e+00)
P_2	2.47e+04 (6.48e−08)	2.75e+04 (8.82e−08)	3.24e+04 (9.93e−08)	2.08e+04 (8.21e−08)	6.39e+05 (1.17e+01)
P_3	1.77e+04 (5.52e−08)	1.98e+04 (4.52e−08)	2.25e+04 (1.35e−07)	1.42e+04 (5.96e−08)	1.71e+05 (6.95e−01)
P_4	1.36e+04 (4.89e−08)	1.55e+04 (5.96e−08)	1.86e+04 (4.36e−07)	1.08e+04 (6.02e−08)	5.43e+04 (1.42e+00)
P_5	1.26e+04 (1.24e−05)	1.86e+04 (7.18e−02)	4.30e+04 (7.83e−01)	2.30e+04 (6.35e−03)	2.07e+04 (2.08e+00)
P_6	3.42e+04 (3.02e+00)	3.73e+04 (1.38e+01)	3.25e+04 (5.13e−01)	3.00e+04 (1.06e−01)	2.16e+04 (1.93e+00)

Fig. 1 Sphere function at $D = 10$

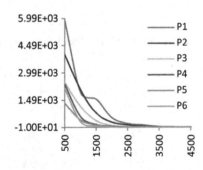

Fig. 2 Hyper-ellipsoid function at $D = 10$

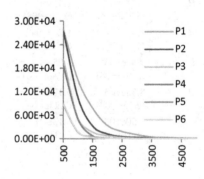

Fig. 3 Ackley function at
$D = 10$

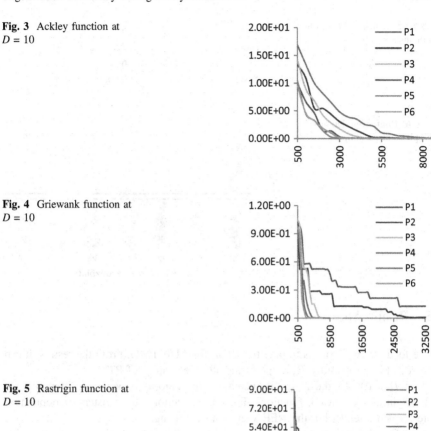

Fig. 4 Griewank function at
$D = 10$

Fig. 5 Rastrigin function at
$D = 10$

Table 4 Ranking based on
Friedman's test

P_{TMO}	Mean rank
P	7.00
P_1	6.00
P_2	5.00
P_3	3.80
P_4	2.80
P_5	1.80
P_6	1.60
CD ($\alpha = 0.05$)	3.604
CD ($\alpha = 0.1$)	3.271

Table 5 Statistics based on Friedman's test

N	5
Chi-square	27.943
df	6
Asymp. sig.	0.000

Fig. 6 Bonferroni–Dunn's graph corresponding to average NFE

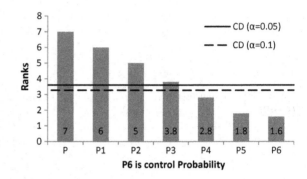

5   Result Analyses

The focus of this work is to tune the P_{TMO} for ADE-TMO. From the results, it can be seen that probability P_6 outperforms all other values of PTMO.

At $D = 10$: As the P_{TMO} is increased, the algorithm behaves differently for Rastrigin and Griewank functions. For these functions, the clusters of population are formed after P_4 but the NFE keeps on decreasing.

At $D = 20$: For every function except Ackley, average NFE is increasing after P_5. Rastrigin and Griewank still form clusters after P_4. SD also varies across all P_{TMO}, and also SD is greater than zero for Rastrigin function.

The convergence graphs also validates that P_6 is best of all values of P_{TMO}. The SR of each function is reported 100%. The Friedman's ranking test also verifies that P_6 probability gives best results for the algorithm.

The Bonferroni–Dunn's graph shows the subsequent significant difference with

- P_6 as control probability.

P_6 gives better results than P, P_1, P_2 and P_3 and at par with P_4 and P_5 at both levels of significance.

6   Conclusion and Future Work

The proposal ADE-TPT seeks for tuning the trigonometric probability for ADE-TMO algorithm in which the mutation operator is modified by embedding trigonometric mutation wherein the convergence rate is improved by using all randomly chosen vectors equivalently along with the target vector for mutation operation. The algorithm is tested over five benchmark functions for different values of trigonometric probabilities. The most efficient probability is validated by comparative result analysis of each probability value considered in the work.

In future, the work will be extended for higher dimensions and will be tested on real-life problems.

References

1. Storn, R., Price, K.: Differential evolution—a simple and efficient adaptive scheme for global optimization over continuous spaces, Berkeley, CA, Technical Report, TR-95–012 (1995)
2. Storn, R., Price, K.: Differential evolution—a simple and efficient heuristic for global optimization over continuous spaces. J. Global Optim. **11**, 341–359 (1997)
3. Rogalsky, T., Derksen, R.W., Kocabiyik, S.: Differential evolution in aerodynamic optimization. In: Proceedings of 46th Annual Conference of Canadian Aeronautics and Space Institute, Montreal, QC, Canada, pp. 29–36, May 1999
4. Ilonen, J., Kamarainen, J.-K., Lampinen, J.: Differential evolution training algorithm for feed-forward neural networks. Neural Process. Lett. **7**(1), 93–105 (2003)
5. Storn, R.: On the usage of differential evolution for function optimization. In: Biennial Conference of the North American Fuzzy Information Processing Society, Berkeley, CA, pp. 519–523 (1996)
6. Zhabitskaya, E., Zhabitsky, M.: Asynchronous differential evolution. In: Adam, G., Busa, J., Hnatic, M. (eds.) MMCP 2011. LNCS, vol. 7125, pp. 328–333. Springer, Heidelberg (2012)
7. Milani, A., Santucci, V.: Asynchronous differential evolution. In: Proceedings of IEEE Congress on Evolutionary Computation, pp. 1210–1216 (2010)
8. Ntipteni, M.S., Valakos, I.M., Nikolos, I.K.: An asynchronous parallel differential evolution algorithms. In: International Conference on Design Optimization and Application, Spain (2006)
9. Zhabitskaya, E.I.: Constraints on control parameters of asynchronous differential evolution. In: Mathematical Modeling and Computational Science, pp. 322–327. Springer, Berlin (2012)
10. Zhabitskaya, E., Zhabitsky, M.: Asynchronous differential evolution with restart. In: International Conference on Numerical Analysis and Its Applications, pp. 555–561. Springer, Berlin (2012)
11. Zhabitskaya, E., Zhabitsky, M.: Asynchronous differential evolution with adaptive correlation matrix. In: GECCO'13, Amsterdam, The Netherlands, July 2013
12. Zhabitskaya, E.I., Zemlyanaya, E.V., Kiselev, M.A.: Numerical analysis of SAXS-data from vesicular systems by asynchronous differential evolution method. Matem. Model. **27**(7), 58–64 (Mi mm3623) (2015)
13. Vaishali, Sharma, T.K.: Asynchronous differential evolution with convex mutation. In: Proceedings of Fifth International Conference on Soft Computing for Problem Solving. Springer, Singapore (2016)
14. Fan, H., Lampinen, J.: A Trigonometric mutation operation to differential evolution. J. Global Optim. **27**, 105–129 (2003)
15. Suganthan, P.N., et al.: Problem definitions and evaluation criteria for the CEC05 special session on real-parameter optimization. Technical report, Nanyang Technological University, Singapore (2005), http://www.ntu.edu.sg/home/epnsugan/index files/CEC-05/CEC05.htm
16. Demšar, J.: Statistical comparisons of classifiers over multiple data sets. J. Mach. Learn. Res. **7**, 1–30 (2006)

17. García, S., Herrera, F.: An extension on statistical comparisons of classifiers over multiple data sets for all pairwise comparisons. J. Mach. Learn. Res. **9**, 2677–2694 (2008)
18. Dunn, O.J.: Multiple comparisons among means. J. Am. Stat. Assoc. **56**(293), 52–64 (1961)
19. Zar, J.H.: Biostatistical Analysis. Prentice-Hall, Englewood Cliffs (1999)

Relevance Index for Inferred Knowledge in Higher Education Domain Using Data Mining

Preeti Gupta, Deepti Mehrotra and Tarun Kumar Sharma

Abstract Optimizing the real-life scenarios facilitate knowledge building. Developing a knowledge model for optimizing certain output criteria enhances the benefits by many folds. Even a non-profit sector like education needs to define knowledge models that optimize their functioning and eventually help in knowledge building. Quantifying the factors determining the academic well-being of the students in any educational organization is of prime importance. The paper exemplifies the implementation of Data Mining Technique to deduce knowledge through classification rules and further assign relevance index to inferred knowledge.

Keywords Higher education · Knowledge · Data mining · Classification

1 Introduction

It is often said that we are drowning in data but starving for knowledge [1]. Extraction of information from data facilitates knowledge building. Information which can be termed as a subset of data stimulates action in an entity, whereas knowledge defines the action of an entity in a particular setting [2]. A number of researchers have classified knowledge on different basis, sometimes defining the manner of codification and occurrence [3], or on the basis of know-what, know-how, know-why and know-when aspect of knowledge [4]. Some have even mapped knowledge in diverse domains [5].

P. Gupta (✉) · T.K. Sharma
Amity University Rajasthan, Jaipur, India
e-mail: preeti_i@rediffmail.com

T.K. Sharma
e-mail: taruniitr1@gmail.com

D. Mehrotra
ASET, Amity University Uttar Pradesh, Noida, India
e-mail: mehdeepti@gmail.com

© Springer Nature Singapore Pte Ltd. 2018
M. Pant et al. (eds.), *Soft Computing: Theories and Applications*,
Advances in Intelligent Systems and Computing 584,
https://doi.org/10.1007/978-981-10-5699-4_27

There are varieties of ways for representing knowledge [6]. Using production rules written in form of IF-THEN rules is one of the most popular approach used for knowledge representation [7]. The IF-THEN rules adopt a modular approach, each defining principally independent and a relatively minor piece of knowledge. A rule-based system will include universal rules and actualities about the knowledge domain covered.

Knowledge building in education domain can be achieved by adopting procedures that optimize their functioning.

The research work is undertaken with an objective of deducing relevance index for inferred knowledge. The case of education sector is taken in particular while inferring knowledge related to student's academic performance in a technical subject at level of higher education. It is important to deduce relevance index to inferred knowledge as it is a clear depiction of the existing system and further helps in decision making.

The paper is organized as follows. Section 2 elaborates the methodology adopted for rule induction and further rule evaluation in the higher education set-up. Finally, the conclusions are drawn and presented in Sect. 3.

2 Adopted Methodology in Higher Education Scenario

Educational organizations strive to achieve the higher academic output for the students. Many researchers have strived hard to predict the factors affecting the academic results of students [8–12]. Identification of such critical parameters, which could improve the academic attainment of students, supports an effective academic planning.

In case the individuals of a population can be separated into different classes, generation of a classification rule is a system in which the individuals of the population are each allocated to one or the other class.

In the study, knowledge is represented through classification rules [13], which exist in the form of IF-THEN rules. The work starts by identifying the variables and collecting the data in the context of these variables. The values of the attributes are then encoded on an 8-level scale. Rule induction is initiated through JRip, which implements a propositional rule learner, repeated incremental pruning to produce error reduction (RIPPER). The rules are then evaluated on the basis of the metrics *Net Benefit* which takes into account both classification and misclassification witnessed by the knowledge rule.

2.1 Variable Identification and Data Collection

This dataset has 5000 records and five independent attributes, all of which are categorical. The independent attribute names in the dataset are as follows:

*ContinuousEvaluationMarks, SGPA_II, Practical_orient, Attendance, Base_Sub_
Marks.*

The independent attributes affect the dependent attribute of *End_Term_Marks*
and are reflected in Table 1.

The attributes were encoded on the 8-level scale, depicted in Table 2.

Table 1 Attributes of the study

Attribute name	Description
ContinuousEvaluationMarks	Performance of the students continuously evaluated by the faculty member with respect to class assignments, marks obtained in class test, performance in viva voce, etc. (maximum marks—30)
SGPA_II	Semester grade point average (SGPA) measures the academic performance of the student in the previous semester on a 10-point scale
Attendance	It reflects the presence of the student in the class of the subject under study
Practical_orient	It reflects the ability of the students to solve the problems related to the subject under the study in a practical manner
Base_Sub_Marks	Performance of the student in physics (base subject) studied in the earlier semester
End_Term_Marks	Performance of the student in the end-term exam of the subject under scrutiny

Table 2 Encoding of the attributes

ContinuousEvaluationMarks (Maximum value 30)		SGPA_II (Maximum value 10)	
Marks range	Encoding	SGPA range	Encoding
0–3	000	0–2	000
4–7	001	2.1–4	001
8–11	010	4.1–5	010
12–15	011	5.1–6	011
16–19	100	6.1–7	100
20–23	101	7.1–8	101
24–27	110	8.1–9	110
28–30	111	9.1–10	111
Base_Sub_Marks (maximum value 100)		**Attendance (maximum value 100%)**	
Marks range	Encoding	Attendance range	Encoding
0–20	000	Below 75%	000
21–40	001	75.1–77%	001
41–50	010	77.1–80%	010

(continued)

Table 2 (continued)

ContinuousEvaluationMarks (Maximum value 30)		SGPA_II (Maximum value 10)	
Marks range	Encoding	SGPA range	Encoding
51–60	011	80.1–83%	011
61–70	100	83.1–85%	100
71–80	101	85.1–90%	101
81–90	110	90.1–95%	110
91–100	111	95.1–100%	111
Practical_orient (maximum value 100)		**End_Term_Marks (maximum value 100)**	
Marks range	Encoding	Marks range	Encoding
0–20	000	0–20	000
21–40	001	21–40	001
41–50	010	41–50	010
51–60	011	51–60	011
61–70	100	61–70	100
71–80	101	71–80	101
81–90	110	81–90	110
91–100	111	91–100	111

2.2   Rule Induction

In the year 1995, Cohen proposed JRip which implemented a propositional rule learner, repeated incremental pruning to produce error reduction (RIPPER) [14].

Error reduction can be witnessed in JRip since the process of incremental pruning examination of the classes is done in the increasing order of their size. The initial ruleset is generated on the basis of incremental reduced error. Initially, JRip (RIPPER) treats all the instances from the training dataset related to a particular judgment as a class and deduces a ruleset that covers all the members of that class. The procedure is repeated for all the classes.

Initialization
Initialize RS = {}, and from each class from the less frequent one to the most frequent one.
Repeat
{

1. *Building phase: Repeat the phases given below, grow phase and prune phase until there are no positive instances or error rate increases more than 50%.*

   1.1 *Grow phase: Follow the greedy approach of adding conditions to the rule until the accuracy of the rule reaches 100%.*

1.2 *Prune phase: Incremental pruning approach should be followed for each rule. The pruning metrics can be measured in terms of 2p/(p + n) − 1, where p—number of positive instances covered in the ruleset and n—number of negative instances covered in the ruleset.*

2 *Optimization Phase: On generation of the initial ruleset {R$_i$}, two variants of each rule are to be generated and pruned from randomized data using procedures Grow and Prune. The generation of the first variant is done from an empty rule, and the next variant is created by adopting a greedy approach of adding conditions to the original rule. The metrics of Description Length (DL) are computed for each variant. The final representation of the ruleset is done by the rule having the minimal DL. After the examination of all the rules in R$_i$, Building phase is again used for generating more rules if there are still residual positives.*

3 *Those rules that increase the DL of the complete ruleset are then deleted from the ruleset, and the final ruleset is added to RS.*

}

In the study, JRip was implemented using Weka 3.8.0 and the following ruleset of 87 rules was generated. A snapshot of the rules and the output achieved is shown in Fig. 1.

2.3 Rule Analysis and Interpretation

For each of the 87 rules acquired by implementing JRip on the dataset, the value of classification (true positive, TP) and misclassification (false positive, FP) was recorded [15].

True positive (TP)— the number of examples satisfying A and C
False positive (FP)—the number of examples satisfying A, but not C

where *A—antecedent of the rule, C—consequent of the rule*

The rules were further evaluated on the basis of *Net Benefit* [16] considering a range of thresholds and calculating the NB across these thresholds. The result was then plotted against Rule Number and Net Benefit. For each threshold P_t, the Net Benefit was calculated as per Eq. 1:

$$\text{Net Benefit (NB)} = \frac{\text{TP}}{\text{N}} - \frac{\text{FP}}{\text{N}}\left(\frac{P_t}{1 - P_t}\right) \tag{1}$$

On evaluating the rules for Net Benefit for different values of P_t, the following observations were met and are depicted through Fig. 2.

On cross-tabulating the rule count for $P_t = 0.1$–0.6, the NB values for all the 87 rules can be witnessed in Table 3.

```
=== Run information ===

Scheme:     weka.classifiers.rules.JRip -F 3 -N 2.0 -O 2 -S 1
Relation:   jrip 1
Instances:  5000
Attributes: 6
            ContinuousEvaluationMarks
            SGPAII
            Practical_orient
            Attendance
            Base_Sub_Marks
            End_Term_Marks
Test mode:  evaluate on training data

=== Classifier model (full training set) ===

JRIP rules:
===========

(Base_Sub_Marks = 110) and (Attendance = 110) and (SGPAII = 010) =>
End_Term_Marks=111 (5.0/1.0)
(SGPAII = 111) and (Base_Sub_Marks = 000) and (Attendance = 000) =>
End_Term_Marks=111 (4.0/0.0)
(Base_Sub_Marks = 110) and (ContinuousEvaluationMarks = 101) and (Prac-
tical_orient = 010) and (SGPAII = 110) => End_Term_Marks=111 (5.0/0.0)
(Base_Sub_Marks = 111) and (SGPAII = 101) and
(ContinuousEvaluationMarks = 011) => End_Term_Marks=111 (5.0/0.0)
(Attendance = 111) and (ContinuousEvaluationMarks = 000) and (Practi-
cal_orient = 011) => End_Term_Marks=110 (7.0/2.0)
.
.
.
.

Time taken to build model: 4.47 seconds
```

Fig. 1 Weka implementation

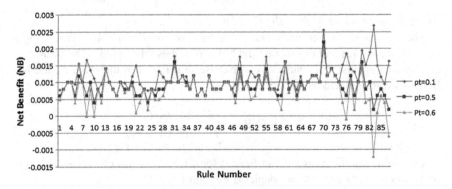

Fig. 2 Consolidated plot across P_t values (0.1, 0.5, 0.6)

Table 3 Analysing NB for the rules

Criteria	$P_t = 0.1$	$P_t = 0.2$	$P_t = 0.3$	$P_t = 0.4$	$P_t = 0.5$	$P_t = 0.6$
Number of rules with NB < 0	0	0	0	0	0	3
Number of rules with NB $\geq$ 0 and NB < 0.0005	0	0	0	0	4	14
Number of rules with NB $\geq$ 0.0005 and NB < 0.001	32	32	33	41	44	36
Number of rules with NB $\geq$ 0.001 and NB < 0.0015	41	44	46	40	35	31
Number of rules with NB $\geq$ 0.0015 and NB < 0.002	12	9	7	5	3	2
Number of rules with NB $\geq$ 0.002 and NB < 0.0025	0	1	1	1	1	1
Number of rules with NB $\geq$ 0.0025 and NB < 0.003	2	1	0	0	0	0
Total rules	87	87	87	87	87	87

The consolidated plot depicting the NB values for all 87 rules across thresholds (P_t = 0.1, 0.5, 0.6), shown in Fig. 2, depict that the Net Benefit of the rule having maximum Net Benefit across all the threshold values of P_t (P_t = 0.1–0.6) decreases as we increase the threshold value (P_t) from 0.1 to 0.6. In fact at P_t = 0.6, some of the rules exhibit the negative NB.

P_t = 0.5 signifies that FP and TP are weighted equally. Hence, maintaining a P_t = 0.1 signifies assigning more weightage to the classification, i.e. true positive (TP), rather than to misclassification, i.e. false positive (FP).

The study selects P_t = 0.1. Maximum NB and distinct peaks are achieved on selecting a P_t = 0.1. It is also observed that NB value decreases as we move from P_t = 0.1 to P_t = 0.6. Moreover, the NB value also shows a negative growth in case of P_t = 0.6. P_t = 0.6 signifies the assignment of more weightage to misclassification rather than to classification.

However, for P_t = 0.1, the rule that acquires the highest benefit is:

Base_Sub_Marks = 010 and Attendance = 001 and ContinuousEvaluation-Marks = 101 => End_Term_Marks = 011

On decoding the rule, it can be stated as:

Base_Sub_Marks is between 41 and 50 and Attendance between 75.1 and 77% and ContinuousEvaluationMarks between 20 and 23 => End_Term_Marks between 51 and 60.

The relevance index assigned to the knowledge rule is on the basis of its Net Benefit (NB), keeping into account the classification and misclassification done by the rule. The Net Benefit (NB) for the above said rule at a threshold value P_t of 0.1 is 0.002689.

The reason for using Net Benefit (NB) to assign relevance index to inferred knowledge is:

1. The prediction model incorporates consequences and hence can be used to infer a decision on the usage of the given model.
2. It can be directly applied to the validation set and does not need any additional information.
3. Even if the model outcome is in binary or continuous form, the method for evaluation is applicable.

3 Conclusion

Rule induction can deduce the relationship existing between the various attributes. The influence of the independent variables on the dependent variable can be observed. Rules with a higher relevance index are much more apt to the system and can be used for appropriate syllabus planning, designing structured lesson plans, structuring criteria for the evaluation of the student's performance and adoption of suitable teaching pedagogy for the improvement in the overall academic performance of the students. The knowledge derived in the form of rules bears relevance in the context of the domain and hence can be added to the knowledge set that can supplement the process of decision making in a knowledge base environment.

References

1. Han, J., Kamber, M.: Data mining: concepts and techniques. Morgan Kaufmann Publishers, Canada (2000)
2. Boisot, M.H.: Knowledge Assets Securing Competitive Advantage in the Information Economy. OUP Oxford New Edition (2006)
3. Polanyi, M.: The Tacit Dimension. Routledge and Kegan Paul, London (1966)
4. Nickols, F.W.: The knowledge in knowledge management. In: Cortada, J.W., Woods, J.A. (eds.) The Knowledge Management Yearbook 2000–2001, pp. 12–21. Butterworth-Heinemann, Boston, MA (2000)
5. Gupta, P., Mehrotra, D., Singh, R.: Achieving excellence through knowledge mapping in higher education institution. Int. J. Comput. Appl. 5–10 (2012)
6. Jong, T.D., Ferguson Hesler, M.G.M.: Types and quality of knowledge. Educ. Psychol. **31**,105–113(1996)
7. Rich, E., Knight, K., Nair, S.B.: Artificial Intelligence. TMH, New Delhi (2010)
8. Kabakchieva, D.: Student performance prediction by using data mining classification algorithms. Cybern. Inf. Technol. **13**, 61–72 (2013)

9. Kumar, S.A., Vijayalakshmi, M.N.: Efficiency of decision trees in predicting student's academic performance. Int. J. Comput. Sci. Inf. Technol. **23**, 335–343 (2011)
10. Sembiring, S., Zarlis, M., Hartama, D., Ramliana, S., Wani, E.: Prediction of student academic performance by an application of data mining techniques. In: International Conference on Management & Artificial Intelligence, vol. 6, pp. 110–114 (2011)
11. Ramanathan, L., Dhanda, S., Kumar, S.D.: Predicting students' performance using modified ID3 algorithm. Int. J. Eng. Technol. (IJET) **5**(3), 2491–2497 (2013)
12. Gupta, P., Mehrotra, D., Sharma. T.K.: Genetic based weighted aggregation model for optimization of student's performance in higher education. In: Advances in Intelligent Systems and Computing, pp. 877–887. Springer, Singapore (2015)
13. Gupta, P., Mehrotra, D.: Effective curriculum development through rule induction in knowledge centric higher education organization. In: Confluence 2013: The Next Generation Information Technology Summit (4th International Conference), Noida. IET Digital Library, vol. 2013, issue 647 CP, pp. 475–480 (2013)
14. Cohen, W.W.: Fast effective rule induction. In: Twelfth International Conference on Machine Learning, pp. 115–123 (1995)
15. Freitas, A.A.: A survey of evolutionary algorithms for data mining and knowledge discovery. In: Advances in Evolutionary Computing, pp. 819–845, Springer, Heidelberg (2003)
16. Vickers, A.J., Elkin, E.B.: Decision curve analysis: a novel method for evaluating prediction models. Med. Decis. Making **26**, 565–574 (2006)

Analytical Study on Cardiovascular Health Issues Prediction Using Decision Model-Based Predictive Analytic Techniques

Anurag Bhatt, Sanjay Kumar Dubey and Ashutosh Kumar Bhatt

Abstract The Healthcare industry has grown tremendously in recent years providing best possible medical facility in such an effective manner, i.e., in terms of time as well as cost. Healthcare industry needs to define the standard and need to bring the analytical approach to next level. In this Paper, we have gone through various researchers ideas that represent their approach to effectively provide the solutions regarding prediction of various cardiovascular health issues at multiple levels. Predictive analytics differ from the descriptive and prescriptive analytics by utilizing patient's medical records and analyze them with various statistical techniques as well as advanced machine learning algorithms. This paper will present recent research using various predictive analytical tools and techniques in order to analyze the cardiovascular health issues and in predicting the future outcome of the analysis with much efficient and effective way. There is a need to analyze and examine possible future work to have a better understanding of applying more hybrid and effective algorithms.

Keywords Data mining · Naïve Bayes algorithm · Data preprocessing
Cardiovascular disease

A. Bhatt · S.K. Dubey (✉)
Amity University, Noida, Uttar Pradesh, India
e-mail: skdubey1@amity.edu

A. Bhatt
e-mail: anurag15bhatt@gmail.com

A.K. Bhatt
Birla Institute of Applied Sciences, Bhimtal, Nainital, Uttarakhand, India
e-mail: ashutoshbhatt123@gmail.com

© Springer Nature Singapore Pte Ltd. 2018
M. Pant et al. (eds.), *Soft Computing: Theories and Applications*,
Advances in Intelligent Systems and Computing 584,
https://doi.org/10.1007/978-981-10-5699-4_28

1 Introduction

Data extraction and pattern recognition is one of the most relevant fields in knowledge extraction process. In Today's world, loads of data are available from numbers of organization showing the record of each activity or event occurred. This numerous amount of data can be processed and used to extract the actual meaning or knowledge by reducing data sizes and by gracing us with immense knowledge. Medical informatics and clinical research is one of the unplowed fields that can yield valuable information to replenish medical services with knowledge for betterment of medical services. Every year according to WHO (world health organization) annual report, more than 12 million deaths occur worldwide and in USA the death rate due to Cardiovascular diseases are increasing day by day [1]. Previously, data mining was used as to discover patterns through unsupervised learning, i.e., without making assumptions about the structure of the data. Heart diseases including sudden cardiac death (SCD) are one of the fatal diseases in India. Applying intelligent data mining algorithms in cardiovascular disease (CVD) datasets can help to develop a system that can predict the future outcomes of cardiovascular disease and can take precise investigation report values and can predict the degree of heart disease (Fig. 1).

Forecasting of results on the basis of predictive analytics is the part of data mining concerned that gives us insight of various futuristic probabilities and analyzed trends.

There are lots of techniques that are used to determine the accuracy of data mining algorithms in medical datasets, but while working with cardiovascular diseases like heart disease, sudden cardiac arrest we need to work with variable like heart rate variability (HRV) [2] based on Echocardiogram dataset [3]. In this paper, we have introduced the differences between predictive data modeling, descriptive, and prescriptive data modeling. The purpose of descriptive analytics is to summarize what happened from the data given, while predictive analytics focuses on providing the forecasting that "what might happen in the future" rather than what

Fig. 1 Relationship between data mining, predictive analytics, and predictive modeling

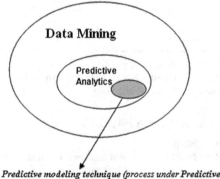

Predictive modeling technique (process under Predictive Analytics)

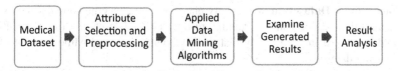

Fig. 2 Steps of medical dataset analysis

will happen in the future because all predictive analytics is based on the probability distribution of events. Prescriptive analytics is the mode of action derived by the results of predictive analytics. Prescriptive analytics basically relies on the "what course of action is required to do the task." The healthcare industry generally relies on doctor's expertise [2, 4] and experience for diagnosis of diseases, while decision support system is required in order to diagnosis of heart disease through prediction. The idea of developing a framework is very simple in order to understand, it proposes the use of sample training set in which we'll run various algorithms of data mining and according to the results shown by it, accuracy of algorithm and predicted value per class will be determined so that we can analyze our results and in the same way, required values will be provided to the decision support system and generated results will determine the possible health vulnerabilities according to provided data (Fig. 2).

2 Literature Review and Problem Discussion

Cardiovascular disease detection is need of the hour as this problem is mushrooming day by day. Data mining application for the detection of these diseases has proved that data mining algorithms can be considered as the most effective approach for these disease detections. Researchers [5] have proposed that using neural network approach for early detection of cardiovascular diseases produces effective results. In [3], we have seen that researcher has shown results of Electro cardiograph (ECG) signal derivative, i.e., HRV and magnetic resonance imaging (MRI) images classification through classification algorithm like support vector machine (SVM) [6, 7] that provides an eye-opening fact about the data and researcher favors statistical analysis over classifiers. Researchers have shown results of ECG signal derivative [8, 9], i.e., HRV that provides an eye-opening fact about the data and researcher favors statistical analysis over classifiers.

Cardiovascular disease detection is divided into various parts through which the results will be generated and these generated results will provide information about data. The whole process of designing optimized framework is divided into several parts. These parts are represented in the form of activity plan which will be followed accordingly.

- **Selection of Dataset**: In this activity phase, we are concerned about selection of dataset that will be using in running results. These datasets must be normalized and organized in such a way that no redundancy is found while running tools into the dataset.
- **Attribute Selection**: Attribute Selection method involves a unique method of selecting attributes that cause variations in the result of data and this variation provides a better approach to deal with results.
- **Data mining algorithm**: In this activity phase, we are concerned about which data mining algorithms we need to apply for better results from data. We need to select the algorithm with high accuracy and precision. Genetic algorithm and associative classification [10] are used by researchers for better results.

Paper [11] has proposed Intelligent Heart Disease Prediction System using data mining techniques using decision tree, Naïve Bayes, and neural network. This paper has demonstrated a prediction system using these classification algorithms (Table 1).

Table 1 Other studies on predicting cardiovascular health issues

Ref.	Database	Features used	Classification method
[4]	Cleveland database	Training the neural network using feed forward neural network	Neural network technique
[11]	Cleveland heart disease database	Data mining extension (DMX) query language	Decision tree, ID3 (iterative dichotomized 3)
[19]	South African medical practitioners	Classification using J48, bayes net	Simple cart and REPTREE (regression tree representative) algorithm
[23]	UCI machine learning repository	Weka tool is used	Naïve Bayes and decision tree
[24]	One minute of ECG Signal	HRV (heart rate variability)	Wigner ville transform
[25]	UCI breast cancer database	Classification, clustering, rule mining	Association rule mining
[26]	Blood-glucose homo monitoring data	Association, classification, subgroup discovery	Best predictor and rules to predict glycemic control
[21]	Physio bank database	Attribute selection and em clustering	Rule base system by clustering techniques

3 Cardiovascular Health Issues Prediction Using Decision Model and Predictive Analytic Approach

Many hospital systems today are designed with hi-tech medical facilities and clinical support stuffed with huge amount of data and these data are required to be analyzed for patterns. In most cases, medical reports of various patients are stored in the database of hospitals that define the various attributes of diagnosis report. Suppose, we have a diagnosis report showing "gender," diagnosis derivatives like "cholesterol," HRV.

Clinical decisions are often based on the doctor's experience and their intuitions that are developed through many years of practice [12, 13]. We want to propose such a framework that helps in defining the accurate prediction based on the clinical data provided to the system [1, 14]. The main concept behind making the framework is to apply data mining algorithms [4, 11], i.e., genetic algorithms, classification algorithms, and techniques like classification based on predictive association rules (CPAR), classification based on multiple association rules (CAMR), C4.5, mesocyclone detection algorithm (MDA), decision tree algorithms, classification and regression tree (CART), but it must have the ability to successful selection of appropriate data mining algorithms. Naïve Bayes algorithm and associative classification can be considered to use datasets to produce effective results [15].

The research will aim at finding solution and proposing a framework that will show the optimal results based on given datasets. In this research, we are focusing on creating a design and framework for cardiovascular disease prediction system [16] that will help heart disease patients or other cardio patients to identify their early symptoms and degree of complexity of the disease [17]. An optimal framework will have certain degree of correctness of data that will define its capability to optimize results from given datasets.

Through this paper, we have used data mining algorithms into consideration for giving better performance and various data analysis tools will be used. Data analysis and pattern finding process will be based on the use of various data mining algorithms as well as predictive analytical techniques, i.e., neural networks, belief networks, cluster analysis, etc. In healthcare delivery environment, predictive analytics can bring surprising results by forecasting various medical outcomes, i.e., cost-effective treatments, doctor's availability, and persistent care facilities (Fig. 3; Table 2).

Fig. 3 Steps involved in attribute selection to compare results

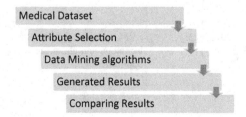

Medical Dataset

Attribute Selection

Data Mining algorithms

Generated Results

Comparing Results

Table 2 Data mining technologies for cardiovascular health issues analysis and prediction

Year	Reference	TOOL/technique/model/dataset	Objective	Analysis
2011	[4]	Open source Cleveland dataset	Result oriented paper using neural network approach to analyze heart disease dataset	Experiment showed that neural network performed better when used with multilayer
2008	[11]	Open source Cleveland dataset	To develop an Intelligent Heart Disease Prediction System using data mining algorithms	Naïve Bayes performed better than decision tree by identifying all important medical predictors
2014	[19]	Data collected from South Africa medical practitioners, Naïve Bayes, REPTREE, cart	Predictive analytics is used to analyze the data using various techniques	REPTREE, simple cart and J48 describes the importance of various predictors for accuracy
2014	[23]	UCI (UC irvine) machine learning repository	Result oriented paper using UCI datasets and providing analytical results	Naïve Bayes provided best result as compared to decision tree, while running in weka
2011	[24]	MIT-BIH database and ECG signal using HRV as feature	Sudden cardiac death prediction using HRV features	SCD prediction is done using HRV as a feature of ECG signal, with KNN (k-nearest neighbor) classifier as a technique
2008	[26]	Classification techniques like CAMR, CPAR, C4.5, MDA	To use cardiovascular disease features to develop a model for better prediction	CPAR, SVM outperformed of all other techniques and HRV helped in cardiac disease detection
2013	[25]	UCI (University of California, Irwin) machine learning repository, Pima Indian breast cancer data warehouse	To analyze the application of data mining in medical image mining using classifiers and rule mining	PCAR (packet delivery conditions aware routing) algorithm is efficient to predict the cancer levels in breast cancer
2011	[21]	SVM, nearest neighbor technique and clustering approach	To predict cardiovascular events using ECG time series data	Anomaly detection framework is used to identify patients

Fig. 4 Steps involved in data analytics

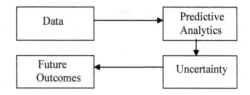

4   Heart Disease Prediction Using Predictive Analytics

Cardiovascular diseases are the leading cause of death in both young and old age group people, leaving behind the traces of events that caused this disease. In this modern society, changing lifestyle, improper nutrition, and mental stress are considered as one of the most effective reasons behind having these cardiovascular diseases, even in young people. We are basically suggesting heart disease prediction system using predictive analytics on the basis of decision based model. In order to predict, we need sufficient amount of knowledge and test cases required to calculate the probability of occurring of these events, which are based on data mining algorithms. Knowledge extraction is one of the clear approaches that allow us to identify those hidden patterns required for finding patterns in the cardiovascular disease data. There are multiple approach to use decision-based model like researcher used genetic algorithm and associative classification in [10], Naïve Bayes in [18]. In order to get the desired result, we also need to consider the performance of the algorithms, on the basis of which efficient algorithm will be associated with particular dataset. Researcher Abhisek Taneja [1] has proposed J48 pruned algorithm and Naïve Bayes algorithm for the datasets taken from UCI Machine Learning Repository. In this paper, researcher has proposed the methodology in which data analysis has been done in two ways: using selected attributes and using all attributes. First, attribute selection method is taken as in data preprocessing phase and attributes are then run through J48 and Naïve Bayes Algorithm with taking both as selected attributes and all attributes (Fig. 4).

5   Heart Disease Prediction Using Attribute Selection

5.1   Using Selected Attributes

Attribute selection is one of the data preprocessing techniques used to generate results. Data preprocessing techniques work in defining the attributes required in the most effective way. Attribute selection process comes under preprocessing where some relevant attributes are considered as efficient as compared to others on the basis of the algorithm applied for preprocessing (Fig. 5).

Fig. 5 Categories of datasets

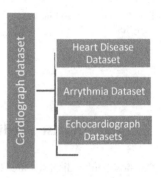

5.2   Using All Attributes

The experimental analysis of the paper can be concluded with both the techniques:
with selected attributes and all attributes. Results are generated totally different from
each other when compared while executing with other algorithms. Attributes
selection plays an important role in generating results, as every attribute plays an
important role in result gathering and analysis. Results in [1] are found different
when attributes run through data preprocessing as the overall performance of the
classifiers are different. Attribute selection algorithms select attributes according to
the effectiveness of the overall attributes in the given subset. Results were found
numerically different when run through selected attributes as compared to all
attributes (Fig. 6).

Naïve Bayes algorithm and J48 algorithm in [1] are compared with the help of
bar representation. In this way, we can easily compute the algorithmic performance
when provided with selected attributes and with all attributes, respectively. Naive
Bayes algorithm outperformed J48 in both the cases, i.e., with selected attributes
and all attributes.

Fig. 6 Result analysis using
all attributes and selected
attributes

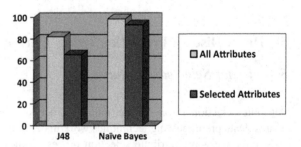

6   Decision Model Based Predictive Analytics

Cardiovascular health issues are constantly affected by the multivariate predictors used in order to predict the cardiovascular disease in general public which is based on their provided data. Predictive analytics in this paper is discussed on the basis of considering multivariate prediction, i.e., multivariate predictive models, tools, and scores are used in order to increase the precision rate of predicting cardiovascular health issue. In the way to multivariate predictive analytics, various factors play an important role, i.e., multivariate predicts help in better analysis of the predictive outcome as it helps in testing the predictive outcome generated by model through local environment as well as external environment. Updating procedures should be used in order to improve the performance of the models generated by considering various predictive factors and the updates in the predictive model must be cross validated with the external environment variables as well as development samples by monitoring the performance of the model generated.

7   Conclusion

In this Paper, this survey has discussed various recent studies that relate the various techniques to identify the hidden pattern form scratch through comparing various predictive analytical methods. Research on using these tools based on predictive analytics and generated decision model is critical and will help for forecasting future outcomes and results for the patients about their cardiovascular health issues, because it requires a great deal of testing and conformation before generating real-world predictions. This whole new dimension will identify the root cause of the cardiovascular disease and will help us to analyze new datasets using results produced by training datasets [19, 20]. Predictive analytical technique will help to generate decision model approach used in order to get the deep insight into cardiovascular health issues and helps in notifying the expected outcome based on the probabilistic aspect of that data arrived from model. Researchers [21, 22, 17] have introduced an effective technique for faster ECG data analysis when provided in compressed form. Predictive analysis plays a vital role in anamoly detection and ensures the effective probability distribution, so that it does not affect the future outcome of the results.

But, on the other hand getting optimal results based on the figures or data provided by us will be highly sensitive to standard environment variables. Changes in these values will cause effects in the integrity and accuracy of the result because we cannot predict the exact result based on the training data [10, 15]. This paper has witnessed various scenarios show promises and provide inspiration about future work, and show the importance of using all accessible levels of data in cardiovascular issues prediction, describing best possible probability of developing a framework that will help to analyze the patient's data and will help develop us with the simple GUI that can determine the prediction using forecasting based on patient's data. This framework will be decision model-based deploying facility that can generate great probabilistic results.

In a nutshell, we can conclude that there is a lot more work is required in the field of machine learning and data mining having specialized branch of predictive analytics which will enable to develop a system that will work in the cellular level to diagnose human body and have a better understanding of applying more hybrid and effective algorithms.

References

1. Taneja, A.: Heart disease prediction system using data mining techniques. Orient. J. Comput. Sci. Technol. **6**(4), 457–466 (2013)
2. Soni, J., Ansari, U., Sharma, D., Soni, S.: Predictive data mining for medical diagnosis: an overview of heart disease prediction. Int. J. Comput. Appl. (0975–8887), **17**(8), 43–48 (2011)
3. Murukesan, L., Murugappan, M., Iqbal, M.: Sudden cardiac death prediction using ECG signal derivative (heart rate variability): a review. In: 2013 IEEE 9th International Colloqium on Signal Processing and Its Application, 8–10 Mac. Kuala Lumpur, Malaysia, pp. 269–274 (2013)
4. Rani, K.U.: Analysis of heart disease dataset using neural network approach. Int. J. Data Min. Knowl. Manage. Process (IJDKP) **1**(5), 1–8 (2011)
5. Chitra, R., Seenivasagan, V.: Review of heart disease prediction system using data mining and hybrid intelligent techniques, ICTACT J. Soft Comput. **03**(04), 605–609 (2013). ISSN 2229-6956
6. Vatankhah, M., Attarzadeh, I.: Proposing an efficient method to classify mri images based on data mining techniques. Int. J. Comput. Sci. Netw. Solut. **2**(8), 38 (2014). ISSN 2345-3397
7. Mao, Y., Chen, Y., Hackmann, G., Chen, M., Lu, C., Kollef, M., Bailey, T.C.: Medical data mining for early deterioration warning in general hospital wards. In: IEEE International Conference on Data Mining Workshops, p. 1042 (2011)
8. Azhim, A., Yamaguchi1, J., Hirao, Y., Kinouchi1, Y., Yamaguchi, H., Yoshizaki, K., Ito, S., Nomura, M.: Monitoring carotid blood flow and ECG for cardiovascular disease in elder subjects. IEEE Eng. Med. Biol. 5496 (2005)
9. Islam, M.R., Ahmad, S., Hirose, K., Molla, M.K.I.: Data adaptive analysis of ECG signals for cardiovascular disease diagnosis, IEEE, 2246 (2010)
10. Jabbar, M.A., Chandra, P., Deekshatulu, B.L.: Heart disease prediction system using associative classification and genetic algorithm. In: International Conference on Emerging Trends in Electrical, Electronics and Communication Technologies-ICECIT 2012
11. Palaniappan, S., Awang, R.: Intelligent heart disease prediction system using data mining techniques, IEEE, pp. 108–115 (2008)
12. Nayak, S.R., Dash, B.K., Mishra, S., Bhutada, T., Jena, M.K.: Sudden cardiac death in young adults. J Indian Acad Forensic Med. **37**(4), 438–440 (2015). ISSN 0971-0973
13. Langara, B., Georgieva, S., Khan, W.A., Bhatia, P., Abdelaziz, M.: Sudden cardiac death in young man. In: Case Report, Dept of Respiratory Medicine, Tameside General Hospital, Ashton Under Lyne, UK, vol. 11. March 2015
14. Sudhakar, K., Manimekalai, M.: Study of heart disease prediction using data mining. Int. J. Adv. Res. Comput. Sci. Softw. Eng. **4**(1), 1157–1160 (2014). ISSN 2277-128X
15. Kaur, B., Singh, W.: Review on heart disease prediction system using data mining techniques. Int. J. Recent Innov. Trends Comput. Commun. **2**(10), 3003–3008 (2014). ISSN 2321-8169

16. Somanchi, S., Adhikari, S., Lin, A., Eneva, E., Ghani, R.: Early prediction of cardiac arrest (code blue) using electronic medical records, ACM, 2119–2126, 11–14 August (2015). ISBN 978-1-4503-3664
17. Herland, M., Khoshgoftaar, T.M., Wald, R.: A review of data mining using big data in health informatics. J. Big Data, p. 35
18. Patil, R.R.: Heart disease prediction system using Naïve Bayes and Jelinek-mercer smoothing. Int. J. Adv. Res. Comput. Commun. Eng. 3(5), 6787–6792 (2014). ISSN 2278-1021
19. Masethe, H.D., Masethe, M.A.: Prediction of heart disease using classification algorithms, In: Proceedings of the World Congress on Engineering and Computer Science. San Francisco, USA, WCECS 2014, pp. 22–24 October (2014). ISSN 2078-0966
20. Manikantan, V., Latha, S.: Predicting the analysis of heart disease symptoms using medical data mining methods. Int. J. Adv. Comput. Theor. Eng. (IJACTE) 2(2), 5–10 (2013). ISSN 2319-2526
21. Syed, Z., Guttag, J.: Unsupervised similarity-based risk stratification for cardiovascular events using long-term time-series data. J. Mach. Learn. Res. 1002 (2011)
22. Sudhir, R.: A survey on image mining techniques: theory and applications. Comput. Eng. Intell. Syst. (Paper), (Online), 2(6), 45–46 (2011). ISSN 2222-1719, ISSN 2222-2863
23. Venkatalakshmi, B., Shivsankar, M.V.: Heart disease diagnosis using predictive data mining. Int. J. Innov. Res. Sci. Eng. Technol. 3(3), 1873–1877 (2014). ISSN 2319-8753
24. Ebrahimzadeh, E., Pooyan, M.: Early detection of sudden cardiac death by using classical linear techniques and time-frequency methods on electrocardiogram signals. J. Biomed. Sci. Eng. (JBiSE), 699–706 (2011)
25. Kavipriyal, A., Gomathy, B.: Data mining applications in medical image mining: an analysis of breast cancer using weighted rule mining and classifiers, IOSR J. Comput. Eng. (IOSRJCE), 8(4), 18–23 (2013). ISSN 2278-0661, ISBN 2278-8727
26. Marinov, M., Mosa, A.S.M., Yoo, I., Boren, S.A.: Data-mining technologies for diabetes: a systematic review. J. Diab. Sci. Technol. 5(6), 1550–1551 (2011)

Automated Sizing Methodology for CMOS Miller Operational Transconductance Amplifier

Pankaj P. Prajapati and Mihir V. Shah

Abstract In the last decade of twentieth century, experts anticipated the expiry of analog circuits as the world was becoming increasingly dependent on digital signals. But most of the digital systems need analog modules for interfacing to the external world. Every high-speed digital circuit needs analog circuit design concepts. So design of analog circuits is of great significant. The manual design procedure of analog circuit is very difficult, time-consuming and cumbersome. Optimization algorithms help to take the mystery out of analog integrated circuit design. In this paper, a simple rail-to-rail CMOS Miller operational transconductance amplifier (OTA) is optimized using particle swarm optimization (PSO) algorithm. PSO algorithm is implemented using C language and interfaced with circuit simulator. This circuit is simulated using BSIM3v3 MOSFET models in 0.35 and 0.18 μm CMOS technology. Simulation results of this circuit are compared to previous works.

Keywords OTA · PSO · Optimization

1 Introduction

Design and testing of digital circuits is relatively systematic, whereas design and testing of analog circuits depends on experience. Small circuits can be designed easily using manual calculations and simulations. But, for complex circuits, this process may become lengthy and time-consuming. Designing a good analog circuit is still a tough work, which usually requires senior designer's understanding and expertise. Also, after optimization, one can not claim the absolutely optimized parameters. So automation is highly preferred in this space.

P.P. Prajapati (✉) · M.V. Shah
L. D. College of Engineering, Ahmedabad, India
e-mail: pankaj@ldce.ac.in

M.V. Shah
e-mail: mihirec@gmail.com

© Springer Nature Singapore Pte Ltd. 2018
M. Pant et al. (eds.), *Soft Computing: Theories and Applications*,
Advances in Intelligent Systems and Computing 584,
https://doi.org/10.1007/978-981-10-5699-4_29

Circuit sizing is an optimization process by its nature, and some approaches are proposed in the past to deal with this task. One of the earliest approaches to solve this problem is convex optimization technique [1]. In [1], convex optimization technique is used to decrease area of the circuit. But this method is more complex. Genetic algorithm (GA), developed by Holland [2], is the most popular optimization algorithm. Many authors have tested GA for different optimization problem in the past. In [3], GA is tested for analog circuit design automation. GA is claimed to be computationally efficient algorithm in this optimization problem. In the past few years, evolutionary algorithms have emerged as powerful optimization tools. Particle swarm optimization (PSO) algorithm, proposed by Kennedy and Eberhart, can be used efficiently to optimize some analog circuits [4–6]. In [7], ant colony optimization algorithm is implemented to find transistors size of some analog circuits to meet the required specifications. Differential evolution algorithm is also tested to design some CMOS-based analog circuits like CMOS voltage divider, CMOS three-stage current-starved voltage-controlled oscillator, CMOS common source amplifier, CMOS differential amplifier with current mirror as load [8].

Operational transconductance amplifier (OTA) is an important building block for various analog circuits and systems. Based on the system requirements, an OTA must fulfill many design parameters. There are different topologies of CMOS Miller OTA available in literature [9, 10]. Here, a simple rail-to-rail CMOS Miller OTA is sizing using PSO algorithm. In Sect. 2, PSO algorithm is described. In Sect. 3, rail-to-rail CMOS Miller OTA structure is explained. The automated sizing methodology using PSO algorithm is presented in Sect. 4. In Sect. 5, simulation results obtained by this work are presented. Conclusions are discussed in Sect. 6.

2 Particle Swarm Optimization Algorithm

Particle swarm optimization algorithm is fundamentally a population-based evolutionary algorithm [5, 6]. It mimics the behavior of birds assembling in search of food. It works on the social activities of particles in the group. So it discovers the global best solution by simply correcting the path of each individual toward its own best location and toward the best particle of the entire swarm at each time step. Consider particle swarm consists of N number of particles with D dimension. The position and velocity of ith particle are characterized by $x_i = [x_i^1, x_i^2, \ldots, x_i^D]$ and $v_i = [v_i^1, v_i^2, \ldots, v_i^D]$, respectively. In the search space, the velocity of ith particle after each iteration is modified as given by the following equation [6, 11].

$$v_{id}^{k+1} = w * v_{id}^k + C_1 * r_1 * (pbest_{id}^k - x_{id}^k) + C_2 * r_2 * (pbest_{id}^k - x_{id}^k). \quad (1)$$

The position vector of the ith particle in $N \times D$ dimension of the search space is given by the following equation [6, 7].

$$x_{id}^{k+1} = x_{id}^{k} + v_{id}^{k+1}. \tag{2}$$

Here, the range of i is $\{1 \dots N\}$, the range of d is $\{1 \dots D\}$, the range of k is $\{1 \dots$ max iteration number$\}$, C_1 and C_2 are two positive constants, r_1 and r_2 are random numbers between 0 and 1, w is inertia weight which is selected less than one and decreased linearly with each repetition. The main steps for PSO algorithm are given in the following flow diagram as shown in Fig. 1 [12].

For this work N, C_1, C_2 and w are selected as 30, 1.47, 1.47 and linearly vary from 0.9 to 0.4 with repetitions, respectively [12].

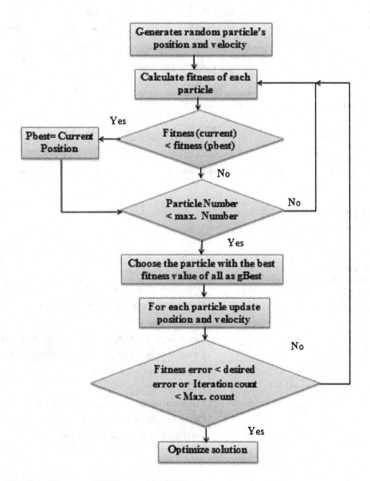

Fig. 1 The flow diagram of PSO algorithm [12]

3 CMOS Miller OTA

The rail-to-rail CMOS Miller OTA is shown in Fig. 2, as proposed in [9]. The main parts of this circuit are bulk-driven differential amplifier and dc level shifters. This circuit has rail-to-rail input and output swing. This circuit works with 600 mV power supply. The desired specifications for this circuit are listed in Table 1.

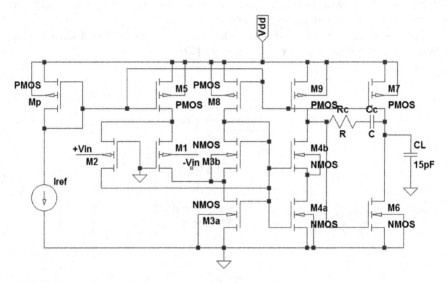

Fig. 2 CMOS Miller OTA [9]

S. No.	Desired specifications
1	Open loop gain $A_v > 75$ dB
2	Unit gain bandwidth (UGB) >25 kHz
3	Phase margin (PM) >45°
4	Power supply rejection ratio (PSSR) (+ve) >60 dB
5	Power supply rejection ratio (PSSR) (−ve) >60 dB
6	Rise slew rate > 15 V/ms
7	Fall slew rate > 15 V/ms
8	Common mode rejection ratio (CMRR) >60 dB
9	Dc power dissipation (P_{diss}) < 10 µW
10	Total transistor area <14,500 µm^2

Table 1 Desired specifications for CMOS Miller OTA

4 Automated Sizing Methodology

The main aim of automated sizing methodology is to find the optimum size of width of NMOS transistors, width of PMOS transistors, value of resistors, value of capacitors, and bias currents of given circuit for required specifications. The flow diagram of automated sizing methodology is given in Fig. 3 [12]. In this work, we have used automated sizing method to design the rail-to-rail CMOS Miller OTA for specifications listed in Table 1. PSO algorithm is implemented using C language and interfaced with circuit simulator tool. Ngspice simulator tool is used to check performance of this circuit. This system receives spice net list of rail-to-rail CMOS Miller OTA as its input. The sizing methodology is implemented on a vector of design variables of this circuit. In this optimization procedure, particle position

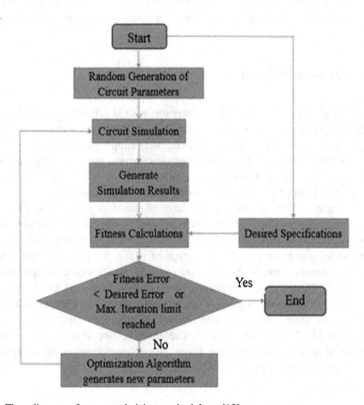

Fig. 3 Flow diagram of automated sizing methodology [12]

vector is channel width of NMOS transistors, channel width of PMOS transistors, value of resistor, value of capacitor, and bias current.

Initially, each particle position vector takes random value. During each iteration, an error function is evaluated based on the spice circuit simulator results. The error function is defined in the following equation [4, 12],

$$F_{error} = \sqrt{\sum_{j=1}^{n} \left(\frac{Spec_{desired} - Spec_{sim}}{Spec_{desired}} \right)^2}, \tag{3}$$

where $Spec_{desired}$ = desired specifications, $Spec_{sim}$ = specifications obtained by a simulator for a specific result delivered by the optimizer. When a termination criterion is fulfilled, the automated sizing procedure will stop. The termination criteria are maximum number of repetitions or minimum value of the error function. For this work, the minimum value of the error function F_{error} is selected as 10^{-6}.

5 Results and Discussion

We have used a core i5, 2.4 GHz processor with 4 GB internal RAM to run complete design setup. This system is running using Ubuntu operating system. In this work, we have added specifications like power supply rejection ratio (+), power supply rejection ratio (−), and common mode rejection ratio as compared to previous works. The obtained simulation results of rail-to-rail CMOS Miller OTA using 0.35 and 0.18 μm CMOS technology are listed in Table 2.

For 0.35 μm CMOS technology, we have obtained at par results. This circuit design requires less area but consumes more power as compared to previous works. We have also tested this circuit with 0.180 μm CMOS technology. We have obtained better results for open loop voltage of 91.37 dB, unit gain bandwidth of 145.55 kHz, common mode rejection ratio of 124.38 dB, power supply rejection ratio (+) of 124.48 dB, power supply rejection ratio (−) of 104.81 dB, rise slew rate of 11.9 V/μs, fall slew rate of 17.61 V/μs, as compared to work reported in [4]. But this design has more total transistor area of 4839.81 μm^2 and more dc power dissipation of 14.92 μW. This design has more open loop voltage gain, more unit gain bandwidth, and less area but more power consumption as compared to 0.35 μm CMOS technology results listed in Table 2.

Table 2 Obtained results and comparison with previous works

S. No.	Desired specifications	Obtained results (0.35 μm)	Results (0.35 μm) [7]	Results (0.35 μm) [4]	Results (0.35 μm) [9]	Obtained results (0.180 μm)	Results (0.180 μm) [4]
1	A_v > 75 dB	77.71 dB	74.82 dB	82.67 dB	73.5 dB	91.39 dB	75 dB
2	UGB > 25 KHz	34.44 KHz	42.27 KHz	48.61 KHz	42.27 KHz	145.55 KHz	58 KHz
3	PM >45°	89.24°	49.56°	58.01°	54.1°	68.87°	59.7°
4	PSSR (+) > 60 dB	71.11 dB	–	–	–	124.48 dB	–
5	PSSR (−) > 60 dB	90.70 dB	–	–	–	104.81 dB	–
6	Rise slew rate > 15 V/ms	12.58 V/μs	–	21.77 V/ms	14.7 V/ms	11.9 V/μs	30.87 V/ms
7	Fall slew rate >15 V/ms	72.73 V/μs	–	23.11 V/ms	14.7 V/ms	17.61 V/μs	16.4 V/ms
8	CMRR >60 dB	85.27 dB	–	–	–	124.38 dB	–
9	P_{diss} 10 μW	9.0 μW	540 nW	545.49 nW	550 nW	14.92 μW	571 nW
10	Total transistor area 14,500 μm²	5033 μm²	–	5586 μm²	20,772 μm²	4839.81 μm²	4598 μm²

6 Conclusion

In this work, PSO-based methodology is used to size rail-to-rail CMOS Miller OTA using 0.35 and 0.18 μm standard CMOS technology parameters. We have added more number of specifications to be optimized in this work. So the objective function of this work is more complicated as compared to previous works. We have obtained results at par as compared to reported works in literature for 0.35 μm CMOS technology. Whereas, we have obtained good results for 0.18 μm CMOS technology. Automated sizing of an analog circuit is easy and effective compared to manual design even though the designer has primary knowledge of a given circuit. Analog circuit performance is changed with process, voltage, and temperature (PVT). For robust optimizer development, PVT variations should be considered during design procedure. Development of optimizer with PVT variations may create an interesting area for future work.

References

1. Sapatnekar, S.S., Rao, V.B., Vaidya, P.M., Kang, Sung-Mo: An exact solution to the transistor sizing problem for CMOS circuits using convex optimization. IEEE Trans. Comput. Aided Des. Integr. Circ. Syst. **12**(11), 1621–1634 (2006)
2. Koza, J.R., Bennett, F.H., Andre, D., Keane, M.A., Dunlap, F.: Automated synthesis of analog electrical circuits by means of genetic programming. IEEE Trans. Evol. Comput. **1**(2), 109–128 (1997)
3. Vaze, A.P.: Analog Circuit design using genetic algorithm: modified. World Acad. Sci. Engi. Technol. **14**, 62–64 (2006)
4. Thakker, R.A., Shojaei Baghini, M., Patil, M.B.: Low-power low-voltage analog circuit design using hierarchical particle swarm optimization. In: 22nd International conference on VLSI Design, IEEE, pp. 427–432 (2009)
5. Kennedy, J., Eberhart, R.C.: Particle swarm optimization.In: Proceeding of IEEE International Conference on Neural Networks, Piscataway, NJ, pp. 1942–1948 (1995)
6. Shi, Yuhui, Eberhart, Russell C.: Empirical study of particle swarm optimization, pp. 1945–1950. Proceedings of the Congress on Evolutionary Computation, IEEE (1999)
7. Gupta, H., Ghosh, B.: Analog circuits design using Ant colony optimization. Int. J. Electron. Commun. Comput. Technol. **2**, 9–21 (2012)
8. Sharma, S.A., Prajapati, P.P.: Design of various analog circuits using differential evolutionary algorithm. Int. J. Sci. Process Res. **23**(3), 159–165 (2016)
9. Ferreira, L. et al.: An ultra-low-voltage ultra-low-power CMOS Miller OTA with rail-to-rail input/output swing. IEEE Circ. Syst.-II: Exp. Briefs., 843–847 Oct (2007)
10. Daoud Dammak, H., Bensalem, S., Zouari, S., Loulou, M.: Design of folded cascode OTA in different regions of operation through gm/ID methodology. World Acad. Sci. Eng. Technol. Int. J. Electri. Comput. Energ. Electron. Commun. Eng. **2**(9), 174–1746 (2008)
11. Vural, R.A., Yildirim, T.: Analog circuit sizing via swarm intelligence. Int. J. Electron. Commun. **66**(9), 732–740 (2012)
12. Prajapati, P.P., Shah, M.V.: Two stage CMOS operation amplifier design using particle swarm optimization algorithm. In: Proceeding of IEEE UP Section conference on Electrical, Computer and Electronics, Dec (2015)

Analytical Review on Image Compression Using Fractal Image Coding

Sobia Amin, Richa Gupta and Deepti Mehrotra

Abstract Compression is the process of reducing the size of an image. Reduction can be achieved through reducing the number of bits and encoding time. Encoding and decoding can be used to achieve compression and decompression. Encoding and decoding can be done through fractal image coding which uses the property of self-similarity between the blocks on the basis of affine transformations. Thus, our main goal is to analyze various techniques of fractal encoding for compressing the image. A systematic study of various fractal encoding techniques is done in this paper and based on the study a comparative analysis is presented.

Keywords Fractal transforms coding · Iterative function system
Contractive mapping · Discrete cosine transform · Quadtree partitioning
Edge property based neighborhood region search · Kernel density estimation

1 Introduction

Image compression has been the most important area of research and plays a vital role in applications such as tele-videoconferencing, video surveillance, image sensing, and many others. Compression is the process of reducing the size through reduction in the number of bits and encoding time of images. Encoding is used for compression, while decoding is used for decompression.

Compression and decompression are used for the storage and transfer of digital images as given by [1]. Thus, for image compression, fractal method can be used.

S. Amin (✉) · R. Gupta · D. Mehrotra
ASET, Amity University, Noida, Uttar Pradesh, India
e-mail: sobiya.amin60@gmail.com

R. Gupta
e-mail: rgupta@amity.edu

D. Mehrotra
e-mail: mehdeepti@gmail.com

© Springer Nature Singapore Pte Ltd. 2018
M. Pant et al. (eds.), *Soft Computing: Theories and Applications*,
Advances in Intelligent Systems and Computing 584,
https://doi.org/10.1007/978-981-10-5699-4_30

Fractal image coding is based on the self-similarity between the blocks through iterated function system (IFS).

Fractal transform was first introduced by Barnsley and Demko [2]. However, image compression using fractals was first practically introduced by Jacquin [3] and Jacqobs et al. [4]. Hence, it was called Jacquin–Fisher method.

The procedure undertaken by Jacquin–Fisher is as follows:

1. Dividing the original image into blocks called as range blocks by the use of 8 * 8 and 4 * 4 levels.
2. Division of blocks into shaded, mid-range and edge blocks.
3. Search of same "code book" for the same class using domain pool of blocks and transformations.

Shaded blocks were coded using gray level, and other blocks were coded using zero-order polynomial.

In Jacquin–Fisher algorithm, two issues have been found one being the cost of search for the particular domain block that was too large and other being the choice of good algorithm for the classification. These issues had been solved by Fisher et al. [5] using the quadtree which divides the image into non-overlapping ranges with the tree structure. The more generalized method was given by [6] using vector quantization which uses first and second domain blocks only, thereby reducing the search necessary to find the appropriate domain block.

Hence, many techniques have been given by many authors with the relative advantages and disadvantages related to them. Thus, the main motive of this paper is to provide the study of various existing techniques. This paper has been organized as introduction in Sect. 1, followed by the literature review in Sect. 2, followed by the analysis of various techniques of fractal encoding for image compression in Sect. 3, and finally, the conclusion and future work in Sect. 4.

2 Literature Review

Many techniques have been used since centuries by many researchers for fractal image compression. All of these techniques have been found to provide a new improvement over others. We have done our review by going through many techniques that have been used by many researchers and formulated many research questions to do the comparison between them. These research questions are as follows.

2.1 Classification of Fractal Image Compression Techniques

In images, there are many such regions which are similar to other regions in the same image; thus, these regions have self-similarity features among them and are said to possess fractal nature. Fractal can be regarded as an image with particular shape or geometry which is subjected to some mathematical equations. Images can be represented by the set of mathematical iterations called as iterated function system. Thus, fractal image compression takes into idea of incorporating iterated function system for an image.

The idea of iterated function system was quiet complex due to the large computation involved in coding as well as processing; hence, iterated function system was enhanced with partitioned iterated function system. In partitioned iterated function system, each image is divided into set of non-overlapping blocks called as range. These non-overlapping blocks are chosen on the basis of self-similarity among other blocks [7–9].

Similarly, the blocks larger than the range blocks or equal to them are chosen called as domain blocks. Each of the range blocks is mapped onto the appropriate domain block on the basis of self-similarity. Mapping each range blocks to a particular domain block adds a lot of computation time. Thus, large computation time was involved in this mapping. To reduce the computation time, affine transformations have been applied to map the particular domain block to the selected range block. Affine transformations involve rotations, translations, resizing of the images. Decoding phase for fractal compression is quite simple. On the basis of domain-range pool self-similarity criteria, fractal image compression has been classified as follows:

a. **Domain-Based Classification**: Fractal image compression has been found to be asymmetric in both time of execution and operations of algorithms. The time required for encoding and decoding has been found to be different. Decoding time has found to be less as compared to encoding time. In order to compress an image, a particular domain block needs to be matched to the range block. In order to improve quality, domain block was considered to be overlapping. In this way, many domain blocks were mapped to the range blocks; thus, increasing the time as well as search for the domain block in the large set of domain becomes complex. Thus, "no-search algorithm" has been proposed to reduce search for the particular domain from the large set of domain pool. Thus, reducing time for encoding as well as searching complexity but the quality has been degraded. In these algorithms, domain blocks have been found to be static; thus, to improve quality as well as reduce encoding time, dynamic domain blocks have been introduced [10, 11].

b. **Dynamic-Based Method**: In previous algorithms, the search for the domain block within the domain pool has been static and involves computational complexity. Thus, to reduce the computational time and to improve quality, dynamic search for domain-range similarity has been proposed. Thus, for dynamically search for domain block local, fractal dimension of the range block has been used [12–14].

Local Fractal Dimension-Based Search: Fractal dimensions are the measures for searching similarities between two or more than two fractals. It determines how densely fractals can be fitted in a given image. Fractal dimensions for gray scale can be determined by many techniques. Similarly, for the color images, fractal dimensions can be calculated by using the fractal dimensions of gray images. Fractal dimensions can be calculated by calculating range block fractal dimension, and then, search for the particular domain block is placed in the domain pool. Fractal distance for most of the images using "differential box counting" has been found to lie in between 2.0 and 3.0. In this method, domain selection for the particular range block has been calculated in advance. To further reduce the computational time, other methods have been proposed [15, 16].

c. **Height-Balanced Tree-Based Search**

For the stable time complexity, height-balanced method is used because insertion, deletion, search, rotation becomes stable. In height-balanced method, balance factor has been used which was calculated on the basis of difference between left subtree nodes and right subtree nodes. During insertion and deletion, balance factor is calculated in intervals thus maintaining the time complexity in each step. Many height-balanced search methods have been used, but AVL has been found to be the best in terms of time complexity and stability. Quadtree method is an example of height-balanced tree-based search which is used to optimize the performance of encoding technique as proposed in [17, 18].

In order to do the literature review, we have formulated some of the research questions which are mentioned below:

RQ2: To identify various fidelity measures/criteria that can be used for fractal encoding.

RQ3: To identify limitations with those existing methods of fractal encoding.

RQ4: To identify the scope of improvement that can be performed on these existing techniques in order to overcome these limitations.

In [19] Bono et al. proposed and examined the methods related to the finite fractals related to the particle size. In the paper proposed by the author, size of particle has been related with the coordination number. The particles with the less coordination number consist of lesser number of voids, thereby decreasing the overall stress. In the papers published earlier, much of the work has been done on the one-dimensional and isotropic normal compression [20–22]. In the paper published by the author, the main stress was put on the particle size and the occurrence of the new particle size when the particle has been crushed as a result of fractal particle size distribution. Thus, a particle size has been stabilized to reduce the generation of new particles which increase gradually to about 20%. It has been proved that the average stress was decreased with the particle size. The lesser the particle size the lesser would be stressed. Coordination number, average strength, and particle stress have also been analyzed. For the particular particle size, average stress has been found to be dependent on the strength. Further, particle which has not been undergone through fractal compression was continuously crushed into smaller particle sizes. Similarly, void space has been established to be lesser for those particles that were having 4–5 lesser contacts and has been regarded as the smallest or the critical particles.

In [23], Sun et al. proposed a method for compressing encrypted image using fractal dictionary and Julia set. To reduce time, the authors of the paper had chosen *M* blocks and BTC queue is used to search domain blocks. Blocks have unique BTC value, and many domain blocks can have same BTC value. Domain blocks sharing same BTC value form a BTC queue, thus using fractal dictionary for the compressing time as well as maintaining quality with good peak signal-to-noise ratio (PSNR). Experimental results have been obtained by performing all the experiments on CORE™ 15 (240 GHZ) personal computer. Experimental results have shown analysis of compression performance, key space analysis, sensitivity analysis, cipher randomness analysis, and speed of encryption and decryption. From the experimental results, encryption process has been found highly sensitive to plain text and key for the small perturbation values. It has been proven experimentally that encryption process is less than 15% of compression process proving no delay in the compression due to the encryption process. Thus, the method used by the authors has been proved to provide real-time compression and encryption. Future work for the authors would be the use of better Julia set and other methods for the further improvement.

In [24] Prashanth et al. proposed fractal encoding fort images with the help of wavelet transform. Quadtree encoding has been employed, and fractal encoding has been done for the images of JPEG, HD, PNG, and BMP images by supplying both with noise and without noise. Median filter has been introduced to eliminate noise to be present in the image. Two types of noises have been presented by authors: one being "SALT and PEPPER" and other being "Gaussian noise." Experiment has been performed on the image size of 1920 * 1080 and 1920 * 1200 pixels for the formats JPEG, PNG, BMP and 8 bits per pixel resolution. Results have been shown for both the noiseless and with the noise simulations. Both the noises are "SALT and PEPPER" and "Gaussian noise." Results have been performed. Encoding, decoding time, mean square error, PSNR for both with filter and without filter have been calculated. Maximum absolute squared deviation of the data from the approximation (MAXERR) and ratio of squared norm of the image approximation to the input image (L2RAT) have been calculated. Thus, the experimental results have evaluated that the proposed method has reduced encoding time to about 95% and had worked well with SD images.

In [25] Padmashree et al. proposed the fractal image compression for medical images. Quadtree partition has been used in the paper proposed by the author. Experiment has been performed on mammogram images and magnetic resonance image (MRI). Compression ratio and peak signal-to-noise ratio (PSNR) have been measured. Authors have used fractal image compression using quadtree partitioning. Embedded block coding optimization truncation (EBCOT) technique has been used. In EBCOT, encoder divides discrete wavelet coefficients in blocks of codes and each coded blocks are encoded in bit-stream-based blocks. Bit stream is truncated to reduce embedded bit stream. For experimental purposes, 80 sets of mammogram images with 256 * 256 images and 20 (mR) images of brain have been used. Technique of fractal image compression has been used both with and without EBCOT on mammograms and MRI. Compression ratio (CR) and peak signal-to-noise ratio (PSNR) have been calculated for all the conditions:

PSNR and CR for mammograms with EBCOT, PSNR and CR for mammograms without EBCOT, PSNR and CR for MRI with EBCOT, PSNR and CR for MRI without EBCOT

Those of fractal image compressions without EBCOT have been found to be 7.6 compression ratio and 26 db PSNR. However, PSNR values for mammogram images have been found to be high as compared to MRI images. But CR has been found to be same. Thus, experimental results had shown that the technique of fractal image compression using quadtree had implemented well with mammograms giving better standard deviation than MRI.

In [26], Xiaoqing et al. proposed a new approach of using kernel density estimation as a statistical method for fractal coding. Texture image retrieval has been used by the author in the paper. Thus, the proposed method has been used by the author for the retrieval of texture image using kernel density estimation. Experimental results have been performed on the images in the library. 512 * 512 grayscale Brodatz texture images have been adopted, and 640 texture images have been created by dividing each image into 16 128 * 128 subimages. Results have shown that KM method has given better results than KS, but joint retrieval method has been found to be the best method. Influence of threshold T on the retrieval rate has also been estimated. Retrieval rate with different thresholds has been estimated using KE and histogram analysis. Results have shown that the retrieval rate had increased with the increase in . Further work for the authors of the paper would be the use of different bandwidth to show the increase in retrieval rates and speed. In addition to the Gaussian kernel function, different other functions would be adopted to obtain the better results

In [27], Hilo discussed the fractal image compression with the better improvements over IFS matching. Stopping search (e) has been used to reduce coding time. Instead of using (Xd, Xy) coordinates for matching of domain blocks (POSI) index position has been used. In order to achieve more compression, (RGB) component of data has been changed to YUV component. Fractal encoding has been used for gray images, but color images have been encoded by considering each color image as a component of gray image (RGB) color components. After (RGB, YUV) has been widely used for coding of colored images and videos where Y represents luminance components and (U, V) represent chromatic parts. Thus, in the paper given by authors, (RGB) components have been changed to (YUV) color spaces. For effective compression, experimental result of the algorithm used by author has been performed using VISUAL BASIC and tested on Pentium IV with 256 RAM. 256 * 256 pixels of color image have been used. PSNR, compression ratio, and encoding time have been used as fidelity criteria for parameters quantization tests, maximum scale and minimum scale tests, block size tests, step size tests and domsize tests. Individual tests have been performed with the comparison of PSNR, compression ratio, and step size for each test. Thus, the experimental results after performing the tests had shown that

Stopping search (e) had reduced time with decrease in PSNR, change of (Xd, Yd) parameters into (POSI) had increased compression ratio, and (U, V) components have also increased encoding time and maintained quality of image.

In [28], Shih et al. performed the image annotation using fractal image coding that extracts fractal properties and uses color and texture for image annotation. Image annotation is a method in which metadata like caption, keywords or other features are provided to the digital image. Fractals are like geometries; thus, fractal features provide information related to geometries that are different from color, shape, size of an image. Thus, the authors of the paper had proposed fractal image annotation combined with color, texture. Experiments have been performed on the public information set of Barcelona urban scenes where 139 JPEG images were used, having resolution of 1600 * 1200 pixels. Four different labels have been used including flora, people, sky, and building, Accuracy has been used as an evaluation measure to determine outcomes. Experiment has been performed over six evaluation measures like:

Selection of color and texture features, co-relation of visual features, accuracy of image annotations, influence of fractal features, influence of thresholds, and influence of labels.

Experimental results have shown that fractal properties possess no correlation with color and texture indicating novel property of fractal features with 5% of error. Thus, results had shown that the fractal features with the combination of scaling parameter and block size had shown better results and accuracy. Fifth experiment has been performed to show the impact of threshold on the accuracy of fractal features. Histogram and variance have been used for this purpose, and the results had shown minimum impact of thresholds on the accuracy. Sixth experiment has been performed to show the impact of labels. Experimental results had shown that the label like flora and people show no variation due to non-fixed shape and size. Thus, for more accurate results, wavelet and shape could be used.

In [29], Lin et al. proposed an edge-based neighborhood region search method in which speed of fractal encoder has been increased. Optimal search was used for the frequency domain. Two discrete transform coefficients have been used to construct coordinate systems. One being lowest vertical coefficient and other being lowest horizontal coefficient. Similarly, range block in case of proposed method given by the author was matched to only 4 transformed domain blocks. Thus, encoding time of proposed method has been reduced to about 1/5 of Duh's classification method [30]. For the experimental results, image size of 256 * 256 has been used with the range block of 8 * 8 and 16 * 16 ha been used, respectively. Results had shown that EP-NRS had performed well with less encoding time and almost same quality of images for both EP-NRS and NRS. Experiment has been performed between Duh's classification [30] and EP-NRS. Results had shown that EP-NRS was having better image quality than Duh's classification [30] and PSNR of about 0.3–0.4 db. Results had shown that the encoding time depends on the mean square error. Thus, the result had shown that the image with more mean square error has more encoding time because of non-uniform distribution of blocks. Experiments have also been performed with the Tseng's PSO-K and PSO-KI methods [31]. Results have shown better encoding time of EP-NRS that uses PSO-K and PSO-KI due to large complexity. Thus, proposed method has been found 2.57 times than Duh's classification [30] but the decay of about 0.9 db.

In [32] Muruganandham et al. proposed a fractal algorithm in which particle swarm optimization has been used. Particle swarm optimization has been used to improve the speed of fractal algorithm in which mean square error (MSE) was used. Mean square between the errors was used as a stopping criterion between the range and domain blocks. Experiment has been performed on the images to compare the encoding time and peak signal-to-noise ratio (PSNR) of full search as well as proposed method. Results have been obtained by showing variations in the PSNR by measuring the changes in the mean square error. Changes in the mean square error have been calculated by changing the percentage of the maximum iterations; thus, for experimental purposes, the authors of the paper have used 10% maximum iterations. After performing the experiment, results have shown that the encoding time and PSNR ratio of the proposed method were better than the full search method, but the loss of 1.2 db in image quality has been experienced.

In [33], Truong et al. proposed an algorithm named fast encoding algorithm for the compression of fractal image. In the paper proposed by the author, redundant computations are removed by calculating the mean square errors of the range block. Similarly, eight dihedral symmetric are obtained for the domain block, thus involving many computations. Fast algorithm used by the authors to reduce computations by the use of discrete cosine transforms. In the experimental setup, fast algorithm has been compared with baseline algorithm. C++ has been used for programming. Gray-level 256 * 256 image has been used. Experimental results show the comparison between fast algorithm and baseline on the basis of time and peak signal-to-noise ratio. It has been proved that the fast algorithm used by the author is six times faster for all coefficients and zonal filter of image. Thus, algorithm has been found to improve speed of decoder with unaffected PSNR.

In [34], Woon et al. discussed the fractal image compression implementation on two satellite images. Image has been first divided into non-overlapping range. Compression ratio and PSNR have been used as fidelity criteria. Experimental results had shown better results for coastal areas having many similarities. Further improvement has been found in fractal compression algorithm used by author of the paper by applying orthogonal wavelet filters. These filters could be used to transform image into frequency domain. During wavelet filter algorithm implementation, image could be decomposed into coarse and detail components. Coarse components represent low-pass-filtered subimage, and detail component represents high-pass-filtered subimage. Adaptive differential pulse modulation method could be used on coarse components, and fractal algorithm could be used on the detail components. Thus, experimental results on applying fractal algorithm have shown that compression ratio of 173:1 and PSNR of 34.9 db has been achieved for coastal areas, and 11.1 and 25.03 db compression ratio and PSNR, respectively, have been achieved for city area.

3 Analysis on Fractal Encoding Techniques

On the basis of analysis, we have formulated a table relating the techniques and fidelity criteria used (Table 1, 2, 3).

Table 1 Analysis of image compression using fractal image coding

S. No.	Techniques used	Fidelity criteria
[23]	Julia set and fractal dictionary	Compression performance, key space analysis, sensitivity analysis
[24]	Wavelet transform, Quadtree encoding	Encoding time, decoding time, mean square error, peak signal-to-noise ratio, MAXERR, L2RAT
[25]	Quadtree and embedded block coding optimization truncation	Compression ratio, peak signal-to-noise ratio
[26]	Kernel density estimation function	Image retrieval rate and speed
[27]	Stopping search, YUV component	Peak signal-to-noise ratio, encoding time
[28]	Image annotations	Texture, color, accuracy measurement
[29]	Edge-based search using neighborhood method	Encoding time, mean square between errors, decay
[32]	Particle swarm optimization	Peak signal-to-noise ratio, encoding time
[33]	Fast encoding algorithm	Encoding time, peak signal-to-noise ratio
[34]	Fractal image compression using affine transformations	Compression ratio and peak signal-to-noise ratio

Table 2 Encoding time and PSNR

S. No.	Encoding time (SEC)	PSNR (db)	Experiment set
[23]	0.234	32.483	MATLAB
[24]			MATLAB
a. Image format without noise	53.84	49.22	
b. Image format with salt and pepper	55.13	44.59	
c. Image format with Gaussian noise	54.76	35.97	
[27]			VISUAL BASICS
E0 = 0.3	128.06	33.29	
[29]			BORLAND C ++
a. DUH [34]	200.95	31.63	
b. PSO-K [30] PSO-KI [30]	60.21 67.21	31.59 31.59	
c. EP-NRS	59.24	31.59	

(continued)

Table 2 (continued)

S. No.	Encoding time (SEC)	PSNR (db)	Experiment set
[32]			MATLAB
a. Full search FIC	09:07:20	35.80	
b. Proposed method	00:15:34	35.03	
[33]			C ++
a. Baseline	22.42	28.90	
b. DCT	6.46	28.93	
c. Zonal filter	3.80	28.18	

Table 3 Compression ratio and PSNR

S. No.	Compression ratio	PSNR	Experiment set
[25]			MATLAB
a. MRI with EBCOT	19.97	28.377	
b. Mammograms with EBCOT	19.97	34.19	
[27]			VISUAL BASICS
E0 = 0.3	9.04 (xd.yd)	33.29	
	9.72 (POSI)	33.29	
[34]			MATLAB
a. Satellite image of city	11.1	25.03	
b. Satellite image of coastal	173.1	34.9	

On the basis of different techniques used, we have analyzed different values of fidelity criteria used by these techniques and formulated a table showing different values of fidelity criteria used.

4 Conclusion and Possible Research Directions

4.1 Conclusion

After doing the analysis, we have found the advantages and limitations of many techniques. It has been found that the decay has been reduced to a greater extend by existing technique, and encryption process has been improved. Image compression has found a greater use in real-time applications by using fractals. Encoding time has been reduced to 95%, and SD images have found to be compressed by using fractals. Standard deviations and mean have been found to improve. Medical images like MRI and mammograms have been compressed with good quality and better encoding time. Satellite image has been optimized with greater efficiencies. Image retrieval rates and speed have been greatly improved with the help of kernel functions. Though many advantages have been incorporated by existing techniques, many limitations have been found. Use of different Julia sets would have been used, different bandwidth would have been used to further reduce the time of encoding, and different color spaces would have been used. Working on all these limitations would be our future work. Thus, we conclude that although a lot of work has been done to improve the compression still some limitations have been found there.

4.2 Possible Research Directions

Nowadays, fractal image compression has been found to be used in the field of mobile telecommunication. In many logical programmable devices like field programmable gate array, fractal has been used because of its simplicity and efficiency as proposed by Gupta et al. [35]. Video telecommunication has found its application because of the reduction in complexity and computation due to the use of fractals. In the medical field, where doctors having problems in clear image of patients using MRI's and mammograms, fractals can help in improving quality of images and survey fields. Image compression using fractals could be applied to the premature atrial contraction and other telemedicine. In order to obtain better allocation of bits in case of images, fractal image compression can be incorporated with adaptive partitioning. In this case, edge block can be shortened, and shaded blocks can be extended. Due to increase in the crime in every field, fractal image compression can help a lot by providing method face detection, biometric, DNAs, etc. Similarly, multi-fractals could be used for image enhancement and coding simplicity. Thus, the use of overlapped blocks, dynamic domain blocks, and the use of more wavelets transform could be further investigated.

References

1. Hassan, T.M., Wu, X.: An adaptive algorithm for improving the fractal image compression (FIC). J. Multimedia Acad. **6**(6), 477–485 (2011)
2. Barnsley, M.F., Demko, S.: Iterated function systems and the global construction of fractals. In: Proceedings of Royal Society London. A, vol. 399, pp. 243–275 (1985)
3. Jacquin, A.E.: Image coding based on a fractal theory of iterated contractive image transformations. IEEE Trans. Image Process. **1**, 18–30 (1992)
4. Jacobs, E.W., Fisher, Y., Boss, R.D.: Image compression: a study of the iterated transform method. IEEE Trans. Signal Process. **40**, 251–263 (1992)
5. Fisher, Y.: Fractal image compression: in theory and application. Springer-Verlag, New York (1994)
6. Ramamurthi, B., Gersho, A.: Classified vector quantization of images: IEEE Trans. Commun. **34**, 1105–1115 (1986)
7. Hu, L., Chen, Q.A., Zhang, D.: An image compression method based on fractal theory. In: The 8th International Conference on Computer Supported Cooperative Work in Design Proceedings, pp. 546–550 (2003)

8. Zhao, Y., Yuan, B.: Image compression using fractls and discrete cosine transform. Int. Lett. **30**(6), 474–475 (1994)
9. Hamzaoui, R.: Decoding algorithm for fractal image compression. Electron. Lett. 32(14), 1273–1274 (1996)
10. Wang, J., Zheng, N.: A novel fractal image compression scheme with block classification and sorting based on Pearson's correlation coefficient. IEEE Trans. Image Process. **22**(9), 3690–3702 (2013)
11. Kodgule, U.B., Sonkamble, B.A.: Discrete wavelet transform based fractal image compression using parallel approach. Int. J. Comput. Appl. **122**(16), 18–22 (2015)
12. Kapoor, A., Arora, K., Jain, A., Kapoor, G.P.: Stochastic image compression using fractals. Inf. Technol. Coding Comput.1–6 (2003)
13. Hamzaoui, R., Saupe, D., Hiller, M.: Fast code enhancement with local search for fractal image compression. In: 2000 International Conference on Image Processing, pp. 156–159 (2000)
14. Jacquin, A.E.: Image coding based on a fractal theory of iterated contractive image transformation. IEEE Trans. Image Process. **1**(1), 18–30 (1992)
15. Hamzaoui, R., Saupe, D., Hiller, M.: Distortion minimisation with fast local search for fractal image compression. J. Vis. Comm. Image Represent. 2(4), 450–468 (2001)
16. Lee, C.K., Lee, W.K.: Fast fractal image block coding based on local variances. IEEE Trans. Image Process. **7**(6), 888–891 (1998)
17. Gupta, R., Mehrotra, D., Tyagi, R.K.: Adaptive searchless fractal image compression in DCT domain. Imag. Sci. J. 1–7 (2016)
18. Revathy, K., Jayamhan, M.: Dynamic domain classification for fractal image compression. Comput. Vision. Pattern Recogn. **4**(2), 95–102 (2012)
19. De Bono, J.P., Mcdowell, G.R.: The fractal micro mechanics of normal compression. Comput. Geotech. **78**,11–24 (2016)
20. Mcdowell, G.R., de Bono, J.P.: On the micro mechanics of one-dimensional normal compression: Geotechnique. **63** 895–908 (2013)
21. Mcdowell, G.R., de Bono, J.P., Yue, P., Yu, H-S.: Micro mechanics of isotropic normal compression. Géotechnique Lett. **3** 166–172 (2013
22. McDowell, G.R., Yue, P., de Bono, J.P.: Micro mechanics of critical states for isotropically over consolidated sand. Powder Technol. **283**, 440–446 (2015)
23. Sun, Y., Xu, R., Chen, L., Hu, X.: Image compression and encryption scheme using fractal dictionary and Julia set. IET Image Proc. **9**, 173–183 (2015)
24. Prashanth, N., Singh, A.V.: Fractal image compression for hd images with noise using wavelet transforms. In: International Conference on Advances in Computing, Communications and Informatics, pp. 1194–1198 (2015)
25. Padmashree, S., Nagapadma, R.: Statistical analysis of objective measures using fractal image compression for medical images. In: IEEE International Conference on Signal and Image Processing Applications, pp. 563–568 (2015)
26. Xiaoqing, H., Qin, Z., Wenbo, L.: A new method for image retrieval based on analyzing fractal coding characters. J. Vis. Commun. Image Represent. 42–47 (2013)
27. Al-Hilo, E.A., George, L.E.: Study of fractal color image compression using YUV components. In: IEEE 36th International Conference on Computer Software and Applications, pp. 596–601 (2012)
28. Shih, C.W., Chu, H.C., Chen, Y.M., Wen, C.C.: The effectiveness of image features based on fractal image coding for image annotations. Expert Syst. Appl. **39**, 12897–12904 (2012)
29. Lin, Y.L., Wu, M.S.: An edge property-based neighbourhood region search strategy for fractal image compression.Comput. Math. Appl. 62(1), 310–318(2011)
30. Duh, D.J., Jeng, J.H., Chen, S.Y.: Speed quality control for fractal image compression. Imag. Sci. J. **56**, 79–90 (2008)
31. Truong, T.K., Kung, C.M., Jeng, J.H., Hsieh, M.L.: Fast fractal image compression using spatial correlation. Chaos, Solitons Fractals. **22**, 1071–1076 (2004)

32. Muruganandhan, A., Banu, R.S.D.W.: Adaptive fractal image compression using PSO. Procedia Comput Sci. **2**, 338–344 (2010

33. Troung, T.K., Jeng, J.H., Reed, I.S., Lee, P.C., Li, A.Q.: A fast encoding algorithm for fractal image compression using the DCT inner product. IEEE Trans. Image Process. **9**, 529–535 (2000)

34. Woon, W.M., Ho, A.T., Yu, T., Tan, S.C., Yap, L.T.: Achieving high data compression of self-similar satellite images using fractal. In: IEEE 2000 International Symposium on Geo Science and Remote Sensing, vol. 2, pp. 609–6111 (2000)

35. Raman, V., Gupta, R.: JPEG multi-resolution decomposition of image compression using integer wavelets. Int. J. Comput. Appl. **95**, 17–20 (2014)

Biological Infrared Antenna and Radar

P. Singh, R. Doti, J.E. Lugo, J. Faubert, S. Rawat, S. Ghosh,
K. Ray and A. Bandyopadhyay

Abstract This paper presents a report on the wasp antenna working in infrared region and the communication between two wasps at 5 m distance resembling radar equation. To the best of our knowledge, this is the first theoretical analysis and simulation to illustrate the presence of radar-like mechanism in living systems.

Keywords Biological physics · Infrared radiation · Antenna · Radar

1 Introduction

The communication between insects is quite a mystery. It does involve not only zoology but also electromagnetics. American biologist Philip Callahan concluded based on his studies on moths' communication that insect's sensory mechanism is

P. Singh · K. Ray (✉)
Amity University Rajasthan, Jaipur, Rajasthan, India
e-mail: kanadray00@gmail.com

P. Singh
e-mail: singhpushpendra548@gmail.com

R. Doti · J.E. Lugo · J. Faubert
Visual Perception and Psychophysics Laboratory, Universite de Montreal, H3T 1P1
Montreal, Canada

S. Rawat
Manipal University Jaipur, Jaipur, India
e-mail: sanyograwat@gmail.com

S. Ghosh
Natural Products Chemistry Division, CSIR-North East Institute of Science & Technology,
Jorhat 785006, Assam, India

A. Bandyopadhyay
Advanced Key Technologies Division, National Institute for Materials Science,
1-2-1 Sengen Ibaraki, Tsukuba 3050047, Japan

© Springer Nature Singapore Pte Ltd. 2018
M. Pant et al. (eds.), *Soft Computing: Theories and Applications*,
Advances in Intelligent Systems and Computing 584,
https://doi.org/10.1007/978-981-10-5699-4_31

both infrared and olfactory. Insects smell odors electronically by tuning into the emitted narrowband infrared radiation.

It has been reported earlier that the structure of insect antennae is similar to antennas in terms of operation [1, 2]. To the best of our knowledge, this is the first theoretical analysis and simulation to explain the communication between wasps developing a natural radio detection and ranging (RADAR) navigation system.

It has been reported that the wasp antenna organs behave as infrared emitter [3, 4] and could be also used for detecting the host from nearly 5 m away [5–7]. Signals involved in insect communication vary considerably. An electroantenno-gram is a technique by which the electronic signal of an insect's antenna is mea-sured [8, 9]. The structure of the Melanophila wasp is shown in Fig. 1.

2   Antenna Design

The design of the antenna consists of substrate with relative permittivity (ε) = 2.5 having height h = 1.59 mm, operating at center frequency of 300 GHz as shown in Fig. 2. The length and the width of the antenna have been derived from eqs. (1), (2) & (3). The wasp-shaped patch has the width of 0.37 mm and length 0.25 mm. The rectangular feed line has dimensions of 15 mm × 0.4 mm. The antenna geometry has finite ground plane of dimension 15 mm × 15 mm.

Fig. 1 Melanophila wasp [3]

Fig. 2 Top view of the antenna

Width of the patch [10, 11]:

$$(W) = (C_0/2f_0) \cdot \sqrt{2}/(\in r + 2). \tag{1}$$

Effective refractive index:

$$(\in r_{\text{eff}}) = \frac{\in r + 1}{2} + \frac{\frac{\in r - 1}{2}}{\sqrt{\left[1 + 12 \cdot \frac{h}{w}\right]}}. \tag{2}$$

Length of the patch:

$$(L) = C_0/(2f_0 \cdot \sqrt{\in r_{\text{eff}}}). \tag{3}$$

The variation of return loss with frequency is shown in Fig. 3a, which clearly reveals that antenna resonates at 390.94 GHz and provides bandwidth of nearly 42.81 GHz, which lies in mid-infraradiation region. The gain of the antenna is more than 16.5 dBi over the entire bandwidth as depicted in Fig. 3b.

Maximum radiation efficiency (Kr) of the proposed antenna at 386.18 GHz is 84.2% approximately. Immediate variation occurs at 395.39 GHz frequency, which is shown in Fig. 3c.

3 Theoretical Result

3.1 Efficiency Measurement

The received power at the proposed antenna is not equal to the input power. The smallest received power that can be detected by the radar is called P_r. Smaller powers than P_r are not usable since they are lost in the noise of the receiver [12]. The minimum received power is detected at the maximum range of r (5 m) as seen from the equation 'as in 4' (Fig. 4).

$$P_r = \frac{P_t G_t}{4\pi r^2} \sigma \frac{1}{4\pi r^2} A_{\text{eff}}, \tag{4}$$

where

P_t (Power transmitted by the radar)	26.2 dB mW (rms)
	26.2 × 1.414 dB mW
	10 log (26.2 × 1.414) mW
	1.568 × 10^{-4} W
G_t (Gain of the radar transmit antenna)	12.3
r (Distance from the radar to the target)	5 m

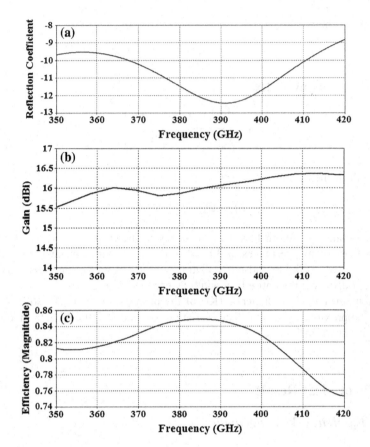

Fig. 3 a Variation of reflection coefficient, **b** variation of gain, **c** radiation efficiency versus frequency

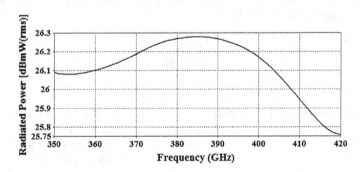

Fig. 4 Radiated power versus frequency

σ (Radar cross section of the target) $7.144 \times 10^{-9} \text{ m}^2$
A_{eff} (effective area of the radar receiving antenna) $1.89 \times 10^{-4} \text{ m}^2$

Put all the value in 'as in (4)'

Power received back from the target to the radar (P_r) $2.6 \times 10^{-16} \text{ W}$
Efficiency (Kt) = $P_t/(P_t + P_r)$ 0.0001568/
                                                                                      $(0.000157 + 2.6 \times 10^{-16})$
                                                                                      $Kt \leq 1$

Here, the transmit power by the radar is 1.568×10^{-4} W and its effective area is $1.89 \times 10^{-4} \text{ m}^2$. So the wasp irradiance is about 0.829 W/m^2, which is approximately same with the Melanophila wasp's threshold energy 60–500.

Watt/cm^2 (0.60–5.0 W/m^2). That is small compared to the sun irradiance (1.344×10^3 W/m^2) and the human irradiance (480 W/m^2).

3.2   Theoretical Host Body Temperature Measurement

The reflected power from the target (wasp-2) to the radar (wasp-1)is calculated by

$$P_{rt} = (Pt.G.\sigma)/(4\pi r^2). \tag{5}$$

After putting the value,

$$P_{rt} = 0.41 \times 10^{-13} \text{ W}.$$

This is the power required at the ideal receiver (target) having the same noise figure as the practical receiver (target). We write the formula as [12, 13]

$$P_{rt} = (F - 1)K.T_0.BW, \tag{6}$$

where

F   (Noise figure)
K   (Boltzmann constant) = 1.38×10^{-23} J/°C
T$_o$ (Standard temperature) = 290 K
BW   (Bandwidth) = 42.81 GHz

After putting the value in Eq. (7),
Noise figure (F) = 1.011.

Relation between noise figure and temperature of the target (noise temperature) is

$$(F) = \left[1 + \left(T_{target}/T_o\right)\right]. \tag{7}$$

After putting the value in Eq. 8,
Target temperature (T_{target}) = 293 K.

3.3 Radiation Pattern

The elevation radiation patterns of proposed antenna at resonant frequencies within the impedance bandwidth region are shown in Fig. 5. At resonant frequency (390.94 GHz), the E and H pattern is tilted by 90° normal to the patch geometry and has most of the radiation in the front direction.

3.4 Radiation's Mechanism

The total surface loss E of a surface is defined as the total radiant energy emitted by the surface in all direction over the entire wavelength per unit surface area per unit time. The amount of the loss E per unit time from area (A) is proportional to the fourth power of its absolute temperature (T) [14].

$$E = A\sigma T^4, \tag{8}$$

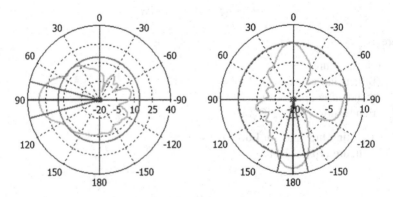

Fig. 5 Variation of E and H plane elevation patterns at the resonant frequency

where

$$\sigma(\text{Stefan} - \text{Boltzmann constant}) = 5.67 \, x \, 10^\wedge - 8 \text{W}/(\text{m}^2.\text{K}^4).$$

By theoretical analysis,

$$\text{Area of the target} = 2.25 \times 10^{-4} \text{m}^2.$$

Temperature of the target = 40 − 45 °C.
Putting the value in equation 'as in 8'

$$E = 49.60 \times 10^{-4} \times 5.67 \times 10^{-8} \times (45 + 273)^4,$$

$$E(\text{total surface losses}) = 0.034 \, \text{W}. \tag{9}$$

3.5 Wien's Displacement Law

The product of absolute temperature and the wavelength is constant at which the surface loss is maximum [14].

$$\lambda.T = 2.898 \times 10^{-3} \text{mK}, \tag{10}$$

where $T = 308{-}318$ K (temperature range); $\lambda = 0.1 - 100$ μm (thermal radiation wavelength band)
Equation (10) indicates (Table 1).

$$\lambda.T = (0.308 - 3.08)10^{-3}\text{mK} \tag{11}$$

Table 1 Antenna parameter comparison

Parameter	Theoretical result	Simulation result
Efficiency	$Kt \leq 1$	$Ks = 0.84$
Temperature	308–313 K	293 K
Radiation		
In both cases, the main lobe field (radiation) direction is normal to the target (90°)		
Total surface loss	0.03494 W	0.062 W
Wien's displacement law		
It is satisfying Wien's displacement law		

4 Matching of Simulation and Theoretical Result

4.1 Radar Cross Section of the Target

When target is in the rectangular form [15],

$$\sigma_{max} = \frac{4\pi w^2 h^2}{\lambda^2}. \tag{12}$$

A definition of the radar cross section found in some texts on the electromagnetic scattering is

$$\sigma = (\text{power reflected toward source/ unit solid angle})/(\text{incident power density}/4\pi)$$
$$= 4\pi R^2 \times [\text{Mod}(Er)^2/\text{Mod}(Ei)^2]$$
$$= 4\pi R^2 \times Mod(\Gamma)^2, \tag{13}$$

where Γ is the reflection coefficient

Comparing both equations, $\frac{w^2 h^2}{\lambda^2} = \text{Mod}(\Gamma)^2$.

After putting the value,

Width W	15 mm
Height H	1.59 mm
Wavelength λ	1 mm
$\text{Mod}(\Gamma)^2$	$= 22.75 \times 10^{-3}$
Γ	0.0047
VSWR	$(1 + \Gamma)/(1 - \Gamma)$

After calculation,

VSWR = 1.0094, which is approximately same with the simulated result 1.75 shown in Fig. 6.

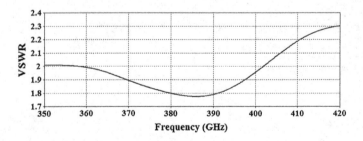

Fig. 6 VSWR/ frequency

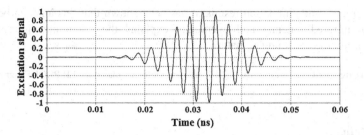

Fig. 7 Variation of excitation signal with respect to time

4.2 Pulse Repetition Frequency

Pulse repetition frequency corresponding to unambiguous range R_{un} is given by [16]

$$f_p = c/2 R_{un}$$
$$\text{pulse repetition time} = 1/f_p = 2 R_{un}/c, \qquad (14)$$

where $T_p = 0.003$ ns (Fig. 7).

By equation 'as in 14,'

$$R_{un} = 4.5 \text{ mm.}$$

Consider the target labeled A which is the unambiguous range interval R_{un}. If the first pulse repetition frequency f_1 has an unambiguous range R_{un1}, and the apparent range measured with pulse repetition frequency f_1 is denoted by R_1 (5 m), then the true range is given by

$$R_{true} = R1, \text{ or } (R1 + R_{un1}), \text{ or } (R1 + R_{un2}), \text{ or } \dots$$

$$R_{true} = 5 \text{ m, or } 5.0045 \text{ m or } 5.090 \text{ m, or} \dots$$

5 Conclusions

The antenna discussed so far explains the natural communication between two wasps by forming radar-like operation and provides a broader bandwidth 42.82 GHz, 16.5 dBi nearly gain and 84.25% efficiency with desired frequency range. The antenna operates in resonance frequency operations in infrared region.

By theoretical and simulation results, it has been shown that this antenna behaves like an infrared antenna and eventually explains the communication process between the insects found in nature.

Acknowledgements J.E. Lugo thanks the magnetophotonics material SEP-PRODEP grant.

References

1. Philip, S.C.: Insects and the battle of the beams. pp. 27–37. (http://wlym.com/basement/fusion/fusion/19850910-fusion.pdf) (1997).)
2. Philip, S.C.\S.: Exploring the spectrum, Acres (1984)
3. Schmitz, H., Bleckmann, H.: The photomechanism infrared receptor for the detection of forest fires in the battle melanophila acuminate. J. Comp. Physiol. A **182**, 647–657 (1997)
4. Campbell, A.L., Rajesh, R., Laura, S., Morley, O.S.: Biological infrared imaging and sensing micron. **33**(2), 211–225 (2002)
5. Gavan, J., Ishay, J.S.: Hypothesis of natural radar tracking and communication direction finding systems affecting hornets flight. J. Electromagn. Waves Appl. **13**(2), 247–248 (2000)
6. Gavan, J., Ishay, J.S.: Hypothesis of natural radar detection and navigation system guiding hornets flight. Int. J. Infrared Millimeter Waves **21**(2), 309–320 (2000)
7. Gavan, J., Ishay, J.S.: Hypothesis stipulating that a natural radar navigational system guides hornet flight. J. Electromagn. Waves Appl. **13**, 1661–25 (1999)
8. Keil, T.A.: Functional morphology of insect mechanoreceptors. Microsc. Res. Techn. **39**, 506–31 (1997)
9. Agmon, I., Litinetsky, L., Gavan, J.: Hypothesis on radar sensing and communication by hornest: comment on their antennal organelles and electrical activity. Physiol. Chem. Phys. **33**, 83–102 (2001)
10. Ebnabbasi, K.: A bio-inspired printed antenna transmission-range detection system. IEEE Antenna Propag. Mag. **55**(3), 01–193 (2013)
11. Delgado, J.A.B., Mera, C.A.B.: A bio-inspired patch antenna array using fibonacci sequences in trees. IEEE Antenna Propag. Mag. **55**(5), 193–03 (2013)
12. Balanis, C.A.: Antenna Theory: Analysis and Design. Wiley-Interscience, New York (2005)
13. Millian, T.A.: Morden Antenna Design. Wiley-Int, New York (2005)
14. Powell, J.L., Crasemann, B.: Quantum mechanics, Oxford & IBH publishing co. 66 Janapath (1961)
15. Skolnick, M. I.: Introduction to radar system, 2nd ed, Artech House. Inc. p 231 (1993)
16. Stewart, W.J., Galvin, K.A., Gillam, L.D., Guyer, D.E., Weyman, A.E.: Comparison pulse repetition frequency and continuous wave doppler echocardiography in the assessment in patients with valvular stenosis and regurgitation. J. Am. College Cardiol. **6**(3), 565–571 (1985)

The Fulcrum Principle Between Parasympathetic and Sympathetic Peripheral Systems: Auditory Noise Can Modulate Body's Peripheral Temperature

J.E. Lugo, R. Doti and J. Faubert

Abstract The Fulcrum principle represents a mechanism that improves our capability to extract information from the external world. It describes the interaction between at least two physiological signals, namely excitatory and facilitation. The excitatory signal can be present in any sensory system, motor, or reflex mechanism and normally is too weak to be detected by the central nervous system and/or to enact a change on the physiological, behavioral, cognitive, etc., state of a human subject. Simultaneously, the facilitation signal can be present in any sensory system, motor or reflex mechanism and it is its energy and frequency content that may create a general activation between the central nervous system and peripheral nervous systems. Consequently, the excitatory signal can be detected and/or the physiological, behavioral, cognitive, etc., state of a human subject can be changed. Herein, we present an example of such principle where auditory noise can induce transitions between sympathetic and parasympathetic nervous systems, which are part of the autonomic nervous system.

Keywords Fulcrum principle · Autonomic nervous system · Body temperature Biofeedback

1 Introduction

The autonomic nervous system is divided in sympathetic and parasympathetic nervous system. The main function of these two systems is regulating the body's unconscious actions. The sympathetic nervous system's main function is to

J.E. Lugo (✉) · R. Doti · J. Faubert
NSERC-Essilor Industrial Research, Faubert Lab, School of Optometry,
Université de Montreal, Montreal H3T 1P1, Canada
e-mail: eduardo.lugo@gmail.com

© Springer Nature Singapore Pte Ltd. 2018
M. Pant et al. (eds.), *Soft Computing: Theories and Applications*,
Advances in Intelligent Systems and Computing 584,
https://doi.org/10.1007/978-981-10-5699-4_32

stimulate the well-known fight-or-flight response. On the contrary, the parasympathetic nervous system is believed to activate the rest-and-digest response. That is, both systems can be seen as being complementary to each other.

It is known that blood gradients perfusing the skin have an impact on the body's peripheral temperature as measured on its extremities. This mechanism depends on the subject's state of sympathetic arousal. When a person is not stressed, their extremities tend to be warm but when a person is stressed; their fingers tend to get colder. Nowadays, there are biofeedback relaxation techniques that help people to learn how to trigger parasympathetic responses to voluntary increase their finger temperature or electrical conductivity (GRS) [1].

The Fulcrum principle [2–13] can be summarized as follows: A subthreshold and/or subconscious excitatory signal (present in one sense, reflex, and/or motor mechanism) that is synchronous with a facilitation signal (present in a different sense, reflex, and/or motor mechanism) can be increased (up to a resonant-like level) and then decreased by the energy and frequency content of the facilitating signal. As a result, the sensation of the signal, and/or the physiological, behavioral, cognitive state of the subject, changes according to the excitatory signal strength an inverted U shape like function. For instance in sensory mechanisms, the excitatory stimulus refers to the signal applied to the sense that we want to study. The facilitation stimulus means the signal applied simultaneously to the same subject, intended to trigger a multisensory integration in a way that facilitates the perception of the excitatory stimulus. In this context, sensitivity transitions represent the change from subthreshold activity to a synchronized firing activity in multisensory neurons. Initially, the energy of their activity (supplied by the weak signals) is not enough to be detected but when the facilitation signal enters the brain, it generates a general activation of multisensory neurons, modifying their original activity. In the past, we have explored different multisensory combinations; facilitation auditory noise versus excitatory tactile, proprioceptive, and visual signals; facilitation auditory deterministic signal versus excitatory tactile signal; and facilitation tactile noise versus excitatory visual signals [2, 3, 9].

In this work, we showed that by using the Fulcrum principle, it is possible to facilitate finger temperature increases by using an effective auditory noise. Therefore, transitions from the sympathetic system to the parasympathetic system occur thanks to the energy and frequency content present in the auditory noise.

The present work is divided as follows: Sect. 2 describes the materials and methods used in this work. Section 3 shows the results we obtained and Sect. 4 wraps up with a discussion and some conclusions.

2 Materials and Methods

2.1 Temperature Measurements

We used a BioGraph Infiniti V4 plus software (TT Ltd), a ProComp Infiniti System amplifier with eight channels (TT Ltd), and a temperature sensor—SA9310 M (TT Ltd). The temperature sensor was strapped to the palmar side of the index finger using the short strip of Velcro that is provided with the sensor.

2.2 Noise Generation

The auditory stimuli were presented binaurally by means of a pair of headphones (Grado Lab SR80) plugged in an amplifier (Rolls RA62b). We used a calibrated high fidelity capacitor microphone (Behringer ECM8000) to verify the frequency response of the headphones inside an acoustically isolated chamber. A computer provided auditory white noise to the amplifier and the intensity range of the noise generator was calibrated to supply white noise with an intensity of 60–95 dBSPL with a systematic error of 3.5 dBSPL. The testing room sound disturbances (e.g., computer fans and low power sounds coming from outside the testing room) were recorded by using the same microphone as before and their cutoff frequency was 2.5 kHz with the intensity of 50 ± 3.5 dBSPL. Although the noise generation band and the electronic amplification bandpass are wider than the auditory spectrum, the acoustic transducer of the headphones drastically modifies the noise spectrum density because of mechanic and electric resonances. The most limiting factors at successive stages are the headphones that cannot reproduce the full white noise spectrum but still have an effective acoustic noise spectrum. Common transducers have a high cutoff frequency between 10 and 12 kHz, but the human auditory noise thresholds are best for a noise spectrum between 5 and 12 kHz (ITU-R 468 noise weighting standard), partially compensating each other. The different processing stages required for the original noise to finally reach the cortex inevitably modify the original white noise spectrum. This implies that the cortex interprets only a limited effective noise band with a cutoff frequency of about 12–15 kHz (where its spectrum is attenuated) instead of a full white noise spectrum. Here we used three different noise levels with values 60, 65, and 70 dBSPL.

2.3 Subjects

Six subjects began the experiments; the first consisted of taking the peripheral temperature baseline for 10 min. In the second part, the experiment was divided into five events. First with a no-noise condition during 2 min, and then we manually increased the noise level. We used three different noise levels, which duration was 2 min as well. Finally, the experiment finished with another 2 min with no-noise again to have 5 events in total. We tracked down peripheral temperature with the equipment described before. The experiments took place in an illuminated room were the subjects sat down comfortably. This study obtained ethics approval from the Institutional Review Board of the University of Montreal.

2.4 Data Analysis

We calculated the integral (surface under the curve) of temperature's time measurements plus the average gradient normalized by its own magnitude; the latter is always a positive unit (positive slope) or negative unit (negative slope). The product of the area and the positive or negative unit gives a figure of merit to evaluate the optimal noise level, which we will call the noise index. Therefore, the optimal noise level should be the one with the highest positive noise index.

3 Results

Figure 1 shows the result for subject 1. As an example, we plotted the result for the baseline. We observe that its behavior is completely different from the experiment where noise levels were manipulated. In fact, by taking the temperature average on each event and compare it with the temperature average for the same period of time in the baseline, we can distinguish that noise level at 70 dBSPL facilitates a temperature increment above the baseline. The result was also obtained from the noise index analysis. For the second experiment, as the test commenced S1's peripheral temperature began to decrease, reflecting the stress level of S1. As soon as the first noise level was applied the peripheral temperature increased and the same thing happened for the next two noise levels to finally decrease once the noise was ceased. From noise index values, the optimal noise level was determined to be 70 dBSPL.

Figure 2 shows the result for subject 2. As the test commenced S2 peripheral temperature began to decrease, again reflecting the stress level of S2. For this subject, as soon as the second noise level is presented the peripheral temperature increased and the same thing happened for the next noise level to finally stay almost constant once the noise was ceased. The optimal noise level was 70 dBSPL.

Figure 3 shows the result for subject 3. As the test commenced S3 peripheral temperature began to decrease as for the other subjects, again reflecting the stress level of S3. As soon as the first noise level is on the peripheral temperature increased for a while and then decreased again. When the second noise level was activated the temperature increased once more. With the third noise level temperature values fluctuated up and down slightly but keeping an approximately constant

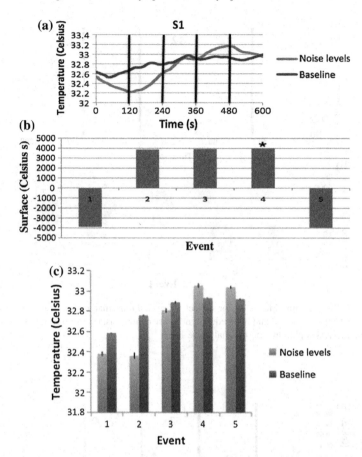

Fig. 1 **a** Temperature time series for subject 1. *Vertical lines* marked the beginning of an event (No-noise, noise level 1, noise level 2, noise level 3 and no-noise). All subjects started and finished with no-noise conditions. During the baseline condition, there was no-noise levels presented. **b** It shows noise index values; the asterisk highlights the optimal noise level. **c** Average temperature per event compared with the baseline for each event. In all cases, both averages were statistically significantly different ($p < 0.05$, t-test, bilateral, variance heteroscedastic)

average value. Clearly at the end, when there was no-noise, temperature values dropped drastically. From the noise index values, the optimal noise level was 70 dBSPL.

Figure 4 shows the result for subject 4. As the test commenced S4 peripheral temperature fluctuates slightly up and down then decreased drastically. This decrement continued even after first noise level was on for the first 40 s then temperature increased rapidly. When the second noise level began, the temperature kept increasing but at a smaller rate and then suddenly started dropping steadily with some fluctuations. Same thing happened with the third noise level and the last no-noise condition. The optimal noise level was 60 dBSPL.

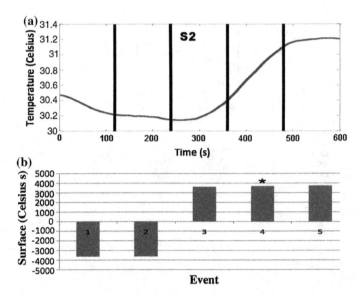

Fig. 2 (*Top*) Temperature time series for subject 2. *Vertical lines* marked the beginning of a noise condition. All subjects started and finished with no-noise conditions. (*Bottom*) it shows noise index values; the *asterisk* highlights the optimal noise level

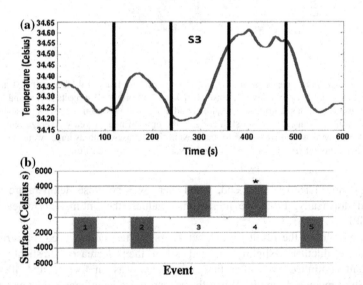

Fig. 3 (*Top*) Temperature time series for subject 3. *Vertical lines* marked the beginning of a noise condition. All subjects started and finished with no-noise conditions. (*Bottom*) it shows noise index values; the *asterisk* highlights the optimal noise level

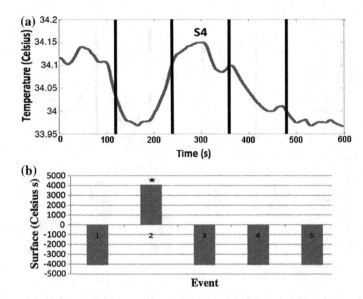

Fig. 4 (*Top*) Temperature time series for subject 4. *Vertical lines* marked the beginning of a noise condition. All subjects started and finished with no-noise conditions. (*Bottom*) it shows noise index values; the *asterisk* highlights the optimal noise level

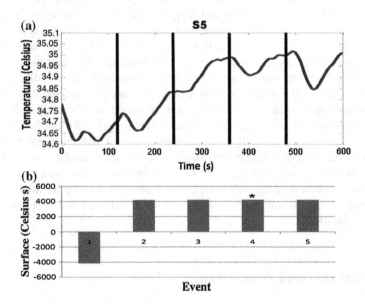

Fig. 5 (*Top*) Temperature time series for subject 5. *Vertical lines* marked the beginning of a noise condition. All subjects started and finished with no-noise conditions. (*Bottom*) it shows noise index values; the *asterisk* highlights the optimal noise level

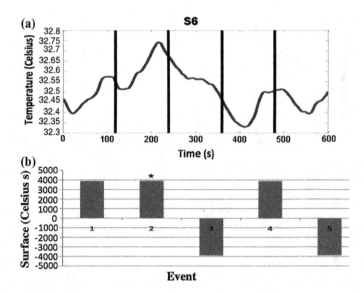

Fig. 6 (*Top*) Temperature time series for subject 6. *Vertical lines* marked the beginning of a noise condition. All subjects started and finished with no-noise conditions. (*Bottom*) it shows noise index values; the *asterisk* highlights the optimal noise level

Figure 5 shows the result for subject 5. The S5 peripheral temperature began to decrease then fluctuated up, down, and then up again. When the first noise level was on temperature dropped for around 30 s and then increased noticeably. This increase kept going even under the influence of noise level 2. When noise level 3 was on temperature dropped again for 40 s approximately and then increased to a level higher than precedent values. The optimal noise level was 70 dBSPL.

Figure 6 shows the result for subject 6. As the test commenced S6 peripheral temperature began to decrease for 30 s then increased rapidly until it decreased again. When noise level 1 was on temperature kept dropping for another 10 s, and suddenly it increased drastically up to a maximum. 30 s before this noise level went off temperature dropped continuously up to 50 s after noise level 3 was on. At that point, temperature increased again but not to preceding values. With the no-noise condition, temperature decreased again for 80 s and then went up for the rest of the experiment. From the noise index values, we can determine that the optimal noise level was 60 dBSPL.

4   Discussion and Conclusions

All subjects presented an initial decreased on peripheral temperature. The time interval of this decrement is subject depend, and it probably reflects the initial stress expressed by the subjects that started even before the test begins. Then, some subjects are capable of reversing this trend as S5 and S6 (although the stabilization mechanism is not that constant). Therefore, in this part of the experiment, two subjects already showed transitions between both autonomic nervous systems.

S1 and S3 responded immediately after the first noise level was on by showing an increase in their peripheral temperature. However, only S1 showed a steady increase. The temperature in S2 continued to drop under this noise level. The rest of the subjects' temperatures fluctuated up and down or down and up showing transitions between both autonomic systems. This is evidence that auditory noise is in fact facilitating such transitions.

When the second noise level was presented, the peripheral temperature increased steadily in subjects S1, S2, S3, and S5. Meaning that at this level, the noise facilitated transitions from sympathetic to parasympathetic systems in a more consistent way. For S4 peripheral temperature fluctuated up and down showing a transit toward the sympathetic system. Clearly, for S6 a transition to the sympathetic system is taking place.

When the third noise was applied, the peripheral temperature on subjects S1 and S2 kept increasing stably. This may be interpreted as both subjects having stronger parasympathetic responses. On the contrary, peripheral temperature on S4 decreased stably showing stronger sympathetic responses. The peripheral temperature of subjects S5 and S6, fluctuated first down and then up and in subject S3 temperature slightly fluctuated following the sequence up-down-up-down-up.

When the auditory noise is off at the end of the experiment, the peripheral temperature in subject S1 decreased immediately with no fluctuations. Suggesting that for this subject the transition from parasympathetic system to sympathetic system took place immediately after the noise level was off. In S3, there is an abrupt decay in peripheral temperature up to levels similar to the ones measured at the beginning of the experiment. This also happened just after the last noise level was off. For S4 peripheral temperature continued to drop up to levels even lower than at the beginning. The peripheral temperature in S5 and S6 first decreased abruptly and then increased with a similar rate. The only difference is the peripheral temperature final levels in both subjects. For S5, they are close to the one obtained when the optimal noise level was presented. On the contrary, for S6, they are similar to the ones measured at the beginning of the experiment. Only S2 maintained constant peripheral temperature values higher than any other preceded experimental condition.

From these results, we can observe that the parasympathetic system state lasted even after all noise levels were off in at least two subjects. On the contrary, the sympathetic state is always present in all subjects at the beginning of the experiments.

Noise index values showed that four subjects out of the six presented 70 dBSPL as optimal level, and in two subjects, the value was 60 dBSPL. In a previous experiment, we also found similar results where, in 70% of the subjects from a group, the 69 dBSPL auditory noise level always facilitated excitatory signal transitions [2].

In conclusion, we can show that the Fulcrum principle is also present within the central nervous system and the autonomic nervous system. Here, transitions from sympathetic to parasympathetic systems are facilitated centrally by means of a noticeable auditory noise.

Acknowledgements This work was partly supported by an NSERC Discovery operating grant.

References

1. Critchley, H.D., Melmed, R.N., Featherstone, E., Mathias, C.J., Dolan, R.J.: Brain activity during biofeedback relaxation: a functional neuroimaging investigation. Brain **124**, 1003–1012 (2001)
2. Lugo, J.E., Doti, R., Faubert, J.: Ubiquitous crossmodal stochastic resonance in humans: auditory noise facilitates tactile, visual and proprioceptive sensations. PLoS ONE **3**, 1–18 (2008)
3. Lugo, J.E., Doti, R., Wittich, W., Faubert, J.: Multisensory integration central processing modifies peripheral systems. Psychol. Sci. **19**, 989 (2008)
4. Harper, D.W.: Signal detection analysis of effect of white noise intensity on sensitivity to visual flicker. Percept. Mot. Skills **48**, 791 (1979)
5. Manjarrez, E., Mendez, I., Martinez, L., Flores, A., Mirasso, C.R.: Effects of auditory noise on the psychophysical detection of visual signals: cross-modal stochastic resonance. Neurosci. Lett. **415**, 231 (2007)
6. Ai, L., Liu, J., Liu, J.: Using auditory noise to enhance the fine motor of human's hand due to cross-modal stochastic resonance. In: International Conference on Biomedical Engineering and Informatics 5305070 (2009)
7. Söderlund, G., Sikström, S., Smart, A.: Listen to the noise: noise is beneficial for cognitive performance in ADHD. J. Child Psychol. **48**, 840 (2007)
8. Wlodzislaw, D.: Neurodynamics and the mind. International Joint Conference on Neural Networks, San Jose, California, pp. 3227–3234 (2011)
9. Lugo, J.E., Doti, R., Faubert, J.: Effective tactile noise facilitates visual perception. Seeing Perceiving **25**, 29 (2012)
10. Haas, C.T.: Vibratory stimulation and stochastic resonance therapy: results from studies in Parkinson's disease and spinal cord injury. Technologies of Globalization Congress, Darmstadt, Germany, pp. 1–16 (2008)
11. Metha, R., Zhu, R.J., Cheema, A.: Is noise always bad? Exploring the effects of ambient noise on creative cognition. J. Consum. Res. **39**, 784–798 (2012)
12. Sawada, H., Egi, H., Hattori, M., et al.: Stochastic resonance enhanced tactile feedback in laparoscopic surgery. Surg. Endosc. **29**, 3811 (2015). doi:10.1007/s00464-015-4124-y
13. Hoskins, R., Wang, J., Cao, C.G.L.: Use of stochastic resonance methods for improving laparoscopic surgery performance. Surg. Endosc. **30**, 4214 (2016). doi:10.1007/s00464-015-4730-8

Fractal Information Theory (FIT)-Derived Geometric Musical Language (GML) for Brain-Inspired Hypercomputing

Lokesh Agrawal, Rutuja Chhajed, Subrata Ghosh, Batu Ghosh, Kanad Ray, Satyajit Sahu, Daisuke Fujita and Anirban Bandyopadhyay

Abstract We propose fractal information theory (FIT) to compute, and it uses a Fractal tape, wherein "every single cell of a Turing tape contains a Turing tape inside." To use this tape, we introduce a geometric musical language (GML). This language has only one letter, a time cycle, a rhythm, a clock, or a unitary operator; on the circle perimeter, multiple singular bursts or "bings" (singularity represented as circles) are located. Time gap or "silence" between the "bings" is adjusted to hold the geometric parameters of structures such as square, triangle. Each time cycle is part of a phase space or a Bloch sphere; hence, information is now a "Bloch sphere with a clocking geometry." Several such spheres self-assemble and expand like a balloon to store and process complex information; "bings" are singularity glue to add clocking Bloch spheres into it; this is the basic of fractal information theory (FIT). The conversion of five sensory signals into geometric shapes and rhythms like music and vice versa is called geometric musical language (GML). New information is integrated as guest into a single ever-expanding host Bloch sphere. The distinction between questions and answers disappears and replaced by "situation," written as geometric shapes and always paired together in a time cycle,

L. Agrawal · R. Chhajed · D. Fujita · A. Bandyopadhyay (✉)
Advanced Nano Characterization Center, National Institute for Materials Science,
1-2-1 Sengen, Tsukuba, Ibaraki 305-0047, Japan
e-mail: anirban.bandyopadhyay@nims.go.jp; anirban.bandyo@gmail.com

S. Ghosh
Natural Products Chemistry Division, CSIR-North East Institute of Science & Technology,
Jorhat 785006, Assam, India

B. Ghosh
Department of Physics, TDB College, Burdwan University, Burdwan 713347, West Bengal,
India

K. Ray
Department of Physics, Amity University Rajasthan, Jaipur, Rajasthan, India

S. Sahu
Nano Bio Systems Science, IIT Jodhpur, Jodhpur, Rajasthan, India

© Springer Nature Singapore Pte Ltd. 2018
M. Pant et al. (eds.), *Soft Computing: Theories and Applications*,
Advances in Intelligent Systems and Computing 584,
https://doi.org/10.1007/978-981-10-5699-4_33

side by side or one inside another. Just like a human brain, FIT-GML hypercomputing does not require algorithm or programming, and it uses the fractal beating, i.e., geometric nesting inside a Hilbert space. FIT reduces to quantum information theory, QIT, if the clocking geometry in the Bloch sphere and virtual poles are removed and the singularity feature of a "bing" is eliminated, which makes it a classical state.

Keywords Fractal information theory · Geometric musical language
Bloch sphere

1 Introduction

The data deluge problem [1] is due to a wrong information theory that has alone ruled for 100 years and still being followed like a religion in every single form of computing and information processing. The unit of information is bits or qubits, 0 and 1, and we add facts as real-world significance to these numbers (Fig. 1a). Existing information theory (EIT), be it classical information theory (CIT) [2–4] or quantum information theory (QIT) [5–7], both, discretizes every single natural phenomenon and events as a sequential sum of elementary decisions "yes" and "no," only to discover eventually that the length of the sequential chain has turned astronomical. This chain of instructions could be read instantly using a quantum computer [8], but both quantum and classical computers have a common problem; they need meticulous details of the path preconceived and a built-in circuit implementing that path, else they do not work. The actual data deluge problem is not the amount of data, but wiring the elementary events as those accurately prevails in its environment at real time [9]. Who would write millions of lines of codes and then build an equivalent circuit at every 200 ms [10] to replicate a human conscious experience? Human brain would not only have to imagine the true "one-to-many and many-to-one" wiring [11, 12] of all events in our environment, but also have to implement the equivalent circuit instantly and in real-time process all of it. If we continue to use our intuition, imagination, and creativity to see meticulous details of the future in our own brain and lay down intelligent algorithms to unearth the hidden mysteries in an incredible amount of data, we would fail. Just imagine, a search engine looks into 0.01% of total Internet; even if we see 100%, still we would be left with the same task, "our super brain would have to find what is it" instantly.

Rejection of EIT is not enough; to replace it with a new information theory, we need to define what the true structure of information is in nature. Current science connects the signal bursts and fits with equations; models hardly care about the silence between the signals. Quantum information theory (QIT) does care about phase or the "silence," but ends up compromising it totally in the real values. The network of "silence" has remained untouched. We suggest developing a new way of writing information, processing it, and memorizing it, by harnessing the network of

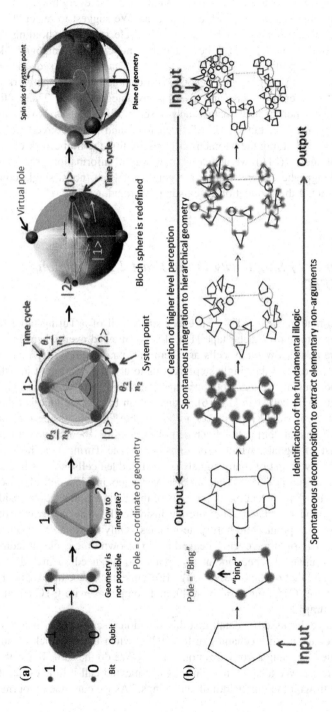

Fig. 1 The basic concept of FIT and GML: a time cycle is a complex unitary matrix as shown in **a**. **b** Shows that any given input would do two things, integrate the "bings" to create smaller and simpler architectures, and at the same time since lots of geometries densely populate the time cycles, they decompose into simpler geometries. Both the process bottom-up and top-down run in parallel

"silence" such that we never need to discretize nature and lose everything, only to rebuild it later by using novel and pathbreaking ideas. We suggest to reject "bits" and "qubits" and to consider information as a time cycle or rhythm holding the basic geometric shapes in a Bloch sphere that has no classical analogue like quantum (Fig. 1a). The philosophy that "Everything that happens is a sequence of events" is replaced by "everything that happens is a change in the 3D network of rhythms that includes environment and user as its integral part." Thus, "the geometry of silence" morphs with nature and observer; if one asks how many qubits, the reply is just one, but no "bits." Even mass and space convert into a network of time cycles holding the geometric shapes. Within the framework of this fractal information theory (FIT), we develop a new way of information processing and decision making, where there is no logic gate, no switch, not a single component that constructed the foundation of existing information theory (EIT) based on probability.

1.1   The Science of Singularity in the Fractal Information Theory (FIT)

In the fractal information theory (FIT), in every single cell of a Turing tape, we insert a Turing tape, to get a Fractal tape [13], this is followed everywhere in the network, so if one asks "how many cells are there in your system?" The reply is "one." That is all we do, yet the single step takes us to a world where not just the entire Turing tape, but also every cell becomes undefined; we cannot assign a definite value to a single cell state [14–16]. If any cell in the Fractal tape is asked "what is your state?", it would reply "I do not know, but the cells inside me, ask them, they might state that definitely." Each cell of the Turing tape makes a "bing" or a singular burst of signals, which represents the whole Turing tape residing within. It is a collective burst of many signals from all hidden cells, so it is like an emission from singularity [17–19]. When an observer enters inside a cell of a used Fractal tape, the "bings" representing all the cells of the Turing tape residing inside a single cell are found correlated. The journey inside cells is never ending, the network starts from a point, but hardly terminates. Only a suitable observer deconstructs the "bing" or "time cycle," depending on its own time cycle structure; yet, we cannot terminate the network at any point, so we need to include the network of nature and that of the observer (Fig. 1a). Time cycle resembles Gödel's closed time-like curve (CTC) proposal of 1949, and computing with CTC equates classical and quantum [20].

Recently, we have discussed in detail that this situation resembles the history of the development of quantum mechanics in the 1920s, when the deadlock of singularity was avoided by using renormalization [21, 22]. We do the same but with a fundamental difference. We take a unitary matrix of time and call it time cycle; its diameter varies to fit with the time width of cell "bings." As we continue to journey inside the cells, the outer diameter of "bing" sets the nodes for the time cycles of cells inside. Thus, the cells residing side by side could be integrated using a time cycle with multiple "bings" on its perimeter. While the cells that reside one inside another would fit inside a single "bing," we replace the "bing" with a network of time cycles inside. So, we put "bing" as a single circle, whose perimeter is empty. As one rotates around the circle, starting from a "bing" there are "silence" between the "bings", a "silence" is an arc of a circle, makes a certain angle to the center, sum of all "silence" is 360°. Since every single geometric shape, be it 1D, 2D or 3D, would fit in a circle or sphere, we use "silence" to encode the angular property of geometric shapes into a time cycle, and diameter is always set by the timescale we operate (Fig. 1b). Encoding geometric information into the time cycle lets us build the geometric musical language, GML. The word musical is used to express the signal produced when system point rotates along the circle perimeter; to an observer, it is a song.

The "silence" cannot be explicitly written on the Turing tape, since the network is 3D and each cell can have millions of connections operating simultaneously. Quantum mechanics allows us to do so. We simply rotate the time cycle and construct a 3D phase space, a Bloch sphere, just like quantum. However, the basic difference is incompleteness [23]. In the world of intermediate states of quantum, we could never make a journey; now, in the network of "bings" one inside another, we have an endless network of those imaginary worlds one inside another, not that infinity we knew thus far [24, 25]. They have their own geometric wiring, links, and interactions. In the current formulations of quantum mechanics, all discrete Hilbert spaces are merged into one. Here, they remain distinct and form geometric connections, i.e., imaginary of an imaginary space. Thus, FIT-GML protocol is best suited to make a journey into the world of singularity and unravel unprecedented features. Since quantum mechanics historically avoided this journey by using renormalization, we suggest that GML produces a better map of nature. This is the reason; we have set morphing of nature by a matrix of the modified Bloch spheres (MBS) as the singular protocol, to learn, decision making, and evolve. Morphing is mimicking unseen dynamics of nature in the MBS, so that all futuristic possibilities are produced when needed.

Decorating the time cycles with the geometric ordering of "bings" only makes the lowest level of the information network, wherein each time cycle could rotate and generate all possible pathways the "bings" could get connected. We find a 3D Bloch sphere for each time cycle. In the MBS, we never store "dead" geometric shapes alone, geometry for any two situations or events, say situation 1 is a triangle and the geometry for correlated event or situation 2 is a square; then, they are always paired in a single time cycle (could be more). On the perimeter, both events are written side by side, with equal status (no identity). This is a nonargument, and since we do not reject, it is not computing. We use morphing instead of computing. These Bloch spheres containing a group of events or situations self-assemble to generate a massive 3D architecture with an ever-increasing complexity (Fig. 2a, b, MBS = Integrated Information Architecture, IIA). Once the network forms, the elementary geometric shape combines and the new time cycles are produced,

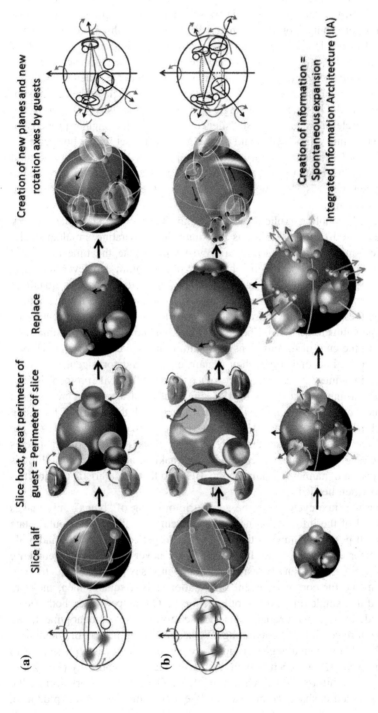

Fig. 2 Self-assembly of Bloch spheres to form integrated information architecture (IIA): **a** two examples are shown how actually on the Bloch sphere new Bloch spheres are added when a "bing" decomposes. Top row shows a case for triangle, and the bottom row shows the same for a square. **b** When several Bloch spheres are added one inside another, then the time cycles must find a space, and while faster time cycles are created, the whole event appears as an expansion to an observer residing at faster time cycles

naturally, so that several correlated geometric shapes eventually become just one geometric shape at the highest level. This hierarchical integration enables MBS to simulate the future and run a perpetual morphing of nature. The two fundamental aspects of MBS are that the processing of information is the evolution of its hardware (imagine a fractal logic gate, [26]). Second, morphing maps the structure of information as it happens in nature, i.e., keeps the network of "silence" intact. EIT breaks and fails to retrieve nature. We make sure that in FIT-GML, nature is morphed as it is, and it is so robust that the observer could change his orientation with the MBS structure in infinite ways and each time he would get a new per- spective. *IIA is made of only singularities, so it is a noncomputable architecture.*

1.2 Ten Reasons Why Existing Information Theory (EIT) that Includes Classical (CIT) and Quantum (QIT) Cannot Morph Noncomputable Nature or Brain

FIT rejects everything in the existing information theory (EIT), i.e., Turing–Russell philosophy, data, bit/qubit, rule, software, switch, lexicon, logic, algorithm, pro- gram, model, memory, switch, processor, subroutine, encoder, decoder, symbol, or buffer, probability, communication, reduction, data transfer all should go. Then, which terminologies would be relevant? Logically, circular, redundant, undefined, singularity, unbounded or divergent, infinite loop or no halting, noise or random- ness, broken link or error, discontinuity, spontaneous reply, etc. *Noncomputability can have architecture in FIT*.

(1) **Relies on the capacity of human intellect**: User measures only what he/she wants to see in nature, not what the system is in totality, and then imagine a theory to link observations, try to rebuild the natural system by adjusting core principles only to fit its prediction, and improve the network of equations that fit the obser- vation. (2) **Rejects nature's original information structure**: It consolidates a faith that everything that happens in nature is an output of a series of events happening one after another. Existing parallelism is simply running several sequential events in parallelism. The science of simultaneity is considered in an ad hoc manner in quantum, nature might have mastered the science of simultaneity, which is not parallelism, and such possibilities are not considered. (3) **Isolates the concept information and structure**, as if information is a property of a structure, what if, events, objects and everything is just information, both sides of the same coin. (4) **Endorses "real" as universal truth; purity of imaginary is compromised**: Though quantum information theory (QIT) explores the phase space, the sensed information is always considered as real and factual truth. Note that phase is always an unseen part in information and quantum always makes sure that there is a projection from the Hilbert space to the real world. (5) **Practices intelligence as magical surprise**: All protocols of artificial intelligence suggest to match output and input by any means, and out-of-the-blue senseless correlations are endorsed as

key to nature (e.g., cancer cell growth = radioactivity). Why are they similar? (6) **Believes hierarchical processing exists only in the real world**: "Deep learning" means keeping several mysterious layers between input and output, and none makes sense, but better fits an observation. When fitting becomes an absolute target, one imaginary world is sufficient because a function of imaginary i supplies the missing part. However, there could be a hierarchical network in the imaginary too. (7) **Believes randomness is fundamental to nature**: Nature is not sequential, not parallel, not simultaneous, and not random alone. Randomness speeds up searching, that is used to depict efficiency of quantum, but the origin of randomness was never formulated scientifically. Fast searching does not ensure higher intelligence; it could be that randomness is derived from something more fundamental. (8) **Endorses two-point mapping of an event**: Self-taught software (STS) is a mere extrapolation of outcomes. An extrapolation is not a prediction. It is a beyond domain value of an optimized function that maps only two choices. The science of equations inherently considers that at any moment, there are only two particles in the universe. The rest of the particles simply disappear. Thus, many-body systems are simple sum of two body interactions. (9) **Believes in making decision by reduction of choices**: In the real life, even if one finds an exact logical solution by thinking but may do something different which was never a logical input during the thought process. One may accept all or reject all. All computing protocols proposed till date believe that the only way to make decisions is to first expand choices by imagination and then reduce it drastically by rejection. (10) **Believes that the speed and the amount of resources can solve any problem of the universe**: The quest for a better computing is a faster speed and more memory/power. What computers do? Send current through circuits as the algorithm instructs, not a single step is taken beyond instruction. Thus, in all measures, the supercontrol of intelligence left permanently to the human's wild imagination.

How materials dynamics would eliminate the utmost need of a user: We need to take every right of decision making from user and give it back to the spontaneous dynamics of a system/material structure. Acquiring information from the environment, interacting back and forth with the environment, autocorrecting errors in detection, learning, etc.; ever-evolving materials dynamics would do everything so that we do not rely on even a single line of programming (we are designing and synthesizing brain jelly for that purpose [26]). GML converts vision, sound, smell, touch, taste all possible sensory signals into a single nested time cycle network of events (as question–answer loses identity) (Fig. 2a, b) at the sensory hardware. Sensors form their own self-assembled Bloch spheres as integrated information architecture. Then, the matrix of Bloch spheres (MBS) or the actual computing jelly (e.g., brain jelly) holds another MBS that absorbs the time cycles as Bloch spheres from outside and transforms jelly organization, and it is part of morphing the nature and external observers. When a new problem is given as input to explore the future possibilities of all routes, the modified MBS attempts to make a reverse morphing, thus change nature and the external observer it delivers a spontaneous reply (answer to the query). FIT-GML inspired new age computer (we call it AjoChhand computing [13], Fig. 3) does not require instructions or even a request to solve its mutual morphing process to accommodate more with nature and user. If not, even attempting to changing them is a universal drive that produces the answer to the computational query. User or observer, IIA, nature, and all integrated architectures of Bloch spheres reside as guests on the most primitive time cycle or host Bloch sphere of nature (Fig. 3).

1.3 Ten Fundamentals of the Fractal Information Theory (FIT)

1. **Use of Fractal tape by replacing the Turing tape**: Every single cell of a Turing tape would have a Turing tape inside [13]. Thus, finite state cells assemble side by side or one inside another to turn undefined. While the side-by-side cells could be analyzed by a Turing tape, the one-inside-another network cannot be done. Simplest device implementing a Fractal tape should harness singularity, i.e., could transit through infinite path across the cells and return a value. There could be only one cell visible at the top and infinite cells inside. Bloch sphere of a cell of a Fractal tape will not have real classical poles like quantum (see Fig. 1a).

2. **Information is a "network phase" beyond dynamic and geometric phase**: Fractal machine is a three-cell Turing tape, wherein each cell is a Fractal tape, one belongs to the immediate environment, then one to the morphing matrix, and the last one belongs to the observer. Information unit is a time cycle made of singularity points ("bing") that holds the geometric parameters, residing as a great circle on its own Bloch sphere. Information holds the geometric connections using phase (arc of silence) with every cells inside, above, and side by side, and the direction of motions of system points on the time cycles, so we call it "network phase." An observer can rotate around the 3D information structure and see the same structure in infinite possible ways. Since entire system has only one unit of information, the quantity of information and the computing (morphing) speed are irrelevant. Apart from geometric and dynamic phase seen in quantum, there is an additional "network phase" generated by the fractal wiring of the phase space spheres, and inside a single phase, sets of several phases reside geometrically to constitute a network phase.

3. **Smallest decision-making unit ("Fractal Machine")**: FIT envisions one machine that would act as a sensor, decision maker, memory element, and learning element, all in one. Each time cycle has at least two correlated events (analogue of a question–answer), and this is the smallest decision-making unit. Several event pairs act as a single point and arranges in the form of a single geometric shape, and it continues. Thus, an entire hierarchical network of decisions from the elementary timescale to the largest is also the smallest decision-making unit. Unlike Turing machine, a Fractal machine's four tuples are performed simultaneously, all at once, continuous, not step by step. The

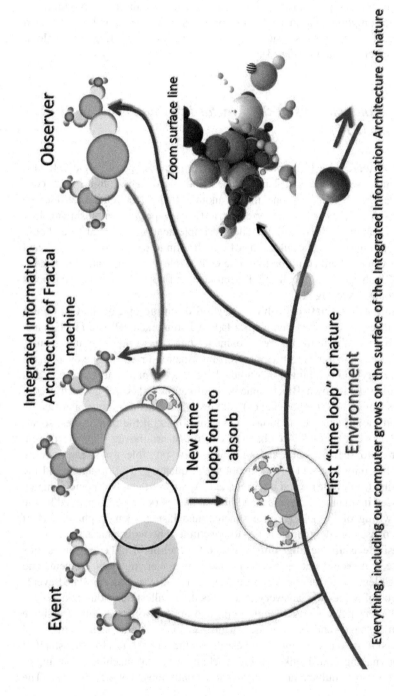

Fig. 3 The supreme host Bloch sphere of the universe and three guests: the arc shows the time cycle embedded in the host Bloch sphere of the universe. The Bloch sphere assembly for the computer (integrated information architecture, IIA), user, or observer and the observed all three primary matrices of Bloch spheres (MBS) are guests on the time cycle of the host Bloch sphere. All discrete events (to be observed) are discretely spaced MBS, residing on the elementary Bloch spheres of nature or universe. The perpetual collective evolution of three MBS systems (IIA, user, observed) is the computing of FIT-GML, and we do not say computing, but say morphing

Fractal machine (i) converts and absorbs time cycles, (ii) expands to find associations, (iii) transforms to integrate, (iv) reply and edit to learn these four steps are taken together, repeated indefinitely, computation never stops.

4. **Morphing drive naturally executes decision making (Five fundamental drives that cause information processing)**: Morphing is the only drive that happens in MBS, i.e., the triangular time cycle network (System–User Environment, SUE), and it generates several subdrives. Morphing means encoding the nested time cycles in S such a manner that the dynamic of S holds specific parts of U and E that take part in synchronization. S first tries to absorb U and E rhythms, if the discrepancies are sorted out by integrating rhythms; S sends suitable time cycles to U and E to manipulate them. How this happens? Density of "bings" is matched between the networks; new time cycles are added in a similar direction of rotation. Thus, phase, geometry, spin, density are all replicated in a loop that ensures morphing. Information includes various kinds of drives that regulate the dynamics; unlike quantum, all these drives are well defined: (i) unitary drive (diameter of time cycle is autoadjusted), (ii) C2 symmetry drive (phase transition of Bloch sphere network to generate symmetry); (iii) Fractal clock drive (system point moves to the fastest clock and then to the slowest and does back and forth); (iv) synchronization drive (triggering self-assembly and disassembly of time cycles); (v) protection drive (long-term memory protects the core of S and short-term memory enables editing).

5. **System point speed, relative phase, number of system points, the role of system points**: The speed of the system point rotating in a time cycle is determined by width of "bing." Its diameter is fixed and time taken by the system point to cross the diameter of a "bing" is the unit of speed. A waveform is a periodic oscillation, represented by a single system point rotating around a circle. If another time cycle sits on it with another system point, then together they make a beating. If we continue to add more, classical beating gets more nested. If we have a time cycle, which oscillates its diameter between zero and maximum, then we get quantum beating. If more than three such oscillatory cycles engage in beating, then we get fractal beating and creation of a new time cycle with a new system point. If many time cycles nest but only one system point activates, then we get a stream of binary pulses, and the nature of binary pulse stream could be edited using the diameter and location of the time cycles. Imagine time cycles touch one inside another, on another, and one touches other from outside.

6. **Infinite nesting, breathing of time cycles, Hilbert space nesting as fractal: origin of uncertainty: converting Bloch sphere of FIT into a Bloch sphere of QIT**: In the quantum information theory (QIT), when the intermediate states of multiple quantum transitions mix, the Hilbert spaces that hold all possible solutions combine into one singular Hilbert space. In FIT, particular Hilbert spaces engage into a geometric relationship, not just at one level, but several hierarchical geometric relationships among Hilbert spaces ($\{H\}$) are created. Several imaginary Bloch spheres integrate using imaginary functional relations

to fit "guest Hilbert spaces inside a single value in the host Hilbert spaces," but the origin of uncertainty is very well defined: (i) Nesting of Hilbert spaces, generating a fractal beating, (ii) an observer sees a single triangle on a sphere from infinite angular positions, (iii) the smaller the mass, the larger the diameter of time cycles, so any point could be a superposition of many such time cycles, (iv) diameters of the time cycles may oscillate as a function of time. Hence, the noise must be intelligent to eliminate a nested Hilbert space, unlike quantum. When (i) the geometric nesting and clocking of Hilbert space disappears, (ii) world of singularity inside a "bing" disappears, (iii) virtual poles become real classical poles, and then, we get Bloch sphere of a qubit in a quantum system.

7. **Harnessing infinity and the philosophy of illogical or nonsense statements**: Truth table (logic gate) for FIT's Bloch sphere would be a fractal [26]; one could zoom any part of the table continuously, only to find a new table inside. We suggest that Turing like arguments will not work; then, if there is no argument, how could a system run? We are not accustomed to think like that. In Sect. 1.8, we explained ten ways, how a correlated event pair evolves spontaneously in an IIA network as nonargument more profoundly than the Turing paradigm.

8. **The glue of "bing" singularity, entropy for Integrated Information Architecture (IIA)**: In basic information structure, the geometrically clocking Bloch sphere has virtual poles as "bing," and these are undefined points, as could be decomposed unlike QIT. As the Bloch sphere expands during learning, it explores, e, π, φ, and i, three infinite series. At different compositions of Bloch spheres, distinct fundamental constants emerge spontaneously in the network. If the total number of cycles participating in synchronization is P and observer syncs with S number of loops, then the product of density of loops and the time bandwidth for P is DenP and the same for S is DenS. Therefore, the ratio is DenP/DenS, the information entropy of communication channel. Entropy depends on location of observer, and it is a function of time. Measures by sync–density ratio, at specific observer location in the network.

9. **No data transmission, no communication, no rejection or reduction of choices. Not parallel, not sequential, not simultaneous, its fractal: The purpose of information processing is morphing (no computing)**: Information is processed simultaneously everywhere in the network which includes the questioner/observer. Clocks residing side by side and above-and-within operate together and hence no data transmission, no communication, no rejection of choices. The purpose of computing is to match the dynamics of morphing matrix, MBS, or S with that of the observer, U and the nature, E.

10. **There is no actual fractal: ordered factor metric of number system and density of primes controls the number of cells and nesting capacity in the tape**: Number of cells in a Fractal tapes are not chosen randomly. Every natural number has ordered factor, except the primes because the number of compositions is zero. Ordered factor suggests in how many different ways a waveform

could be composed in other words (waveform = perimeter of a time cycle), and boundary values of number of tapes and their characters are determined by density of primes. Since there is no universal pattern, a true fractal does not exist in a Fractal tape.

1.4   Ten Fundamentals of the Geometric Musical Language (GML)

We developed a language where a circle is the only one letter; natural numbers are its only variable. Variation of natural numbers means number of waveform n in a cavity; language is to connect circles to represent cavity solution. Varying n varies circle perimeter or length, and we edit rhythms (length = time). Property of constructing a single circle is the only grammar rule to couple millions of such system points spontaneously, the growth propagates by making system points using doublets and triplets, etc. Any perturbation to the nested rhythm network beyond fractal boundary enables it to adjust rhythms, and it is a spontaneous drive to maintain periodicity—key to the functioning of GML. Cyclic oscillation between multiple frequencies is called music, the relation between multiple frequency peaks if encodes geometric properties then it is geometric musical language. Integrated information architecture (IIA) made of geometrically clocking Bloch spheres runs a large number of clocks, and when system points touch "bings," IIA plays a composition of music, representing a complex network of geometric architecture.

1. **The definition of a time cycle: Why circular geometry is advantageous**: A time cycle is a pixel in the perimeter of its host time cycle, and each pixel on its parameter holds a time cycle inside. Initially, a time cycle perpetually oscillates between its two limits, equaling as large as its host, and decreases to become one of its pixel, with time, it fixes its diameter, become static (long-term memory, and short-term memory). Two types of time cycles, static and oscillating, construct a matrix of morphing system, MBS. For purity, a circle tries to minimize correlated event pair (analogue of question–answer or if-then pair) clusters on its perimeter by decomposing into a network and hold only one geometric shape or only one associative nonargument; the shape or event pair nonargument is generated when system point moves along its perimeter. Silence between "bings" holds the geometric parameters. **Use of circle that uses frequencies is superior to logic gates that use bits, because** (1) a system point running along the perimeter of a circle is itself a memory that need not to be refreshed ever. At the same time, it makes a decision; (2) System point is never lost at a far distant place in the operational circuit, its distributive, no central control system is required to manage system points, and no external user is thus required. Due to natural property of a circle; (3) a circle is a rhythm, and hence, automated clocks maintain time, over entire information architecture

spontaneously; (4) a circle is a fractal harmonic oscillator and hence can spontaneously self-assemble, and creation of new logical arguments, i.e., learning and simulating future, is done by simple self-assembly of such nested circles. Because it is an oscillator; (5) it can spontaneously reply by synchrony and desynchrony so we do not need circuit connection to address and require minimum resources; (6) we do not need logic gate for rejection and hence enormous power saving. A circle made of frequencies in a rhythm (7) is a natural sensor, and we do not need to design any; (8) it is a natural filter as unmatched signals are rejected, and noise is eliminated naturally. Some incredible feature of circular assembly is that (9) a circle can activate or cross-check infinite loops at once via escape time (ET, one inside another) and IFS (iterative function system, side by side) routes simultaneously; (10) it holds any geometric shape or symmetry in 1D, 2D, and 3D, using only diameter and ratio of arc gaps and hence could emulate any "form of argument" as "geometric shape." Simply by self-assembling the material representing the MBS, one can create an evolving fractal circuit that can sense, decide, respond, and evolve by itself (a programmable matter, or brain jelly [26]).

2. **Time is fundamental, mass, space, and everything else are its derivative**: Time cycle = Rhythm = Complex unitary operator: GML has only one letter, a circle, only variables are the natural numbers, and primes are the only constraints that determine the boundary conditions at various timescales. The language GML considers mass = length = time, i.e., there is only one dimension (in the dimension analysis if one puts $M = L = T$, dimensions of everything in the universe are $...T^{-3}, T^{-2}, T^{-1}, T, 0, T^1, T^2, T^3...$). The diameter and arc gap between two "bings" (or frequency points or coordinates of geometric shapes) are only two variables. The magnitude of frequency is the diameter of the circle, and gap between two frequency points is the parameter that decides geometric shapes and synchronization criterion. While sensing, the ratio of arc gaps between two frequency points of two circles is compared. Length of the perimeter of circle is the period of a rhythm T; if a circle is made of eight frequencies or local circles (8 waves inside one wave), the frequencies are played at a time gap dt on the perimeter, and sum of dt is T. The energy transmission from a pure circle is a sinusoidal wave. Perimeter of a circle = wavelength = relative diameter is the amplitude of output signal. Nested network circle inside a circle holds period as T, $T/8$, $T/16...$ $T/8n$.

3. **Nesting of beats, fractal mechanics, a perfect nesting creates a system point**: A circle is never left alone. A simulation of the "future" is natural in MBS via beating, as if what one user is going to hear next as soon as we have heard a word or sentence or even a letter. Two circles with two system points are connected to convert a sinusoidal wave into a rhythm; we call it "nesting"; if they have just one system point on the guest, we get binary pulse stream. For one system point, nesting three circles/rhythms only fine-tunes the feature, but if all have different system points, nesting goes complex. If nesting uses "baby" or "just born" time cycles (oscillating diameter), then we get quantum beating; if more than three time cycles participate, we get fractal beating. When sum of

the wavelengths of the guest waves is the wavelength of the host wave, then the system point on the cycle moves automatically. It means two waves are naturally synchronized, binding the nested waves. Thus, a perfect nesting creates a system point.

4. **The letters to make words and sentences: IIA learns means big bang like balloon expands**: There are 1D, 2D, and 3D basic structures, which combine to form an information replica of any natural signal. Time cycle is a circle, and we have selected a set of linear shapes L, U, V, S, M, N, O and 2D geometric shapes triangle, square, pentagon, hexagon to be used as elementary geometries or letters of geometric musical language (GML). Time cycles hold a Bloch sphere like quantum, but "bings" are not real (qubits are real points), so each "bing" holds a sphere, so more the information a MBS learns, due to spontaneous repulsion of the spheres, it expands. Time cycles can self-assemble to fuse inside or at any point on the expanding balloon like structure, imaginary worlds nested inside another maps Hilbert space inside a Hilbert space.... It is not quantum but a world of fractal mechanics.

5. **Fractal decomposition of event seeds and expansion**: When an image, sound, touch, taste, or smell information is converted into time cycles of geometric shapes, they combine into one nested time cycle. This geometric structure holds an association between various signals, and this is question and answer pairing. It is an event, and the concept of question and answer in FIT does not exist here. The information irrespective of its nature is converted into the fractal seed of geometry, by spontaneous decomposition of "bings" in the absorbed time cycle and splitting of composition of networks of correlated event pair (analogue of question–answer or if-then pair) into single event pair time cycles. Since the system point moves an event is robust, we get a spin direction, a phase, creation of a clock that keeps time, holding a memory state and making a decision all happens with a single step. If a seed time cycle is placed in a network, it would get connected at a proper location and the singular drive of morphing would generate several new time cycles and that we call as expansion of the seed to the higher level.

6. **Spontaneous self-assembly for information integration following ordered factor metric: the drive of the prime**: Objective of self-assembly is to develop an equivalence of an argument, i.e., a nonargument. Now, instead of "rejection," self-assembly of cycles enables the MBS to have all possible "associations"—search happens without searching. A circle or system of circles could be part of one or many circular assemblies; simultaneously, are possibilities infinite, random? When morphing drive triggers several subdrives, it essentially grows the time cycles following the ordered factor metric, which is not an artificially induced metric; it is a universal relationship what happens to rhythms if we increase the number of waveforms one by one. During self-assembly of time cycles, prime numbers work as boundary points because number of choices one could reorder prime number of waveforms is zero. Hierarchical network of Bloch spheres spontaneously evolves following the ordered factor of natural numbers. The number of geometric corners and total

count of geometric points in a nested network and thus the limits of "bings" in a circle during self-assembly are regulated by ordered factor metric. Hence, it leads to a perpetual expansion of the entire network following strict mathematical regulations.

7. **When a circle activates and when it gets a silence: when it is born and when it disappears**: Geometric circles are activated by association, only if equal or more "bing" points in its perimeter activate. In a network of associated circles, the system uses its resources, energy, and time to sustain networks of maximum matching. Thus, associative circles with a large diameter spontaneously get activated even if there is no connection or wiring, i.e., it is purely wireless processing. If there is no longer a match, time cycle silences. Fractal beating gives births to a new time cycle and stops giving birth to a new one, when the oscillatory time cycles that make fractal beating become permanent. If activation stops before a time cycle becomes permanent, then the time cycle disappears.

8. **Creation of rhythm is a choice of system point selection and ratio of diameters**: Clockwise rotation and anticlockwise rotation of system points generate similar results. Interestingly, if the system points of two nested circles rotate in the same direction or in opposite direction, they produce similar pulse streams or rhythms. Only variable is the relative angular velocity between the two circles and the starting phase difference between the two circles. They control output rhythm; hence, by absorbing output, other nested rhythms get these two information, $D1/D2 = V2/V1$, and ratio of diameter determines relative angular speed. This is the only communication analogue of Turing systems. Phase difference in rotation is not random or users choice; it is born by strict rule. The difference between the common "bing" locations between two circles (to be nested) locks their system points, and it is the initial phase difference. Note that if the number of nested circles increases, the resultant sinusoidal wave produced is not effectively edited into the output rhythm. Eight circles are found to be the best choice to edit/construct a rhythm. Eight smaller constituent circles are "peaks," the larger encompassing circle is the linear "band," and hence, an optimized band is an octave.

9. **Management of Rhythms: 1D, 2D, and 3D geometric shapes are mere selection of system points and mode of output acquisition**: Triplet/three nested circles make a rhythm with one system point. If two triplets have small common frequency distribution on cycles perimeter (common frequency = "bing") (at least three common), instead of 360° rotation, the smaller cycle flips between two points on the perimeter of the larger one, and we call it locking. Then, the output rhythm/signal is discrete linear pulses (U, L, V, S, C, etc.). Three types of bonding are there between two cycles, IN, ON, and OUT (IN if bonds with faster rhythms, OUT if bonds with slower triplet, ON if diameter of two triplets similar, i.e., conservation required). For in-nest and out-nest, only one contact point, for on-nest, two contact points. Two triplets isolate if longer common overlap is available for cycles in vicinity. To match the external input, some guest cycles in IIA even die so that the host time cycles

make a better match while morphing. A single triplet could generate cascade of other triplets for generating larger triplets to meet an external demand of morphing.

10. **Continued fraction geometric alzebra** (CFGA): Two cycles nest in three ways, in-nest (reduce, IN), on-nest (conserve, ON), and out-nest (amplify signal, OUT). If 6 or 8 rhythms/cycles/frequencies/bings reside on the cycle perimeter to constitute a geometric shape, the 6/8 time cycles expand into a fractal distribution of frequencies. Geometric constants e, π, φ and imaginary index i build unique relationships to map a specific infinite series that represent a local nested rhythm network, and the relation also provides a geometric constant that is essential to build any linear relation or variation in the network. We can do entire complex mathematics of infinite series, simply by drawing the nested cycles; this is called continued fraction geometric alzebra (CFGA). Fractal distribution of 8 fundamental frequencies is the key for the locks that come from other systems, and ordered factor metric delivers the frequency values (ordered factor fractal is a unique pattern made of 8, 24, 40… frequencies $8(2n + 1)$, and this pattern is the only self-similarity that exists in the FIT-GML protocol). In a continued fraction system, the layers of this fractal are for nesting. When two time cycles or circles fuse, particular layers in the ordered factor metric get active. The material that makes a time cycle generates fractal resonance peaks, and their interactions make nested time cycles; in 66% cases, IN, ON and OUT triplets can make infinite fractal net alone.

1.5　Ten Rules of Self-assembly of Time Cycles

The self-assembly of the time cycles residing in the clocking Bloch spheres replaces the need to programming in the fractal information theory (FIT) and its prime component Geometrical Musical Language (GML). FIT-GML does not require a user to imagine the logic and frame instructions, only a singular physical process, self-assembly of time cycles in a brain-like jelly enables it to learn, processes information and reply—do everything by self-assembly but do not use a single line of code. Prior to learning, the Fractal tape computer's core is made of zillions of time cycles or Bloch spheres and its materials equivalence would feel like a jelly (brain jelly). The jelly changes its structure to as the Bloch sphere self-assembly. A host Bloch sphere grows by hosting the learned Bloch spheres as shown in Fig. 2b. The time cycles continue to self-assemble and undergo successive ten primary transformations noted below.

(i) **Transform by phase transition and symmetry breaking**: Spontaneous phase transition and symmetry breaking of various local time cycle clusters continue in a network of time cycles, i.e., our system MBS, to regenerate symmetry.

(ii) **Create and destroy time cycles generating new shorter routes**: Creating or deleting new time cycles or even local network of time cycles to sync with environment and observer and often only to simplify the communication route.

(iii) **Copy and paste unknown time cycles**: The system MBS replicates a local pattern or cluster of time cycles, at a distant domain in the network for nesting. For nesting, and applied learning, often we need to generate redundant time cycles, and it prepares the system for emergency response.

(iv) **Reorient and rewrite geometric information in existing time cycles**: Runs a perpetual synchronization of time cycles with the observer O (paradox: if all observers see an object together, then?) and the environment E (OUE, SUE). Vibrations with wavelength far beyond the maximum time cycle of the system S.

(v) **C2 symmetry drive**: Hierarchical network of oscillating time cycles (two types of time cycles, one perpetually oscillating and another static) triggers changes in the network as the entire geometry tries to build a C2 symmetry. C2 drive is precursor for hierarchical learning.

(vi) **Morphing to mimic evolutionary dynamics of environment**: Dynamics of the time cycles is replicated at a seed level so the evolution of dynamics runs similarly like external environment even if the system S isolates itself from external connection.

(vii) **Protection drive: Long-term and short-term memory**: Time cycle network (matrix of Bloch spheres, MBS) restructures to follow the fundamental learning drives as a part of morphing, which also acts a filter to the learning process. However, the time cycles those activate for a longer time stop oscillating, and they fix their diameter. Only the temporary ones oscillate as soon as they are born. This feature protects the learned core.

(viii) **Rule extraction**: The system changes and transforms the time cycles that certain rules are automatically extracted from the system that operates at any hierarchical levels.

(ix) **Expands**: The 3D network of Bloch spheres continuously expands to hold major new time cycles in a giant Bloch sphere that appeared at the beginning and holds everything. Due to repulsion, entire system expands proportionally.

(x) **Rule of evolution is mathematics of ordered factor**: Time cycles strictly follow the ordered factor of numbers while executing every single change in the network of time cycles.

1.6 Ten Physical Conditions that Trigger the Self-assembly of Time Cycles

(i) **Time cycles bond only under a certain specific condition**: Time cycles should stay in touch for longer than the longest time cycles of the cluster of time cycles we want to add to a network of system S. If shorter, it may get connected at the wrong location, and later, the system might need to defragment the memory and the learning.

(ii) **Matching the difference in the density of time cycles turn perpetual**: In order to match the difference in the density of time cycles, empty time cycles without "bings" are added at particular angle to match the density. Normally, three nested time cycles take part in the density matching process, the environment, the observer, and the system that computes.

(iii) **The geometric information encoding process** is identical to the density matching process. "Bings" with a "silence" hold the geometric information, but a "bing" is the "bing" location, spin orientation, and density of time cycles. Perfect match and equilibrium are never reached in the system of the time cycle network.

(iv) **Without cross-check, there is no abrupt formation of a new time cycle**: 3D orientation of the time cycle made Bloch spheres may bring two or more very different Bloch spheres nearby, during the course of self-assembly. This situation would trigger the formation of new time cycles and unwanted integration mixing distinct pathways, destroying the true essence of the geometry encoded. The largest and the oldest time cycle-based nested time cycles expand to avoid such rouge links. However, if the accidentally met pair of nested time cycles finds significant similarities, they would synchronize and form a bond by creating a time cycle.

(v) **Fusion and fission of the tiny time cycles**: If there are giant and tiny time cycles at a high density in a particular domain of the time cycle network, and the "bings" of the tiny cycles are not heavy, means do not contain large number of "bings" inside, then the small time cycles unlock from a certain point and adsorbs to some other points in the network. This process avoids major defragmentation of the network bypassing the structural phase transitions. (The same mathematical deduction under simulation would show that a single time cycle may act like a bomb to a network).

(vi) **Matching the spin direction for the time cycles**: Every single time cycle, each storing a geometric shape, forms the great perimeter in its exclusive Bloch sphere. Several such Bloch spheres are embedded on the surface of a host Bloch sphere. When the system point rotates along the great circle of the Bloch sphere, there are poles and the direction of rotation matters, if not the geometric information made by connecting the "bings" would change. These, rotational directions or spins also need to match during synchronization.

(vii) **Phase synchronization run in parallel to the geometric synchronization**: We are aware about the synchronized oscillation of a pair of oscillators; normally by synchronization, we understand that both the oscillators would vibrate in phase. They might also adjust the sphere diameter to achieve coherent vibrations.

(viii) **Creating a mirror image from the phase space hierarchical network by fractal route**: Often, the time cycle network undergoes a heavy loss due to structural defect or otherwise recreates a major part of the lost section by generating a fractal extension. As soon as a memory is added, during the integration process, a large number of hierarchical time cycles are produced as a cascade effect. They reproduce the lost information.

(ix) **Time cycle network expands and continuously tries to produce longer than the longest time cycles**: A fundamental primary drive of the time cycle network is to produce longer and longer time cycles, integrate them, and grow perpetually. The oscillation of the time cycle diameter is the root cause and elementary factor governing this drive. This drive is so fundamental that the system compromises with the minor changes in the "bing" location, spin orientation, and density of time cycles. Perfect match and equilibrium is never reached in the system of the time cycle network.

(x) **Prime frequency wheel drive**: While generating and expanding the time cycles, we should note that ordered factor for a particular number is the number of all possible routes a waveform could be made of smaller waveforms. This means, the ordered factor tells us, allowed superposition of nested time cycles. The ordered factor for a prime number is zero; hence, we find a boundary value. Perimeter of a time cycle is time, inverse of time is frequency, and nested time and nested frequencies are equivalent. We create an alternate frequency wheel, for any time cycle network, which tells us that the growth cannot be random, rather restricted by primes and its factors that regulate the ordered factor values of numbers.

1.7 Ten Grammar Rules How Correlated Event Pair (Analogue of Question–Answer or if-then Pair) Spontaneously Evolve in a Time Cycle Network

(i) **No question is born without an answer; they are born together, live together, and reply spontaneously**: Elementary time cycles self-assemble but never break apart. There are no true and no false statements, all are true; there is no question or answer; all are correlated events or situations. Correlated events pair or form a group. They always remain as a pair in a time cycle (we say event pair). When one sees an event pair forming a higher level, only two "bings" are heard, one represents situation 1 (triangle) and the other situation 2 (square). If one geometry is triggered, all associated others play spontaneously, even if they reside at different time cycles. (ii) **Nested clock deconstructs truth into a set of new event pair network: not sequential, not parallel, not simultaneous, but fractal**: If the geometry of one event activates, a cascade of activation continues in the integrated Bloch sphere architecture (Fig. 3). The activated pattern acts as a single unit, since a higher-level geometry integrates them and that higher-level geometry activates similar patterns. Correlated events are analogues or previous if-then statements or arguments. The fractal decomposition of correlated events happens at all timescales at once. It is neither sequential nor parallel. Along with sequential, parallel, and simultaneous, we introduce the fourth category, fractal. (ii) **Hierarchical arguments are progeny; they are born by ever-changing connectivity of elementary arguments**: Intermediate states of quantum cannot process arguments, but when each intermediate state has many intermediate states inside, then quantum beating is replaced by fractal beating. All hierarchical nonargument pathways born out of single elementary geometric information coexist as separate identity in any set of arguments. So, in fractal beating, distantly located (timewise) and discretely activated event pairs during vibration are nested and higher level correlated event time cycles are born. Unpredictability is not the sign of randomness, but a single observation of nested rhythm could originate from several alternate situations, which are fundamentally different. (iii) **Incompleteness does not generate error here; it is fractal route to infinite connections**: No event is absolute (like no answer or question is complete), and there is always a question that invalidates the answer, fixing that to evolve answer is a journey toward completeness. Similarly, the facts are relatively static points, but its connections are emergent feature as a geometric shape could be part of various distinct arguments in the network at the same time. (iv) **Logically circular or circular logic** is the unit of a single nonargument to enter inside a time cycle and go above. Discrete time cycles self-assemble into an unit cycle, its a circular logic. As the system point rotates, each defines the other, individually remain undefined. (v) **Astronomically large possible contexts are included**: As soon as a time cycle for a correlated event is born, every single other time cycles in the network are connected in all possible ways. Since some time cycles oscillate, consequently a large number of hierarchical network of time cycles are born, and thus, all possible contexts which were never coded are born. (vi) **The motive of correlated event doublet (analogue of an argument) is to make the geometry more symmetric and integrating the geometries into one singular geometry at the highest level**. Every single set of arguments forms a geometric shape, and a complex pattern of composition of basic geometries should form to compose a complete nonargument; the shape would be symmetric and follow the rules of geometries. (vii) **A correlated event network cannot be stated without suggesting the angular orientation or location of an observer on the Bloch sphere network**, i.e., perspective of an observer sets the entire network of nonargument: Many correlated event networks for a single correlated event coexist simultaneously. Even though a fractal arrangement argues that at the top there should be one nonargument, but even that would require at least three perspectives, one from the arguments below that makes it, one from above or the nonargument it constructs with others, and one the perspective of an observer. (viii) **"Bing" does not contain information but a "silence" does: flow of time**: Silence gives circle diameter and phase from coordinates of bing, from which we get amplitude, mass, space all essentials to construct geometric shape and its dynamics. In a nested time cycle network, time does not have any direction, time does not flow, but for an observer who is moving in any time cycle in the network past present future like perceptions appear. There is only "phase" in the information network, dynamic and geometric phases. (ix) **Expansion of an optimal 12 note frequency wheel is the best way to expand a nonargument**. In a time cycle, one can put several correlated event pairs. It can act like a seed; sustaining its parameters, we can expand them, adding multiple time cycles to its network with longer and shorter time cycles. This is how a correlated event wheel perfects its time cycle structure, and it is essential to morphing or learning and embedding a dynamics (autocorrect a nonargument). (x) **Morphing with the universal geometries is a drive for the nonargument**: Morphing is a singular drive for learning, and it gives rise to several other local drives. We have described this point in this article in other contexts.

1.8 The Construction of a Nonargument for a Noncomputable Decision Making

Ten reasons we say that FIT-GML decision making does not have any argument: All properties of an argument are missing here. We find (i) no if-then statement, (ii) no rejection, (iii) no extrapolation, (iv) no either-or statement, (v) no probability, (vi) no finite statement, (vii) no specific task or instruction, (viii) no sequence of events, (ix) no additive or multiplicative statements, (x) no idioms.

Twelve ways of making a nonargument: Every natural event has a set of geometric arrangement of shapes associated with each other; this association is essential and sufficient to replace a correlated event cycle (nonargument). Our noncomputing (morphing) does not rely on an external user and the practice of argument, so it accurately morphs with the nature's geometrical arrangement in real time. As a result, all possible situations that might arise in future are already encoded as a seed into the matrix of the Bloch spheres (MBS). We follow twelve protocols to optimize a nonargument naturally in the time cycle network, or there are twelve ways of making a nonargument in the FIT-GML decision making. These steps will replace a user, so these are very crucial; we plan to match human intelligence, intuition, and future simulation with these few steps.

1. **Negating all the most bright choices continuously**: A query/geometric shape spontaneously activates most matched associated geometries in the network spontaneously, and the system ignores that as an absolute and finds for the

nearest other associated choices. It repeats until the choice fails to find any taker in the query matrix.

2. **Relocating observer, repeat**: On the reality sphere change, the point of an observer or change perspective of an observer and repeat the cycle.

3. **Extracting fractal geometric seeds at the most fundamental level**: Every event when extracted to a network of geometries and time cycle dynamics, the system naturally tries to find self-similarity in the dynamics in terms of geometries, because that is the only thing the matrix of Bloch spheres (MBS) can do. The hierarchical network of geometries that it finds is a continued fraction series e, π, φ, and i; this relation identifies the fractal seed of basic dynamics; more we repeat it at different timescales, more new dynamics are generated, which were never there during morphing.

4. **A cascade of generating slower rhythms encodes all possible future events**: Time cycles periodically oscillate under certain condition, and the same physical phenomenon helps generating ultra-long time cycles (due to quantum beating or classical beating) and its hierarchical network. It means this oscillation encodes all possible future of an event seed.

5. **Simultaneity as a tool to extract superposed possibilities**: Evolution of 3D geometry teardrop and ellipsoid shapes is embedded in the number system, and we get the philosophy of a pair of pole division from a single pole spontaneously. Hence, bifurcation of arguments, opening up routes that estimates correlation of correlated event vibrations from geometric equilibrium, it sets a safe route to choose one of many choices.

6. **The orientation of the spin flips to trigger phase transition**: Spin flips if an expansion of the time cycle network is such that certain geometric information could not be hold reliably. This is an essential and sufficient requirement to fine-tune morphing. Spin direction is reversed only to keep the geometric shapes intact during transformation and expansion of the time cycles. This ensures reliability.

7. **Defragmentation of time cycles**: Often during the formation of a correlated event pair in a time cycle, several associated event components (read several distinct geometries) are bonded together. Therefore, in course of time with learning, the objective of the matrix of time cycles is to deconstruct such that all such mixed assemblies are eventually made of time cycles that have only two events. This is spontaneous factorization of a messy logic in an illogical manner.

8. **Eight variability in the time cycles and a maximum 12 imaginary states within are estimated for any decision making**: The system cross-checks to find eight domains of times operating simultaneously side by side (frequency wheel) and 12 layers of time cycles one inside another. Two sets of fractal clocks are equivalent, and thus, matrix of time cycle system has only one nonargument; the system adjusts time fluctuations at various time domains.

9. **Twelve network features used for constructing a nonargument: Four classes of synchronizations run in parallel**: All time cycles belonging to the observer, the environment, and the matrix of time cycle system continuously

merge with each other. The structure of a nonargument is framed with 12 kinds of network features: (i) hierarchical, (ii) spin-foam replication, (iii) exchange-able time cycles, (iv) filtering a complex net, (v) rejecting redundant time cycles, (vi) sensing all possible time cycles within limit and beyond, (vii) key geometric points, (viii) 3D orientation symmetry of time cycle network, (ix) creating mirror replica of time cycles, (x) transmission lines via time cycles, (xi) reverberation of time cycles, a little activation at one point of network expands all over through a favorable route, and (xii) classified rule search. These 12 classes of time cycle networks engage into four classes of sponta-neous synchronization run in the system. The hierarchical class, exchange class, split-join class, and mimic class are the four classes of spontaneous editing of rhythms. Thus, the formation of nonarguments is not a random selection of geometric structures.

10 **The umbrella protocol**: There are two types of time cycles, one that is learnt by the matrix of time cycles from environment in the presence of a user. The other one is a hierarchical network of time cycles produced by the matrix. The hierarchical network simplifies the geometry and thus integrates geometries or correlated event pairs far beyond the pairs under concern. Thus, there is an umbrella like processing by which higher-level associations are accurately underpinned and intricate reconfigurations are made.

11. **Use of ordered factor metric**: The universal geometric relationship evolved in the number system is emergent and nonrepeating. However, that regulates the evolution of nonargument and self-assembly and disassembly of correlated event doublets. A nonargument always has one less variable than the fixed truth, and total number of variable and truth should compose like our number system metric; hence, the network of nonargument could be constructed. If there is a little change in the most static nonargument, all nested arguments would change simultaneously, strictly following the ordered factor metric ("everything is connected to everything"). FIT-GML's world of nonsensical nonarguments suggests that doublet or binary logic could be right for the 50% of the time. Hence, our contribution of nonsensical nonargument is to complete the world of knowledge that were left out i.e. 50%, here we try to see the universe of triplets pentate to couplet of 37 arguments putting together that makes sense.

12. **Ten limits of a nonargument originating in the mathematics of number system and geometry**: (1) Though nonargument is a singularity (undefined), one could enter inside a nonargument 12 times to generate 12 different spin directions. We need to change the starting nonargument after 12 times. (2) A nonargument holds only in a fractal cavity resonator, and it nests packets of beating oscillations in slower rhythms. (3) Frequency wheels of only the first 12 primes make elementary structure of a nested nonargument. (4) A true nonar-gument is one that has a pair of simultaneous events (12 triplets). (5) 10^{11} truths $(2.3.5.7.11.13.17.19.23.29.31.37 \sim 10^{11})$ or pattern of 12 primes (up to 37) is the limiting length of a nonargument. (6) There is 64 matrices of teardrop to

ellipsoid shape transition, which configures the shape of a nonargument in the IIA. (7) Triplet of triplet fractal forms the boundary of any information content. (8) Nested phase is the only variable of a nonargument. (9) Only 10 geometric dimensions are feasible for 12 spin directions in a nonargument. (10) IIA runs infinite series of e, π, and φ as a function of i, while playing any geometry in a nonargument.

1.9 Comparison Between Classical Information, Quantum Information, and Fractal Information

Classical information	Quantum information	Fractal information
A real state	An imaginary state sphere, with real poles, 0 and 1	Virtual poles, each pole ("bing") is a coordinate of a geometric shape on a time cycle; the time cycle is located on an imaginary sphere. Information is expressed using e, Pi and Phi as a function of i Wiring of oscillators (hardware of information) = network of "silence" (information of object/event) = Geometric phase (information of fractal seeds) = Perimeter and arc (location of information)
Information is always real, and phase is a separate real information	Real part of information comes by projection from the Hilbert space, which contains superposition of infinite number of intermediate states. The uncertainty is undefined so noise can terminate quantum information discretely	Several imaginary Bloch spheres integrate, but the origin of uncertainty is very well defined: (i) nesting of Hilbert spaces, generating fractal beating, (ii) observer sees a single triangle on a sphere from infinite angular positions, (iii) the smaller the mass, the larger the diameter of time cycles, so any point could be a superposition of many such time cycles, (iv) diameter of time cycles may oscillate as a function of time. Hence, noise must be intelligent to eliminate nested Hilbert space

(continued)

(continued)

Classical information	Quantum information	Fractal information
Needs logic gate, circuit, algorithm to integrate information at the behest of user. Spontaneous self-assembly does not exist	The concept of phase space exists, but there is no way they could self-assemble, because the qubits have no geometric inherent property	Self-assembly of Bloch sphere follows ten basic rules and with learning continuously expands, undergoes symmetry breaking and phase transitions. At least four imaginary phase spaces exist to construct the smallest unit of information
Only dynamic phase, if a particle completes a cycle, no one can tell it surely	Dynamic and geometric phase, if a particle completes a cycle; along with dynamic, there is a geometric phase two	Apart from geometric and dynamic phase, there is a network phase too generated by the network of phase space spheres. The hierarchical network of phase overshadows the dynamic phase, and hence, the roles of observer and nature are also included.
Information is a static state	Information has dynamics; the numerous intermediate states keep it dynamic, sustaining the possibility of change until it is measured	Information includes various kinds of drives that regulate the dynamics; unlike quantum, all these drives are well documented: (i) unitary drive (diameter of time cycle is autoadjusted), (ii) symmetry drive (phase transition of Bloch sphere network to generate symmetry), (iii) Fractal clock drive (system point moves to the fastest clock and then to the slowest and does back and forth), (iv) synchronization drive (triggering self-assembly and disassembly of time cycles)
Classical static phase generates classical beating to create "bing"	Quantum evolution of phase for the infinite elementary states which creates additional beating than classical and hence appearance of unpredicted "bing" that edits "silence" on the time cycle of information	At various levels of the network, the oscillating time cycles generate new very long duration time cycles and they also interact and create new. Thus, an evolving network of "silence" phases is created, not just one like quantum

(continued)

(continued)

Classical information	Quantum information	Fractal information
Maximum information transmits through single-channel use. Source entropy $H = -\sum(p(x)\log(p(x)))$, measures by bits/minute or actions per minute	Unassisted quantum capacity $Q <= C$; assisted quantum capacity $Q2 >= Q$. Source entropy $-\text{Tr}(d \log d)$	If total number of cycles participating in synchronization is P and observer syncs with S number of loops, then the product of density of loops and the time bandwidth for P is DenP and the same for S is DenS. Therefore, the ratio is DenP/DenS, the information entropy of communication channel. Entropy depends on location of observer and it is a function of time. Measures by sync–density ratio, at specific observer location in the network
Turing–Russell philosophy, data, bit/qubit, rule, software, switch, lexicon, logic, algorithm, program, model, memory, switch, processor, subroutine, encoder, decoder, symbol, or buffer, probability, communication, reduction, data transfer, etc	Same as classical information theory, except we add unpredictability and superposition	Rejects all terminologies of classical and quantum information theory. Includes logically circular, redundant, undefined, singularity, unbounded or divergent, infinite loop or no halting, noise or randomness, broken link or error, discontinuity, spontaneous reply, etc.
Information is processed "step-by-step," observer measures object	Information is processed "simultaneously," but encoded instructions step by step	Information is processed simultaneously everywhere in the network which includes the questioner/observer and follows side-by-side and above-and-within protocols together and hence no data transmission, no communication, no rejection of choices. Questions couple with answers in a single time cycle as event from the very beginning of the learning process
Classical information does not have any information about the object or event, and it is merely a data without any meaning. Classical	Quantum information does not have any information about the object or event, and it is merely a data without any meaning. Quantum mechanics	Uses geometric musical language and fractal mechanics to generate information architecture that is a filter. It carries out fractal decomposition of any event

(continued)

(continued)

Classical information	Quantum information	Fractal information
mechanics needed to process such information	needed to process such information	happening in the environment of the morphing architecture. So "event seed" converts into "geometric seed," which is processed as "seeds" but when spontaneously replied back to the observer, the original fractal is regenerated. Time cycle = an event = periodic oscillation of an oscillator

2 Conclusions

Why do we need global database of biological rhythms: The time cycle in GML is like a complex unitary matrix, and it has a virtual spin, thus, Bloch sphere made self-assembled architecture (integrated information architecture) is an equivalent of a spin-foam (no electron or mass). GML promises a grand unification of intelligence protocols that would include any object, any event, and any sensory signal. All incredible features of the brain originate from ever-expanding big bang like balloon, i.e., host Bloch sphere expands on which the guest spheres grow, the connecting perimeter between two spheres run, and if it happens over the entire architecture, we can hear a music. This is why the IIA of FIT's special Bloch sphere is tagged as GML. This is not possible with the existing "black box" model. In FIT-GML, every single feature of brain-like intelligence is achieved by rewiring. We are building the database of nested rhythms of every single part of the brain and determine exactly what the rhythms are; at the same time, we convert sensory signals in terms of a few geometric shapes, to improve the grammar for the GML. Digitization destroys crucial information; conventional analogue processing is only about increasing resolution. Our language of circles captures higher-level correlation (intelligence) from the input data—never captured before. Weak signals get importance, never neglected as noise; all credit goes to fractal mechanics, a generic form of quantum mechanics that provides "fractal beating"—the foundation of FIT-GML.

Acknowledgements Authors acknowledge the Asian office of Aerospace R&D (AOARD) a part of United States Air Force (USAF) for the grant no FA2386-16-1-0003 (2016–2019) on electromagnetic resonance-based communication and intelligence of biomaterials. Authors acknowledge Ben Goertzel and Roger Penrose for critical comments.

References

1. Klimke, W.: The problem: genome annotation standards before the data deluge. Stand Genomic Sci. **5**(1), 168–193 (2011)
2. Shannon, C.E.: A mathematical theory of communication. Bell Syst. Tech. J. **27**, 379–423 (1948). (pp. 623–656)
3. Landauer, R.: IEEE.org.: Information is Physica. In: Proceedings of Workshop on Physics and Computation Phys Comp. vol. 92, pp. 1–4 (1993) (IEEE Comp. Sci. Press, Los Alamitos, 1993)
4. Khinchin, A.I.: Mathematical foundations of information theory. Dover, New York (1957). ISBN 0-486-60434-9
5. Bennett, C.H., Shor, P.W.: Quantum information theory. IEEE Trans. Inf. Theory **44**, 2724–2742 (1998)
6. Nielsen, M.A., Chuang, I.L.: Quantum computation and quantum information. Cambridge University Press (2000). ISBN 0-521-63235-8
7. Brukner, C., Zeilinger, A.: Information and fundamental elements of the structure of quantum theory. In: Castell, L., Ischebeck, O. (eds.) Time, Quantum, Information. Springer, Berlin (2003)
8. Shor, P.W.: Polynomial-time algorithms for prime factorization and discrete logarithms on a quantum computer. SIAM J. Sci. Stat. Comput. **26**, 1484 (1997)
9. Mayer-Schönberger, V., Cukier, K.: Big Data: A Revolution That Will Transform How We Live, Work, and Think, 1st edn., 256 p. Hardcover, 5 Mar 2013. Eamon Dolan/Houghton Mifflin Harcourt. ISBN 10: 0544002695; ISBN 13: 978-0544002692
10. Gomes, G.: The timing of conscious experience: a critical review and reinterpretation of Libet's research. Conscious. Cognit. **7**, 559–595 (1998)
11. Bandyopadhyay, A., Acharya, A.: A 16 bit parallel processing in a molecular assembly. Proc. Natl. Acad. Sci. USA **105**, 3668–3672 (2008)
12. Bandyopadhyay, A., Pati, R., Sahu, S., Paper, F., Fujita, D.: A massively parallel computing on an organic molecular layer. Nat. Phys. **6**, 369 (2010). Highlight Nat. Phys. **6**, 325 (2010)
13. Ghosh, S., Sahu, S., Fujita, D., Bandyopadhyay, A.: Design and operation of a brain like computer: a new class of frequency-fractal computing using wireless communication in a supramolecular organic, inorganic systems. Information **5**, 28–99 (2014)
14. Chen, W., et al.: Anomalous diffusion modeling by fractal and fractional derivatives. Comput. Math Appl. **59**(5), 1754–1758 (2010)
15. West, B.J., Bologna, M., Grigolini, P.: Physics of Fractal Operators. Springer, New York (2003)
16. Kolwankar, K.M., Gangal, A.D.: Fractional differentiability of nowhere differentiable functions and dimensions. Chaos: Interdiscip. J. Nonlinear Sci. **6**(4), 505–513 (1996)
17. Dan, A.L., Xue, W., Guiqin, W., Zhihong, Q.: Methodological approach for detecting burst noise in the time domain. World Acad. Sci. Eng. Technol. Int. J. Electr. Comput. Energ. Electron. Commun. Eng. **3**(10) (2009)
18. Kuramoto, Y.: Each singularity generates a signal burst" Chaos and Statistical Methods. In: Proceedings of the Sixth Kyoto Summer Institute, Kyoto, Japan, p. 273, 12–15 Sept 1983, 6 Dec 2012, Springer. ISBN 9783642695599
19. Mallat, S., Hwang, W.L.: Singularity signal detection and processing with wavelets. IEEE Trans. Inf. Theory **38**, 617–643 (1992)
20. Watrous, J., Aaronson, S.: Closed time like curves make quantum and classical computing equivalent. In: Proceedings of the Royal Society A: Mathematical, Physical and Engineering Sciences, vol. 465, No. 2102, p. 631 (2009)
21. Jaynes, E.T.: Foundations of radiation theory and quantum electrodynamics. In: Barut, A. (ed.) Quantum Beats, pp. 37–43. Springer, Berlin (1980)
22. Abbot, L.F., Wise, M.B.: Dimension of a quantum mechanical path. Am. J. Phys. **49**, 37 (1981)

23. Goldstein, R.: Incompleteness: The Proof and Paradox of Kurt Gödel. W.W. Norton & Company (2005). ISBN 0-393-05169-2
24. Ziegler, M.: Computational power of infinite quantum parallelism. Int. J. Theor. Phys. **44**(11), 2059–2071 (2005)
25. Hamkins, J.H., Lewis, A.: Infinite time Turing machines. J. Symb. Log. **65**(2), 567–604 (2000)
26. Ghosh, S., Fujita, D., Bandyopadhyay, A.: An organic jelly made fractal logic gate with an infinite truth table. Sci. Rep. **5**, 11265 (2015)

Design and Analysis of Fabricated Rectangular Microstrip Antenna with Defected Ground Structure for UWB Applications

Sandeep Toshniwal, Tanushri Mukherjee, Prashant Bijawat, Sanyog Rawat and Kanad Ray

Abstract This research paper proposes to design and fabrication of MS patch antenna with defected ground structure. In the proposed design, the geometry operates from 3.2 GHz to 5.06 GHz and provides impedance bandwidth of 45.3%, having stable pattern characteristics over the entire range. Antenna is fabricated on a FR-4 epoxy substrate ($h = 1.59$ mm), and IE3D simulation software is used.

Keywords MS patch antenna · Bandwidth · Defected ground structure VSWR · Smith chart

1 Introduction

UWB systems have been used extensively, because of their intrinsic advantages like minute dimension, greater data rate, high bandwidth, easy to integrate, and less power consumption. UWB utilizes the frequency spectrum ranging from 3.1 to 10.6 GHz allocated by the FCC [1–8]. In this bandwidth, a number of additional licensed systems exist for which the ultra-wide band systems cause the interference. [9–14].

S. Toshniwal · T. Mukherjee
Department of ECE, Kautilya Institute of Engineering and Technology, Jaipur, India
e-mail: toshniwal.sandeep@gmail.com

P. Bijawat
Department of ECE, MNIT, Jaipur, India

S. Rawat
Department of ECE, Manipal University, Jaipur, Rajasthan, India
e-mail: sanyograwat@gmail.com

K. Ray (✉)
Amity School of Applied Science, Amity University Rajasthan, Jaipur, India
e-mail: kanadray00@gmail.com

© Springer Nature Singapore Pte Ltd. 2018
M. Pant et al. (eds.), *Soft Computing: Theories and Applications*,
Advances in Intelligent Systems and Computing 584,
https://doi.org/10.1007/978-981-10-5699-4_34

373

The **defected ground structure** (DGS) is a recent method, where the metal ground plane of a microstrip patch is purposely customized to improve antenna performance.

MS patch antennas have various advantages like low profile, less weight, inexpensive, and easy fabrication. For the handheld wireless devices like cellular phones, pagers microstrip antennas are extremely compatible with embedded antennas. But patch antenna has many disadvantages as less gain and bandwidth. Some other problems which will occur while using microstrip patch antennas are surface waves in the substrate layer. Due to the surface waves, excitation losses the gain and BW of antenna will decrease. So to overcome that entire drawback, there have been inventions of the new technique called **defected ground structure** [15–23].

2 MS Patch Antenna Design (with Defected Ground Plane)

The design of a rectangular MS patch antenna with defected ground plane is shown in Fig. 1. The MS patch antenna is fabricated on the FR-4 dielectric substrate ($h = 1.59$ mm and tan $\delta = 0.02$). A radiating patch (10 mm × 12 mm) and a feed of size (1.9 mm × 8 mm) are printed on the same surface of the FR-4. The antenna performance (BW and gain) is enhanced by taking defected ground plane of dimension of 30 mm × 60 mm. Using defected ground, a bandwidth of 1.8 GHz and gain 4 dBi is achieved.

Fig. 1 Fabricated MS patch antenna with finite ground plane

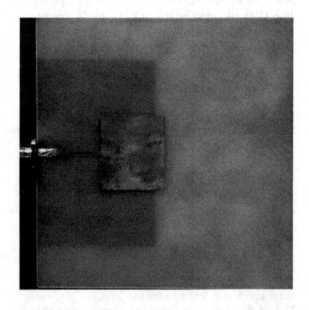

3 Results and Discussion

Measured results of design are presented in this section, Fig. 2 represents the return loss (S11) curve for the fabricated design. The antenna is efficiently operating from 3.23 to 5.061 GHz. The proposed antenna exhibits bandwidth of 1.8 GHz (45.3%).

Figure 3 represents the VSWR with frequency curve for fabricated design. The voltage standing wave ratio falls below 2 for the preferred band (Fig. 4).

Figure 5 represents the Smith chart for fabricated MS patch antenna.

Fig. 2 Return loss (S_{11}) curve for fabricated design-I

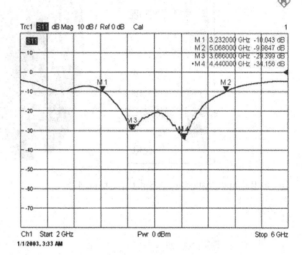

Fig. 3 VSWR with frequency curve for fabricated design-I

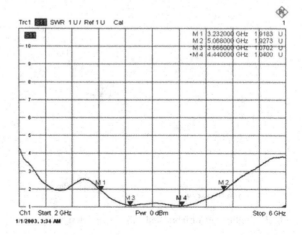

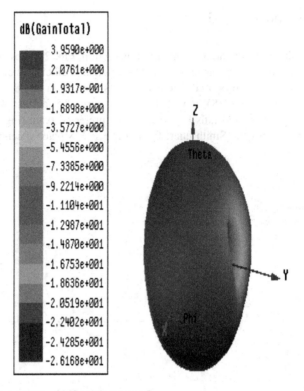

Fig. 4 Radiation pattern of MS patch antenna-I

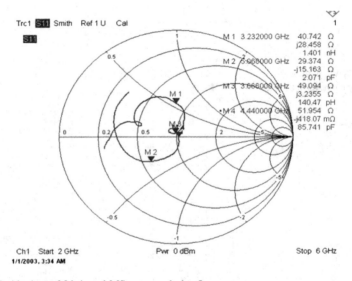

Fig. 5 Smith chart of fabricated MS antenna design-I

4 Conclusions

The proposed design showed a broad bandwidth MS patch antenna can be fabricated with **defected ground** plane with rectangular patch. A BW of 2.36 GHz (45.3%) is achieved. The bandwidth of antenna with defected ground with simple patch increases up to 45.3% and in antenna with defected ground with modified patch increases up to 47%. Also, the gain of modified patch has been amplified up to 4 dBi. The modified antenna can be used for ultra-wide band applications of wireless communication.

References

1. James, J.R., Hall, P.S.: Handbook of Microstrip antennas. Peter Peregrinus
2. Constantine, A. Balanis: Antenna Theory, Analysis and Design. Wiley
3. Ansoft Designer, www.ansoft.com
4. FCC, First report and order, revision of part 15 of the commission's rules regarding ultra-wideband transmission systems FCC (2002)
5. di Benedetto, M.-G., Kaiser, T., Molisch, A.F., Oppermann, I., Politano, C., Porcino, D. (eds) UWB Communications Systems: A Comprehensive Overview. Hindawi (2006)
6. Allen, B., Dohler, M., Okon, E.E., Malik, W.Q., Brown, A.K., Edwards, D.J. (eds.): Ultra-Wideband Antennas and Propagation for Communications, Radar and Imaging. Wiley, London (2006)
7. Mailloux, R.J.: Phased Array Antenna Handbook, 2nd edn. Artech, Boston (2005)
8. Kim, Y.M.: Ultra wide band (UWB) technology and applications. technical report, NEST group The Ohio State University, July 10, 2003.(a) Batra et al.: Multi-band OFDM physical layer proposal. Document IEEE 802.15-03/267r2 (2003)
9. Win, M.Z., Scholtz, R.A.: On the energy capture of ultra-wide bandwidth signals in dense multipath environments. IEEE Comm. Lett. **2**(9), 245–247 (1998) (a) Molisch, F.: Ultrawideband propagation channels—theory, measurement, and modeling. IEEE Trans. Veh. Technol. **54**(5), 1528–1545 (2005)
10. Rawat, S., Sharma, K.K.: A compact broadband microstrip patch antenna with defected ground structure for C-band applications. Central Eur. J. Eng. **4**, 287–292 (2014) Springer
11. Nerguizian, C., Despins, C., Affes, S., Djadel, M.: Radiochannel characterization of an underground mine at 2.4 GHz wireless communication. IEEE Trans. Wireless Commun. **4**(5), 2441–2453 (2005)
12. Foschini, G., Gans, M.: On limits of wireless communications in a fading environment when using multiple antennas. Wireless Pers. Commun. **6**(3), 311–335 (1998)
13. Rawat, S., Sharma, K.K.: Stacked elliptical patches for circularly polarized broadband performance. In: International Conference on Signal Propagation and Computer Technology (ICSPCT 2014), pp. 232–235 (2014)
14. Rawat, S., Sharma, K.K.: Stacked configuration of rectangular and hexagonal patches with shorting pin for circularly polarized wideband performance. Central Eur. J. Eng. **4**, 20–26 (2014). (Springer)
15. Huang, H.-F., Hu, Y.-H.: A compact dual-band printed monopole antenna for WiMAX/WLAN applications. Prog. Electromagnet. Res. Lett. **49**, 91–97 (2014)
16. Rachmansyah, A.I., Benny Mutiara, A.: Designing and manufacturing microstrip antenna for wireless communication at 2.4 GHz. Int. J. Comput. Electr. Eng **3**(5) (2011)

17. Singh, G., Singh, J.: Comparative analysis of microstrip patch antenna with different feeding techniques. In: International Conference on Recent Advances and Future Trends in Information Technology (iRAFIT2012) Proceedings published in International Journal of Computer Applications® (IJCA)

18. Anchit Bansal, P.G.: A compact microstrip-fed dual-band coplanar antenna for WLAN applications. In: International Conference on Recent Trends in Engineering & Technology (ICRTET2012) ISBN: 978-81-925922-0-6

19. Breed, G.: An introduction to defected ground structures in microstrip circuits. From November 2008 High Frequency Electronics Copyright © 2008 Summit Technical Media, LLC

20. Parui, S.K., Das, S.: A new defected ground structure for different microstrip circuit applications. Radio Eng. **16**(1) (2007)

21. Dua, R.L., Singh, H., Gambhir, N., 2.45 GHz microstrip patch antenna with defected ground structure for bluetooth. Int. J. Soft Computing Eng. (IJSCE), **1** (6) (2012) ISSN: 2231-2307

22. Biswas, S., Biswas, M., Guha, D., Yahia, M.M. Antar: New defected ground plane structure for microstrip circuits and antenna applications. Centre of Advanced Study in Radio Physics and Electronics, University of Calcutta, India and the Natural Sciences and Engineering Research Council of Canada

23. Kaur, P., Nehra, R., Kadian, M., Dr. Asok De, Dr. S.K. Aggarwal: Design of improved performance rectangular microstrip patch antenna using peacock and star shaped DGS. Int. J. Electron. Sig. Syst. (IJESS) **3**(2) (2013) ISSN: 2231-5969

Improved Clustering Algorithm for Wireless Sensor Network

Santar Pal Singh and S.C. Sharma

Abstract Wireless sensor network (WSN) is comprised of miniature devices with limited energy resources. In some applications, the sensor nodes are unreachable so the backup of energy resource is not possible. Therefore, energy competence is a crucial issue that is inevitably to be improved to extend the network lifetime. Usually, clustering is used to improve the energy proficiency of network routing. In this paper, we focus on the cluster-based or hierarchical routing algorithms for sensor networks. We perform the analysis of popular hierarchical routing algorithm low-energy adaptive clustering hierarchy (LEACH) and focus on how to choose the next hop nodes during data transmission phase. In this article, we propose an improved clustering algorithm for WSN. Simulation analysis confirms that proposed clustering algorithm outperforms the LEACH algorithm.

Keywords WSN · Clustering algorithms · Energy · Network lifetime · LEACH

1 Introduction

Latest advancement in embedded systems and communication technologies allows the microautonomous system comprised of small tiny devices known as sensors. These sensors can detect, compute, and communicate via suitable sensor technology that gives birth to wireless sensor network [1, 2]. Deployment ease and low-cost sensors make wireless sensor network suitable for many applications like health care, transportation, smart building, disaster management, and environmental

S.P. Singh (✉) · S.C. Sharma
Electronics and Computer Discipline, DPT, Indian Institute of Technology Roorkee,
Roorkee, India
e-mail: sps78dpt@iitr.ac.in

S.C. Sharma
e-mail: scs60fpt@iitr.ac.in

© Springer Nature Singapore Pte Ltd. 2018   379
M. Pant et al. (eds.), *Soft Computing: Theories and Applications*,
Advances in Intelligent Systems and Computing 584,
https://doi.org/10.1007/978-981-10-5699-4_35

monitoring. Since the sensor nodes have limited energy supply, optimization of energy must be considered as the important aim in sensor network design [3]. Clustering is a process used to handle the energy usage effectively [4]. In this scheme, each and every group of sensors has a head node recognized as cluster head (CH) that performs data fusion and data forwarding toward the base station (BS) or sink node [5]. Numerous cluster-based routing algorithms have been reported in text to extend the lifetime of network [6–8]. The LEACH is most popular routing protocol due to its simplicity and efficiency [9]. LEACH divides the complete network into a various clusters, and execution time of system is partitioned into several rounds. In every round, the member node contends to be CH in accordance with some specified criterion. However, this protocol does not guarantee that the preferred amount of CHs is selected in the network. Some enhancement on LEACH is reported in the literature [10–12].

In this article, we proposed an improved clustering algorithm to increase network lifetime. The main initiative in proposed algorithm is CH election with consideration of residual energy. The proposed method is described in Sect. 3. Simulation results analysis is presented in Sect. 4. At last, Sect. 5 concludes the work.

2 Related Work

Here, in this part or work, we concentrate on the similar work reported in the literature. Routing is a process of selecting best path in the network. Routing protocols are liable for discovering and managing efficient routes in the networks [6, 7].

2.1 Classification of Routing Protocols in WSN

WSN routing protocols can be classified on the basis of some measures like the way of constituting the routing paths, structure of network, protocol operation, originator of communications, how a protocol selects a next hop on route of forwarded message [5–7]. The categorization of routing protocols is given in Fig. 1.

On the basis of structure of network, routing protocols are categorized as flat, hierarchal (cluster-based), and location-based protocols. In flat-structured routing, each sensor node plays similar role, while in cluster-based routing, sensor nodes have dissimilar roles. Due to network scalability and transmission effectiveness, cluster-based routing is the best choice [13, 14]. LEACH is the most influential protocol of this category [9]. So, we focus on LEACH protocol and its drawbacks.

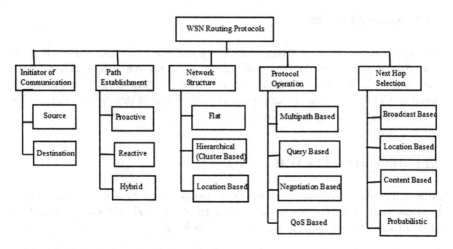

Fig. 1 Classification of WSN routing protocols

2.2 LEACH Protocol

LEACH is hierarchy-based routing protocol developed by Heinemann et al. [9]. The execution of LEACH consists of several rounds. Every round is made of setup phase and steady-state phase.

2.2.1 Setup Phase

Initially when clusters are being formed, each node makes a decision either to become a CH for present round. This decision is performed by a node in choosing a random number between 0 and 1. If chosen number is less than a threshold $T(n)$, the node becomes a CH for present round.

$$T(n) = \begin{cases} \frac{p}{1-p \times [r \bmod (1/p)]} & n \in G \\ 0 & \text{otherwise} \end{cases} \tag{1}$$

where p represents the ratio of CHs in network (generally p is 5%), r is the present round number. G is the set of nodes which have not been the CH in running round.

Using the above threshold, nodes those have been CH cannot become CH for subsequent time for p rounds. Afterward, every node has a chance of $1/p$ becoming a CH and attach to that cluster. The CHs merge and compact the data and advance it to sink; thus, it extends the network lifetime.

2.2.2 Steady-State Phase

The formation of a CH in every cluster at some stage in setup phase gives an assurance intended for data communication in steady-state phase. Intra-cluster transmission is based on TDMA schedule. The inter-cluster communication is based on CDMA schedule. After certain amount of time in steady-state phase, CHs are again elected during the setup phase.

2.2.3 Drawbacks of LEACH

Although cluster-based routing protocols act in fine way, they also experience many troubles. Here, we focused on the drawbacks of LEACH protocol [9, 15]. This protocol suffers from the following weaknesses like:

(i)   CH's election is random and does not consider the node's remaining energy.
(ii) Reclustering frequency is high so the some energy is wasted.
(iii) It covers small area.
(iv) Non-uniform distribution of cluster heads (CHs).

3 Proposed Algorithm

We propose a new clustering algorithm to extend the network lifetime. The proposed method is given in Fig. 2.

The proposed method consists of two phases such as:

3.1 Cluster Formation Phase

During the period of the cluster creation, to avoid the node with small remaining energy to be chosen as CH; take the consideration of the remaining energy when we define the threshold $T(n)$

$$T(n) = \begin{cases} \frac{p}{1-p\times[r\bmod(1/p)]} \times \frac{E_{residual}}{E_{initial}} & n \in G \\ 0 & \text{otherwise} \end{cases} \qquad (2)$$

where p, r, G are same as in Eq. (1). $E_{residual}$ and $E_{initial}$ are the remaining energy and initial energy of node correspondingly.

Fig. 2 Flowchart of
proposed method

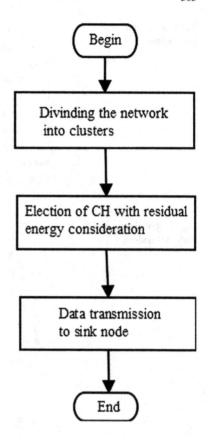

3.2 Data Transmission Phase

In this phase, the intra-cluster communication uses TDMA schedule similar to
LEACH. In LEACH, every CH accomplishes direct communication with base
station, regardless of the distance among CH and BS. Here, in proposed method, the
communication between CHs and BS has two types: one-hop communication and
multi-hop communication. The objective of the proposed method is to reduce the
transmission loss. The proposed algorithm described here chooses optimal path and
follows multi-hop communication between CH and BS.

4 Simulation Results

We simulated the proposed improved algorithm and analyzed with the original
LEACH in this section. Simulation parameter setting is as per Table 1.

Table 1 Simulation parameters

Parameters	Values
Number of nodes	100
Area	200 m × 200 m
Initial energy	2 J
Base station position	[100, 175]
Electronics energy (E_{elec})	50 nJ/bit
Energy for data aggregation (E_{da})	5 nJ/bit/signal
Packet size	25 bytes
Number of CH	5
Simulation time	600 s

Fig. 3 Remaining energy with 100 nodes

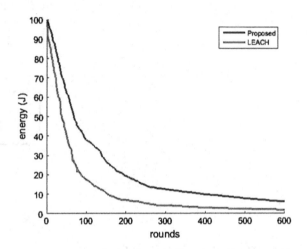

The performance analysis of our improved algorithm with original LEACH is shown in Figs. 3, 4, and 5.

Figure 3 shows the comparative presentation of remaining energy of LEACH with improved clustering algorithm. From the above presentation, we make out that improved algorithm has more residual energy in comparison with LEACH. Figure 4 shows the performance of first node dead (FND) means after how much round the first node dies. From the results, we know that our improved algorithm takes more rounds for FND. Figure 5 shows the performance of half node dead (HND) means after how much round the half nodes die. From the results, we know that our improved algorithm takes more rounds for HND. So, from the simulation result analysis, it is understandable that improved clustering algorithm extends the lifetime of network in comparison with LEACH protocol.

Fig. 4 Time of first node dead

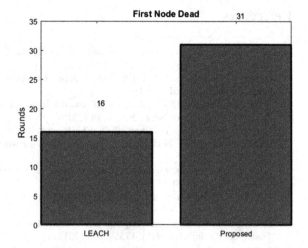

Fig. 5 Time of half nodes dead

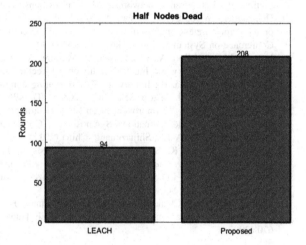

5 Conclusion

The improved clustering algorithm for wireless sensor network is described in this paper. This proposed improved algorithm uses residual energy for CH election process and follows multi-hop approach among cluster and base station. It is clear from simulated result analysis that proposed algorithm outperforms the LEACH protocol. Future works include the implementation of load balancing approaches among cluster head nodes to further improve energy efficiency.

References

1. Akyildiz, I.F., Su, W., Sankarasubramaniam, Y., Cyirci, E.: Wireless sensor networks: a survey. Comput. Netw. **38**(4), 393–422 (2002)
2. Romer, K., Friedemann, M.: The design space of wireless sensor networks. IEEE Wirel. Commun. **11**(6), 54–61 (2004)
3. Singh, S.P., Sharma, S.C.: A survey on research issues in wireless sensor networks. Open Trans. Wireless Sensor Netw. **2**(1), 1–18 (2015)
4. Naeimi, S., Ghafghazi, H., Chow, C.O., Ishi, H.: A survey on the taxonomy of cluster-based routing protocols for homogeneous wireless sensor networks. Sensors **12**(6), 7350–7409 (2012)
5. Abbasi, A.A., Younis, M.: A survey on clustering algorithms for wireless sensor networks. Comput. Commun. **30**, 2826–2841 (2007)
6. Liu, X.: A survey on clustering routing protocols in wireless sensor networks. Sensors **12**, 11113–11153 (2012)
7. Liu, X., Shi, J.: Clustering routing algorithms in wireless sensor networks: an overview. KSII Trans. Internet Inf. Syst. **6**(7), 1735–1755 (2012)
8. Kumar, D.: Performance analysis of energy efficient clustering protocol for maximizing lifetime of wireless sensor networks. IET Wireless Sens. Syst. **4**(1), 9–16 (2014)
9. Heinzelman, W.R., Chandrakasan, A., Balakrishnan, H.: Energy-efficient communication protocol for wireless micro-sensor networks. In: Proceedings of 33rd Hawaii International Conference on System Sciences, Hawaii, USA (2000)
10. Yassein, M.B., Al-Zou, A., Khamayseh, Y., Mardini, W.: Improvement on LEACH protocol of wireless sensor network. Int. J. Digital Content Technol. Its Appl. **3**(2), 132–136 (2009)
11. Zhao, F., Xu, Y., Lia, R.: Improved LEACH routing communication protocol of wireless sensor network. Int. J. Distrib. Sens. Netw. Article ID 649609
12. Xiaowen, M., Xiang, Y.: Improvement on LEACH protocol of wireless sensor network. In: Proceedings of the 2nd International Symposium on Computer, Communication, Control and Automation (ISCCCA-13), Shijiazhuang, China (2013)
13. Singh, S.P., Bhanot, K., Sharma, S.: Critical analysis of clustering algorithms for wireless sensor networks. Adv. Intell. Syst. Comput. (AISC) **436**(1), 783–792 (2016)
14. Lai, W.K., Fan, C.S., Lin, L.Y.: Arranging cluster sizes and transmission ranges for wireless sensor networks. Inf. Sci. **183**(1), 117–131 (2012)
15. Heinzelman, W.B., Chandrakasan, A.P., Balakrishnan, H.: Application specific protocol architecture for wireless micro-sensor networks. IEEE Trans. Wireless Commun. **1**(4), 660–670 (2002)

Performance Measurement of Academic Departments: Case of a Private Institution

Sandeep Kumar Mogha, Alok Kumar, Amit Kumar
and Mohd Hussain Kunroo

Abstract This paper evaluates the relative performances of academic departments of The NorthCap University, Gurgaon, for the academic year 2014–2015. The study is carried out with two inputs, namely number of academic staff and number of non-academic staff, and four outputs, namely total enrolled students, total pass students, students placed for jobs, and research index. Data envelopment analysis (DEA)-based dual CCR model is used for the efficiency evaluation. The results show that out of 7 academic departments, four are technically efficient with average efficiency score 0.899. The rest three inefficient departments are operating on increasing returns to scale.

Keywords Data envelopment analysis · Dual CCR model · NCU

S.K. Mogha (✉)
Department of Mathematics, Faculty of Physical Sciences, SGT University,
Gurgaon 122505, India
e-mail: moghadma@gmail.com

A. Kumar
Department of Mathematics, Faculty of Science, Subharti University,
Meerut 250005, India
e-mail: alokkumar1784@gmail.com

A. Kumar
Department of Mathematics, BITS, Pilani 333031, Rajasthan, India
e-mail: amitk251@gmail.com

M.H. Kunroo
Institute of Economic Growth, Delhi University Enclave, New Delhi 110007, India
e-mail: mhkunroo@gmail.com

© Springer Nature Singapore Pte Ltd. 2018
M. Pant et al. (eds.), *Soft Computing: Theories and Applications*,
Advances in Intelligent Systems and Computing 584,
https://doi.org/10.1007/978-981-10-5699-4_36

1 Introduction

Education and research is the main contributor of the overall development of any nation. There are several government and private institutions in India for the scientific and technical education which works for the development of the nation. There are several names in the list of best government institutions. In counterpart of the government higher education, there are also several private institutions which also provide the scientific and technical education. The first name in the list of private institution/university in NCR region comes into insight is The NorthCap University (previously known as ITM University Gurgaon).

The NorthCap University was founded in 1996. Initially, it was known as Institute of Technology and Management. In the year 2009–2010, the institute gained its status as an autonomous university as ITM university. In August 2015, it has been renamed as The NorthCap University (NCU).

This paper measures the relative efficiencies of academic departments of NCU based on multiple input and multiple output criteria. The second objective of the paper is to measure the comparative efficiency of the departments, and the third objective is to identify the efficient and inefficient performers.

There are two approaches to measure the efficiencies of DMUs. One is parametric, and the other is nonparametric. In parametric approaches, there are two methods, namely ordinary least square method and the stochastic frontier method, in which functional form is required, while data envelopment analysis (DEA) and free disposal hull are nonparametric approaches which do not require any functional form. The most frequently and most suitable technique for measuring the performance of non-profit organizations is DEA. Therefore, this paper applies DEA-based methodology for the performance measurement of the departments. It is, on the whole, appropriate technique for the researcher in investigating the efficiency of DMUs that exchange multiple inputs into multiple outputs.

In recent years, DEA has become very popular among researchers for measuring the performance of DMUs. DEA is extensively used in several areas including education, banking, health, drugs and pharmaceutical industries across the world to measure the efficiencies. Some recent studies are reviewed here.

Tyagi et al. [1] use DEA-based CCR and BCC models to measure the efficiencies of 19 academic departments of IIT Roorkee (India). Rayeni and Saljooghi [2] used DEA to determine the relative efficiency scores of academic departments of a university and set benchmarks for the departments. Aziz et al. [3] use DEA to measure the relative efficiency of 22 academic departments of a public university in Malaysia. Aristovnik [4] used DEA to measure the relative efficiency in public education expenditures in the EU and OECD countries. Barra and Zotti [5] used DEA to assess technical efficiency in a big public university. Puri and Yadav [6] measure the efficiencies of 27 public and 22 private sector banks for the year 2009–2010 using DEA-based CCR and BCC models. Mogha et al. [7] assess the technical and scale efficiencies of 50 private sector hospitals in India for the period 2004–05 to 2009–10 using CRS- and VRS-based DEA models. Mahajan et al. [8] estimate technical efficiencies, slacks, and input/output targets of 50 large Indian pharmaceutical firms for the year 2010–11. Mogha et al. [9] used DEA-based CCR model to assess the efficiency of 36 public sector hospitals of Uttarakhand, state of India, for the calendar year 2011.

The rest of the paper is organized as follows: Sect. 2 contains methodology; Sect. 3 describes the research design, input/output variables, DEA model used in the study and data; Sect. 4 gives efficiency results followed by the conclusion in the last.

2 Methodology

The application procedure of DEA analysis includes three stages: The first stage is to select the DMU(s) departments (in this case). Then, the input and output variables are to be selected for the efficiency analysis. Finally, model is selected to analyze the data. The DEA model is described later in the paper.

2.1 Selection of DMUs

We have selected all seven departments of NCU for the efficiency evaluation, since all the departments have similar objective, i.e., academic orientation (teaching and research), and therefore similar in structure, funded by the same governing body, and thus could be taken for the DEA study.

2.2 Selection of the Inputs and Outputs

As per the availability of data, cross-sectional analysis is based on data of 7 departments for the academic year 2014–2015. The descriptions of used variables are given below.

2.2.1 Academic Staff

Academic staff is a mandatory factor for any academic institution for teaching and research purpose. This includes professors, associate professors and assistant professors, lecturers, and research scholars. These are the main employees of the institution to create enrollments by convincing the students from schools and also to create the research outputs. Thus, the selection of this variable as an input variable is appropriate in this study.

2.2.2 Non-academic Staff

Each institute/university has its construction, staff rooms, and laboratories for each department. So, for non-academic work like laboratory attendant and other file handling work, each department has particular manpower. So, this factor can also be used as an input variable for the efficiency evaluation.

2.2.3 Total Enrolled Students

There are three programs: undergraduate (UG), postgraduate (PG), and doctoral program (Ph.D.) in each of the department of the university. For making a variable of the total enrolled students, we assign weights to the number of students of all the levels. By this, we get a new variable "total enrolled students" as a new variable:

$$\text{Total enrollment of the students} = \text{Number of undergraduate students}$$
$$+ 1.3\,(\text{no. of post graduate students})$$
$$+ 2\,(\text{no. of doctoral students})$$

Since all the undergraduate students in first-year courses are taught by the APS department faculty members, in this study, the number of students in APS department is taken as the number of students of B.Sc., M.Sc., and the total first year of UG students. This is used as an output variable.

2.2.4 Total Passed Students

In this, we include the number of graduate and postgraduate degrees conferred. Students have to pass in all the subjects of their concern department. By adding number of undergraduate, postgraduate and doctoral degree awarded, a new output variable "total passed students" is formed.

2.2.5 Total Students Placed for Jobs

This variable includes the number of the students get offer from any company or organization (either they joined or not). This variable is used as an output variable.

2.2.6 Research Index

This is the main activity in the growth of an academic department or any institution. This index is formed with the combination of research papers published in journals and conferences with proceedings, presented in the conferences and research projects carried out by the department. About the quality of research, there are some contradictions between the researchers. In this context, the appropriate term for the variable is "research index." Thus:

$$\text{Research Index} = \text{No. of research papers published in journals}$$
$$+ 0.5 \text{ (papers published in conferences)}$$
$$+ 1.2 \text{ (no. of research project)}$$

2.3 Selection of the Model

Firstly, DEA was developed in [10] by Charnes et al. which was extended in [11] by Banker, Charnes, and Cooper. These models evaluate the performance based on the criteria of multiple inputs and outputs. There are two types of orientations in DEA literature. One is input-oriented measure which aims to minimize inputs with existing level of outputs, and other is output-oriented measure which aims to maximize outputs with existing levels of inputs.

In DEA applications, where the non-flexible inputs are used, the output orientation is more suitable. If the outputs variables are fixed by the organizations, an input orientation is used (Ramanathan [12]).

In the present study, two inputs, namely academic staff and non-academic staff are used. In most of the cases, these variables are unbendable for a year. Also, the variables such as "progress" and "research index" are also not decided at advance bases. In some of the cases, total enrollment can be fixed by the organization. So, the output orientation is proper for the current analysis.

In this study, dual CCR model, also known as CCR envelopment model, is used and is given by:

$$\left.\begin{array}{ll} \text{Max } Z_k = \theta_k + \varepsilon\left(\sum_{r=1}^{s} s_{rk}^+ + \sum_{i=1}^{m} s_{ik}^-\right) & \\ \text{subject to} & \\ \sum_{j=1}^{n} \lambda_{jk} y_{rj} - s_{rk}^+ = \phi_k y_{rk} & \forall r = 1,2,3,\ldots,s \\ \sum_{j=1}^{n} \lambda_{jk} x_{ij} + s_{ik}^- = x_{ik} & \forall i = 1,2,3,\ldots,m \\ \lambda_{jk} \geq 0 & \forall j = 1,2,3,\ldots,n \\ s_{ik}^-, s_{rk}^+ \geq 0 & \forall i = 1,2,3,\ldots,m \\ & r = 1,2,3,\ldots,s \\ \theta_k \text{ is unrestricted in sign} & \end{array}\right\} \quad (1)$$

where s_{rk}^+ denotes the slack of the kth DMU in the rth output, s_{ik}^- denotes the slack of the kth DMU in the ith input, and λ_{jk}'s are the dual variables. θ_k is the (proportional) decrease functional concurrently to all inputs and results in a radial movement forward to the efficiency frontier. This is properly identified as CCR

envelopment model. The explanation of the outcome of the envelopment model
(1) can be summarized as:

The DMU_k is fully efficient if

(a) Optimal value of θ is 1.
(b) All the slack variables becomes zero.

The DMU_k is weakly efficient if

(a) Optimal value of θ is 1.
(b) $s_{rk}^{+*} \neq 0$ and/or $s_{ik}^{-*} \neq 0$ for some r and i in some alternate optima.

The nonzero value of slacks and/or the optimal value of θ less than 1 identify the
sources and sum of any inefficiency that may be present in the DMU. If the optimal
value λ_{jk}^* of λ_{jk} is nonzero, then the jth DMU represents the reference set (peer) of
the kth DMU and the value for these reference set elements is the coefficients used
to construct the benchmark. The reference set can be defined as the collection of
DMUs used to construct the virtual DMU as benchmark which shows about
increment in outputs and decrement in inputs to make the kth DMU efficient. The
results of the study are calculated by using DEA Solver software.

3 Data

Table 1 demonstrates the explanatory statistics of the data collected from different
sources. The data for input variables are taken from HR admin office of the university. Total enrolled student's undergraduate, postgraduate, and doctoral records
are taken from annual report and VC report of NCU (2014–2015). Details of got
offered students (job) (anywhere) are taken from controller of records (COR).

Number of research papers published in journals and conference of the department and research project in number of taken by the department all are taken from
newsletters of NCU from July 2014 to June 2015. The collected data of the
departments for all the factors are given in Table 1.

Table 1 Descriptive statistics in input and output variables

	Academic staff	Non-academic staff	Total enrolled students	Total passed students	No. of students placed for job	Research index
Max	39	27	676.3	248	180	423.42
Min	10	1	40.6	21	1	59.22
Average	24.43	10	237.37	105.71	55.71	209.19
SD	10.15	9.41	198.77	83.47	61.32	123.61

4 Findings

The efficiency score (OTE) for 7 academic departments of NCU is estimated for year 2014–2015. The results are calculated on the basis of output orientation. The efficiency scores calculated by CCR output-oriented model along with reference set, peer count, slacks, and targets are discussed.

4.1 Overall Technical Efficiency (OTE)

DEA calculates the efficiency of academic departments that create a production leading edge. The calculated efficiency scores indicates that departments whose efficiency is equal to 1 are on the efficiency leading edge under CRS assumption and those having efficiency score less than 1 are inefficient compared to the departments on the efficiency frontier. Table 2 evinces that out of 7 academic departments, only 4 are overall technically efficient and form the efficiency frontier. The remaining 3 departments are inefficient. The efficient departments APS, CSE and IT, ECE and EE, and management form the "reference sets" for the departments which are not efficient. Figure 1 shows the regression line of OTE.

4.2 Slacks and Targets

For inefficient departments, DEA calculate slack in inputs and outputs and set targets so that they can be efficient. These targets can be calculated by the relations given below:

For outputs:

$$\overline{y_{rk}} = \phi_k^* y_{rk} + S_{rk}^{+*}$$

Table 2 Efficiency scores, reference set, peer weight, and peer count

Departments	DMU	OTE	Reference set	Peer weight	Peer count	Rank
APS	D1	1	D1	1	0	1
ME	D2	0.786	D3	0.641	0	5
CSE and IT	D3	1	D3	1	2	1
ECE and EE	D4	1	D4	1	1	1
Civil	D5	0.758	D3, D6	0.045, 0.466	0	6
Management	D6	1	D6	1	2	1
Law	D7	0.748	D4, D6	0.382, 0.044	0	7
	Mean	0.899				

Fig. 1 Regression line for
OTE score

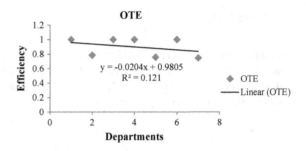

Table 3 Inputs and outputs slacks for inefficient departments

Departments	Academic staff	Non-academic staff	Total enrolled students	Total passed students	No. of students placed for job	Research index
ME	0	1.69	37.01	0	51.6	34.18
Civil	0	2.32	4.69	0	20.06	0
Law	0	0	30.93	0	5.54	33.58
Mean	0	1.34	24.21	0	25.73	22.59

For inputs:

$$\overline{x_{rk}} = x_{ik} - S_{ik}^{-*}$$

where $\overline{y}_{rk}(r = 1, 2, 3, 4)$ and $\overline{x}_{ik}(i = 1, 2, 3)$ are the output and input targets, respectively, for the kth DMU; y_{rk} and x_{ik} are the actual rth output and ith input, respectively, of the kth DMU; ϕ_k^* is the optimal efficiency score of the kth DMU; s_{ik}^{-*} and s_{rk}^{+*} are the optimal input and output slacks of the kth DMU for $(i = 1, 2, 3)$ and $(r = 1, 2, 3, 4)$. The optimal input and output slacks for inefficient DMUs are given in Table 3.

Table 4 represents the input and output target values for the departments, which are not efficient, along with percentage diminution in inputs and expansion in outputs. The results expose that on average a department has no considerable scope to shrink inputs, but it has a significant scope to expand outputs compared to the best performing department. On average 59.61% of total enrolled students, 22.09% total passed students, 149.46% of students placed for jobs and 49.00% research index should be increased with 16.75% reduction in non-academic staff and full utilization of academic staff, if all the inefficient departments run at the efficient level.

Table 4 Target for inefficient departments under CCR output-oriented model

Departments	Academic staff	Non-academic staff	Total enrolled students	Total passed students	No. of students placed for job	Research index
ME	25 (0.00)	17.31 (8.89)	198.14 (56.39)	158.97 (27.18)	115.39 (130.78)	271.42 (45.50)
Civil	12 (0.00)	1.68 (58.00)	104.95 (38.09)	47.49 (31.92)	25.34 (533.50)	138.78 (31.92)
Law	10 (0.00)	1 (0.00)	85.22 (109.90)	38.78 (33.72)	18.91 (89.10)	112.77 (90.43)
Mean	15.67 (0.00)	6.66 (16.75)	129.43 (59.61)	81.74 (29.09)	53.21 (149.46)	174.32 (49.00)

5 Conclusion

In this paper, we measure the relative efficiencies of 7 academic departments of The NorthCap University, Gurgaon (India), using the DEA-based dual CCR model. The study finds that out of total 7 departments, only 4 (57.14%) departments are found to be efficient as they scored the maximum level of efficiency. Average overall technical efficiency of the departments is found to be 89.90%, which indicates that on average, the inefficient departments are not using their potential by 10.10%; that is, these departments have the scope of producing more outputs with lesser inputs than their existing levels. The target setting results show that only one input have some scope to reduce but all the outputs have significant scope to augment. The results also suggests that on average, 59.61% of total enrolled students, 22.09% total passed students, 149.46% of students placed for jobs and 49.00% research index should be increased with 16.75% reduction in non-academic staff if all the inefficient departments operate at the level of efficient departments.

Acknowledgements The present study is just for research purpose. The data used in the study are not for commercial purpose. The data are collected from different sources of NCU with the permission of vice-chancellor of the university.

References

1. Tyagi, P., Yadav, S.P., Singh, S.P.: Relative performance of academic departments using DEA with sensitivity analysis. Eval. Prog. Plan. **32**, 168–177 (2009)
2. Rayeni, M.M., Saljooghi, F.H.: Benchmarking in academic departments using data envelopment analysis. Am. J. Appl. Sci., Iran **11**, 1464–1469 (2010)
3. Aziz, N.A.A., Janor, R.M., Mahadi, R.: Comparative departmental efficiency analysis within a university: a DEA Approach. In: 6th International Conference on University Learning and Teaching, vol. 22, pp. 212–219 (2012)

4. Aristovnik, A.: Relative efficiency of education expenditures in eastern Europe: a non-parametric approach. J. Knowl. Manage., Econ. Inform. Technol, Slovenia **3**, 1–13 (2013)
5. Barra, C., Zotti, R.: Measuring efficiency in higher education: An empirical study using a bootstrapped Data envelopment analysis. Int. J. Adv. Econ. Res. **21**, 11–33 (2016)
6. Puri, J., Yadav, S.P.: Performance evaluation of public and private sector banks in India using DEA approach. Int. J. Oper. Res. **18**, 91–121 (2013)
7. Mogha, S.K., Yadav, S.P., Singh, S.P.: Estimating technical and scale efficiencies of private hospitals using a non-parametric approach: case of India. Int. J. Oper. Res. **20**, 21–40 (2014)
8. Mahajan, V., Nauriyal, D.K., Singh, S.P.: Technical efficiency analysis of the Indian drug and pharmaceutical industry: a non-Parametric approach. Benchmarking: Int. J. **21**, 734–755 (2014)
9. Mogha, S.K., Yadav, S.P., Singh, S.P.: Estimating technical efficiency of public sector hospitals of Uttarakhand (India). Int. J. Oper. Res. **25**, 371–399 (2016)
10. Charnes, A., Cooper, W.W., Rhodes, E.: Measuring the efficiency of decision-making units. Eur. J. Oper. Res. **2**, 429–441 (1978)
11. Charnes, A., Cooper, W.W.: Programming with linear fractional functional. Naval Res. Logistics Q. **9**, 181–186 (1962)
12. Ramanathan, R.: An Introduction to Data Envelopment Analysis. Sage Publication India Pvt. Ltd., New Delhi (2003)
13. Avkiran, N.K.: Investigating technical and scale efficiencies of Australian Universities through data envelopment analysis. Socio. Econ. Plan. Sci. **35**, 57–80 (1999)
14. Coelli, T.: A Guide to DEAP Version 2.1: a data envelopment analysis (computer) program. Available at: http://www.owlnet.rice.edu/∼econ380/DEAP.PDF (1996)
15. Coelli, T., Rao, D.S.P., Battese, G.E.: An Introduction to efficiency and productivity analysis. Kluwer Acadmic Publisher, London (1998)
16. Ramanathan, R.: Comparative risk assessment of energy supply technologies: a data envelopment analysis approach. Energy **26**, 197–203 (2001)
17. Singh, S.: Measuring the performance of teams in the Indian premier league. Am. J. Oper. Res. **1**, 180–184 (2011)
18. Tomas, R.: Using data envelopment analysis: a case of universitie. J. Rev. Econ. Perspect. **14**, 34–54 (2014)
19. Yousfat, A., Boukmiche, L., Yahiaoui, N.: Measurement of the relative efficiency in the Algerian University: evidence for Adrar University based on DEA method. Eur. J. Bus., Econ. Acc. **3**, 78–91 (2015)

Application of Shuffled Frog-Leaping Algorithm in Regional Air Pollution Control

Divya Prakash, Anurag Tripathi and Tarun Kumar Sharma

Abstract Shuffled frog-leaping algorithm (SFLA), a memetic algorithm modeled on the foraging behavior of natural species called frogs. SFLA embeds the features of both particle swarm optimization (PSO) and shuffled complex evolution (SCE) algorithm. It is well documented in the literature that SFLA is an efficient algorithm to solve nontraditional optimization problems. In this study, SFLA has been applied to a real-world problem of air pollution control. The results are at par with the state of the art of algorithm.

Keywords SFLA · Air pollution control · Global optimization
Shuffled frog-leaping algorithm

1 Introduction

The problem of optimization arises in every sphere of human life. As the level of problem increases in terms of complexity, dimension, limited knowledge about the problem domain, feasible region, and alike, it becomes difficult to solve such multifaceted problems using traditional optimization methods (e.g., gradient search methods, etc.). SFLA, introduced in 2003 by Eusuff and Lansey [1, 2], is a recent member of memetic algorithm family. SFLA, like other memetic algorithms, enthused by natural foraging behavior of species. In SFLA, these natural species are frogs. Since introduction, many problems of versatile domain have been successfully

D. Prakash (✉) · T.K. Sharma
Amity School of Engineering and Technology, Amity University Rajasthan,
Jaipur 303007, India
e-mail: divya1sharma1@gmail.com

T.K. Sharma
e-mail: taruniitr1@gmail.com

A. Tripathi
Arya Institute of Engineering and Technology, Jaipur, India
e-mail: anuragtripathi1305@gmail.com

© Springer Nature Singapore Pte Ltd. 2018
M. Pant et al. (eds.), *Soft Computing: Theories and Applications*,
Advances in Intelligent Systems and Computing 584,
https://doi.org/10.1007/978-981-10-5699-4_37

solved using basic SFLA and its extended variants. The same can be witnessed from the literature. SFLA embeds the features of both particle swarm optimization (PSO) and shuffled complex evolution (SCE) algorithm. PSO helps in performing local search, while SCE helps in global search. In SFLA, like other memetic algorithms, colony of frogs is initialized randomly and in second step the colony is divided into subcolonies or memeplexes.

SFLA performs both exploration by dividing the randomly initialized population of frogs into number of memeplexes and exploitation by performing evolution process in each memeplexes.

In this study, SFLA has been applied to solve a real-world problem of air pollution control. Air pollution is a biggest challenge throughout the globe. The control over this pollution can help or raise an alert for human health protection, pollution in atmosphere, and ensures safety for flight. This study is focused on optimization air pollutants.

The structure of the study is as follows: Sect. 2 describes SFLA in brief followed by the problem discussion in Sect. 3. Experimental settings and computational results are presented in Sect. 4. Section 5 presents the conclusions drawn and future work plan.

2 An Overview of SFL Algorithm

SFLA, introduced in 2003 by Eusuff and Lansey [1], has been successfully implemented in solving many world optimization problems. SFLA is performed at par when compared with GA, PSO, and ant colony optimization (ACO) algorithms [1, 2]. Researchers have proposed many improved variants of SFLA and applied to many applications of science, engineering, and management [2–16].

SFLA is a memetic algorithm that is inspired from foraging behaviors of natural species called frogs. SFLA has the abilities to perform local search process as in PSO algorithm and at the later stage information sharing between memeplexes. The next paragraph details the working principle of SFLA.

Like in other memetic algorithms, a population of frogs (solutions) is randomly initialized in feasible search space. Then, frogs are distributed into small colonies called memeplexes. This is done by calculating fitness values using objective function. Now, each memeplex contains frogs from different cultures and each frog is supposed to perform local search process in order to modify worst frog position (in terms of fitness value). This process, within each memeplexes, is performed till some fixed criterion or generations. This process also helps in optimization. Later, the information or idea hold by the frogs in each memeplex is then exchanged among other memeplexes through shuffling process. Both local and information exchange process are performed till the fixed number of iterations. In general, there are four steps in basic SFLA and are as follows:

A. Initialization of Frog Population

The population of frogs is randomly initialized between fixed upper and lower bounds, i.e., (ub_j) and (lb_j), using Eq. (1): Set of frogs is presented by F.

$$x_{ij} = lb_i + rand(0, 1) \times (ub_i - lb_i) \tag{1}$$

where rand $(0,1)$ is random number between zero and one which is uniformly distributed; $i = 1, 2,..., F; j = 1, 2,..., D$ (dimension).

B. Sorting and Division of Frogs in Memeplexes

The fitness of each frog is calculated and then sorted in descending order. Then, division of sorted population is done into m memeplexes such that $F = m$ (memeplexes) x n (number of frogs in each memeplex). The division of frogs is done such that the frog with higher fitness value will be the part of first memeplex, accordingly the next frog into second memeplex, and the following fashion frog $m + 1$ will again move to the first memeplex and so on.

C. Local Searching Process

Frog with global best (X_{global}), best (X_{best}), and worst (X_{worst}) based on fitness values in each memeplex is noted. Now, an evolution process is implemented to update the position of worst frogs only using the following Eqs. (2) and (3):

Worst frog's updated position is

$$\begin{aligned} X_{new} &= \text{current position}(X_{worst}) + Mov_t \\ &- Mov_{max} \leq Mov_t \leq Mov_{max} \end{aligned} \tag{2}$$

where

$$Mov_t = rand(0, 1) \times (X_{best} - X_{worst}) \tag{3}$$

where t, Mov_t represents the generations $(1, 2, ..., N_{gen})$ and movement of a frog, respectively. Mov_{max} is the maximum acceptable frog movement in the feasible region. If, using above equations, the worst frog position shows significant improvement, then the position of worst frog is updated; otherwise, the same process using Eqs. (2) and (3) is repeated with respect to the global best frog (i.e., X_{global} replaces X_{best}). If process shows no improvements, then a new solution is randomly generated using Eq. (1) to update the worst frog. This is an iterative process till the maximum number of generations (N_{gen}).

D. Information Exchange or Shuffling Process

Information exchange is performed among memeplexes, or the frogs are sorted and shuffled again to perform the evolution process. This process again continues till the fixed number of iterations.

The searching process and basic SFLA algorithm is depicted in Fig. 1.

Algorithm – Basic SFLA

Step 1: Initial population of frog size F is generated and evaluated (fitness value)

Step 2: Distribute the population of frog in m memeplexes where each memeplex
        contains n fog members such that $F = m$ x n.

Step 3: Start Evolutionary process

Step 4: Set $i = 1$

Step 5: while $i \leq i_{max}$

Step 6: Identify the global best (x_g) position of frog in the population.

Step 7: Identify the worst (x_{worst}) and best (x_{best}) frog in each memeplex.

Step 8: Apply eq. (2) & (3) to generate new frog position and evaluate.

Step 9: **If** $f(x_{new}) < f(x_{worst})$

Step10: Set $x_{worst} = x_{new}$ and go to Step 13.

Step 11: $x_{best} = x_{global}$. Repeat Step 8 & Step 9.

Step 12: Random position is generated and replaced with x_{worst}.

Step 13: $i = i + 1$

Step 14: End While

Step 15: Population of frog is shuffled.

Step 16: If fixed termination criterion (N_{gen}) is achieved then exit else go to Step 2.

Fig. 1 SFL algorithm

3 Modeling of Air Pollution Control

It has been analyzed that there lies three major pollutants, namely aerosol, acid rain formation, and hydrocarbon in the atmosphere. These pollutants may be the result of industrial waste, pollution through vehicles, etc. It is not feasible and even possible to completely remove these pollutants from the atmosphere but their emission can be optimized using some control measures. Let us define the three above-discussed pollutants as $P1$, $P2$, and $P3$ and can be represented as a matrix, i.e.,

$$P = [P1, P2, P3]^{T}.$$

Further, the pollutant monitoring area is defined in two parts, i.e., fixed path for the flights, and second is the surrounded area of the path in the space (rectangular). Let us name them as area $A1$ and $A2$, respectively. Three different measures to control the pollutants have been applied in these two areas, and they are

- usage of selective adsorbent,
- artificial sedimentation, and
- pollutant source reduction.

The efficiency of the above-stated controlling measures varies on each pollutant. The quantities of pollutants' reduction are presented in Table 1. Since the three controlling measures are independent and do not affect the other one, they can be summed up together to formulate total reduction quantity.

Table 1 Quantities of pollution reduction using different measures

	Controlling measure-1 (CM1)		Controlling measure-1 (CM1)		Controlling measure-1 (CM1)	
	A1	A2	A1	A2	A1	A2
P1	q_{11}	q_{12}	q_{13}	q_{14}	q_{15}	q_{16}
P2	q_{21}	q_{22}	q_{23}	q_{24}	q_{25}	q_{26}
P3	q_{31}	q_{32}	q_{33}	q_{34}	q_{35}	q_{36}

Table 2 Associated cost for each controlling measure

	A1	A2
CM1	x_1	x_2
CM2	x_3	x_4
CM3	x_5	x_6

Table 3 Reduction of pollutants

	A1	A2
CM1	r_1	r_2
CM2	r_3	r_4
CM3	r_5	r_6

Let $Q = [q_{ij}]$ represents a matrix, where $i = 1, 2, 3$ and $j = 1, 2... 6$. The associated cost for each measure is discussed in Table 2. This is also considered as a matrix and represented as $C = [x_1 x_2 x_3 x_4 x_5 x_6]^T$.

As all the three controlling measure are independent and can be used alone. But practically, this may limit their impact. It may result effectively if these controlling measures would be applied proportionally. This may also reduce the cost. Table 3 presents the effect of the controlling measures when applied in proportionate to reduce pollutants. Again this can be written in form of matrix as $R = [r_1 r_2 r_3 r_4 r_5 r_6]$.

Based upon this, expenditure function model is formed as:

$$\text{minimize } Z = RX \tag{4}$$

with respect to the constraints

$$QX \geq P.$$

4 Experimental Parameter Settings

The population of frog is fixed to 200 with 20 memeplexes. There will be 10 frogs in each memeplexes performing local search. Ten generations are performed in each memeplexes to explore the local region. The maximum movement step is fixed to 0.5 times of the search area. Twenty-five runs are performed to evaluate mean function value and standard deviation. The maximum numbers of function evaluations (NFE) are fixed to 1000. The simulations are performed on Dev C++ with the following machine configurations: Intel(R) CPU T1350@1.86 GHz with 1 GB of RAM.

In the present case (problem considered), the data is consulted from the literature [17] and is follows:

$$P = \begin{bmatrix} 30 \\ 75 \\ 63 \end{bmatrix}, P = \begin{bmatrix} 6 & 4 & 13 & 10 & 8 & 7 \\ 17 & 20 & 9 & 15 & 28 & 24 \\ 18 & 27 & 14 & 12 & 15 & 10 \end{bmatrix} \quad \text{and } R = \begin{bmatrix} 4 & 5 & 3 & 3 & 6 & 4 \end{bmatrix}.$$

After putting the above information in Eq. 4, the following results are drawn. $(x_1, x_2, x_3, x_4, x_5, x_6) = (0.66299, 0.709199, 0.475399, 0.9440, 0.336399, 0.904199)$ with $Z = 16.09811$. It took only 416 NFE to evaluate the results. These results are at par with that of genetic algorithm, i.e., $(x_1, x_2, x_3, x_4, x_5, x_6) = (0.6633, 0.7092, 0.4754, 0.9439, 0.3364, 0.9042)$ with $Z = 16.1$.

5 Conclusions and Future Direction

In this study, basic SFL algorithm has been implemented in optimizing air pollution control problem. There are major three types of pollutants available in the atmosphere, and also three controlling measures are used to reduce these pollutants. A model is discussed to reduce the cost of the method to reduce these pollutants from the atmosphere. The area is divided into two regions. The simulated results show that the pollutant decreasing capability of three controlling measures is 66.299, 47.539, and 33.639% in the fixed path of flight where as the pollutant decreasing capability in the surrounding region, i.e., rectangular is 70.919, 94.40, and 90.419% with the minimal annual expenditure of 161,000 for air pollution. In future, we will concentrate on theoretical analysis of problem with more constraints or multi-objective formation of the problem.

References

1. Eusuff, M.M., Lansey, K.E.: Optimization of water distribution network design using the shuffled frog leaping algorithm. J. Water Resour. Plan. Manage. **129**, 210–225 (2003)
2. Eusuff, M., Lansey, K., Pasha, F.: Shuffled frog-leaping algorithm: a memetic meta-heuristic for discrete optimization. Eng. Optim. **38**, 129–154 (2006)
3. Ahandani, M.A., Alavi-Rad, H.: Opposition-based learning in the shuffled differential evolution algorithm. Soft. Comput. **16**, 1303–1337 (2012)
4. Li, J., Pan, Q., Xie, S.: An effective shuffled frog-leaping algorithm for multi-objective flexible job shop scheduling problems. Appl. Math. Comput. **218**, 9353–9371 (2012)

5. Tang-Huai, F., Li, L., Jia, Z.: Improved shuffled frog leaping algorithm and its application in node localization of wireless sensor network. Intell. Autom. Soft Comput. **18**, 807–818 (2012)
6. Ahandani, M.A., Alavi-Rad, H.: Opposition-based learning in shuffled frog leaping: an application for parameter identification. Inf. Sci. **291**, 19–42 (2015)
7. Sharma, T.K., Pant, M.: Shuffled artificial bee colony algorithm. Soft Comput. doi:10.1007/s00500-016-2166-2 (2016)
8. Sharma, T.K., Pant, M.: Identification of noise in multi noise plant using enhanced version of shuffled frog leaping algorithm. Int. J. Syst. Assur Eng. Manage, Springer. doi:10.1007/s13198-016-0466-7 (2016)
9. Sharma, S., Sharma, TK., Pant, M., Rajpurohit, J., Naruka, B.: Centroid mutation embedded shuffled leap frog algorithm. Elsevier Procedia Comput Sci. **46**, 127–134 (2015)
10. Rajpurohit, J., Sharma, T.K., Nagar, A.: Shuffled frog leaping algorithm with adaptive exploration. In: Proceedings of Fifth International Conference on Soft Computing for Problem. Solving Volume 436 of the series Advances in Intelligent Systems and Computing, pp 595–603 (2015)
11. Liu, C., Niu, P., Li, G., Ma, Y., Zhang, W., Chen, K.: Enhanced shuffled frog-leaping algorithm for solving numerical function optimization problems. J. Intell. Manufact uring. 1–21 (2015)
12. Aungkulanon, P., Luangpaiboon, P.: Vertical transportation systems embedded on shuffled frog leaping algorithm for manufacturing optimisation problems in industries. Springer Plus (2016). doi:10.1186/s40064-016-2449-1
13. Liu, H., Yi, F., Yang, H.: Adaptive grouping cloud model shuffled frog leaping algorithm for solving continuous optimization problems. Comput. Intell. Neurosci. **2016**(2016). Article ID 5675349
14. Dalavi, A.M., Pawar, P.J., Singh, T.p.: Tool path planning of hole-making operations in ejector plate of injection mould using modified shuffled frog leaping algorithm. J. Comput. Des. Eng. **3**(3), 266–273 (2016)
15. Lei, D., Guo, X.: A shuffled frog-leaping algorithm for hybrid flow shop scheduling with two agents. Expert Syst. Appl. **42**(23), 9333–9339 (2015)
16. Jadidoleslam, Morteza, Ebrahimi, Akbar: Reliability constrained generation expansion planning by a modified shuffled frog leaping algorithm. Int. J. Electr. Power Energy Syst. **64**, 743–751 (2015)
17. Zhang, Dayong, Zhang, Xiang, Zhongping, Wu: Air pollution control simulation based on genetic algorithm. J. Eng. (JOE) **1**(4), 74–76 (2013)

Estimating Technical Efficiency of Academic Departments: Case of Government PG College

Imran Ali, U.S. Rana, Millie Pant, Sunil Kumar Jauhar
and Sandeep Kumar Mogha

Abstract This paper estimates the technical efficiencies of 20 departments of a government PG college Gopeshwar, Chamoli, Uttarakhand (India), for the academic year 2012–2013 through DEA technique. Number of academic staffs, number of taught course students, average students' qualification, departmental operating cost (DOC) are taken as input variables, and number of graduates from taught courses, average graduates' results (%), and graduate rate (%) are taken as output variables. Input–output data are collected from the NAAC report 2014 and annual SSR report academic year 2012–2013 and from the examination section of the government PG college, Gopeshwar. Some data are collected from the departments of the college and college annual book Madhuri. The study concludes that overall efficiency of departments is 87.8%, which indicates that the departments are not using their potential by 12.20%.

Keywords DEA · Efficiency · LP · Departments · Uttarakhand

I. Ali (✉)
Department of Mathematics, Government PG College,
Gopeshwar, Chamoli, Uttarakhand, India
e-mail: ali.imranmath@gmail.com

U.S. Rana
Department of Mathematics, DAV PG College, Dehradun,
Uttarakhand, India
e-mail: drusrana@yahoo.co.in

M. Pant · S.K. Jauhar
Department of Applied Sciences and Engineering, Indian Institute
of Technology, Roorkee, India
e-mail: millidma@gmail.com

S.K. Jauhar
e-mail: suniljauhar.iitr@gmail.com

S.K. Mogha
Department of Mathematics, SGT University, Gurgaon, India
e-mail: moghadma@gmail.com

© Springer Nature Singapore Pte Ltd. 2018
M. Pant et al. (eds.), *Soft Computing: Theories and Applications*,
Advances in Intelligent Systems and Computing 584,
https://doi.org/10.1007/978-981-10-5699-4_38

1 Introduction

We know that mostly the development of any country is based on the education and research. Presently in India among all the postgraduates degree colleges of higher education in Uttarakhand, state (India), government PG college, Gopeshwar (Chamoli), Uttarakhand is the best college as it is received an A-grade given by national assessment and accreditation council (NAAC), Bangalore, in the year 2014. Gopeshwar, GPGC [1].

The college has several academic programs such as B.Sc., BA, Bachelor of Business Administration (BBA), Bachelor of Commerce, Bachelor of Education Programs, and M.Sc., M.Com., MA degree programs in different disciplines and also has Ph.D. programs. In total, twenty-two academic departments offer these programs, but we select here only twenty departments; out of these 20 departments, 17 departments run UG, PG, and Ph.D. programs and 3 departments (Bachelor of Education, Home Science, Bachelor of Business Administration) run only UG degree courses.

Our objective here to find the comparative performance of 20 academic departments of Government PG college Gopeshwar, Uttarakhand (India), for the academic year (2012–2013) using DEA. The other purpose is to measure how efficiently the departments work in the college and to identify the efficient and inefficient departments. In the degree colleges, emphasis is given more on teaching. So finally, we measure the performance of the departments for the activities of teaching only.

There are several methods for measuring the teaching efficiency of the departments of postgraduates degree college like parametric methods such as stochastic frontier method, Izadi et al. [2] and nonparametric method such as data envelopment analysis (DEA) Johnes and Yu [3], Bougnol and Dula [4]. Each method has its merits and demerits. When there is only one input and output handling data, then parametric method is used, but in case of several input and output handling data, nonparametric methods are applicable Charne's [5] (Ramanathan [6]).

In this paper, Sect. 2 comprises of a brief survey of literature of DEA in education field. Sections 3 and 4 describe the DEA model and problem description. Section 5 describes the model for teaching efficiency. Result analysis and conclusions are given in Sects. 6 and 7, respectively.

2 Literature Survey

Recently, more than one studies have been undertaken to measure the efficiency for academic departments in higher educational institutes such as degree colleges or universities. Every study differs in its scope and meaning. Some studies are briefly reviewed as given below. Bessent et al. [7] used DEA in assessing the comparatively efficiency of higher educational institute or colleges in USA. Other important studies that utilized DEA to evaluate the relative efficiencies among universities include Ahn et al. [8], Abbott and Doucouliagos [9], Avkiran [10], Johnes [11], Fandel [12], and Breu and Raab [13].

There is no fixed rule to select the inputs and outputs for efficiency measurement in higher educational institutes like a university or degree colleges or departments; e.g., Ahn et al. [8] have selected faculty salaries, research fund given by state, etc., as inputs variables and total enrolled students in UG and PG, total credit hours in a semester, etc., as a outputs variables. Johnes and Johnes [14], Johnes [11], Beasley [15], and Ali et al. [16] have given previous studies on DEA application in the context of university departments or courses. Jauhar et al. [17, 18] have given previous study on DEA in the context of Indian higher educational institute departments. For further details, interested readers may refer Cook and Seiford [19] and Kuah et al. [20]. The author verified that how DEA could be used in getting better programs, starting new program, discontinuous the deficient programs.

3 DEA Model

3.1 DEA Fundamentals

DEA was first developed by Charnes et al. [21] and extended by Banker et al. [22]. It is also known as CCR and BCC models. The main role of DEA is to find the comparative efficiency assessment of the firms or departments or DMUs for multiple inputs and outputs. A DMU can be defined as a body it transforms inputs into outputs.

Let us consider n decision-making units such that DMU_1, DMU_2, DMU_3,..., DMU_n, all decision making units DMU_j ($j = 1,2,3,4,5,...,n$) use n inputs α_{ij} ($i = 1,2,3,4,5,...,m$) and s outputs β_{rj} ($r = 1,2,3,4,5,...,s$). Let us consider a set of inputs and outputs weights u_i ($i = 1,2,3,4,5,...,m$) and v_r ($r = 1,2,3,4,5,...,s$) that maximize the efficiency score selected as variables for all the DMU.

Let us consider the decision-making units to be calculated on any trial be designated as DMU_o ($o = 1,2,3,2,4,5,...,n$). The efficiency of all DMUs e_o is obtained by solving the LP multiplier form of DEA which is given below.

General form LPPs of CCR and BCC DEA Model

Mathematically, the general form of CCR model is given by:

$$e_o = \text{Max} \sum_{r=1}^{s} v_r \beta_{ro} \tag{1}$$

$$\text{Subject to} \sum_{i=1}^{m} u_i \alpha_{io} = 1, \tag{2}$$

$$\sum_{r=1}^{s} v_r \beta_{ro} - \sum_{i=1}^{m} u_i \alpha_{io} \leq 0 \tag{3}$$

$$v_r, u_i \geq 0. \tag{4}$$

Another version of DEA is BCC model given by Banker et al. [22] who extended the idea of constant return to scale (CRS) in CCR model for the variable return to scale (VRS). They added a separate variable u_{0o} to the CCR multiplier model to evaluate whether operation was conducted in regions of increasing, constant, or decreasing returns to scale (RTS). The model is given below. They examined return to scale locally at a point on the frontier, related to the value of the term u_{0o} as (Mogha et al. [23]):

$u_{0o} < 1$ imply increasing RTS;
$u_{0o} = 0$ imply constant RTS;
$u_{0o} > 1$ imply decreasing RTS.

Mathematically, the general form of BCC model is given by:

$$e_o = \text{Max} \sum_{r=1}^{s} v_r \beta_{ro} + u_{0o} \tag{5}$$

$$\text{Subject to} \sum_{i=1}^{m} u_i \alpha_{io} + u_{0o} = 1, \tag{6}$$

$$\sum_{r=1}^{s} v_r \beta_{ro} - \sum_{i=1}^{m} u_i \alpha_{io} + u_{0o} \leq 0, \tag{7}$$

$$u_i, v_r \geq 0 \text{ and } u_{0o} \text{ is unrestricted in sign} \tag{8}$$

The model executes n-times to justify the efficiency scores of each DMU. The efficiency of all DMUs varies in the range between 0 and 1: If efficiency of a DMU is unity, then it is efficient; otherwise, it is called inefficient.

There are no certain criteria to choose the inputs or outputs for the efficiency measurement of postgraduate college departments; e.g., Johnes [11] has taken as a input variables total number of UG and PG students for teaching and research, expenditure on plants and library and first division undergraduates (UG), postgraduates (PG), research graduates, and research grants as outputs.

4 Problem Descriptions

4.1 Methodology

The DEA methodology is mainly used to calculate the efficiency of non-profit entities. DEA provides relative efficiency of a DMU. DMU may comprise of banks, educational institutes, private organizations, transportation firms, etc. The efficiency of DMUs is obtained by the general formula which is given by the ratio of output weighted and weighted input.

Data envelopment analysis gave us some important and useful idea that may help to improve the efficiency of a DMU. With the help of DEA, we can evaluate input slacks and output slacks. DEA also identifies the reference set called peer group (Tyagi. et al. [24]).

4.2 Selection of DMUs

To analyze the efficiency study, we select the DMUs. There are two major factors for a selection of DMUs—homogeneity and number of DMUs. They should perform similar tasks and should have same target. Here, we choose the departments of government PG college, Gopeshwar (Chamoli), Uttarakhand. They are homogeneous since they perform the similar task and have same target.

4.3 Selection of Inputs and Outputs

There is no fixed rule to select the inputs and outputs. In this study, we select the inputs and outputs to measure the teaching efficiency of the departments of government postgraduates degree college, Gopeshwar, Chamoli, Uttarakhand (India), which is given below.

4.3.1 Inputs

Number of Academic Staffs (α_1)

Number of academic staff is must for all the departments for both research and teaching activities. The number of academic staffs is also considered as inputs (Tayagi. et al. [24], Kuah and Wong [25], Johnes and Johnes [14]).

Number of Taught Course Students (α_2)

Teaching is the initial work of all departments and is connected to the academic staff. In most of the department, two types of students are enrolled, (1) under-graduates (UG) and (2) postgraduates (PG), while in some departments, only UG students are enrolled. So we take total enrolled students in UG and PG as a input variable. This input is also used as a variable (Kuah and Wong [25]):
    Number of taught course students = Total enrolled students in UG and PG.

Average Students Qualification (α_3)

To calculate the academic progress of the students at pass-out level, we take average students qualification as an input variable at entering level. This is also used as an input variable in Kuah and Wong [25].

Departmental Operating Cost (DOC, α_4)

For the development of all activities of the departments, departmental operating cost (DOC) is necessary. So we take DOC as an input variable. This input is also used as a variable in Basely [15], Abbott and Doucouliagos [9], Tyagi. et al. [24].

4.3.2 Outputs

The outputs of teaching activities are focused on: (1) number of graduates from taught courses (β_1), (2) average graduates results in % (β_2), and (3) graduate rate in %(β_3) of departments of the degree college which are associated with the academic quality of graduates. Therefore, we are taking these as output variables for teaching efficiency (Kuah and Wong [25]).

5 Model for Teaching Efficiency Measurement of College Departments

Let us assume n departments of a college DMU_1, DMU_2, DMU_3,...,DMU_n. Each department j, DMU_j ($j = 1,2,3,...,n$) uses four inputs α_{ij} ($i = 1,2,3,4$) to produce 3 outputs β_{rj} ($r = 1,2,3$) from its teaching activities.
    Let us consider p to be the proportion of expenditure on teaching activities and let input and output weight u_i ($i = 1,2,3,...,m$) and v_r ($r = 1,2,3,...,s$) be variables. Let DMU_j to be calculated on any trial be designated as DMU_o ($o = 1,2,3,...,n$).
    The teaching efficiency of the DMU_o is denoted by T_o and defined by:

$$T_0 = \sum_{r=1}^{3} v_r \beta_{ro} / \sum_{i=1}^{3} u_i \alpha_{io} + p(u_4 \alpha_{4o}) \tag{9}$$

$$\text{Subject to } \sum_{i=1}^{3} v_r \beta_{rj} - \sum_{i=1}^{3} u_i \alpha_{ij} - p(u_4 \alpha_{4o}) \leq 0, \quad \text{for all } j, \tag{10}$$

$$0.3 \leq p \leq 0.7 \tag{11}$$

$$u_i, v_r \geq \in \tag{12}$$

Constraint condition (10) is to limit the relative efficiency T_o of all decision-making units to be within 1. Constraint condition (11) is to avoid zero proportion of the costs on either function or $\in = 0.01$ and is a small and non-Archemedean number, and the efficiency model is executed up to n-times in identifying the relative efficiency scores of all DMUs. If the efficiency score of the DMUs is 1, then it means that it is considered as more efficient decision-making unit, while less than 1 efficiency score DMUs are considered as inefficient (Kuah and Wong [25]).

5.1 Choice of DEA Model

In this study, we have taken four inputs and 3 outputs; except the input total number of taught course students (UG and PG), all the inputs and outputs are inflexible since it cannot be decided in advance. Therefore, CCR and BCC output-oriented DEA models are used for our study.

5.2 Input–Output Data for Efficiency Study

Input–output data are collected from the NAAC reports 2014, staff statement register (SSR) of the academic year 2012–13, and the examination section. Some data are collected from the departments of the college and college annual book Madhuri of the academic year 2012–2013. Input and output data are given in Table 1 in Appendix

Table 1 Inputs and outputs data for efficiency analysis of 20 departments Govt. PG College Gopeshear, Chamoli (for the academic year 2012–2013)

Characteristics	Inputs				Outputs		
	Inputs (α_1, α_2, α_3, α_4)				Outputs (β_1, β_2, β_3)		
	α_1	α_2	α_3	α_4	β_1	$\beta_2(\%)$	$\beta_3(\%)$
Max.	12	512	66.9789	3600	289	100	880.6452
Min.	0	10	48.5735	50	08	72	30.3887
Avg.	2.90	168.40	54.9685	1542.50	106	88.4780	95.6659
S.D.	2.1213	195.1615	4.2463	2262.7420	15.2584	4.1587	30.9145

6   Results and Discussion

Teaching efficiency for 20 departments of government PG college, Gopeshwar, Chamoli (Uttarakhand), is estimated for the year 2012–2013 using CCR and BCC output-oriented model. The efficiency scores are described as follows.

6.1   Overall Technical Efficiency (OTE)

The results are given in Table 2; from the result, it is clear that out of all the 20 departments, 6 departments (30%) are technically efficient since the OTE scores of these departments are 1. The names of these departments are Hindi, Sanskrit, political science, military sciences, BBA, and mathematics. The remaining 14 (70%) departments are comparatively less technical efficient since the OTE scores of these 14 departments are less than 1. The average OTE score is 0.878. Seven departments English, sociology, commerce, physics, chemistry, botany, and zoology have score less than the average efficiency score. The efficient departments GCD1, GCD2, GCD6, GCD9, GCD14, and GCD15 form the 'reference set' for inefficient departments. Since 0.536 (53.6%) is the least efficiency from all the above 14 inefficient departments, English department (GCD3) is found the most inefficient department and we can say that its overall performance is very poor. From the above 14 inefficient departments, 7 departments have scored less than average efficiency and 7 departments have scored above the average efficiency score. The average overall technical efficiency score 0.878 reveals that on an average, departments have to increase their output by 12.2% by maintaining the existing level of inputs.

Maximum number of peer count indicates the extent of robustness of the departments compared with other efficient departments. Here, we see that political science department represents peer for maximum number of departments, so it is the most technical efficient department.

Table 2 Efficiency scores of 20 departments of Govt. PG College Gopeshwar, Chamoli (UK), by using CCR model for the year 2012–2013

Code	Name of department	OTE (CRS score)	Peer department (R. SET)	Peer weight Peer count
GCD1	Hindi	1	GCD1	1,1
GCD2	Sanskrit	1	GCD2	1,6
GCD3	English	0.536	GCD2, GCD6	0.086, 0.986 0
GCD4	Geography	0.976	GCD1, GCD6	0.241, 0.811 0
GCD5	History	0.969	GCD2, GCD6, GCD15	0.393,0.102,0.417 0
GCD6	Political science	1	GCD6	1,14
GCD7	Economics	0.946	GDC2, GCD6	0.038, 0.994 0
GCD8	Sociology	0.812	GCD2, GCD6, GCD15	0.265,0.399,0.343 0
GCD9	Military science	1	GCD9	1,0
GCD10	Education	0.964	GCD2, GCD6, GCD15	0.132,0.404,0.441 0
GCD11	Home science	0.970	GCD6, GCD15	0.042,1.0 0
GCD12	Commerce	0.653	GCD6	1.031 0
GCD13	B. Ed	0.922	GCD6	1.095 0
GCD14	BBA	1	GCD14	1.0,0
GCD15	Mathematics	1	GCD15	1.0,4
GCD16	Physics	0.872	GCD6	1.129,0
GCD17	Chemistry	0.56	GCD6	1.298,0
GCD18	Geology	0.948	GCD2, GCD6	0.050,1.0 0
GCD19	Botany	0.670	GCD6	1.0330 0
GCD20	Zoology	0.758	GCD6	1.148 0

Average 0.878

6.2 Pure Technical Efficiency (PTE)

As we know that CCR model works on constant return-to-scale assumption, Table 2 shows DEA results by using CCR model while BCC model work is based on VRS assumption (Banker at el. [22]). Table 3 represents data envelopment analysis results obtained by using CCR and BCC models. It is clear from Table 3 that out of above 20 departments, 6 departments are overall technical efficient ($TE = 1$), and 10 departments are pure technical efficient and they form the VRS efficient frontier, since the VRS efficiency of these departments is equal to one, i.e., in the above efficient departments, there is no scope of improvements in the outputs (maintaining the same input label). The remaining 10 departments, namely English, Economics, Sociology, Education, Commerce, Physics, Chemistry, Geology, Zoology, and Botany, are inefficient since their VRS efficiency score is less than one.

It can be inferred from the results shown in Table 3, that some of the departments namely, Geography, History, Home science and B.Ed. are CRS inefficient since

Table 3 OTE (CRS score) PTE, Peer dept., Peer counts, and RTS of 20 departments Govt. PG College Gopeshwar, (UK) for the year 2012–13

Code	Name of Dept.	OTE	PTE (VRS score)	Peer department	Peer count	SE	RTS
GCD1	Hindi	1	1	GCD1	0	1	–
GCD2	Sanskrit	1	1	GCD2	2	1	–
GCD3	English	0.536	0.573	GCD6	0	0.935	DRS
GCD4	Geography	0.976	1	GCD4	5	0.976	DRS
GCD5	History	0.969	1	GCD5	1	0.969	IRS
GCD6	Political Science	1	1	GCD6	8	1	–
GCD7	Economics	0.946	0.975	GCD6	0	0.97	DRS
GCD8	Sociology	0.812	0.817	GCD2, GCD6, GCD15	0	0.993	DRS
GCD9	Military science	1	1	GCD9	0	1	–
GCD10	Education	0.964	0.971	GCD2, GCD5, GCD6, GCD15	0	0.933	IRS
GCD11	Home Science	0.970	1	GCD11	3	0.970	DRS
GCD12	Commerce	0.653	0.671	GCD6, GCD11, GCD14	0	0.973	DRS
GCD13	B.Ed	0.922	1	GCD4, GCD14	0	0.922	DRS
GCD14	BBA	1.0	1	GCD14	5	1	–
GCD15	Mathematics	1.0	1	GCD15	2	1	–
GCD16	Physics	0.872	0.975	GCD4, GCD14	0	0.895	DRS
GCD17	Chemistry	0.56	0.721	GCD4, GCD6, GCD14	0	0.777	DRS
GCD18	Geology	0.948	0.986	GCD6, GCD11	0	0.962	DRS
GCD19	Botany	0.670	0.688	GCD4, GCD6, GCD14	0	0.974	DRS
GCD20	Zoology	0.758	0.861	GCD4, GCD14	0	0.88	DRS

Average: 0.878 (OTE), 0.912 (PTE), 0.959 (SE)

there CRS efficiency score is less than one, but VRS efficient as there VRS efficiency score is one. This means that all these four departments are capable to reduce these inputs in to outputs efficiently, but its technical efficiency is low due to its scale size.

Increasing Return to Scale (IRS)

If we observe RTS, GCD5 and GCD10 operate at IRS and GCD3, GCD4, GCD7, GCD8, GCD11, GCD12, GCD13, D16, GCD17, GCD18, GCD19, and CD20 at DRS. In the BCC model, political science departments have a maximum number of peer counts so it is the most pure technical efficient departments.

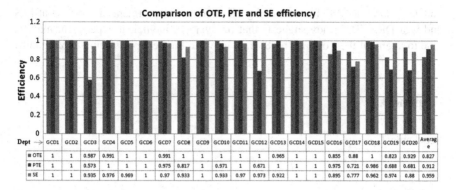

Fig. 1 Comparison of all three TE, PTE, and SE efficiencies (see online version for colors)

Dept →	GCD1	GCD2	GCD3	GCD4	GCD5	GCD6	GCD7	GCD8	GCD9	GCD10	GCD11	GCD12	GCD13	GCD14	GCD15	GCD16	GCD17	GCD18	GCD19	GCD20	Average
OTE	1	1	0.987	0.991	1	1	0.991	1	1	1	1	1	0.965	1	1	0.855	0.88	1	0.823	0.929	0.827
PTE	1	1	0.573	1	1	1	0.975	0.817	1	0.971	1	0.671	1	1	1	0.975	0.721	0.986	0.688	0.681	0.912
SE	1	1	0.935	0.976	0.969	1	0.97	0.933	1	0.933	0.97	0.973	0.922	1	1	0.895	0.777	0.962	0.974	0.88	0.959

Scale Efficiency (SE): Out of the 20 departments only 6 departments, namely Hindi, Sanskrit, Political Science, Military Science, BBA, and Mathematics, are scale efficient since their scale efficiency score is 1. So, all these departments are acting at the maximum scale. Other 14 departments are scale inefficient, due to less than one scale efficiency score. Chemistry department has the least **(0.777)** scale efficiency score (Fig. 1).

7 Conclusion

This paper has evaluated the OTE, PTE, and SE of 20 academic departments of Government PG College, Gopeshwar, Chamoli, in Uttarakhad through DEA techniques.

The study reveals that out of all the above 20 departments, only 6 departments (30%) are efficient. The average overall technical efficiency score is 0.878, i.e., 87.8% represents that 12.2% of the technical potential of department is not in use, implying that these departments have the scope of producing 12.2% more outputs with same level of inputs. The academic departments GCD1, GCD2, GCD6, GCD9, GCD14, and GCD15 form efficiency frontier as their OTE = 1.

From the above efficient departments GCD6 is the most efficient since they form the reference set (i.e., peer department) for the maximum number (14) of academic departments, and also GCD6 has the maximum number of peer count (8) so it is most efficient. In the study of CCR DEA model, GCD3 is the less efficient due to least OTE (=0.536) and it is to be increased by increasing the scale size of the department.

From the BCC model results, it is clear that out of 20 departments, 10 (50%) departments are pure technical efficient out of 10 PTE (VRS), 6 CRS and 3 DRS and 1 working at IRS. The remaining 10 inefficient departments 1 working at IRS and 9 at DRS. The study suggests that on average, inefficient department may be able to augment their outputs by 8.8% relative to the best-performing departments.

Appendix

See Table 1.

References

1. Gopeshwar, GPGC: Chamoli Uttarakhand India, NAAC (2014)
2. Izadi, H., Johns, G., Oskrochi, R., Crouchley, R.: Stocastic frontier estimation of a CES cost function: the case of higher education Britain. Econ. Educ. Rev. **21**, 63–71 (2002)
3. Johnes, J., Yu, L.: Measuring the research performance of chinese higher education institutions using data envelopment analysis. China Econ. Rev. **19**, 679–696 (2008)
4. Bougnol, M.L., Dula, J.: Validating DEA as a ranking tools an application of DEA to assess performance in higher education. Ann. Oper. Res. **145**, 339–365 (2006)
5. Charnes, A., Cooper, W.W., Lewin, A.Y., Seiford, L.M.: Data envelopment analysis: Theory, methodology, and applications. Kulwer Acadmic Publishers, Massachusetts USA (1994)
6. Ramanathan, R,: An introduction to data envelopment analysis. Sage Publication India Pvt. Ltd., New Delhi (2003)
7. Bessent, A.M., Bessent, E.W., Charnes, A., Cooper, W.W., Thorogood, N.C.: Evaluation of educational program proposals by means of DEA. Educ. Admin. Q. **19**(2), 82–107 (1983)
8. Ahn, T., Arnold, V., Charnes, A., Cooper, W.W.: DEA and ratio efficiency analysis for public instructions of higher learning in Texas. Res. Gov. Non-profits Acc. **5**, 165–185 (1989)
9. Abbott, M., Doucouliago's, C.: The efficiency of Australian's Universities a data envelopment analysis. Econ. Educ. Rev. **22**, 89–97 (2003)
10. Avkiran, N.K.: Investigating technical and scale efficiencies of australians university through data envelopment analysis. Socio Econ. Plan. Sci. **35**, 57–80 (2001)
11. Johnes, J.: Data envelopment analysis and its application to the measurement of efficiency in higher education. Econ. Educ. Rev. **25**, 273–288 (2006)
12. Fondle, G.: On the performance of University in North Rhine, Westphalia, Germany: government redistribution of funds judge using DEA efficiency measures. Euro. J. Oper. Res. **176**, 521–533 (2007)
13. Breu, T.M., Rabb, R.L.: Efficiency and perceived quality of the nation's "Top 25" National University and National Liberal Arts Colleges: an application of data envelopment analysis to higher education. Socio-Econ. Plan. Sci. 28, 33–45 (1994)
14. Johnes, J., Johnes, G.: Research funding and performance in U.K. University department of economics: a frontier analysis. Econ. Educ. Rev. **14**, 301–314 (1995)
15. Beasley, J.: Determining teaching and research efficiencies. J. Oper. Res. Soc. **46**, 441–452 (1995)
16. Ali, I., Rana, U.S., Pant, M., Jauhar, S.K., Mogha, S.K.: DEA for Measuring the academic performance of a higher educational institude of Uttarakhand, India. Int. J. of Computer. Info. System and Ind. Manage. Applications. **9**, 206–217 (2017)
17. Jauhar, S.K., Pant, M., Dutt, R. (2016). Performance measurement of an Indian higher education institute: a sustainable educational supply chain management perspective. Int. J. Syst. Ass. Eng. Manage. 1–14. doi:10.1007/s13198-016-0505-4

18. Jauhar, S.K., Pant, M., Nagar, A.K.: Sustainable educational supply chain performance measurement through DEA and differential evolution: a case on Indian HEI. J. Comput. Sci. (2016). doi:10.1016/j.jocs.2016.10.007
19. Cook, W.D., Seiford, L.M.: Data envelopment analysis (DEA)-thirty years on. Europeans J. Oper. Res. **192**, 1–17 (2009)
20. Kuah, C.T et al.: A review on data envelopment analysis (DEA). In: Fourth Asia International Conference on Mathematical Analytical Modelling and Computer Simulation, pp. 168–173 (2010)
21. Charnes, A., Cooper, W.W., Rhodes, E.L.: Measuring the efficiency of decision making units. Euro. Gen. Oper. Res. **2**, 429–444 (1978)
22. Banker et al.: Some models for the estimation of technical and scale inefficiency data envelopment analysis. Manage. Sci. **30**(9), 1078–1092 (1984)
23. Mogha S.K. et al: Estimating technical efficiency of public sector hospitals of Uttarakhand (India). Int. J. Oper. Res. **25**(3) (2016)
24. Tyagi, P., et al.: Relative performance of academic departments using DEA with sensitivity analysis. Eval. Prog. Plan. **32**, 168–177 (2009)
25. Kuah, C.T., Wong, K.Y.: Efficiency assessment of universities through data envelopment analysis. Proc. Comput. Sci. **3**, 499–506 (2011)

Fuzzy-based Probabilistic Ecological Risk Assessment Approach: A Case Study of Heavy Metal Contaminated Soil

Vivek Kumar Gaurav, Chhaya Sharma, Rakesh Buhlan and Sushanta K. Sethi

Abstract The ecological risk assessment tools, viz. index of geoaccumulation (I_{geo}) and enrichment factor (EF), are the classical models for the assessment of risk related to the soil and sediment contamination. These models are well classified into several classes for the assessment of contamination risk. The vagueness in the estimation of risk associated with these models creates voids for computational approaches. In the present study, fuzzy-based probabilistic model was developed to restrict the vagueness of risk estimation. Both I_{geo} and EF were taken as input variables for aggregate risk determination. The linguistic attributes were assigned for risk estimation and qualitative scale presented as trapezoidal fuzzy number. For the validation of methodology, a case study was taken into the account for risk determination. The fuzzy-based aggregate risk assessment revealed high risk of cadmium and arsenic toxicity in the study area with the risk score of 0.751 and 0.698, respectively. The fuzzy-based risk assessment is a conceptual methodology that restricts the vagueness in the estimation of risk for better decision-making approach.

Keywords Ecological risk assessment · Index of geoaccumulation Enrichment factor · Fuzzy logic · Heavy metal · Soil contamination

V.K. Gaurav (✉) · R. Buhlan
Environmental Research Lab (DPT), Indian Institute of Technology Roorkee, Saharanpur
Campus, Saharanpur, India
e-mail: vkgaurav.iit@hotmail.com

C. Sharma
Department of Paper Technology, Indian Institute of Technology Roorkee, Saharanpur
Campus, Saharanpur, India

S.K. Sethi
Department of Polymer and Process Engineering, Indian Institute of Technology Roorkee,
Saharanpur Campus, Saharanpur, India

© Springer Nature Singapore Pte Ltd. 2018   419
M. Pant et al. (eds.), *Soft Computing: Theories and Applications*,
Advances in Intelligent Systems and Computing 584,
https://doi.org/10.1007/978-981-10-5699-4_39

1 Introduction

The fuzzy set theory can be viewed as a qualitative assessment approach that can be used to develop and evaluate a hierarchical model for assessment of soil health risk based on parameters such as organic content, heavy metal deposition, enzymatic activity [1, 2]. The concept of fuzzy logic had been introduced by Zadeh [3] in 1965 is used today in several fields. Initially, this concept applied the human way of thinking to modify the concept of binary logic and abandoning the probabilistic approach of either true or false.

Fuzzy logic is a computational approach which is based on the 'degree of truth', rather than the probabilistic approach of true or false. The truth value in fuzzy logic varies extremely from 0 to 1.

Initially, the concept of fuzzy logic was mainly used to control means of transport, electrical appliances and industrial processes. But today with the advancement of computing technologies, it have broaden the involvement of fuzzy logic in the environment, analysis and diagnosis of HTR nuclear power plant, phosphoric acid production, selecting layered manufacturing techniques, ophthalmology, machine monitoring and diagnostics, hazard identification connected to oil drilling waste, hydraulics or water management, the transportation of hazardous materials on roads, ducts, hydrogeology and for identification of the vulnerability of aquifers [4–10].

Over the past decade, Japan, USA and Europe have achieved numerous application of the fuzzy system in the areas of pattern recognition, robotics, image/signal processing, medicine/biomedical, data mining, healthcare, drug administration, banking and financial decision system, household electronic appliances such as washing machines and televisions, nuclear power stations and handwriting recognition, photocopy machines and elevators.

Today, fuzzy sets have transformed into a number of concepts, models and techniques in order to deal with complex mechanisms which otherwise hard to dissect with classical methods of probability theory. The industrial and atmospheric electrostatics have successfully applied the concept of fuzzy logic in fuzzy-automated diagnosis, fuzzy fault tree analysis and fuzzy risk evaluation [11].

Agricultural and food system have also successfully introduced the concept of fuzzy models in their analytical approaches [12]. In Industrial work environment, there is always involvement of exposing workers to high-risk-associated tasks. Therefore, a decisional instrument to assess the risk has been designed with the help of fuzzy logic as it allows treating uncertain and qualitative data [13].

Fuzzy logic has also been implemented in determining the environmental impact assessment. This system helps in the evaluation of expected change in the physical, social, biological and economic systems due to implementation of an activity or project [14].

Further application of fuzzy logic can be introduced in the area of soil contamination [15]. Soil contamination is among the world's fastest growing global environmental problem. This indeed posing a great threat to human health and environment quality. Over the last few decades, heavy metal soil pollutants have received larger attention. The accumulation of heavy metals in soil can directly affect its physical and chemical properties, reducing its biological activity and nutrient availability. Along with these, it can also pose indirect affects to soil in terms of environmental, human and animal health [16]. Increased level of heavy metals in soil can threaten the food safety via plant uptake [17].

The main challenge faced in heavy metal contaminated soil [18] is to accurately predict the contaminated soil due to complex spatial pattern of heavy metals and high coefficient of variation. Therefore, there is need of innovative scientific methodology to successfully and precisely predict the percentage of heavy metal contaminated soil from the non-contaminated soil areas.

Therefore, the approach of fuzzy logic has been introduced in this work to determine the percentage of heavy metal contaminated soil, which is cost-effective, which quickly and precisely identifies the most serious threat to environment.

The parameters used for defining the soil health in terms of heavy metals are geoaccumulation index (I_{geo}) [19] and enrichment factor (EF) [20]. For each parameter, membership function has been defined in fuzzy system, thus performing the fuzzification. After this, fuzzy rules have been defined in order to identify the soil health of each site in terms of index. Thus, the main aim of fuzzy analysis was to determine the soil health index, which is then further used to define the soil health level based on the basic data obtained from the sampling site.

2 Case Study

Materials and methods

2.1 Fuzzy Logic

To determine the percentage of soil heavy metal contaminate of a sampling site, the application of fuzzy logic has been introduced in this work. It is a mathematical structure having capability to operate along linguistic terms. This concept is totally unrelated from classical Boolean logic as it provides a given statement an intermediary level of truth rather than being either false or true. This provides a better resemblance with the human way of analysing. The final result of fuzzy logic provides an analytical-deductive process which is naturally consistent with a priori set of principles, unlike standard human reasoning which often affected by various considerations and are neither consistent nor realistic.

A fuzzy set can be seen as an extension of a classical set. If U is a functional set and its elements are denoted by x, then we can define a fuzzy set A as a set of ordered pairs

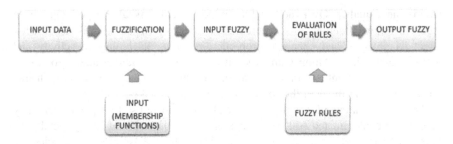

Fig. 1 Basic steps in fuzzy logic system

$$A = \{x, \mu_A(x) | x \in U\} \tag{1}$$

Where $\mu A(x)$ is the membership function of x in A. This membership function provides mapping of each element of U to a membership value between 0 and 1.

The fuzzy logic was facilitated through implementation of a fuzzy interference system, in which there is defined process of connection, mostly made from the data collected from the field. The main steps in fuzzy implementation include (Fig. 1): defining membership functions, fuzzification, inference system, and fuzzy output [21].

The defining of membership function is the most important part on which all other subsequent operations are based. These functions represent the fuzzy sets and can have different shapes (triangular, Gaussian, trapezoidal, etc.); according to the situations, values can be taken between 0 and 1.

2.2 Ecological Risk Assessment of Heavy Metal Contaminated Site

2.2.1 Analysis of Heavy Metal Concentration

In the present study, aggregate ecological risk of heavy metal contamination of soil is assessed. Soil samples from suburban areas of Saharanpur district (India) were collected, and heavy metal concentrations were analysed for metals, viz. Cd, Zn, Cu, Pb and As. The location of sampling areas with their geographical coordinates is mentioned in Table 1.

Composite sampling was performed, and totally, 10 samples were taken with the help of post-hole soil auger from upper horizon (0–20 cm) from the study areas in March 2016; Samples were stored in polyethylene bag, and within 3 h from sampling, samples were delivered to the laboratory. Till complete analysis, samples were kept at 4 °C. For heavy metal analysis, firstly soil samples were digested using microwave digestion oven (MarsXpress) using USEPA 3052 (USEPA 1996)

Table 1 Sampling areas with their geographical coordinates

S. No.	Area	Coordinates
1	Sheikhpura quadeem	29° 55′ 30″N, 77° 34′ 21″E
2	Tapri	29° 55′ 19″N, 77° 35′ 40″E
3	Lakhnaur	29° 53′ 45″N, 77° 36′ 50″E
4	Paragpur	29° 55′ 43″N, 77° 35′ 20″E
5	Kapasa	29° 55′ 0″N, 77° 36′ 28″E

Table 2 Classes of geoaccumulation index (I_{geo})

Class	Value	Soil quality
I_0	$I_{geo} \leq 0$	Practically uncontaminated
I_1	$0 < I_{geo} < 1$	Uncontaminated to moderately contaminated
I_2	$1 < I_{geo} < 2$	Moderately contaminated
I_3	$2 < I_{geo} < 3$	Moderately to heavily contaminated
I_4	$3 < I_{geo} < 4$	Heavily contaminated
I_5	$4 < I_{geo} < 5$	Heavily to extremely contaminated
I_6	$2 < I_{geo}$	Extremely contaminated

methodology. Further, digested solutions were diluted, and heavy metal concentration was analysed using ICP-OES.

2.2.2 Ecological Risk Assessment

The index of geoaccumulation (I_{geo}) estimates the contamination by comparing the present heavy metal concentration with pre-industrial heavy metal concentration originally referred to the bottom sediments [22]. It is expressed as:

$$I_{geo} = \log_2 \frac{C_n}{1.5B_n} \tag{2}$$

In this expression, C_n is the total concentration of element 'n' in the surface layer of soil analysed, and B_n refers to the geochemical background value of the element 'n' in the Earth's crust [23]. A factor of 1.5 refers to the possible natural fluctuation in the base level of the metal estimation in the environment and as a correction factor for any small anthropogenic influences. Six classes of geoaccumulation index (Table 2) have been classified [24].

2.2.3 Enrichment Factor (EF)

Enrichment factor for soil relies on the standardization of an analysed element against the reference element. The reference soil can be any soil free from

Table 3 Classes of
contamination categories of
enrichment factor (EF)

Class	Value	Soil quality
EF_0	<2	Deficiency to minimal enrichment
EF_1	2–5	Moderate enrichment
EF_2	5–20	Significant enrichment
EF_3	20–40	Very high enrichment
EF_4	>40	Extremely high enrichment

contamination, and reference element can be characterized by minor occurrence variability. In the present work, aluminium (Al) is taken as the reference element. The enrichment factor was estimated using modified formula based on the expression

$$EF = \left[\frac{C_n(\text{Sample})}{C_{\text{ref}}(\text{Sample})}\right] \Big/ \left[\frac{B_n(\text{Background})}{B_{\text{ref}}(\text{Background})}\right] \tag{3}$$

where C_n(Sample) indicates the content of analysed element in the study area, C_{ref}(Sample) is the concentration of reference element in the study area, B_n(Sample) is the concentration of examined element in the reference environment and B_{ref}(Sample) is the concentration of reference element in the reference environment. On the basis of the value of enrichment factor, five contamination categories were classified [23] (Table 3).

2.3 The Fuzzy Approach to Determine Aggregate Ecological Risk

Our study was to focus on the development of methodology based on fuzzy for the aggregate ecological risk assessment of heavy metal contaminated sites. Therefore, since the information available for risk analysis of the contaminated sites was not complete and in few cases illustrated just by estimations and not through surveys.

Foremost approach to apply fuzzy logic is to define input data (Fig. 3) in the form of membership functions and defuzzification method selection for obtaining the fuzzy output; in the present case, it is aggregate ecological risk at contaminated sites. The sensitivity was analysed by using the fuzzy approach to solve the uncertainty problem associated with the input data and the described fuzzy model. The fuzzy inference model has been deployed for facilitating the kind of aggregated risk quantification, by using fuzzy membership functions and fuzzy rules (Table 4).

The conceptual diagram in Fig. 2 shows fuzzy inferences, so as to manage the algorithm easily (Fig. 3).

The aggregation waste determines the contaminants depending on the heavy metal content. Therefore, with the increase of geoaccumulation index or enrichment

Table 4 Fuzzy rules for determination of soil risk assessment

Geoaccumulation index	Enrichment factor	Soil risk assessment
I_0	EF_0	No risk
I_0	EF_1	Low risk
I_0	EF_2	Medium risk
I_0	EF_3	Medium risk
I_0	EF_4	High risk
I_1	EF_0	Low risk
I_1	EF_1	Low risk
I_1	EF_2	Medium risk
I_1	EF_3	High risk
I_1	EF_4	High risk
I_2	EF_0	Low risk
I_2	EF_1	Medium risk
I_2	EF_2	Medium risk
I_2	EF_3	High risk
I_2	EF_4	High risk
I_3	EF_0	Medium risk
I_3	EF_1	Medium risk
I_3	EF_2	High risk
I_3	EF_3	High risk
I_3	EF_4	High risk
I_4	EF_0	Medium risk
I_4	EF_1	High risk
I_4	EF_2	High risk
I_4	EF_3	Extreme high risk
I_4	EF_5	Extreme high risk
I_5	EF_0	High risk
I_5	EF_1	High risk
I_5	EF_2	High risk
I_5	EF_3	Extreme high risk
I_5	EF_4	Extreme high risk

factor, the aggregation risk will increase. The procedure to determine the aggregated risk combines the results inferred by two fuggy diagrams (Figs. 4 and 5).

The final result of the process was obtained by the defuzzification (Fig. 6) by providing a numerical value between 0 and 1 and represents the risk index.

In order to ease the methodology, we used the MATLAB® software passing through the graphical interface by means of creating three different graphical interfaces to describe the three planned variable sets. Our attention focussed on the aggregation-based risk assessment.

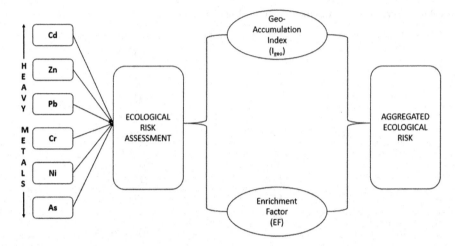

Fig. 2 Conceptual diagram of implemented fuzzy logic model

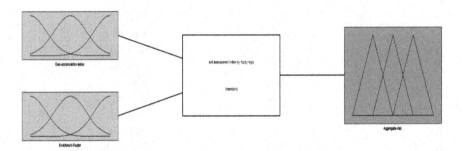

Fig. 3 Fuzzy inference engine

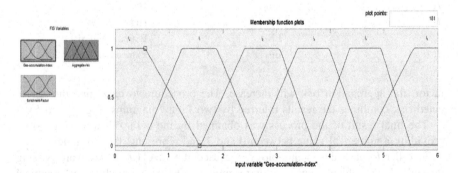

Fig. 4 Fuzzy rule-based input variable of geoaccumulation index

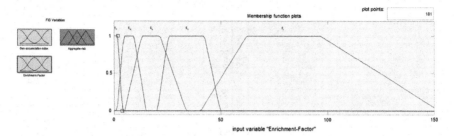

Fig. 5 Fuzzy rule-based input variable of enrichment factor

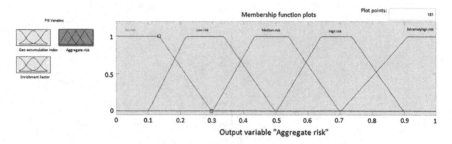

Fig. 6 Fuzzy rule-based output variable of aggregate risk

3 Result and Discussion

The work described dealt with the study of ecological risk assessment of heavy metal contamination. High risk of heavy metal accumulation was observed in suburban areas of Saharanpur district. Geometric mean of heavy metals observed in the study areas is depicted in Table 5.

The fuzzy inferences are shown in conceptual diagram in Fig. 5 so as to manage the algorithm easily.

Geoaccumulation and enrichment factor of metals in the study area were assessed on the basis of equation and classification proposed by Muller (1981) and Buat et al. (1979), respectively. High geoaccumulation of cadmium (Cd) was observed in the study area followed by the arsenic (As) contamination. Samples were moderately contaminated by lead (Pb) and nickel (Ni), whereas zinc (Zn) has practically uncontaminated status in study area.

Extremely high enrichment of cadmium (Cd) and arsenic (As) was observed in the study area, whereas lead (Pb), chromium (Cr) and nickel (Ni) has moderate enrichment in the study area. Moderate enrichment of zinc (Zn) was observed in the study area.

Table 5 Geometric mean of heavy metal concentration in study areas

Metal	Cd	Zn	Pb	Cr	Ni	As
Geometric mean of metal concentration (mg/kg)	2.24 ± 0.48	99.58 ± 40.62	22.45 ± 5.74	84.34 ± 46.28	256.88 ± 46.24	9.38 ± 4.23

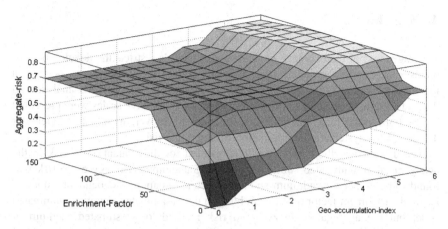

Fig. 7 Fuzzy risk assessment index represented three dimensionalities with enrichment factor and geoaccumulation index

Table 6 Fuzzy inference of aggregate risk assessment

Metal	Index of geoaccumulation (I_{geo})	Enrichment factor (EF)	aggregate risk
Cd	3.92	128.49	0.71
Zn	−0.23	7.17	0.499
Pb	1.9	17.14	0.499
Cr	0.68	13.55	0.499
Ni	0.98	16.66	0.497
As	2.5	47.98	0.701

Fuzzy logic uses linguistic type of variables which are defined by continuous range of truth values or FMFs in the interval [0, 1] instead of the strict binary (true or false) decisions. The linguistic input and output associations when combined with an inference procedure constitute the fuzzy-based system. The membership functions represent the degrees to which the specified concentration belongs to the fuzzy sets. The trapezoidal membership function was used to perform the simulations which are in the agreement with the input data trend.

A surface diagram (Fig. 7) approach is presented for the identification of aggregation risk concentration (Table 6) feature based on a set of rules by considering the methodology variables such as geoaccumulation index and enrichment factor. A fuzzy rule system approach is used because of the imprecise, insufficient, ambiguous and uncertain data available. In this manner, without mathematical formulations, expert maximum risk assessment system is identified.

4 Conclusion

The vagueness of the classical, ecological risk models restricts the development of precise decision-making sense, thus the idea behind the work was to develop a fuzzy-based methodology for aggregate ecological risk analysis. To analyse and illustrate linguistic attributes of risk estimation, the qualitative scales were presented as the trapezoidal fuzzy membership function. Moreover, hierarchal model was developed for systematic determination of aggregate risk.

The designed methodology was implemented for the evaluation of risk in the heavy metal contaminated suburban areas of Saharanpur. The aggregate risk was found to be higher for cadmium and arsenic contamination. Independent values of I_{geo} and enrichment factor of Zn were lying in uncontaminated to low contaminated zone, but the aggregative fuzzy-based risk methodology estimated medium or moderate risk from Zn contamination. Similarly, despite significant enrichment of lead (Pb) and moderately contaminated status of geoaccumulation, the aggregate risk from the designed methodology was found to be lying in medium risk zone. This reduction in vagueness of risk estimation accounts for the fuzzy-based risk assessment methodology.

References

1. Rodríguez, E., et al.: Dynamic quality index for agricultural soils based on fuzzy logic. Ecol. Indic. **60**, 678–692 (2016)
2. Liu, K.F.-R., et al.: Using fuzzy logic to generate conditional probabilities in Bayesian belief networks: a case study of ecological assessment. Int. J. Environ. Sci. Technol. **12.3**, 871–884 (2015)
3. Zadeh, L.A.: Fuzzy sets. Inf. Control **8**(3), 338–353 (1965)
4. Kitowski, J., Książek, E.: Fuzzy logic applications for failure analysis and diagnosis of a primary circuit of the HTR nuclear power plant. Comput. Phys. Commun. **38**(2), 323–327 (1985)
5. Ketonen, M.: A fuzzy logic application in phosphoric acid production. Control Eng. Pract. **1**(2), 273–281 (1993)
6. Khrais, S., Al-Hawari, T., Al-Araidah, O.: A fuzzy logic application for selecting layered manufacturing techniques. Expert Syst. Appl. **38**(8), 10286–10291 (2011)
7. Sokołowski, A.: On some aspects of fuzzy logic application in machine monitoring and diagnostics. Eng. Appl. Artif. Intell. **17**(4), 429–437 (2004)
8. Sadiq, R., Husain, T.: A fuzzy-based methodology for an aggregative environmental risk assessment: a case study of drilling waste. Environ. Model Softw. **20**(1), 33–46 (2005)
9. Bonvicini, S., Leonelli, P., Spadoni, G.: Risk analysis of hazardous materials transportation: evaluating uncertainty by means of fuzzy logic. J. Hazard. Mater. **62**(1), 59–74 (1998)
10. Gemitzi, A., et al.: Assessment of groundwater vulnerability to pollution: a combination of GIS, fuzzy logic and decision making techniques. Environ. Geol. **49**(5), 653–673 (2006)
11. Balog, E., Berta, I.: Fuzzy solutions in electrostatics. J. Electrostat. **51**, 409–415 (2001)
12. Center, B., Verma, B.P.: Fuzzy logic for biological and agricultural systems. Artif. Intell. Rev. **12**(1–3), 213–225 (1998)

13. Murè, S., Demichela, M.: Fuzzy application procedure (FAP) for the risk assessment of occupational accidents. J. Loss Prev. Process Ind. **22**(5), 593–599 (2009)
14. Boclin, A.D.S.C., de Mello, R.: A decision support method for environmental impact assessment using a fuzzy logic approach. Ecol. Econ. **58**(1), 170–181 (2006)
15. Caniani, D., et al.: Application of fuzzy logic and sensitivity analysis for soil contamination hazard classification. Waste Manag. **31**(3), 583–594 (2011)
16. Järup, L.: Hazards of heavy metal contamination. Br. Med. Bull. **68**(1), 167–182 (2003)
17. Sharma, R.K., Agrawal, M., Marshall, F.: Heavy metal contamination of soil and vegetables in suburban areas of Varanasi. India. Ecotoxicol. Environ. Saf. **66**(2), 258–266 (2007)
18. Alloway, B.J.: Introduction. In: heavy metals in soils. Springer, pp. 3–9 (2013)
19. Zhiyuan, W., et al.: Assessment of soil heavy metal pollution with principal component analysis and geoaccumulation index. Procedia Environ. Sci. **10**, 1946–1952 (2011)
20. Kaushik, A., et al.: Heavy metal contamination of river Yamuna, Haryana, India: assessment by metal enrichment factor of the sediments. J. Hazard. Mater. **164**(1), 265–270 (2009)
21. Wang, L.-X.: Fuzzy systems are universal approximators. In: IEEE International Conference on Fuzzy Systems. IEEE (1992)
22. Muller, G.: Index of geoaccumulation in sediments of the Rhine river (1969)
23. Taylor, S.R., McLennan, S.M.: The geochemical evolution of the continental crust. Rev. Geophys. **33**(2), 241–265 (1995)
24. Muller, G.: The heavy metal pollution of the sediments of Neckars and its tributary: a stocktaking. Chem. Ztg **105**, 157–164 (1981)

Performance Evaluation of DV-HOP Localization Algorithm in Wireless Sensor Networks

Vikas Gupta and Brahmjit Singh

Abstract Nowadays, wireless sensor networks (WSNs) have become a key area of research. Localization of sensor nodes in WSN is extremely important for transmission and reception of the data. Localization is used to determine the geographical location of the sensor nodes which is required in routing of data. In the present work, the existing DV-HOP algorithm which is a range-free localization algorithm has been studied and evaluated in terms of localization error. The localization error has been calculated by varying the number of unknown nodes, communication range, and the number of anchor nodes. All simulations have been performed in MATLAB.

Keywords Wireless sensor networks · Localization · Anchor node · DV-HOP

1 Introduction

With the advancements in micro-electro-mechanical systems (MEMS) technology, the wireless sensor networks (WSNs) have shown a remarkable growth. The deployment of small, inexpensive, low-power, distributed sensor nodes which are capable of processing the information locally has found many applications in the field of military, agriculture, disaster management, etc. [1]. Localization in wireless sensor networks is the physical determination of coordinates of sensor nodes spread in the area of interest. Without having the location information, the raw data would not be useful at the base station, and in addition, the location information also

V. Gupta (✉)
Electronics and Communication Engineering Department,
JMIT, Radaur, Yamuna Nagar, Haryana, India
e-mail: vikasgupta2k11@gmail.com

B. Singh
Electronics and Communication Engineering Department, National Institute
of Technology Kurukshetra, Kurukshetra, Haryana, India
e-mail: brahmjitsingh@nitkkr.ac.in

© Springer Nature Singapore Pte Ltd. 2018
M. Pant et al. (eds.), *Soft Computing: Theories and Applications*,
Advances in Intelligent Systems and Computing 584,
https://doi.org/10.1007/978-981-10-5699-4_40

provides interesting possibilities like geographic routing, target tracking. A large number of localization algorithms have been proposed by the researchers in the field of WSN. Broadly, these algorithms can be divided in two categories, namely range-based and range-free. Range-based techniques use absolute point to point estimates of distance or angle estimates in location calculation. Commonly used range-based localization techniques are time of arrival (TOA) [2], time difference of arrival (TDOA) [3], angle of arrival (AOA) [4], and received signal strength measurement (RSSI) [5]. The main advantage of these techniques is high accuracy in the localization of sensor nodes, but the hardware used is expensive [6]. On the other hand, range-free localization techniques use only connectivity information to localize the entire network. A large number of range-free techniques have also been proposed by the researchers. Bulusu et al. [7] proposed a centroid-based range-free localization technique which determines the unknown node location by calculating the centroid of the beacons or anchor nodes those are equipped with GPS-like arrangement. Niculescu et al. proposed DV-HOP algorithm [8] in 2003, in which coordinates of unknown nodes are estimated on the basis of a classical distance vector exchange. He et al. [6] proposed approximation point in triangulation test (APIT) which uses triangles formed by three beacon nodes, and the position of the unknown node is computed by calculating the centroid of the intersection of the beacon triangles. Doherty et al. [9] proposed a range-free localization algorithm which considers localization problem as a convex optimization problem based only on connectivity-induced constraints and uses a semi-definite program (SDP) to solve the problem. Shang et al. [10] proposed a localization method based on multidimension scaling (MDS-MAP) which uses the connectivity information who is in the range of whom and derives the locations of the nodes in the network. Multidimensional scaling (MDS) is a data analysis technique that converts the proximity information into a geometric one. MDS is well suited to node localization in communication networks, where the task is to use the distance information between nodes to determine the coordinates of nodes in a 2D or 3D space. The research is being carried out by the researchers to improve the above-stated range-based and range-free localization algorithm using various soft computing techniques like fuzzy logic systems (FLS), particle swarm optimization (PSO), genetic algorithm (GA), or artificial neural networks (ANN) [11–14]. In the present paper, the performances of DV-HOP algorithm have been evaluated and the effect of variation in the number of nodes, communication range, and the number of anchor nodes (beacon) on localization errors has been studied using simulation work.

The structure of this paper is organized as follows: In Sect. 2, the DV-HOP algorithm has been discussed, and Sect. 3 discusses the steps involved in the implementation of the algorithm. Results and discussion on the same have been presented in the Sect. 4. Section 5 contains the conclusion and future work.

2 DV-Hop Algorithm

This is the one of the basic techniques of localization which is comprised of mainly three stages [15–17]. In the first stage, it uses a classical distance vector exchange method to get distances in hops from all the unknown nodes to the anchor nodes. Each node maintains a table (X_i, Y_i, h_i), where X_i, Y_i are the coordinates of anchor i, and h_i is the hop count. The node exchanges these updates only with its neighbors. In the second stage, average size of one hop is estimated by calculating the distances to other anchor nodes or beacons. The average hop size is used as correction for the nodes in its neighborhood. The average hop size can be calculated using the following formula:

$$\frac{\sum \sqrt{(X_i - X_j)^2 + (Y_i - Y_j)^2}}{\sum h_i}, \quad i \neq j, \text{ for all anchors } j \tag{1}$$

where (x_i, y_i), (x_j, y_j) are coordinates of anchor i and j, h_i is the hops between beacon i and beacon j. After receiving this correction, any arbitrary node then may estimate distances to the beacons which can be used to find the unknown coordinates using trilateration, which forms the third stage of the DV-HOP localization algorithm. As shown in Fig. 1, the coordinates (x_u, y_u) of unknown node (D) can be calculated using the coordinates of the nodes A, B, and C those are already known. The coordinates of beacons A, B, and C are (x_a, y_a), (x_b, y_b), and (x_c, y_c), respectively.

The lengths of these three beacon nodes from node D are d_a, d_b, and d_c, and these lengths can be calculated using formulas given in Eq. (2).

Fig. 1 Trilateration technique [12]

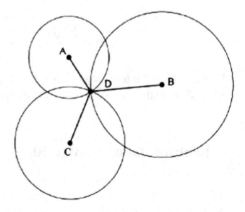

$$(x_1 - x_u)^2 + (y_1 - y_u)^2 = d_1^2$$
$$(x_2 - x_u)^2 + (y_2 - y_u)^2 = d_2^2$$
$$\vdots \tag{2}$$
$$(x_n - x_u)^2 + (y_n - y_u)^2 = d_n^2.$$

Equation (2) can be expanded as

$$\left\{ \begin{array}{l} 2(x_1 - x_n)x_u + 2(y_1 - y_n)y_u = x_1^2 - x_n^2 + y_1^2 - y_n^2 - d_1^2 + d_n^2 \\ 2(x_2 - x_n)x_u + 2(y_2 - y_n)y_u = x_2^2 - x_n^2 + y_2^2 - y_n^2 - d_2^2 + d_n^2 \\ \vdots \\ 2(x_{n-1} - x_n)x_u + 2(y_{n-1} - y_n)y_u = x_{n-1}^2 - x_n^2 + y_{n-1}^2 - y_n^2 - d_{n-1}^2 + d_n^2 \end{array} \right\} \tag{3}$$

These equations can be modeled in the form of linear equation given by $Ax = B$, and the coordinates of unknown node D are calculated using matrix formulas. The formulas are given in Eq. (4).

$$A = 2 \begin{bmatrix} x_1 - x_n & y_1 - y_n \\ \vdots & \vdots \\ x_{n-1} - x_n & y_{n-1} - y_n \end{bmatrix} \tag{4}$$

The coordinates (x_u, y_u) of the unknown node D are further computed by using matrix formula $(x_u, y_u) = (A^T A)^{-1} A^T B$. The localization error (LE) has been calculated using the formula given in Eq. (5)

$$LE = \frac{\sum \sqrt{(X_t - X_e)^2 + (Y_t - Y_e)^2}}{n - m} \tag{5}$$

where (X_t, Y_t) is the true location, and (X_e, Y_e) is the estimated location of unknown node and, m and n are the number of unknown and anchor nodes.

3 Implementation of Algorithm

Step 1: A uniformly distributed random topology within a square area is drawn.
Step 2: The total number of unknown nodes and the anchor nodes amount is decided.

Step 3: Distances in hops are estimated from all the unknown nodes to the anchor nodes. A hop matrix is created which is initialized with 0, and the number of hops to each node is calculated using shortest path algorithm. The average hop distance from unknown nodes to the anchor nodes is calculated using the formula given in Eq. (1) in Sect. 2.

Step 4: Coordinates of the unknown nodes have been determined using the least square method. The necessary mathematical equations have been presented in the DV-HOP localization algorithm explained in Sect. 2.

Step 5: Localization error has been calculated by subtracting the true locations from the estimated location of the unknown nodes. The formula to calculate the localization error (LE) is given in Eq. (5).

4 Results and Discussion

All the simulation results have been obtained using MATLAB. All the parameters considered for the simulation of DV-HOP algorithm and the values of these parameters have been presented in Table 1 showing the summary of the simulation environment. The localization error has been calculated using different values of anchor nodes, unknown nodes, and communication range of the nodes.

Localization Error versus Number of Nodes for Different Values of Beacons: The localization error versus number of nodes for different values of beacons or anchor nodes in DV-HOP algorithm has been shown in Fig. 2. The simulation has been done by varying the number of unknown nodes from 50 to 300 and taking the communication range equals to 50. The results show that the error is

Table 1 Summary of simulation environment

S. No.	Parameters	Values
1	Area	$100 \times 100 \text{ m}^2$
2	Unknown nodes	50–300
3	Anchor nodes or beacons	5–25
4	Communication range	10–100 m

Fig. 2 Plot showing localization error versus number of nodes for different values of beacons

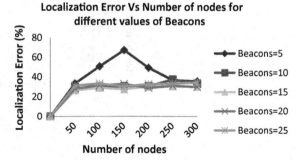

large if the numbers of beacons are very small. And the error is very small if the numbers of beacons are 15 for 50–150 nodes. The error starts increasing with the increase in the number of nodes for all types of beacons.

Localization Error versus Anchor Node Ratio for Different Values of Communication Range: The localization error versus anchor node ratio for different values of communication range has been shown in Fig. 3. The simulation has been done by taking the total number of nodes equal to 100. The results show that for an anchor ratio of about 0.1 to 0.2 and communication range of 50 m, the localization error is small. This can also be seen from the graph that if the range is small, then anchor ratio should be increased for low error but this would be uneconomical as having more beacons costs more.

Localization Error versus Communication Range: The localization error versus communication range has been shown in Fig. 4. The simulation has been done by taking the total number of nodes equal to 100 and beacons amount equal to 10. The results show that localization error is small for communication range of 20–50 m. Below this range, the error is very large, and for above of this range, the error tends to increase. So, the value of communication range should be between 20 and 50 m for good results.

Fig. 3 Plot showing localization error versus anchor node ratio for different values of communication range

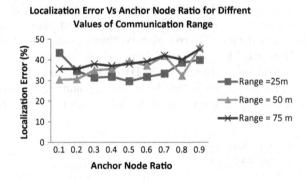

Fig. 4 Plot showing localization error versus communication range

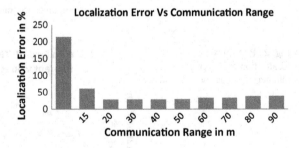

5 Conclusion

In the present work, performance of the basic DV-HOP localization algorithm has been analyzed on the basis of localization error with respect to total number of nodes, anchor ratio, and communication range. The results show that the algorithm is best suitable if the beacon nodes vary from 10 to 20 for 50 to 150 numbers of nodes, because the localization error is small for this range. Localization error is also very small in the basic DV-HOP method for the anchor ratio of 0.1 to 0.2. Simulation results also show that the algorithm is well suited if the communication range is between 20 and 50 m. Although the algorithm works well for different values of anchor ratio and communication range, but still there is a scope of improvement in the DV-HOP algorithm. The performance of DV-HOP localization algorithm in terms of reduction in the localization error can be further enhanced by using some soft computing technique like fuzzy logic systems (FLS), particle swarm optimization (PSO), genetic algorithm (GA), or artificial neural networks (ANN).

References

1. Akyildiz, I.F., Su, W., Sankarasubramaninam, Y., Cayirci, E.: A survey on sensor networks. IEEE Commun. Mag. **40**(8), 102–114 (2002)
2. Voltz, P.J., Hernandez, D.: Maximum likelihood time of arrival estimation for real-time physical location tracking of 802.1 1 a/g mobile stations in indoor environments ad-hoc positioning system. In: IEEE Conference, Position Location and Navigation Symposium (PLNS) (2004)
3. Kovavisarruch, L., Ho, K.C.: Alternate source and receiver location estimation using TDoA with receiver position uncertainties. In: IEEE Conference, ICASSP (2005)
4. Niculescu, D., Nath, B.: Ad hoc positioning system (APS) using AOA. In: IEEE Conference, INFOCOM (2003)
5. Kumar, P., Reddy, L., Varma, S.: Distance measurement and error estimation scheme for RSSI based localization in wireless sensor networks. In: IEEE Conference, Wireless Communication and Sensor Networks (WCSN) (2009)
6. He, T., Huang, C., Blum, B., Stankovic, J., Abdelzaher, T.: Range-free localization schemes for large scale sensor networks. In: MobiCom' 03, pp. 81–95, ACM Press (2003)
7. Bulusu, N., Heidemann, J., Estrin, D.: GPS-less low cost outdoor localization for very small devices. IEEE Pers. Commun. Mag. **7**(5), 28–34 (2000)
8. Niculescu, D., Nath, B.: DV based positioning in ad hoc networks. Telecommun. Syst. **22**, 267–280 (2003)
9. Doherty, L., Pister, K.S., Ghaoui, L.E.: Convex position estimation in wireless sensor networks. In: IEEE Conference ICC' 01, vol. 3, pp. 1655–1663 Anchorage, AK (2001)
10. Shang, Y., Ruml, W., Zhang, Y., Fromherz, M.: Localization from connectivity in sensor networks. IEEE Trans. Parallel Distrib. Syst. **15**(10) (2004)
11. Gong, L., Sun J., Xu, W., Xu, Jian.: Research and simulation of node localization in WSN based on quantum particle swarm optimization. In: IEEE International Symposium on Distributed Computing and Application to Business, Engineering and Science (2012)

12. Chagas, H.S., Martins, J., Oliviera, L.: Genetic algorithms and simulated annealing optimization methods in wireless sensor networks localization using ANN. In: IEEE International Midwest Symposium on Circuit and Systems (2012)
13. Rahman, M.S., Park, Y., Kim, K.: Localization of wireless sensor networks using ANN. In: IEEE International Symposium on Communication and Information Technology (2009)
14. Katekaew, W., So-In, C., Rujirakul, K., Waikham, B.: H-FCD: hybrid fuzzy centroid & DV-hop localization algorithm in wireless sensor networks. In: IEEE International Conference on Intelligent System, Modeling and Simulation (2014)
15. Cui, H., Wang, Y., Liu, L.: Improvement of DV-HOP localization algorithm. In: IEEE International Conference on Modeling, Identification and Control (ICMIC) (2015)
16. Zhipeng, X., Chunwen, L., Huanyu, L.: An improved hop size estimation for DV-hop localization algorithm in wireless sensor networks. In: IEEE International Conference (CCDC) (2015)
17. Ji, W., Liu, Z.: An improvement of DV-hop algorithm in wireless sensor networks. In: IEEE International Conference, Wireless Communication, Networking and Mobile Computing (2006)

Performance Comparisons of Four Modified Structures of Log Periodic Three Element Microstrip Antenna Arrays

Abhishek Soni and Sandeep Toshniwal

Abstract This paper presents performance comparisons of four different structures of log periodic structures of microstrip antenna. Log periodic structure is used for enhancement of bandwidth. In the military aviation platform, the requirements of antenna are lightweight, low volume, low profile with wide bandwidth. Hence in this paper, four log periodic structures of square, rectangle, triangle, and circle shaped are analyzed and simulated using HFSS software tool, and their performances are compared. Proposed antennas find their applications in satellite communications, radio broadcasting, and wireless-fidelity applications, etc.

Keywords Micro strip patch antenna · Impedance bandwidth
HFSS log periodic antenna · Reconfigurable antenna

1 Introduction

Log periodic antennas have a highly directional, wideband, multiple elements, narrow beam characteristics. The radiation characteristics and impedance performances of log periodic antenna repeat as a logarithmic of frequency [1]. The log periodic antenna was invention of Dwight E'Isbell, Raymond Duhamel, and others [2]. Isbell and Mayes–Carrel antennas were patented by University of Illinois at Urbana-Champaign and licensed the design as a package exclusively to JFD electronics in New York [3]. Log periodic antennas are mostly known and usable due to their independence of frequency. They are preferred for wideband arrays because of end-fire structure; their physical areas (perpendicular to directions of main beam) are lesser than the other structures [1]. The antenna is connected to

A. Soni (✉) · S. Toshniwal
Department of ECE, Kautilya Institute of Engineering and Technology,
Jaipur, India
e-mail: absoni19@gmail.com

S. Toshniwal
e-mail: toshniwal.sandeep@gmail.com

© Springer Nature Singapore Pte Ltd. 2018
M. Pant et al. (eds.), *Soft Computing: Theories and Applications*,
Advances in Intelligent Systems and Computing 584,
https://doi.org/10.1007/978-981-10-5699-4_41

source at the nearest to the smaller dipole element, and components of frequencies of the signal are traveling down the antenna structure until near resonance means they reach a dipole [4]. They also required traveling back with the same distance as radiated fields before they can combine with the high-frequency components radiated from small dipoles near the excitation point. It results in a time delay which is approximately proportional to wavelength component. Major advantage of these antennas is the simple design procedure which has their engineering applications. It is not suitable solutions that multiple antennas supporting different wireless bands, because of major requirements of small size, highly efficient, optimum power consumption, and cheap cost [5]. For applications of future technologies of wireless communication like cognitive radio, ground penetrating radar, and RFID, log periodic antenna was mostly preferred by researchers [6, 7].

In paper [8], a frequency reconfigurable annular slotted antenna is discussed. In which three different frequencies can be modified and controlled by the matching stubs placed on the opposite side of antenna. In paper [9], Koch fractal geometry is employed in low-cost multiband printed circuit board (PCB) antenna, and it finds applications in Wi-fi, 3G, WiMAX, and some portion of ultra-wideband.

Four different modified structures of microstrip log periodic antenna arrays are described in this paper. By using the principle of log periodic antenna, proposed antennas of three elements are designed. The log periodic antennas are mostly used for wideband applications because of easy selection of required band from wideband when each element provides radiation at different frequency bands. The HFSS simulation tool is used to design the log periodic antennas. Several parameters such as return loss, bandwidth, and resonant frequencies are analyzed and compared for proposed antennas.

2 Antenna Design and Simulated Results

2.1 Square-Shaped Log Periodic Antenna

The geometrical structures of log periodic microstrip antennas of three elements are shown in Fig. 1. Firstly, microstrip feed line of 50 Ω is made on upper side of the substrate then different-shaped patches are connected with feed lines to form a log-periodic array. Substrate of FR4 material is used which has relative permittivity of 4.4, height of 1.6 mm, and loss tangent of 0.025.

For the designing of a broadband antenna, the log periodic microstrip antenna [10] is mostly preferred. Patches size and spacing between them increase in order to a log periodic manner. Scaling of dimensions is done in the specific way for designing of log-periodic wideband antenna so that performance of antenna is periodic with the logarithm function of frequency [11]. All the parameters like length of patch (L), width of patch (W), side length (S), and diameter(d) of patch are chosen in a specific way which is related to the scaling factor (τ), specified in Eq. (1).

Fig. 1 Geometry of
square-shaped log periodic
antenna

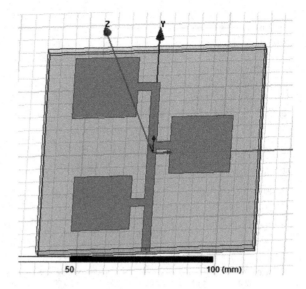

$$\tau = \frac{L_{m+1}}{L_m} = \frac{W_{m+1}}{W_m} = \frac{S_{m+1}}{S_m} = \frac{d_{m+1}}{d_m} \qquad (1)$$

Resonant frequency of first patch is 3 GHz and by a scaling factor of 1.05, the second patch dimensions can be calculated. Similarly, the third patch dimensions are once again obtained by a scaling factor of 1.05 to calculate second patch diameter. For minimizing mutual coupling effect and providing forward-fire radiation pattern, a half-wavelength spacing between each patches is chosen.

The area of patches are 22.98 * 22.98 mm^2, 21.885 * 21.885 mm^2, and 20.843 * 20.843 mm^2, respectively, and distances from feed point are 65.5063, 42.5733, and 20.759 mm, respectively.

Simulation result obtained by square-shaped log periodic antenna is shown in Fig. 2. It is observed that antenna is tuned at 1.75, 2.75, and 3.25 GHz and provides 60, 10.90, and 10.93% bandwidth, respectively.

Current distribution plot of square-shaped log periodic antenna is shown in Fig. 3.

2.2 Rectangular-Shaped Log Periodic Antenna

Figure 4 shows the geometry of rectangular-shaped log periodic antenna. The area of patches are 30.15 * 22.98 mm^2, 28.71 * 21.885 mm^2, and 20.843 * 20.843 mm^2, respectively, and distances from feed point are 65.5063, 42.5733, and 20.759 mm, respectively. Other parameters are same as previous arrangements.

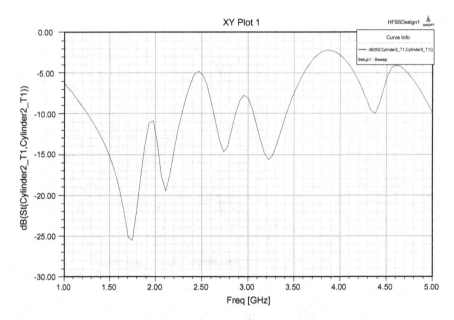

Fig. 2 Plot between return loss and frequency

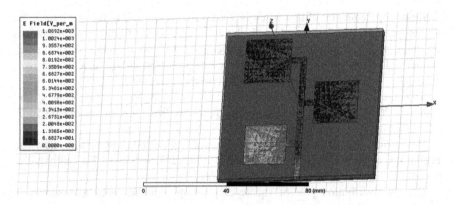

Fig. 3 Current distribution plot at 3 GHz

The simulated return loss versus frequency plot is shown in Fig. 5. It is found that proposed antenna is tuned at 1.7, 2.7, and 3.2 GHz and provides 61.76, 9.25, and 12.5% bandwidth, respectively.

Current distribution plot of rectangle-shaped log periodic antenna is shown in Fig. 6.

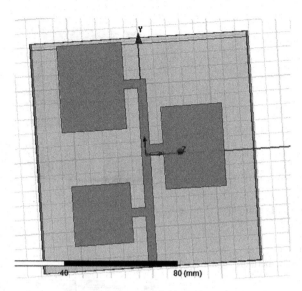

Fig. 4 Geometry of rectangular-shaped log periodic antenna

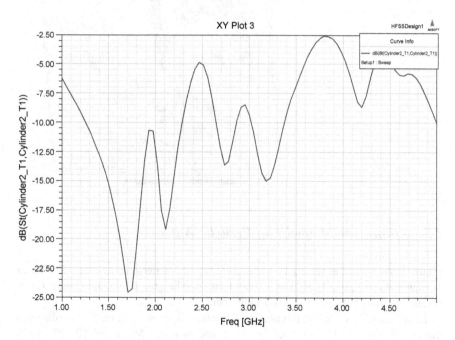

Fig. 5 Plot between return loss and frequency

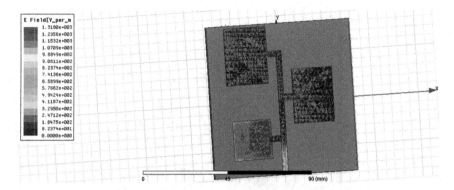

Fig. 6 Current distribution plot at 3 GHz

Fig. 7 Geometry of
triangular-shaped log periodic
antenna

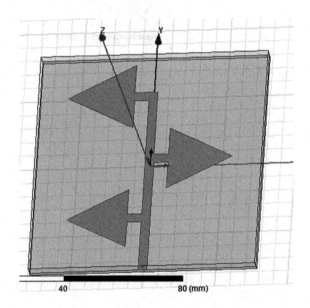

2.3 Triangular-Shaped Log Periodic Antenna

Figure 7 shows the geometry of triangular-shaped log periodic antenna. The bases
of triangles are 22.98, 21.885, and 20.843 mm and heights are equal to bases. Other
parameters are same as previous arrangements.

In the return loss versus frequency plot, which is shown in Fig. 8, it is observed
that antenna is tuned at 1.9 and 2.93 GHz and provides 56.31 and 10.23% bandwidth, respectively.

Current distribution plot of triangle-shaped log periodic antenna is shown in
Fig. 9.

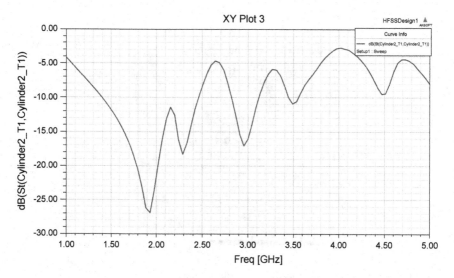

Fig. 8 Plot between return loss and frequency

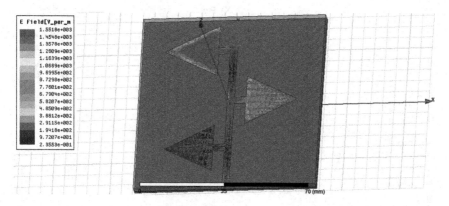

Fig. 9 Current distribution plot at 3 GHz

2.4 Circular-Shaped Log Periodic Antenna

Figure 10 shows the geometry of circular-shaped log periodic antenna. The radiuses of Circles are 13.81, 13.15, and 12.52 mm. Other parameters are same as previous arrangements.

In the return loss versus frequency plot, which is shown in Fig. 11, it can be observed that antenna is tuned at 1.75, 2.85, and 3.55 GHz and provides 68.57, 12.22, and 8.4% bandwidth, respectively. The calculated impedance bandwidth is 68.57% which is greater than other geometries.

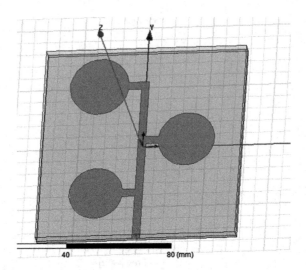

Fig. 10 Geometry of circular-shaped log periodic antenna

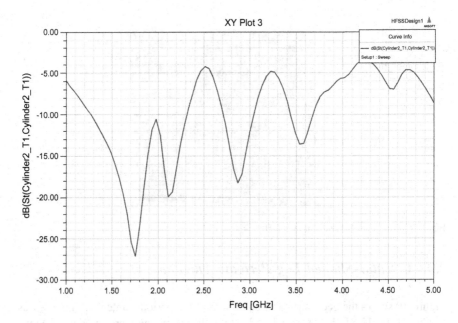

Fig. 11 Plot between return loss and frequency

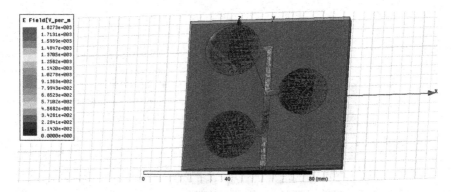

Fig. 12 Current distribution plot at 3 GHz

Table 1 Comparison results of return loss parameter

Arrangement	Bandwidth (%)	Resonant frequency (GHz)	Return loss (dB)
Square-shaped log periodic antenna	60	1.75	−25.5
	10.90	2.75	−15
	10.93	3.25	−15.7
Rectangular-shaped log periodic antenna	61.76	1.7	−24
	9.25	2.7	−13
	12.5	3.2	−15
Triangular-shaped log periodic antenna	56.31	1.9	−27
	10.23	2.93	−17
Circular-shaped log periodic antenna	68.57	1.75	−27
	12.22	2.85	−18
	8.4	3.55	−14

Current distribution plot of triangle-shaped log periodic antenna is shown in Fig. 12.

2.5 Comparative Analysis of Bandwidth and Return Loss

Bandwidth is one of the important antenna performance parameter. It is observed that for 1.75 and 2.85 GHz frequencies, circular-shaped log periodic antenna provides maximum bandwidth and minimum return loss while rectangular-shaped log periodic antenna is suitable for 3.2 GHz frequency.

Comparative results are more clearly shown in Table 1.

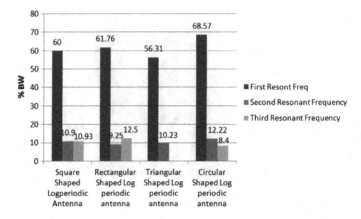

Fig. 13 Bandwidth comparison chart

The graph shown in Fig. 13 represents circular-shaped log periodic antenna having greater bandwidth and minimum losses.

3 Conclusion

Simulations of different-shaped log-periodic microstrip antenna arrays of square, rectangular, circular, and triangular patch elements were presented in this paper. It is analyzed that the lower cut-off frequency of the antenna arrays depends on the choice of shapes and dimensions of the patch. It is observed that the antenna array using circular patch provides the best performance, followed by the rectangle, square, and triangle patch shapes, respectively, in terms of impedance bandwidth. In terms of the return loss, the circular patch has minimum return loss followed by triangle, square, and rectangle. These antennas have applications in satellite communications, radio broadcasting, and Wireless-fidelity applications, etc.

References

1. Balanis, Antenna Theory—Analysis and Design. Wiley, London (1997)
2. Ding, X., Wang, B.-Z., Zang, R.: Design and realization of a printed microstrip logperiodic antenna. iWEM 2012 Proceedings
3. Ismail, M.F., Rahim, M.K.A., Zubir, F., Ayop, O.: Log-periodic patch antenna with tunable frequency. In: Proceedings of the 5th European conference on antennas and propagation (EUCAP), Rome, pp. 2165–2169, 11–15 Apr 2011
4. Hamid, M.R., Gardner, P., Hall, P.S.: Frequency reconfigurable log periodic patch array. Electron. Lett. **46**, 25 (2010)
5. Pozar, D.M., Schaubert, D.H.: Microstrip Antennas, The Analysis and Design of microstrip Antennas and Arrays. IEEE Press (1995)

6. Nickolas, K., Dimitrios, E.A., Manos, T., John, P.: RF MEMS sequentially reconfigurable sierpinski antenna on a flexible, organic substrate with novel dc biasing technique. IEEE/ASME J. Micro Electromechanical Syst. **16**(5), 1185–1192 (2007)

7. Sai Rajanarendra, Nakkeeran, R.: A comparitive study of four different shaped frequency reconfigurable log periodic microstrip antenna arrays. Int. J. Microwaves Appl. **2**(2) (2013)

8. Isbell, D.E.: Log periodic dipole array. IRE Trans. Antennas Propagat. **AP**-S:260–267 (1960)

9. Nasir, S.A., Arif, S., Mustaqim, M., Khawaja, B.A.: A log-periodic microstrip patch antenna design for dual band operation in next generation wireless LAN applications. 2013 IEEE 9th international conference on emerging technologies (ICET), pp. 1–5 (2013)

10. Hall, P.S.: Multioctave bandwidth log-periodic microstrip antenna array. IEE Proceedings H—Microwaves, Antennas and Propagation **133**(2) (2015)

11. Rahim, M.K.A., Gardner, P.: Design of nine element quasi microstrip log periodic antenna. In: Proceedings of RF and microwave conference, Malaysia, pp 132–135, 5–6 Oct 2004

Optimization of Compressive Strength of Polymer Composite Brick Using Taguchi Method

Nitesh Singh Rajput, Dipesh Dilipbhai Shukla, Lav Ishan and Tarun Kumar Sharma

Abstract Taguchi design of experiments (DOE) method was employed to determine the optimum value of controllable factors of the process. It minimizes the number of experimental runs and maintains the quality of the desired product. In this paper, high density Polyethylene and poly propylene/marble dust composite (HDPE-PP/MDC) was prepared using single screw extruder machine. Marble dust is used as a filler as it is waste material of marble industries. The loadings of filler, binder, plasticizer and lubricant were varied from 14–30, 2–4, 2–5 to 2–6%, respectively. The effects of different loadings on the compressive strength of HDPE-PP/MD composite were studied, and the results were optimized by using Taguchi's design of experiment (DOE) method. It is concluded that the filler and the binder loading play significant role in the enhancement of compressive strength of the composite.

Keywords Taguchi method · Polymer composite · Extrusion · Compressive strength

N.S. Rajput (✉) · D.D. Shukla · L. Ishan · T.K. Sharma
Amity School of Engineering and Technology, Amity University Rajasthan,
Jaipur 303007, India
e-mail: niteshthakur72@yahoo.com

D.D. Shukla
e-mail: dipeshdshukla@yahoo.com

L. Ishan
e-mail: amity.lavishan@gmail.com

T.K. Sharma
e-mail: taruniitr1@gmail.com

© Springer Nature Singapore Pte Ltd. 2018
M. Pant et al. (eds.), *Soft Computing: Theories and Applications*,
Advances in Intelligent Systems and Computing 584,
https://doi.org/10.1007/978-981-10-5699-4_42

453

1 Introduction

Many optimization techniques like Taguchi's methods, artificial neural networking, genetic algorithm have been used for determining the most appropriate final result. These techniques not only optimize the result but also help in decreasing the experimental runs and saving time by giving the most appropriate compositions, experimental set-up method, etc., in less runs.

Out of these the Taguchi's design, optimization technique is a cost effective and easy method for improving the quality of product and processes by optimizing set of controllable parameters. For example, as [1] shows that the Taguchi method helps in improving the quality if plastic by optimizing the controllable factors. The quality of product obtained by the Taguchi method may be different from the desired quality. The Taguchi method can be used for making high-quality as well as low-quality products depending upon the need of the customer. The Taguchi method is adopted by number of researchers in the area of thermoplastic polymer composites for optimizing and improving the final results [2–5]. Polymer composite can be made by injection moulding, compression moulding and extrusion method. [6] used the continuous extrusion process for the formation of short wool fibre-PP composite and further used Taguchi design of experiment to find out the most significant factors for preparation of the composite. Similarly [7] reported the use of Taguchi experimental design approach to finding out the effect of denaturants on the mechanical properties on soy meal-based biodegradable bends. Not only for the finding out the most significant factors like filler composition, base polymer composition for the formation of the composite the Taguchi method can also be used to determine the processing parameters for the formation of composite and being able to find out the best set-up condition to optimize the result shown by Campos Rubio et al. [8] and Khosrokhavar et al. [9].

Materials like marble dust, saw dust, waste rubber, bagasse fibre, clay are used as fillers in polymer composites and have significant effect on the mechanical properties of the composites. Similarly [10] used waste ground rubber as the filler. These fillers not only increases the mechanical property if the base composite but are also cheap. References [11, 12] shows the effect of clay as the fillers on the mechanical properties on the formation of the Nano-composites. These fillers play an important role when cost and environmental factors are considered. This paper focuses on the utilization of marble slurry (waste of marble industry) as filler in the polymer-based composite.

The purpose of this work is to investigate the effect of marble dust on the mechanical property of HDPE and PP/marble dust composite. In this paper, extrusion moulding method is used. The processing parameter like screw speed was kept 15 rpm and temperature of feed, transition and the metring section were kept 130, 190 and 230 °C, respectively, using bend heater and maintained throughout.

2 Experimental

2.1 Materials

High Density Polyethylene and Poly Propylene pallets were obtained from local polymer industry having density in between 0.93–0.97 and 0.895–0.92 g/cm³, respectively, and the compressive strength is 40 MPa for PP and 20 MPa for HDPE. The marble dust was obtained from the marble stone industry. Glycerol was used as plasticizers and purchased from Merck Chemicals Ltd.

2.2 Design of Experiment

In this experiment, the polymer composite was made by extrusion process. Five important factors were taken into account to enhance the compressive strength of the composite; these factors are HDPE and PP as resin, marble dust as filler, binder, Glycerol as plasticizer and lubricant. The experiment where large number of parameters were involved, out of those if one of the factors is chosen to be varied then the number of experimental runs will be increased. To avoid this situation design of experiment method is used. Here, we have used Taguchi's method which helped us in reducing the experiments and also maintained the quality of the data.

In design of experiment method, proper selection of factors and their levels is mandatory to get best results. For the present work resin, filler, binder, plasticizers and lubricant were taken as five factors and their values were changed up to level 5 as shown in Table 1.

As discussed, we have selected four factors each having five levels is selected for Taguchi's design and a standard orthogonal array of L25 was designed and shown in Table 2, analysed and S/N (signal by noise) ratio is plotted according to the result.

S/N ratio (1) is used as a transformed response to show the changes in response due to variation of the selected factors in Taguchi's method. From the above table, it can be observed that the optimum value is obtained at 64 wt% fibre, 26% filler, 3 wt% binder, 2 wt% plasticizer and 6 wt% lubricant loading. In the present work,

Table 1 Selected factors and their respective levels

Factors	Symbol	Level 1	Level 2	Level 3	Level 4	Level 5
Resin	A	60	64	68	72	76
Filler	B	30	26	22	18	14
Binder	C	2	3	4	3	2
Plasticizer	D	5	3	2	5	2
Lubricant	E	4	4	4	2	6

Table 2 Taguchi L25 orthogonal array of designed experiments based on the coded levels

Experimental run	A	B	C	D	E	S/N ratio
Run 1	1	1	1	1	1	25.8451
Run 2	1	2	2	2	2	26.1926
Run 3	1	3	3	3	3	26.2773
Run 4	1	4	4	4	4	25.5292
Run 5	1	5	5	5	5	25.2964
Run 6	2	1	2	3	4	26.0639
Run 7	2	2	3	4	5	26.4444
Run 8	2	3	4	5	1	25.8007
Run 9	2	4	5	1	2	25.4368
Run 10	2	5	1	2	3	25.2490
Run 11	3	1	3	5	2	26.0639
Run 12	3	2	4	1	3	24.6599
Run 13	3	3	5	2	4	24.3497
Run 14	3	4	1	3	5	24.2438
Run 15	3	5	2	4	1	24.1903
Run 16	4	1	4	2	5	24.5062
Run 17	4	2	5	3	1	24.4543
Run 18	4	3	1	4	2	24.2969
Run 19	4	4	2	5	3	24.3497
Run 20	4	5	3	1	4	24.1903
Run 21	5	1	5	4	3	24.5577
Run 22	5	2	1	5	4	24.4022
Run 23	5	3	2	1	5	24.2969
Run 24	5	4	3	2	1	24.2438
Run 25	5	5	4	3	2	24.0279

the S/N ratio is used to analyse the effect of the factors on the compressive strength of the composite brick. The S/N ratio can be expressed mathematically as

$$
S/N = \frac{-10 * \log\left[\frac{1}{y_1^2} + \frac{1}{y_2^2} + \frac{1}{y_3^2} + \cdots + \frac{1}{y_n^2}\right]}{n},
\tag{1}
$$

where y is the experimental measurement of the compressive strength and n is number of sample per trials.

2.3 Experimental Procedure

HDPE/PP pellets and marble dust as filler were dumped in the hopper of extruder machine. The temperature of the feed, transition and metring section were kept and regulated throughout to 130, 190 and 230 °C, respectively, using band heaters and

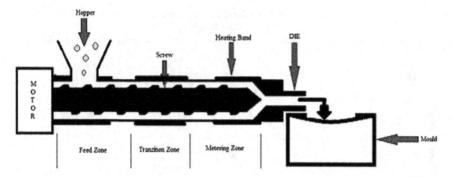

Fig. 1 Schematic diagram of extruder machine

their controllers. The motor was started and its speed was set at 1500 rpm. This rotational power was transferred by a gear box to the extruder screw. The RPM of the extruder was fixed at 15 rpm. Due to high temperature the polymer becomes soft in the feed zone, comes in plastic state in transition zone and in metering zone it melts completely. The extruder prepares homogeneous mixture inside the barallel. The mixture is forced to pass through an opening of die. The extruded flow rate depends on screw rotation and the lubricant is added to the mixture. In the present study, screw rotation is kept fixed so that the weight per cent of lubricant changes the extrudate flow rate. The extrudate was collected in a metallic brick shape mould and was air cooled. The composite was taken out of the mould after cooling and characterization was done (Fig. 1).

3 Result and Discussion

The compressive strength of the composite is influenced by the addition of the fillers, binders and the lubricant in the mixture. The addition of filler in the mixture has the maximum effect on the compressive strength of the composite brick. The composition of the filler used is considered between 14 and 30%. As the amount of filler increases it increase the compressive strength of the sample. The effect due to the addition of binder and lubricant have less effect on the increment of the compressive strength of the brick as compared to the filler and these are taken between 2–4 and 2–6%, respectively. S/N response analysis shows that on increasing the weight per cent of filler and binder increases the compressive strength significantly and on increasing the content of plasticizer the compressive strength decreases. S/N curve is shown in Fig. 2.

Main Effects Plot for Means
Data Means

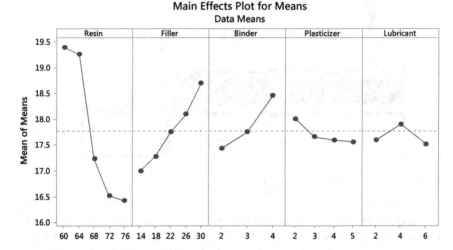

Fig. 2 Effect of factors on compressive strength

4 Conclusion

Taguchi method was used to determine the dependency of compressive strength on filler, binder, plasticizer and lubricant loading. It was found that marble plays a key role in the enhancement of compressive strength of the composite. Also binder proved to be significant contributor to the compressive properties of the composite but the compressive strength decreases with increase in plasticizer loading. The compressive strength can be achieved by mixing High Density Polyethylene and PolyPropylene resin with 22 wt% filler and 3 wt% binder.

References

1. Gu, F., Hall, P., Miles, N.J., Ding, Q., Wu, T.: Improvement of mechanical properties of recycled plastic blends via optimizing processing parameters using the Taguchi method and principal component analysis. Mater. Des. **62**, 189–198 (2014)
2. Esnaashari, C., Khorasani, S.N., Entezam, M., Khalili, S.: Mechanical and water absorption properties of sawdust-low density polyethylene nanocomposite. J. Appl. Polym. Sci. **127**, 1295–1300 (2013)
3. Hong, G.B., Su, T.L.: Statistical analysis of experimental parameters in characterization of ultraviolet-resistant polyester fiber using a TOPSIS-Taguchi method. Iran. Polym. J. (English Ed.) **21**, 877–885 (2012)
4. Chieng, B.W., Ibrahim, N.A., Yunus, W.M.Z.W.: Optimization of tensile strength of poly (lactic acid)/graphene nanocomposites using response surface methodology. Polym. Plast. Technol. Eng. **51**, 791–799 (2012)

5. Lashgari, S., Garmabi, H., Lashgari, S., Mohammadian-Gezaz, S.: Improving the interfacial adhesion of highly filled PP–bagasse composites designed by Taguchi method. J. Thermoplast. Compos. Mater. **24**, 431–446 (2011)
6. Kim, N.K., Lin, R.J.T., Bhattacharyya, D.: Extruded short wool fibre composites: mechanical and fire retardant properties. Compos. Part B Eng. **67**, 472–480 (2014)
7. Reddy, M.M., Mohanty, A.K., Misra, M.: Optimization of tensile properties thermoplastic blends from soy and biodegradable polyesters: Taguchi design of experiments approach. J. Mater. Sci. **47**, 2591–2599 (2012)
8. Campos Rubio, J.C., Silva, L.J., Da Leite, W.D.O., Panzera, T.H., Filho, S.L.M.R., Davim, J. P.: Investigations on the drilling process of unreinforced and reinforced polyamides using Taguchi method. Compos. Part B Eng. **55**, 338–344 (2013)
9. Khosrokhavar, R., Naderi, G., Bakhshandeh, G.R., Ghoreishy, M.H.R.: Effect of processing parameters on PP/EPDM/organoclay nanocomposites using Taguchi analysis method. Iran. Polym. J. **20**, 41–53 (2011)
10. Xin, Z.X., Zhang, Z.X., Pal, K., Byeon, J.U., Lee, S.H., Kim, J.K.: Study of microcellular injection-molded polypropylene/waste ground rubber tire powder blend. Mater. Des. **31**, 589–593 (2010)
11. Majdzadeh-Ardakani, K., Nazari, B.: Improving the mechanical properties of thermoplastic starch/poly(vinyl alcohol)/clay nanocomposites. Compos. Sci. Technol. **70**, 1557–1563 (2010)
12. Majdzadeh-Ardakani, K., Navarchian, A.H., Sadeghi, F.: Optimization of mechanical properties of thermoplastic starch/clay nanocomposites. Carbohydr. Polym. **79**, 547–554 (2010)
13. Patnaik, A., Satapathy, A., Chand, N., Barkoula, N.M., Biswas, S.: Solid particle erosion wear characteristics of fiber and particulate filled polymer composites: a review. Wear **268**, 249–263 (2010)
14. Mu, Y., Zhao, G., Wu, X., Zhang, C.: An optimization strategy for die design in the low-density polyethylene annular extrusion process based on FES/BPNN/NSGA-II. Int. J. Adv. Manuf. Technol. **50**, 517–532 (2010)
15. Wang, J., Mao, Q.: Methodology based on the PVT behavior of polymer for injection molding. Adv. Polym. Technol. **32**, 474–485 (2012)

Application of Unnormalized and Phase Correlation Techniques on Infrared Images

Himanshu Singh, Millie Pant, Sudhir Khare and Yogita Saklani

Abstract Accurate and unambiguous measurement of relative displacement between two images is important in a large number of practical applications. Comparison of a real-time captured image with a stored reference image is one such example. Comparison of two sensed images acquired within a short and accurately known time interval can be used to measure the velocity to height ratio of a moving imaging system. Accurate superposition of two images acquired at different times and their consequent, point by point subtraction, calls for attention to changes which have occurred over time in the real-time scenario. All such applications require measurement of the displacement vector to accuracies within a small fraction of pixel. The phase correlation method is based on the fact that most of the information about the relative displacement vector is contained in the phase of their cross-power spectrum.

Keywords Unnormalized correlation · Phase correlation · Infrared

1 Introduction

Correlation is matching of two signals to see how much similarity exists in between the two. The correlation operation is entirely dependent on the correlator used. The correlator should produce a correlation surface with a strong response at the location of the target object image while suppressing the response of the false target. This correlation can be performed in the spatial domain or in the frequency

H. Singh (✉) · S. Khare · Y. Saklani
Instruments Research & Development Establishment, DRDO, Dehradun, India
e-mail: singh_himanshu18@rediffmail.com

S. Khare
e-mail: sudhir_khare@hotmail.com

M. Pant
Department of Applied Science and Engineering, IIT Roorkee, Roorkee, India
e-mail: millifpt@iitr.ernet.in

© Springer Nature Singapore Pte Ltd. 2018
M. Pant et al. (eds.), *Soft Computing: Theories and Applications*,
Advances in Intelligent Systems and Computing 584,
https://doi.org/10.1007/978-981-10-5699-4_43

domain. The main aim of this paper is to apply the unnormalized and phase correlation techniques over live infrared images and find the correlation pattern. Further, we compare the two techniques as to which is better for application in tracker and subpixel registration applications. The paper is organized as follows: a brief introduction about this paper is given first, and then an extensive literature survey is reported. In it, several relevant papers are refereed and their findings are condensed. The next section is devoted to the mathematical formulation of unnormalized correlation, phase correlation and the relation between the two. Further, image analysis by means of unnormalized correlation and phase correlation for applications like template matching and registration is elaborated. Since, the application is much easier in frequency domain so ensuing discussions are in frequency domain. Next section shows the pseudo-algorithm in MATLAB and simulation results. Finally, conclusion is drawn based on the simulation results.

2 Literature Review

Extensive literature survey has been carried out. A paper by Cozzellan et al. [1] presented a Hartley transform algorithm for accurate and fast elaboration of phase-only correlation instead of 2D FFT. An important paper by Lamberti et al. [2] published in Optical Society of America journal adds flavour by implementing phase correlation method for peak wavelength detection of fibre Bragg grating sensor. A novel subpixel phase correlation method using singular value decomposition (SVD) and the unified random sample consensus (RANSAC) algorithm was proposed by Xiaohua et al. [3]. In the proposed method, SVD theoretically converts the translation estimation problem to one dimension for simplicity and efficiency and the unified RANSAC algorithm acts as a robust estimator for the line fitting, in this case for the high accuracy, stability and robustness. A short paper by Lewis [4] shows that unnormalized cross-correlation can be efficiently normalized using precomputing integrals of the image and image over the search window. In another paper by Goshtasby et al. [5], two stage templates matching using cross-correlation were reported. An important paper by Tomasi et al. [6] details track feature selection and monitoring method that can detect occlusions, disocclusions and features that do not correspond to the points in the world. A tracking algorithm was suggested which extended the Newton-Raphson search method to work under affine transformation. A paper by Seelen et al. [7] reports an adaptive correlation method that has been developed which selectively updates the correlation template in response to scale changes due to zoom lens in an infrared image sequence. Next paper by Jan et al. [8] implements phase correlation method along with edge extraction method to achieve registration in multisensor platform for real-time applications. One popular approach using phase correlation was mentioned in [9]. In this paper, an analytic expression for the phase correlation of down-sampled images is derived. Paper claims that for down-sampled images the signal power in the phase correlation is not concentrated in a single peak, but rather in several coherent peaks mostly adjacent to each other. So, they use the technique for subpixel registration. In all these papers, a comparative analysis and simulation of unnormalized correlation and phase correlation was not carried out. Moreover, applicability of phase correlation on applications like tracking and subpixel registration was not specifically carried out. So, there is a need to study the applicability of unnormalized correlation and phase correlation from practical implementation point of view.

3 Unnormalized Correlation

3.1 Spatial Correlation

Let reference image be described by the vector Y and the section of the current scene imaged by the vector S_{ij}. Then, the correlation at each location (i, j) in the scene image is simply the dot product of the two vectors and is given by:

$$C(i,j) = S * Y. \tag{1}$$

When the reference image is large, it may be more efficient to perform correlation in frequency domain. By applying a Fourier Transform to reference image Y and the scene image S into their frequency domain equivalents and $\hat{S}$ can be computed. Given the complex conjugate of $\hat{S}$ is $\hat{S}^*$ and F^{-1} is the inverse Fourier, then the correlation matrix is given by:

$$C(i,j) = \left(S_{ij} \cdot Y_{ij}\right) / \left(\| S_{ij} \| \cdot \| Y_{ij} \|\right). \tag{2}$$

Normalized spatial correlation as described by the above equation has some useful properties. The normalization removes the bias of the algorithm to match with the intense objects and ensures that correlation value is between zero and one, with small values a poor match, and one being a perfect match.

However, the response of normalized spatial correlation has fairly broad peaks. A more powerful correlation method is symmetric phase-only correlation given by:

$$C = F^{-1}\left(\hat{S} \times \hat{Y}\right) / \left(\left|\hat{S}^*\right| \cdot \left|\hat{Y}\right|\right). \tag{3}$$

Using this equation, a very clean correlation surface is obtained. False matches are suppressed, leaving a distinct peak marking the location of the correct match. Traditional correlation systems have difficulty robustly detecting an object when the object is partially obscured by clutter. In particular case of correlation matching and reference image update subsystem, the system behaves as follows:

1. The reference image is correlated with the current frame of the scene image to produce a correlation surface.
2. The location of the target object is extracted from the scene image; the location corresponds to the scene of the correlation surface.
3. The object is extracted from the scene image. The sequence is repeated for every frame.

Such systems are robust in presence of noise; however, this is not the case when object is obscured by clutter. Updating the reference image without introducing noise or clutter is critical to the performance of the correlation system. Use of an adaptive window for selecting the reference image will reduce the amount of surrounding clutter from being introduced; it will not stop obscuring clutter from being introduced in the reference image [7].

3.2 Phase Correlation Method

The discrete phase correlation function is obtained by first computing the discrete two-dimensional Fourier transforms, G_1 and G_2, of the two sampled images, forming the cross-power spectrum, and at each spatial frequency extracting the phase.

$$e^{j\theta} = \frac{G_1 G_2^*}{|G_1 G_2^*|} \tag{4}$$

and then computing the inverse Fourier transform of the phase array:

$$d = F^{-1}\{e^{j\theta}\}. \tag{5}$$

The reasoning which underlies this approach is that all of the information about the translation of two images reside in the phase of their cross-power spectrum. In particular, in the idealized case of images cyclically shifted by a vector translation L, the above equation yields a delta function located at the position L.

3.3 Relation Between Phase and Unnormalized Cross-Correlation

Phase correlation and unnormalized cross-correlation are related in the following manner. The cross-correlation function, c, can either be expressed as the direct cross-correlation of the images as mentioned as g_1 and g_2 in the spatial domain:

$$C = g_1 * g_2 \tag{6}$$

or, using the convolution theorem, as the inverse Fourier transform of the cross-power spectrum:

$$C = F^{-1}\{G_1 G_2^*\}. \tag{7}$$

Thus,

$$C = d * F^{-1}\{G_1 G_2^*\} \tag{8}$$

where G_1 and G_2 are the Fourier transforms of g_1 and g_2, respectively. These techniques are traditionally used for image matching applications, since it is optimum for signals degraded by white noise [8].

3.4 Filtered Phase Correlation

The above algorithm can be generalized by introducing an arbitrary weighting function (filter) $H(f)$ in the spatial frequency domain giving

$$d_H = F^{-1}\{H * e^{j\theta}\}. \tag{9}$$

In practice, it is useful to choose filters which are scene-independent functions, since this simplifies the analysis of performance. This approach follows from the fact that:

1. Perspective changes in the image domain have a direct analog in the Fourier domain.
2. Regions of the spatial frequency domain which are displayed by more than one resolution element give rise to random phase differences which contribute only noise to the correlation function.

This results in the correlation computation of a spatial frequency filter which has unit amplitude at zero spatial frequency filter and decreases to zero at the boundary of the random phase region. The basic approach, then, is to use filtered phase correlation function with the filter function chosen so as to maximize the information content of the computed correlation function. In this manner, the widest possible set of image noise characteristics can be handled effectively, including low contrast region devoid of significant cultural features.

3.5 Image Analysis by Means of Correlation and Phase Correlation in Frequency Domain

To compare between the correlation and phase correlation techniques, let us consider two related image analysis tasks: image detection and image registration. Detection or template matching is concerned about the determination of presence or absence of objects suspected of being in an image. Registration involves the spatial alignment of a pair of view of a scene.

3.5.1 Template Matching

One of the most fundamental means of object detection within an image field is by template matching in which a replica of an object of interest is compared to all unknown objects in the image field. If the template match between the unknown object and the template is sufficiently close, the unknown object is labelled as the template object. A template match is rarely ever exact because of image noise, spatial and amplitude quantization effects, and a priori uncertainty as to the exact shape and structure of an object to be detected.

As shown in Fig. 1, computation of the numerator is equivalent to raster scanning the template $T(p, q)$ over the search area $S(u, v)$ such that the template always resides within $S(u, v)$, and then forming the sum of products of the template and the search area under the template. For large size templates, it may be computationally efficient to perform the convolutions indirectly by Fourier domain filtering.

3.5.2 Image Registration

In many of the image processing applications, it is necessary to form a pixel by pixel comparison of two images of the same object field obtained from different sensors, or of two images of an object field taken from the same sensors at different times. To form this comparison, it is necessary to spatially register the images, and thereby, to correct for the relative translational shifts, rotational differences, scale

Fig. 1 Relation between template region and search area

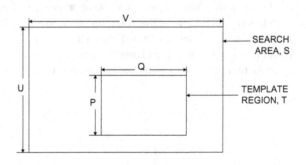

differences and perspective view differences. Often, it is possible to eliminate or minimize many of these sources of misregistration by proper static calibration of an image sensor. The classical technique for registering a pair of images subject to unknown translational differences is to:

1. Form the normalized cross-correlation function between the image pair.
2. Determine the translational offset co-ordinates of the correlation function peak.
3. Translate one of the images with respect to the other by the offset co-ordinates.

3.6 Basic Correlation Function (in Frequency Domain)

Let $F_1(j, k)$ and $F_2(j, k)$, for $1 \leq j \leq J$ and $1 \leq k \leq K$, represent two discrete images to be registered. $F_1(j, k)$ is considered to be the reference image, and

$$F_2(j,k) = F_1(j - j_0, k - k_0) \qquad (10)$$

is a translated version of $F_1(j, k)$ where (j_0, k_0) are the offset co-ordinates of the translation. The normalized cross-correlation between the image pair is defined by a formulation, which is a generalization of the template matching cross-correlation expression, and it utilizes an upper left corner-justified definition for all the arrays. The dashed line rectangle of the further shown Fig. 2 specifies the bounds of the correlation function region over which the upper left corner of $F_2(j, k)$ moves in space with respect to $F_1(j, k)$. The bounds are thus specified and indicated by the shaded region. This region is called the *window region* of the correlation function computation. The computation is often restricted to a constant-size window area less than the overlap of the image pair in order to reduce the number of calculations. This constant-size window region is called a template region. The dotted lines in Fig. 2 specify the maximum constant-size template region which lies at the centre of $F_2(j, k)$.

Fig. 2 Computational bounds in correlation

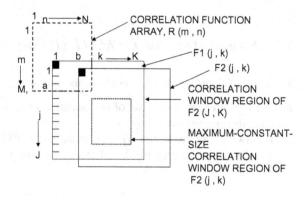

3.6.1 Phase Correlation Method (in Frequency Domain)

Consider a pair of continuous domain images

$$F_1(x, y) = F_1(x - x_0, y - y_0) \tag{11}$$

that are translated by an offset (x_0, y_0) with respect to another. By the Fourier transform shift property, the Fourier transform of the images is related by

$$F_2(\omega_x, \omega_y) = F_1(\omega_x, \omega_y) e^{\left\{ -i\left(\omega_x x_0 + \omega_y y_0\right)\right\}}. \tag{12}$$

The exponential phase shift factor can be computed by the cross-power spectrum of the two images given by

$$G(\omega_x, \omega_y) = \frac{F_1(\omega_x, \omega_y) F_2^*(\omega_x, \omega_y)}{\left| F_1(\omega_x, \omega_y) F_2(\omega_x, \omega_y) \right|} = e^{\left\{ i\left(\omega_x x_0 + \omega_y y_0\right)\right\}}. \tag{13}$$

Taking the inverse Fourier transform of the above equation yields the spatial offset

$$G(x, y) = \delta(x - x_0, y - y_0) \text{ in the spatial domain.}$$

The cross-power spectrum approach can be applied to discrete images by utilizing discrete Fourier transform in place of the continuous Fourier transforms [9].

The basic problems associated with the basic correlation function are:

1. The correlation function is rather broad, making detection of peak difficult.
2. Image noise may mask the peak correlation.

Such problems do not arise in case of phase correlation. The peak is sharply defined thus giving the exact location of object or the target in the image. Moreover, there is no image noise problem that could result in masking of the peak correlation function. Thus, we use the phase correlation function to determine the correlation method.

3.7 Pseudo-Algorithm for MATLAB Simulation

The computational sequence consists of the following sequence steps:

1. The input sequence consists of two sampled images g_1 and g_2 which have the same dimensions say (N by M).
2. The two-dimensional fast Fourier transform (FFT) is taken of each image, resulting in two complex N by M element arrays, G_1 and G_2.

3. The phase difference matrix is derived by forming the cross-power spectrum, $G_1 G_2^*$, and dividing by its modulus, i.e.

$$e^{j\phi} = e^{j(\phi_1 - \phi_2)} = e^{j\emptyset} = \frac{G_1 G_2^*}{\left| G_1 G_2^* \right|}. \tag{14}$$

4. If filtering is required, the phase matrix is multiplied by a weighting function H (f).
5. The phase correlation function, d, is then obtained as a real N by M element array by taking the inverse FFT of the weighted phase matrix:

$$d = F^{-1} \left\{ H * e^{j\emptyset} \right\}. \tag{15}$$

6. The correlation function is searched to determine the location of the highest peak, and the non-integer part of the peak location is estimated by interpolation.

4 Simulation Results

The simulation of unnormalized correlation and phase correlation was implemented in frequency domain. The codes were written in MATLAB language. The applicability and comparison of the two techniques were carried out on infrared video. The precise location of correlation peak is obtained using phase correlation technique. The subpixel precision obtained can be adjudged as shown in Figs. 3, 4, 5 and 6.

Fig. 3 Single frame in input video

Fig. 4 Target object template

Fig. 5 Unnormalized correlation peak

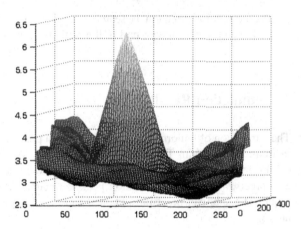

Fig. 6 Phase correlation peak at (120, 184)

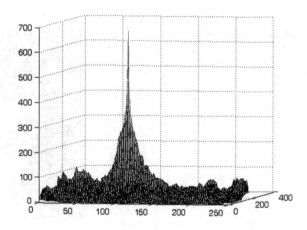

5 Conclusion and Future Work

The two techniques of correlation—unnormalized correlation and phase correlation have been studied and simulated using MATLAB in frequency domain. The simulation of the infrared images from recorded video of target and its template has been carried out using these techniques. The correlation peak due to unnormalized technique is very broad with side lobes and thus cannot be used to precisely locate the spatial position of the target in applications such as tracking of target where the precise location of target is of utmost importance. While the phase correlation technique gives a precise location of the target with subpixel accuracy, it is of vital importance in tracking and sub pixel registration applications. Hence, out of the two techniques **phase correlation** is much suited for applications like tracking, image registration. In future, the algorithm/method can be used for implementation of infrared tracker.

References

1. Cozzellan, L., Giuseppe, S.: Phase-only correlation function by means of Hartley transform. JSM Math. Stat. **1**(1) 2014
2. Lamberti, A., Vanlanduit, S.B., De Pauw, Berghmans, F.: A novel fast phase correlation algorithm for peak wavelength detection of Fiber bragg grating sensor. OSA Optics Express, **22**(6), 7099 (2014)
3. Tong, X. et al. A novel subpixel phase correlation method using singular value decomposition and unified random sample consensus. IEEE Trans. Geosci. Rem. Sens. **53**(8) (2015)
4. Lewis J.P.: Fast template matching. vision interface. pp. 120–123 (1995)
5. Goshtasby A., Gage S. H., Bartholic J.F.: A two-stage cross-correlation approach to template matching. IEEE Trans. Pattern Anal. Mach. Intell. **6**(3), 374–378 (1984)
6. Shi J., Tomasi C.: Good features to track. In: Proceedings of the IEEE Conference on Computer Vision and Pattern Recognition (1994)
7. Cahn von Seelen, UM., Bajcsy, R.: Adaptive correlation tracking of targets with changing scale. Technical Reports (CIS), Department of Computer & Information Science, Univ. of Pennsylvania, June 1996
8. Klimaszewski Jan., Kondej, M., Kawecki, M., Barbara, P.: Registration of infrared and visible images based on edge extraction and phase correlation approaches. In: Image Processing and Communications Challenges 4, vol. 184 of the series Advances in Intelligent Systems and Computing, pp. 153–162
9. Foroosh, H., Zerubia J.B., Berthod M.: Extension of Phase Correlation to Subpixel Registration. IEEE Trans. Image Process. **11**(3), 188–200 (2002)
10. Mitra, S.K., Kaiser, J.F.: Handbook for digital signal processing. Wiley, New York (1993)
11. Oppenheim, A.V., Schafer, R.W.: Digital signal processing. Englewood Cliffs, New Jersey, Prentice-Hall (1975)
12. Pratt, W.: Digital image processing. John Wiley, New York (1978)

Modified Least Significant Bit Algorithm of Digital Watermarking for Information Security

Devendra Somwanshi, Indu Chhipa, Trapti Singhal and Ashwani Yadav

Abstract In today's information age, the technologies have developed so much that most of the users prefer Internet to transfer data from one end to another across the world. So privacy in digital communication is the basic requirement when confidential information is being shared between two users. The attacks on the Internet and Internet-attached systems have grown. Digital watermarking technique is used to hide the information inside a signal. In this paper, least significant bit (LSB) algorithm is discussed. The basic LSB algorithm is easy to decrypt because the watermark image is embedded into a single bit position of an original image. To remove this limitation of basic LSB algorithm, the modified version of LSB algorithm using XOR gate is used and the experimentation results are shown. Performance analysis of the algorithm is also done.

Keywords Digital watermarking · Least significant bit (LSB) algorithm Peak signal-to-noise ratio (PSNR) · Mean squared error (MSE)

1 Introduction

The Internet is global in scope, but this global inter-network is an open and inse-cure. The Internet is today a widespread information infrastructure and a medium for collaboration and interaction between the individuals, without regard to geo-graphical locations. Watermarking is a technique used to hide data or identifying information within digital multimedia. The best-known watermarking technique is

D. Somwanshi (✉) · I. Chhipa · T. Singhal
School of Engineering and Technology, Poornima University, Jaipur, India
e-mail: imdev.som@gmail.com

I. Chhipa
e-mail: indu.chhipa21@gmail.com

A. Yadav
Amity University, Jaipur, India
e-mail: ashwaniy2@gmail.com

© Springer Nature Singapore Pte Ltd. 2018
M. Pant et al. (eds.), *Soft Computing: Theories and Applications*,
Advances in Intelligent Systems and Computing 584,
https://doi.org/10.1007/978-981-10-5699-4_44

473

digital watermarking which is becoming popular, especially for adding undetectable identifying marks, such as author or copyright information [1]. Digital water-marking is used to hide the information inside a signal, which cannot be easily extracted by the third party or unauthorized person [2]. The digital watermarking process embeds a signal into the media without significantly degrading its quality and also process to embed some information called watermark into different kinds of media called cover work. Digital watermarks are inside the information so that ownership of the information cannot be claimed by third parties [3].

Previously, lot of work has been done in the field of watermarking using LSB. Authors had done work using second LSB [1], LSB and inverse bit [2], SVD-based watermarking [4], different bit positions [5], etc. This work shown in this paper proposed modified least significant bit (LSB) algorithm. LSB algorithm was first proposed by Kurah and Mc Huges in 1992 [6]. Many researchers used basic least significant bit (LSB) algorithm using different techniques, but it was found that basic LSB has high PSNR and MSE value but has low security, as in this water-mark was embedded into single bit of original image [7, 8].

The paper is organized as follows. Section 2 gives brief introduction of basic LSB algorithm. Section 3 presents the proposed work. Demonstration of process embedding is described in Sect. 4. Section 5 presents the experimental results and performance analysis. Section 6 gives the conclusion. Future work is given in Sect. 7.

2 Basic LSB Algorithm

The simplest spatial-domain image watermarking technique is to embed a water-mark in the least significant bits (LSBs) of some randomly selected pixels of the cover image. In this watermarking scheme, firstly it embeds watermarks in the least bit of pixel of the original image [9]. All images taken in the form of pixels and each pixel represented by 8-bit value (least and most significant bit) and then the watermark is embedded in the least significant bit and most significant bit of selected pixel of the image, and later watermark image is extracted. There is less chance for degradation of the original image. More information can be stored in an image. The basic algorithm only replaces one bit of original image by corre-sponding value of watermark. In proposed algorithm, watermark is embedded in two bits according to condition. Thus, it provides more security [10].

3 Proposed Work

The simplest spatial-domain image watermarking technique is to embed a water-mark in the least significant bits (LSBs) of some randomly selected pixels of the cover image. In this watermarking scheme, firstly it embeds watermarks in the least bit of pixel of the original image. All images taken in the form of pixels and each pixel represented by 8-bit value (least and most significant bit) and then the watermark is embedded in the least significant bit and most significant bit of selected pixel of the image, and later watermark image is extracted [11]. There is

Fig. 1 Flowchart of modified LSB

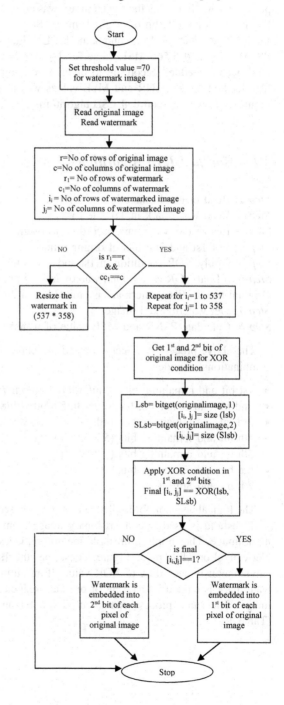

less chance for degradation of the original image. More information can be stored in an image. The design of embedding watermark in a specific first bit position in pair or other higher bits is taken based on logic output of XOR condition for which first two bits are used as input. The performance parameters were calculated for all possible pairs. In this method, watermark was embedded into one bit or other higher bits instead of single bit of original image. So, TIF, GIF, PMG, BMP file fetched same PSNR value of 47.26 and same MSE value of 1.220. JPEG file has 40.675 PSNR value and 5.566 MSE value for 1st and 2nd bit position. Similarly, watermark was embedded according to other bit combinations. TIF, GIF, PNG, BMP files fetched better PSNR and MSE values was as compared to JPEG image file. Figure 1 shows the modified LSB algorithm.

3.1 Modified LSB Algorithm

Step 1: Read original image in any format.
Step 2: Read watermark in TIF file format.
Step 3: Resize the watermark according to original image.
Step 4: Get 1st and 2nd bits of original image.
Step 5: Apply XOR condition in 1st and 2nd bits.
Step 6: If final XOR = = 1, then watermark is embedded into 1st bit of all pixel of Original image. Else watermark is embedded in 2nd bit.
Step 7: Get watermarked image.
Step 8: Calculate PSNR and MSE value of watermarked image.

The above process has been repeated by embedding the watermark in other bit combinations as under;

- 1st bit and remaining bits (2nd/3rd/4th/5th/6th/7th/8th)
- 2nd bit and remaining bits (3rd/4th/5th/6th/7th/8th)
- 3rd bit and remaining bits (4th/5th/6th/7th/8th)
- 4th bit and remaining bits (5th/6th/7th/8th)
- 5th bit and remaining bits (6th/7th/8th)
- 6th bit and remaining bits (7th/8th)
- 7th bit and 8th bit

The logical decision for implementation was based on 1st and 2nd bit value only.

Threshold is used to convert binary image from a gray scale image for watermark image. The simplest threshold method replaces each pixel in an image with a black pixel or white pixel. Figure 1 describes the 1st and 2nd substitution process simultaneously according to XOR value. If an output of XOR condition is 1, then watermark is embed into 1st bit of original image else embed into 2nd of original image. The same process can be used for 1st and remaining bit combinations (Table 1).

4 Demonstration of Process

4.1 Input Process

In this process, two images are taken for experiment.

Step 1: In this step, two images are used as input file. First as an original image and other watermark image this is embedded into original image.
Step 2: Then convert original image and watermark into pixel form.
Step 3: After this, convert pixel form into binary form.

4.2 Working Process

XOR condition: In this pixel value of original image was taken. Every pixel value has 8 bits. According to original image pixel value of 1st row and 1st column is 00000001. 1st and 2nd bits are 1 and 0, respectively. After selection of value, apply XOR condition into selected bit. Proposed algorithm embeds the watermark into one bit or multiple bits (Table 2).

1st and Remaining Bits Substitution: According to XOR output, watermark is embedded into original image. If output is 1, then watermark is embedded into 1st bit of original image else embedded into remaining bits. In this output is 1 so watermark is embedded into 1st bit position of original image (Table 3).

Similarly according to output of XOR condition, watermark is embedding into 1st bit or remaining seven bits value of every pixel. This process can be used for other bit combinations. This process is explained with example in Table 4.

Table 1 Comparison b/w original and watermarked image

Original image	Watermarked image
Original Grayscale Starting Image	Image to be Hidden

Table 2 XOR condition and LSB implementation for 1st and remaining bits

1st bit of original image	2nd bit of original image	Output	Apply watermark in original image
1	0	1	1st bit

Table 3 Working process of modified LSB using 1st and remaining bits combinations

	1			1
1	0 0 0 0 0 0 0 **1**		1	**0**

First Pixel Value of Original Image	Corresponding Watermark

	1	
	1	0 0 0 0 0 0 0 **0**

Table 4 Working process of modified LSB algorithm with example

S. No.	1st bit of original image	2nd bit of original image	Output	Apply watermark in original image	Image	Working process of
1	1	0	1	1st bit		Modified LSB using 1st and remaining bits combinations
2	1	0	1	2nd bit		Modified LSB using 2nd and remaining bits combinations
3	1	0	1	3rd bit		Modified LSB using 2nd and remaining bits combinations
4	1	0	1	4th bit		Modified LSB using 3rd and remaining bits combinations
5	1	0	1	5th bit		Modified LSB using 4th and remaining bits combinations
6	1	0	1	6th bit		Modified LSB using 5th and remaining bits combinations
7	1	0	1	7th bit		Modified LSB using 6th and remaining bits combinations

5 Experimental Results and Performance Analysis

Table 5 shows the results of TIF, GIF, PNG, BMP, and JPEG images. In this method, watermark was embedded into one bit or other higher bits instead of single bit of original image. So, TIF, GIF, PMG, BMP file fetched same PSNR value of 47.26 and same MSE value of 1.220. JPEG file has 40.675 PSNR value and 5.566

Table 5 Performance of modified LSB for different pairs of bit positions

Bit position	File format			
	TIF, BMP, GIF, PNG		JPEG	
	PSNR	MSE	PSNR	MSE
1st and 2nd	47.26	1.220	40.675	5.566
1st and 3rd	42.57	3.596	39.873	6.694
1st and 4th	38.344	9.518	37.945	10.43
1st and 5th	32.88	33.49	33.36	29.99
1st and 6th	26.74	26.74	26.69	139.20
1st and 7th	21.04	21.04	20.85	534.41
1st and 8th	14.50	2304.9	14.42	2345.9
2nd and 3rd	41.625	4.472	39.473	7.340
2nd and 4th	37.962	10.395	37.689	11.06
2nd and 5th	32.76	34.37	33.37	30.55
2nd and 6th	26.71	138.60	26.69	139.17
2nd and 7th	21.03	512.57	26.84	534.87
2nd and 8th	14.49	2307.9	14.42	2348.1
3rd and 4th	37.106	12.660	36.999	12.97
3rd and 5th	32.49	36.64	32.99	32.64
3rd and 6th	26.64	140.87	26.62	141.35
3rd and 7th	21.01	514.84	20.83	536.70
3rd and 8th	14.49	2310.0	14.41	2350.6
4th and 5th	31.72	43.66	32.05	40.49
4th and 6th	26.43	147.49	26.45	147.22
4th and 7th	20.95	521.86	20.77	543.81
4th and 8th	14.48	2317.2	14.40	2358.4
5th and 6th	25.68	175.50	25.70	174.80
5th and 7th	20.73	549.47	20.57	569.58
5th and 8th	14.42	2344.8	14.35	2386.6
6th and 7th	19.87	669.48	19.74	689.05
6th and 8th	14.21	2464.8	14.13	2506.8
7th and 8th	13.49	2906.3	13.44	2944.5

MSE value for 1st and 2nd bit position. Similarly, watermark was embedded according to other bit combinations. TIF, GIF, PNG, BMP files fetched better PSNR and MSE values was as compared to JPEG image file.

Further, it can be seen that when the modified LSB was applied for first two bits the quality is best and when applied to 7th and 8th bits, the quality is worst [5].

5.1 Performance Parameters Used

- **PSNR value**: This parameter defines the accuracy and quality of watermarked image. It is calculated by using the following formula. The phrase peak signal-to-noise ratio (PSNR) is most commonly used as a measure of quality of reconstruction in image compression [6].

$$\text{PSNR} = 10 * \log 10 \frac{(\text{MAX}_1^2)}{\text{MSE}}$$
$$= 20 * \log 10 \frac{(MAX_1)}{\sqrt{\text{MSE}}}$$

where MAX_1 is the highest pixel of watermarked image which is always 255.MSE is mean square error of watermarked image. So, first we calculate the MSE value.

- **MSE value**: Mean squared error (MSE) is one of the earliest tests that were performed to test if two pictures are similar. This parameter also calculates error of watermarked image using this formula [6].

$$\text{MSE} = \frac{1}{m * n} \sum_{i=0}^{m-1} \sum_{j=0}^{n-1} [I(i,j) - K(i,j)]^2$$

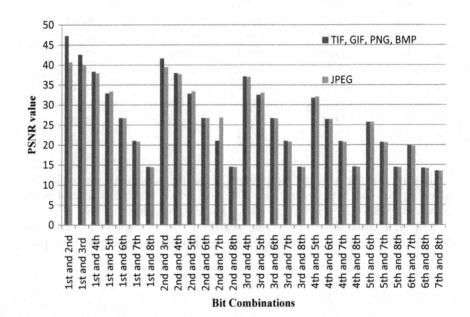

Fig. 2 Graphical analysis of PSNR value for TIF, PNG, GIF, BMP and JPEG file formats

Fig. 3 Graphical analysis of MSE value for TIF, PNG, GIF, BMP and JPEG file formats

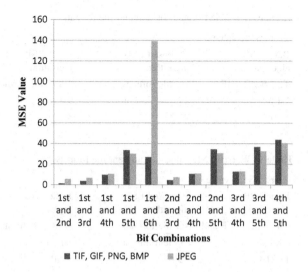

where *m* is number of rows, *n* is number of columns, *I* is original image, and *K* is reconstructed image.

Figure 2 shows the analysis of PSNR value for different image file formats. It shows TIF, BMP, GIF, PNG have same PSNR value and JPEG has low PSNR value for every two bit combinations.

Figure 3 shows the analysis of MSE value for different image file formats. It shows TIF, BMP, GIF, and PNG have same MSE value and JPEG has high of MSE value for every bit combination.

5.2 Comparison between Basic LSB and Modified LSB Algorithm

In this section, the comparison analysis of performance of basic LSB and modified LSB for best and worst cases is discussed.

Tables 6 and 7 show the comparison for best-case PSNR and worst-case PSNR, respectively. From the tables, it can be seen that the best value of PSNR for basic LSB algorithm was found when the watermark is embedded in 1st bit of every pixel of original image of TIF, PNG, GIF, and BMP files and its value could be 51.08, whereas in case of modified LSB, the best value of 47.26 could be obtained for the case when the watermark is embedded into 1st bit or 2nd bit of every pixel of the same files according to the output of XOR condition. The results were different in case of JPEG format and were as 40.25 for basic LSB and 40.675 for modified LSB which is slightly better in case of modified LSB.

Table 6 Best PSNR value for different file formats

Best PSNR value			
Basic LSB algorithm	Proposed LSB algorithm	Basic LSB algorithm	Proposed LSB algorithm
TIF, BMP, GIF and PNG file formats		JPEG file format	
1st bit	1st and 2nd bits	1st bit	1st and 2nd bits
51.08	47.26	40.25	40.675

Table 7 Worst PSNR value for different file formats

Worst PSNR value			
Basic LSB algorithm	Proposed LSB algorithm	Basic LSB algorithm	Proposed LSB algorithm
TIF, BMP, GIF and PNG file formats		JPEG file format	
8th bit	7th and 8th bits	8th bit	7th and 8th bits
11.21	13.49	11.19	13.44

Table 8 Best MSE value for different file formats

Best MSE value			
Basic LSB algorithm	Proposed LSB algorithm	Basic LSB algorithm	Proposed LSB algorithm
TIF, BMP, GIF and PNG file formats		JPEG file format	
1st bit	1st and 2nd bits	1st bit	1st and 2nd bits
0.50	1.22	6.13	5.56

Table 9 Worst MSE value for different file formats

Worst MSE value			
Basic LSB algorithm	Proposed LSB algorithm	Basic LSB algorithm	Proposed LSB algorithm
TIF, BMP, GIF and PNG file formats		JPEG file format	
8th bit	7th and 8th bits	8th bit	7th and 8th bits
4920.415	2906.3	4941.57	2944.5

For worst case, the PSNR values for TIF, PNG, GIF, and BMP files with basic LSB has been 11.21 and for modified LSB 13.49. For JPEG file, these values were 11.19 and 13.44, respectively.

Similarly, Tables 8 and 9 show the best-case and worst-case performance of basic LSB and modified LSB with respect to MSE value. It can be seen that the MSE in best case for basic LSB could be 0.5 while with modified LSB 1.22 for TIF, BMP, GIF, PNG files. For JPEG file, these values were 6.13 and 5.56, respectively. The worst-case values for TIF, BMP, GIF, PNG files were 4920.415 and 2906.3 and for JPEG file was 4941.57 and 2944.5 for basic LSB and modified LSB, respectively.

6 Conclusion

Modified LSB algorithm was implemented and used for watermarking for different file formats such as: TIF, BMP, GIF, PNG, and JPEG. The performance of this algorithm was checked through PSNR and MSE values after embedding the watermark into original image. The experimental analysis could fetch the best PSNR values of 47.26 for TIF, BMP, GIF, PNG files and 40.67 for JPEG file with modified approach. The best MSE value of 6.13 was obtained for TIF, BMP, GIF, PNG files and 5.56 for JPEG file. Whereas with basic LSB the PSNR value was 51.08 for TIF, BMP, GIF, PNG files and 40.25 for JPEG file. The best MSE with basic LSB was 0.50 for TIF, BMP, GIF, PNG files and 1.22 for JPEG file. The experimentation work suggests that the PSNR and MSE values are not good for the JPEG images as compared to other file formats. The PSNR and MSE were quite acceptable, and picture quality was also good. The future work includes further improvement in the LSB algorithm or used of other approaches to improve PSNR and MSE values.

7 Future Work

The research work that is planned for the future is as under:

- The algorithm can further be modified for better PSNR and MSE values for different image file formats.
- This method can be used to set a threshold value of PSNR and MSE on a large dataset of different images.
- The performance of the algorithm can be improved by applying pre-processing and post-processing on image.
- Filters can be used after watermarking process and relation between watermark process and noise can be obtained.

References

1. Singh, A., Jain, S., Jain A.: Digital watermarking method using of second least significant bit (LSB) with the inverse of LSB. Int. J. Emerg. Technol. Adv. Eng. (IJETAE), 3(2) (2013)
2. Bamatraf, A., Ibrahim, R., Mohd, N., Salleh, M.: A new digital watermarking algorithm using combination of least significant bit (LSB) and inverse bit. J. Comput. 3(4) (2011)
3. Bamatraf, A., Ibrahim, R., Salleh, M.N.B.M.: Digital watermarking algorithm using LSB. In: International Conference on Computer Applications and Industrial Electronics (ICCAIE), 2010, pp. 155–159, 5–8 Dec 2010
4. Ansari, I.A., Pant, M., Ahn, C.W.: Int. J. Mach: SVD based fragile watermarking scheme for tamper localization and self-recovery. Learn. and Cyber. (2015). doi:10.1007/s13042-015-0455-1

5. Bansal, N., Deolia, V.K., Bansal, A., Pathak P.: Digital image watermarking using least significant bit technique in different bit positions.In International Conference on Computational Intelligence and Communication Networks (CICN), 2014, pp. 813–818, 14–16 Nov 2014

6. Joshi, K., Yadav R., Allwadhi S.: PSNR and MSE based investigation of LSB. In: International Conference on Computational Techniques in Information and Communication Technologies (ICCTICT), New Delhi, pp. 280–285. (2016). doi:10.1109/ICCTICT.2016. 7514593

7. Garg, M., Gupta S., Khatri, P.: Fingerprint watermarking and steganography for ATM transaction using LSB-RSA and 3-DWT algorithm. In: International Conference on Communication Networks (ICCN), Gwalior, pp. 246–251. (2015). doi:10.1109/ICCN.2015.48

8. Bansal, N., Bansal, A., Deolia, V.K., Pathak, P.: Comparative analysis of LSB, DCT and DWT for digital watermarking. In: 2nd International Conference on Computing for Sustainable Global Development (INDIA Com), pp. 40–45, 11–13 Mar 2015

9. Lakhdar, L., Merouani, F.H., Smain, M.: A new binary similarity metric for two LSB Steganalysis. In: International Conference on Information Technology for Organizations Development (IT4OD), Fez, pp. 1–7. (2016). doi:10.1109/IT4OD.2016.7479313

10. Chi-Kwong Chan, Cheng, L.M.: Hiding data in images by simple LSB substitution. Int. J. Inf. Sci. Tech. (IJIST), 2012, 2(4), 469–474 (2012)

11. Joshi, K., Yadav, R.: A new LSB-S image steganography method blend with Cryptography for secret communication. In: Third International Conference on Image Information Processing (ICIIP), Waknaghat, pp. 86–90. (2015). doi:10.1109/ICIIP.2015.7414745

Megh: A Private Cloud Provisioning Various IaaS and SaaS

Tushar Bhardwaj, Mohit Kumar and S.C. Sharma

Abstract Cloud computing is a collection of heterogeneous computing resources (both hardware and software) that provide various types of services over the Internet on pay-per-use basis. Therefore, number of users and service requests are increasing day by day in cloud environment. Cloud service provider runs the user application (request) parallel so that user gets the response in minimum time, and resource utilization should be maximum. In this paper, the authors have developed a private cloud computing environment called *Megh*, using OpenNebula, that is capable of hosting various IaaS and SaaS services for the end users. For now, Megh is delivering two types of SaaS (i) *Cloud-WBAN*: A pervasive healthcare system that delivers SaaS in terms of analysis service on the sensory data collected from wireless body area networks (WBAN) and (ii) *High-performance computing* (*HPC*): It is the use of parallel processing for running advanced application programs efficiently, reliably, and quickly. *Virtual Machine Provisioning*: Virtual machine provisioning privileges control activities related to deploying and customizing virtual machines. Authors have tested Megh with a case study of high-performance computing scenario. We run several instances of an application on conventional server as well as cloud environment, and computational results show that cloud environment executes the application with minimum response and makespan time.

Keywords High-performance computing · Virtual machine · Cloud computing
Task scheduling

T. Bhardwaj (✉) · M. Kumar · S.C. Sharma
Wireless & Cloud Computing Lab, IIT Roorkee, Roorkee, India
e-mail: tushariitr1@gmail.com

© Springer Nature Singapore Pte Ltd. 2018
M. Pant et al. (eds.), *Soft Computing: Theories and Applications*,
Advances in Intelligent Systems and Computing 584,
https://doi.org/10.1007/978-981-10-5699-4_45

485

1 Introduction

1.1 Overview of Cloud Computing

Cloud computing is a model for enabling convenient, on demand network access to a shared pool of configurable computing resources (e.g., Networks, servers, storage, applications, and services) that can be rapidly provisioned and released with minimal management effort or service provider interaction [1] by NIST, USA.

1.1.1 Deployment Models

The deployment models are briefly described below:

- Public Cloud: Public cloud (off-site and remote) is owned by large organization group and is available for public use. Example: Amazon Web Service, IBM Blue cloud, Sun cloud, Google App Engine, Microsoft Azure Service platform.
- Private Cloud: Private cloud (on-site) is owned and operated for the exclusive use of a particular organization. The cloud may be either on or off premises, but in most cases, it is developed on the premise for security reasons.
- Hybrid Cloud: A hybrid cloud is the combination of public, private, or community cloud. Therefore, it can be also seen as some portion of computing resources on-site (on premise) and off-site (public cloud).

1.1.2 Service Models

The service model offers various kinds of services, which take the form of XaaS, of "<Something> as a Service." Primarily, there are three kinds of service models that have been universally accepted:

- Infrastructure as a Service (IaaS): It provides virtual machines, virtual storage, and virtual hardware infrastructure to the end user for provisioning. Examples: Amazon Web Services (AWS), Microsoft Azure, Google Compute Engine (GCE).
- Platform as a Service (PaaS): It provides the virtual machines, application services, operating system, application development framework, and control structures. Examples: OpenShift, Google App Engine, Cloud Foundry.
- Software as a Service (SaaS): It is a complete set of operating environment and applications that are used by the end users. Examples: Google Apps, Salesforce, Workday, Concur, Citrix's GoToMeeting, Cisco WebEx.

Fig. 1 Services offered by Megh

1.2 Our Contributions

In this paper, the authors have developed a private cloud computing environment called *Megh* using OpenNebula [2] that is capable of hosting various IaaS and SaaS services for the end users, Fig. 1. For now, Megh is delivering two types of SaaS (i) *Cloud-WBAN*: A pervasive healthcare system that delivers SaaS in terms of analysis service on the sensory data collected from wireless body area networks (WBAN) and (ii) *High-performance computing* (*HPC*): It is the use of parallel processing for running advanced application programs efficiently, reliably, and quickly. We run the different application parallel to different virtual machine to test our scheduling algorithm response time and makespan time. *Virtual Machine Provisioning*: Virtual machine provisioning privileges control activities related to deploying and customizing virtual machines.

The rest of the paper is structured as follows. Section 2 shows the related work about private cloud deployment using OpenNebula and task scheduling algorithms. Section 3 describes the system design and deployment of Megh cloud. Section 4 showcases the comparison between high-performance computing using conventional server and HPC. Finally, Sect. 5 includes conclusive remarks and directions of future work.

2 Related Works

There are several task scheduling algorithms that have been proposed in last decade, some of them were static and other were dynamic algorithm. Tripathy and Nayak [3] implement a deadline-based scheduling algorithm considering the parameter average resource utilization ratio and number of job accepted or rejected using the private cloud OpenNebula, and in some cases, backfilling algorithm does not work well; therefore, authors use the analytic hierarchy process (AHP) as a decision-maker in the backfilling algorithm to choose the possible best lease from the given best effort queue in order to schedule the deadline sensitive lease. Hossain [4] implement a load-balancing algorithm at OpenNebula considering the execution time as a parameter. Author allocates the job to the virtual machine and calculates the load on each virtual machine, if any virtual machine is overloaded then transfer the task to other virtual machine. Parmar and Champaneria [5] compare and analyze the performance of different private cloud like Eucalyptus, OpenStack, OpenNebula and CloudStack for different services such as scalability, hypervisor, storage, networking, authentication. The main function of OpenNebula, Eucalyptus, and Nimbus private cloud is to manage the virtual machine at the time of provisioning IaaS services. Authors compared the performance of each system [6] and discussed common challenges in these systems like fair job scheduling in the absence of preemption, priority, deadline, cost, etc., network issues, user security, internal security, etc. Samal and Mishra proposed round-robin technique considering the parameter response time and resource utilization to solve the problem of load-balancing in cloud environment [7], but algorithm did not minimize the response time and makespan time. Lakra and Yadav [8] proposed a multi-objective algorithm to reduce the turnaround time and cost. Trade-off always occurs between cost and time parameter, but authors proposed an optimal task scheduling that improved the throughput and reduced the cost without violating the service-level agreement for an application in SaaS. Proposed algorithm will not work well after considering some qualities of service parameter such as deadline, priority of task. Thomas et al. [9] proposed an improved task scheduling algorithm after analyzing the traditional algorithm that was based on user task priority and task length. In this paper, authors does not give any special importance to high-priority task when they arrive, and consider all the important factor for task scheduling to reduce makespan time, but unable to minimize makespan time for large number of task. It is easy to implement the static and dynamic algorithm in simulation environments, but difficult to implement in real cloud environment like OpenNebula, Eucalyptus, OpenStack, Nimbus, etc. In this paper, we deployed the private cloud and run the algorithm parallel to all the virtual machine and find out the results.

3 Megh: A Private Cloud

3.1 OpenNebula: IaaS Deployment

OpenNebula is a private cloud that provides the resource with the help of Haizea as a lease manager. It contains many number of characteristics which make it a unique software for cloud resource management. One of the unique functionality of

OpenNebula is to group the host in clusters and allows the segmentation of servers according to their characteristics. One more important point regarding the job scheduling in cloud environment is that it have powerful matchmaking algorithm that is useful to balance the allocation of virtual resources in a private cloud, specially, in conjunction with clusters. Cloud is turning into a suitable computing infrastructure for delivering services-oriented computing. Many open-source cloud solutions are available for the same. Open-source software platforms offer an enormous amount of adaptability without immense licensing charges. Accordingly, open-source programming is generally utilized for deploying and provisioning cloud, and private cloud is being assembled progressively in the open-source way. "*OpenNebula* is a *cloud computing* platform for managing heterogeneous distributed *data center* infrastructures. The OpenNebula platform manages a data center's virtual infrastructure to build private, public, and hybrid implementations of infrastructure as a service. OpenNebula is free and open-source software, subject to the requirements of the Apache License version 2" [3].

3.2 Experimental Setup

We have deployed Megh on 3 HP Z620 Workstation Servers, as shown in Fig. 2. It comprises of 36 cores and 3 TB of virtualized storage and currently has over 10 virtual machines commissioned for high-performance computing delivering IaaS and SaaS services. The data center's characteristics are shown in Table 1, and virtual machine's characteristics are shown in Table 2.

Fig. 2 Megh: data center laboratory

Table 1 Data center characteristics

Item	Properties
Industry standard architecture (ISA)	×86
Operating system	Linux
Virtual machine monitor (VMM)	KVM
Storage capacity	3 Terabyte
Number of CPU	3 with 12 cores per CPU
Memory capacity	48 GB

Table 2 Virtual machine characteristics

Item	Properties
Industry standard architecture (ISA)	×86
Operating system	Linux
Virtual machine monitor (VMM)	KVM
Storage capacity	10 GB
Number of CPU	1 with 1 core per CPU
Memory capacity	1 GB

4 Case Study: High-Performance Computing

User submit the job J1, J2,...Jn through the graphical user interface or Web interface with service requirement in terms of quality of service parameter such as deadline, cost, priority of task. When user submits the application to cloud, it is accepted by cloud resource controller. Cloud resource controller checks whether that request is coming from legitimate user or not; if request is coming from legitimate user, then it allocates the resource or provides the service otherwise it discards the request. Cloud resource controller contains the information about all the virtual machine and all the application (tasks) as shown in Fig. 3.

After that, cloud resource controller checks the following parameter:

- Upcoming request is priority-based or non-priority-based.
- Calculate the total number of request and requirement of resource (number of processing element, memory, bandwidth, etc.) by request for processing.
- Upcoming request has a deadline or not.
- User-required parameter (response time, makespan time, execution time, etc.)

After checking all the parameter, cloud resource controller stores the request in a queue based upon required parameter. If application is high-priority or deadline-based then cloud resource controller allocates such type of resource to the application that executes in minimum time, i.e., before the deadline. If application is based on lower priority then it allocates a simple virtual machine to execute them because time or deadline is not crucial factor in this case.

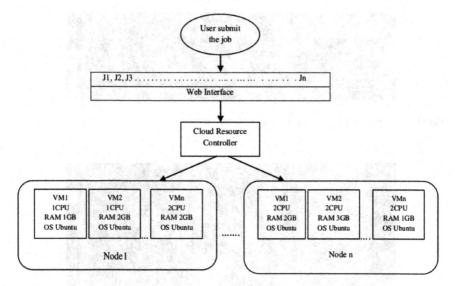

Fig. 3 Architecture of cloud

To test the functionality of Megh, authors have tested it with a case study of high-performance computing scenario. The main aim of this case study is to authenticate the credibility of using parallel processing using our private cloud Megh. The performance metric comprises of response and makespan time. Response time is the total time it takes to respond to a process request. If we ignore the transmission time, then response time is just the sum of service and waiting time. On other hand, the *makespan time* is the total time elapse from start till the end of a process execution. The above two parameters justifies the characteristics of a task to get executed in different computing environments.

For experimental purpose, we have taken a simulated application, which takes *n* seconds (depending upon its complexity and input data set) to get executed on a dedicated server. The application and scheduling algorithms are implemented using bash scripting. Firstly, we have executed several instances of that application on a machine with (4 GB RAM, 4 Cores, Ubuntu 14.04 OS) using the bash script given in Fig. 4 and calculated the response and makespan time. When such an application runs on a single machine, there will be two criteria: context switching and non-context switching. As our application is non-parallel, we opted for non-context switching mode. Moreover, context switching takes more time and leads to per-formance degradation. In this mode, the second application instance gets the CPU when the first instance gets completely executed.

Secondly, we have utilized HPC SaaS, offered by Megh and executed the same applications on four machines in parallel with (1 GB RAM, 1 Core, Ubuntu 14.04 OS) and calculated the response and makespan time. Here, we have scheduled the application instances to the virtual machines on first-come-first-serve (FCFS) basis —see the bash script given in Fig. 5. Here, the virtual machines can be identified by

```
#!/bin/bash
num=numberofapplicationInstances
for((i=1; i<=num; i++))
  do
      ssh root@172.23.1.223 'bash -s' < task$i.bash
  done
```

Fig. 4 Bash script for scheduling on conventional server

```
#!/bin/bash
num=numberofapplicationInstances
for((i=1; i<=num; i++))
  do
      ssh root@172.23.1.219 'bash -s' < task$i.bash &
      if [ $i -eq $num ]
      then
      break
      fi
      i=`expr $i + 1`
      ssh root@172.23.1.220 'bash -s' < task$i.bash &
      if [ $i -eq $num ]
      then
      break
      fi
      i=`expr $i + 1`
      ssh root@172.23.1.221 'bash -s' < task$i.bash &
      if [ $i -eq $num ]
      then
      break
      fi
      i=`expr $i + 1`
      ssh root@172.23.1.222 'bash -s' < task$i.bash &
      if [ $i -eq $num ]
      then
      break
      fi
  done
```

Fig. 5 Bash script for FCFS Scheduling on HPC

the IP addresses starting from 172.23.1.219 to 172.23.1.222. This can be scaled up and down with respect to the user demand and SLA requirements. The application instances are given by task$i.bash.

Each virtual machine has its own buffer that stores the tasks for further execution. Figure 6 shows the comparative analysis of response time between the two approaches. It can be clearly seen that using parallel approach the response time is quite improved. On the other hand, Fig. 7 showcases the variation in the makespan time of application instances running on conventional server and HPC. Using Megh's HPC an average speedup factor of 2 has been observed in makespan time.

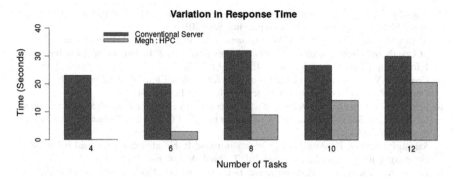

Fig. 6 Response time comparison

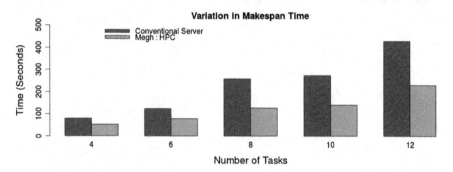

Fig. 7 Makespan time comparison

5 Conclusion

There exist various open-source cloud platforms such as OpenNebula, Eucalyptus, Nimbus, OpenStack. We develop a private cloud environment 'Megh' using the OpenNebula because it is easy to deploy the software in OpenNebula, and it is more flexible compared to all other private cloud. We test the task scheduling algorithm at Megh and compare the computational results (as shown in Figs. 6 and 7) with the algorithm running at conventional server. Our controller node distributes the task in parallel to all running virtual machines, so that all the tasks will get executed in minimum time as compared to conventional server. We will try to implement dynamic task scheduling algorithm and auto-scaling mechanism at Megh in future.

References

1. Mell, P., Grance, T.: The NIST definition of cloud computing. http://csrc.nist.gov/publications/nistpubs/800-145/SP800-145.pdf (2011)
2. OpenNebula. https://en.wikipedia.org/wiki/OpenNebula

3. Tripathy, C., Nayak C.: Deadline sensitive lease scheduling in cloud computing environment using AHP. J. King Saud Univ. Comput. Inf. Sci. (2016)
4. Moniruzzaman, A.B.M., Nafi, K.W., Hossain, S.A.: An experimental study of load balancing of OpenNebula open-source cloud computing platform. In: 2014 International Conference on Information. Electronics & Vision (ICIEV). IEEE (2014)
5. Chavan, S.V., Tilekar, S.K., Ladgaonkar, B.P.: International Journal of Advanced Research in Computer Science and Software Engineering. Int. J. 5(12) (2015)
6. Sempolinski, P., Thain, D.: A comparison and critique of eucalyptus, openNebula and nimbus. In: 2010 IEEE Second International Conference on Cloud Computing Technology and Science (Cloudcom). IEEE (2010)
7. Samal, P., Mishra, P.: Analysis of variants in round Robin algorithms for load balancing in cloud computing. Int. J. Comput. Sci. Inf. Technol. 4(3), 416–419 (2013)
8. Lakra, A., Yadav, D.: Multi-objective tasks scheduling algorithm for cloud computing throughput optimization. Procedia Comput. Sci. 48, 107–113, 2015
9. Thomas, A., et al.: Credit based scheduling algorithm in cloud computing environment. Procedia Comput. Sci. 46, 913–920 (2015)

Predicting the Calorific Value of Municipal Solid Waste of Ghaziabad City, Uttar Pradesh, India, Using Artificial Neural Network Approach

Dipti Singh, Ajay Satija and Athar Hussain

Abstract The municipal solid waste (MSW) can be used as renewable energy resource. The amount of energy generated depends upon calorific value of the MSW. The aim of present study is to discuss about the chemical composition and corresponding calorific value of 39 samples of municipal solid waste of Ghaziabad City, Uttar Pradesh (India). In this study, artificial neural network (ANN) technique has been applied to predict the calorific value of MSW of the city. The elements such as carbon, hydrogen, oxygen, nitrogen, sulfur, phosphorus, potassium, and ash obtained from chemical analysis have been used as input data points to predict the calorific value of MSW in ANN model. The developed ANN model has been validated with the help of minimum value of mean squared error [0.003703 (training), 0.03760 (validation), and 0.2269 (testing operations)] and optimized value of coefficient of correlation (0.9088) between observed and predicted calorific values of MSW samples. The proposed ANN model has shown better predictive results.

Keywords Municipal solid waste · Calorific value · Artificial neural network

D. Singh
School of Vocational Studies & Applied Sciences, Gautam Buddha University,
Gautam Budh Nagar, Uttar Pradesh, India
e-mail: diptipma@rediffmail.com

A. Satija (✉)
Department of Applied Science & Humanities, Inderprastha Engineering College,
Ghaziabad, Uttar Pradesh, India
e-mail: aajaysatija@rediffmail.com

A. Hussain
Department of Civil Engineering, Ch. B.P. Govt. Engineering College, Jaffarpur,
New Delhi, India
e-mail: athariitr@gmail.com

© Springer Nature Singapore Pte Ltd. 2018
M. Pant et al. (eds.), *Soft Computing: Theories and Applications*,
Advances in Intelligent Systems and Computing 584,
https://doi.org/10.1007/978-981-10-5699-4_46

1   Introduction

The MSW is the result of daily used items discarded by the society. The MSW has various hazardous impacts on the society. The municipal authorities can maintain the health of society by proper waste collection and disposal techniques. This is observed that landfilling is a conventional waste disposal technique in the world. But this waste disposal technique has some drawbacks. The hazardous gases like methane generally are emitted from the landfills. These gases enhance air pollution and have adverse effect on health of mankind. The Leachate formation from landfill area pollutes groundwater and hence creates water pollution. The lands are very costly in various metropolitan cities [1, 2]. The existing landfill areas are insufficient for waste disposal in these cities [3]. The bomb calorimeter is conventional tool to evaluate the calorific value of MSW. The related experimental work requires expertise attention, cost, and time. To reduce such experimental cost and time, an ANN model has been developed to estimate the calorific value of Ghaziabad City, India. The ANN models have high learning capabilities from the previous target data (actual data) and have high degree of nonlinear relationship between input and hidden nodes, hidden nodes and output nodes. But multiple linear regression (MLR) models do not have such learning capabilities [4]. In the literature, there are various MLR models that have been developed for predicting the calorific value of MSW. The physical component analysis, ultimate analysis, and proximate analysis of MSW have been performed in various case studies of various municipalities to predict the calorific value of the waste [5–11]. The solid waste components analysis has been carried out in December 2008 by an environment organization "M/s ECOPRO Environmental services" in Ghaziabad City. Our study is based on their project report. In their survey, three types of analysis (physical component analysis, ultimate analysis, and proximate analysis) of 39 samples of MSW have been performed. In the proposed ANN model, the amount of carbon (C), hydrogen (H), nitrogen (N), oxygen (O), sulfur(S), phosphorous (P), potassium (K), and ash estimated from ultimate analysis of the report has been used as input parameters to predict the calorific value of MSW. The model has been validated by estimating the optimized value of mean squared error (MSE) and coefficient of correlation R.

2   Materials and Methods

2.1   Case Study Area

The study area of our research is Ghaziabad City, Uttar Pradesh (India). Ghaziabad is the fastest growing, highly urbanized, highly industrialized metropolitan city of National Capital Region, India. Delhi is located in its west, Bulandshahr district is in southeast, Gautam Buddha Nagar is in southwest, and Baghpat district is in northwest. Hapur district is in east, and Meerut is in north. The people of the city

Fig. 1 Ghaziabad City Map

enjoy the six seasons such as spring, summer, monsoon, autumn, pre-winter, and winter. The land of the city is fertile. Hindan River passes through the city. Ghaziabad is located in Latitude 28.6691565 (28° 40' 8.96"N) North and Longitude 77.45 (77° 27' 13.53"E) East. Figure 1 illustrates about Ghaziabad City Map.

2.2 Analysis of Composition of Waste

The municipal corporation has no records regarding the composition of MSW generated in Ghaziabad. The Asian Development Bank has directed Wilbur Smith Associates to generate Master Plan for Solid Waste Management for Ghaziabad City in 2008. The concerned report was prepared in January 2009. In this report, the MSW collected from five zones (Mohan Nagar, Kabir Nagar, Vijay Nagar, Vasundhara, and City Zones) has been considered for analyzing the composition of solid waste. Overall, the 39 solid waste samples have been collected from the residential areas, commercial and market areas, slum areas, hospitals, slaughter

Variables	Minimum value	Maximum value
Input variables		
Carbon (%)	18.32	59.92
Hydrogen (%)	3.91	9.87
Oxygen (%)	8.33	55.54
Nitrogen (%)	0.87	1.68
Sulfur (%)	0.18	1.24
Phosphorus (%)	0.35	1.39
Potassium (%)	0.44	1.53
Ash (%)	2.91	41.35
Output variable		
Calorific value	0.4587 (Kcal/gm)	2.2574 (Kcal/gm)

Table 1 Analysis of chemical composition of municipal solid waste of 39 samples of Ghaziabad City, India

houses, hospitals, hotels, restaurants, etc. Table 1 illustrates the ultimate analysis (chemical composition) of various samples of MSW in which maximum as well as minimum values of C, H, O, N, S, P, potassium, ash and corresponding calorific value (Kcal/gm) have been calculated.

2.3 Artificial Neural Network

ANN approach has been successfully applied to predict the calorific value of MSW in various case studies of municipalities of various countries [1, 2, 12, 13]. Now, in the proposed artificial neural network Model, we have used elements (during ultimate analysis) C, H, O, N, S, P, K, ash (percentage on dry basis) of 39 samples of MSW as input variables and corresponding calorific value (Kcal/gm) as output (target) variable. Real-valued tan-sigmoid transfer function has been used between input and hidden layer neurons, while linear transfer function has been used between hidden and output layer neurons. The best optimum ANN structure with 10 neurons in hidden layer has been estimated by trial and error method. Levenberg–Marquardt backpropagation learning algorithm (trainlm) has been used for learning the weights and to keep minimized difference between output calorific value and predicted calorific value of MSW. The 39 samples' data sets have been subdivided into 27 training data sets (70% of 39 samples), six testing data sets (15% of 39 samples), and six validating data sets (15% of 39 samples). Each ANN model initially gets trained. The network model is then validated and finally tested for randomly selected data.

3 Result and Discussion

MATLAB 7.12.0 (2011a) software has been used for simulating the results. Various ANN models have been examined by changing the hidden number of neurons from 10 to 20 by error and trial method. On the basis of minimum value of MSE 0.003703 Kcal/gm (during training), 0.03760 Kcal/gm (during validation) and 0.2269 Kcal/gm (during testing) operations and maximum value of Coefficient of correlation R (for all) 0.9088 between target and output values, the model 8-10-1* has been found as best ANN Model. Table 2 illustrates the best ANN structure (8-10-1) with eight input variables (C, H, O, N, S, P, K, and ash), 10 hidden neurons, and one output as calorific value.

Basically, there is no algorithm available in the literature to determine the exact number of neurons in hidden layer. If we choose fewer nodes in hidden layer, then this may lead to large error in our system. Our system may become under-fit. If we choose large number of nodes in hidden layer, this may over-fit. Our training and generalization will not be good. We have decided to choose 10 hidden neurons to get rid of under-fitting as well as over-fitting. Figure 3 illustrates this. Up to epoch 3, the training error, validation error, and testing errors are nearly equal, but after epoch 4, this difference grows more, which means over-fitting increases now. The number of neurons for this network is 10. Finally the ANN model 8-10-1* has been considered as best fit model because training (learning) error 0.003703 Kcal/gm and validation (generalizing) error 0.003760 Kcal/gm are very small. In the literature, some authors suggest that the number of hidden layer neurons must lie within the limit: twice of input layer neurons. So to get rid of under–fitting, fewer number of neurons have not been chosen and experimentation has been done up to 20 hidden layer neurons (beyond the limit of twice of input layer neuron). But in each case, over-fitting has been seen due to larger difference between training error and validation error. The coefficient of correlation (between observed and predicted

Table 2 Results of training, validation and testing operations of ANN models

ANN model structure	MSE (training)	Regression R			
		Training	Validation	Testing	All
8-10-1*	**0.0037**	**0.9907**	**0.9226**	**0.8632**	**0.9088**
8-11-1	0.0281	0.9037	0.6692	0.6305	0.7444
8-12-1	0.0045	0.9848	0.6673	0.6151	0.8227
8-13-1	0.0088	0.9664	0.7909	0.5455	0.7586
8-14-1	0.04143	0.8884	0.08045	0.5595	0.7606
8-15-1	0.0142	0.9751	0.8055	0.8607	0.8792
8-16-1	0.0427	0.8616	0.7627	0.7608	0.8182
8-17-1	0.0083	0.9856	0.7501	0.8609	0.9087
8-18-1	0.0290	0.9074	0.5486	0.9519	0.6954
8-19-1	0.0225	0.9237	0.6804	0.7683	0.7909
8-20-1	0.0062	0.9705	0.9236	0.5846	0.6013

calorific values) varies in different ANN models. In 8-10-1*, MSE has been observed as smallest and R be highest. Hence, 8-10-1* has been considered as best ANN structure.

Figure 2 illustrates the values of coefficient of correlation during training (0.9907), validation (0.9226), and testing (0.8632) and for all time steps (0.9088) between target calorific value and output calorific value from the proposed ANN model.

Figure 3 represents the variations of the mean squared errors of training, validation, and test time steps. The best performance regarding MSE during training process is 0.003703 on iteration 4. MSE of validation and testing operation is 0.03760 and 0.2269, respectively. MSE of training operation during 2 and 3 epochs is decreasing continuously. After epoch 4 MSE of training operation drastically gets decreased. On the other hand, during validation and testing operations MSE tends to increase.

Figure 4 illustrates the comparison between target data (actual calorific values) and output data (predicted calorific values). This is observed that both values are very close. This shows that the results from proposed ANN model are consistent. The ANN models have good learning capability from previous results, and in present study, better relationships have been established between eight input waste constituents elements (C, H, O, N, S, P, K, and ash) and corresponding calorific

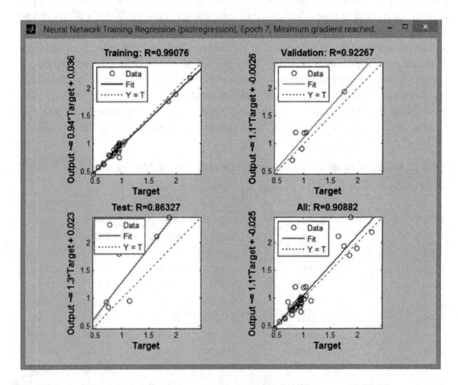

Fig. 2 Regression analysis between target and output data (ANN structure 8-10-1*)

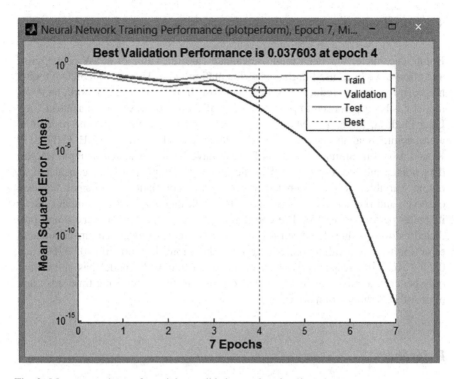

Fig. 3 Mean squared error for training, validation, and testing time steps

Fig. 4 Comparison between actual and predicted calorific values of waste samples

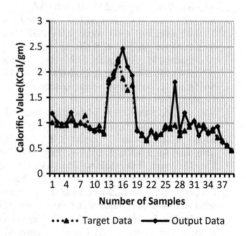

value (Kcal/gm). When artificial neural network (ANN) and linear regression modeling approaches are compared, then multiple regression approach is observed much transparent than ANN model [14]. In the literature, various regression equations like Dulong model, Steuer model have been successfully used to interpret the calorific values.

4 Conclusions

Bomb calorimeter has been used to evaluate calorific value of MSW. Time, cost, and effort have been used during experimentation work. Hence, a predictive model has been required to reduce such time and cost. In present research work, an ANN model has been proposed to predict the calorific value of MSW of Ghaziabad City, Uttar Pradesh, India. The MSW samples have been collected from the residential areas, commercial and market areas, slum areas, etc. The elements C, H, O, N, S, P, K, and ash from ultimate analysis of MSW have been considered as input (target) data points, and corresponding calorific value (Kcal/gm) has been considered as output variable. If the estimated value of the correlation coefficient between observed and predicted values is greater than 0.9, then the predictive model shows its better performance [15]. The correlation coefficient between observed and predicted calorific values has been found as 0.9088 in our model. The mean squared error for training, validation, and testing operations is 0.003703, 0.03760, and 0.2269 Kcal/gm, respectively. Hence, the proposed ANN model has shown not only better predictive results but will also prove useful for reducing time, cost, and effort of experimentation work.

References

1. Dong, C., Jin, B., Li, D.: Predicting the heating value of MSW with a feed forward neural network. Waste Mgmt. **23**, 103–106 (2003)
2. Shu, H.Y., Lu, H.C., Fan, H.J., Chang, M.C., Chen, J.C.: Prediction for energy content of Taiwan municipal solid waste using multilayer perception neural networks. J. Air Waste Mgmt. Assoc. **56**, 852–858 (2006)
3. Abu-Qudais, M., Abu-Qdais, H.A.: Energy content of municipal solid waste in Jordan and its potential utilization. Energy Convers. Mgmt. **41**, 983–991 (2000)
4. Swingler, K.: Applying neural networks: a practical guide, pp. 21–39. Academic Press, London, UK (1996)
5. Wilson, D.L.: Prediction of heat of combustion of solid wastes from ultimate analysis. Environ. Sci. Tech. **6**, 1119–1121 (1972)
6. Khan, A.M.Z., Abu Ghararah, Z.H.: New approach for estimating energy content of municipal solid waste. J. Environ. Eng. **117**, 376–380 (1991)
7. Liu, J.I., Paod, R.D., Holsen, T.M.: Modeling the energy content of municipal solid waste using multiple regression analyses. J. Air Waste Mgmt. Assoc. **46**, 650–656 (1996)
8. Cooper, C.D., Kim, B., MacDonald, J.: Estimating the lower heating values of hazardous and solid wastes. J. Air Waste Mgmt. Assoc. **49**, 471–476 (1999)
9. Abu-Qudais, M., Abu-Qdais, H.A.: Energy content of municipal solid waste in Jordan and its potential utilization. Energy Convers. Mgmt. **41**, 983–991 (2000)
10. Chang, Y.F., Lin, C.J., Chyan, J.M., Chen, I.M., Chang, J.E.: Multiple regression models for the lower heating value of municipal solid waste in Taiwan. J. Environ. Mgmt. **85**, 891–899 (2007)
11. Meraz, L., Domínguez, A., Kornhauser, I., Rojas, F.: A thermochemical concept-based equation to estimate waste combustion enthalpy from elemental composition. Fuel **82**, 1499–1507 (2003)

12. Ogwueleka, TCh., Ogwueleka, F.N.: Modeling energy content of municipal solid waste using artificial neural network. Iran. J. Environ. Health. Sci. Eng. **7**(3), 259–266 (2010)
13. Akkaya, E., Demir, A.: Predicting the heating value of municipal solid waste-based materials: an artificial neural network model. Energy Sour. **32**, 1777–1783 (2010)
14. Jahandideh, S., Jahandideh, S., Asadabadi, E.B., Askarian, M., Movahedi, M.M., Hosseini, S., Jahandideh, M.: The use of artificial neural networks and multiple linear regression to predict rate of medical waste generation. Waste Manage. **29**, 2874–2879 (2009)
15. Coulibaly, P., Baldin, C.K.: Non-stationary hydrological time series forecasting using non-linear dynamic methods. J. Hydrol. **307**, 164–174 (2005)

Advertisement Scheduling Models in Television Media: A Review

Meenu Singh, Millie Pant, Arshia Kaul and P.C. Jha

Abstract For promoting products in a highly competitive market, it is imperative for firms to develop appropriate promotions. Advertising is one of the major components of promotional mix which contributes in communicating the distinct product features to customers. Multiple media, namely television, radio, mobile phones, newspaper, magazines, Web sites, are available to the firms for advertising. Although over the years there have been many new media that have been included in the media mix of firms, yet television has remained important to advertisers. This importance is owing to its features such as large reach to the audience as well as effective communication with the audience. This paper presents a review of literature in the field of scheduling of advertisements in television.

Keywords Advertising · Television · Optimization · Recent advances

1 Introduction

Television (TV) as a medium has always been an important part of the media mix of firms and over the years has maintained its prominent position. For instance in India, as per the report by IBEF, TV is the largest and fastest growing media segment from those available in India and is expected to grow at compound annual

M. Singh (✉) · M. Pant
Department of Applied Science and Engineering, Indian Institute
of Technology Roorkee, Saharanpur Campus, Roorkee, India
e-mail: msingh.dase.iitr@gmail.com

M. Pant
e-mail: millifpt@iitr.ernet.in

A. Kaul · P.C. Jha
Department of Operational Research, University of Delhi, New Delhi, India
e-mail: arshia.kaul@gmail.com

P.C. Jha
e-mail: jhapc@yahoo.com

© Springer Nature Singapore Pte Ltd. 2018
M. Pant et al. (eds.), *Soft Computing: Theories and Applications*,
Advances in Intelligent Systems and Computing 584,
https://doi.org/10.1007/978-981-10-5699-4_47

growth rate (CAGR) of 15.5% and reach US\$ 15.2 billion in 2019.[1] The audio-visual nature of this medium makes it an effective medium for reaching a large audience. Promotion of different products is carried out by placement of advertisements in various forms. The aim of the paper is to put forward the various studies that have considered the advertising through television medium by considering optimization of the objective of the TV network.

While considering advertising through the television medium, it must be noted that there are two major participants, namely television network (or channels/stations in some problems) and advertisers. The television network is a telecommunications network that consists of groups of TV stations aligned to distribute unified television programming content across the country. It is termed as an entertainment programmer that provides various TV channels via a central operation or geographically distributed television station. The television station is a single independent programmer that is equipped to broadcast the affiliated programming received from the network to the local market through the airwaves using broadcasting licensed issued by Government [1, 2]. On the other hand, an advertiser can be a person, company, or an intermediary such as advertising agency, whose product or service is placed in the advertisement in order to provide information to the target the audience. An advertising agency can be internal department or an internal agency that handles the advertising on behalf of their clients and imparts their effort in promoting and selling of their client's product or services [1, 2]. Advertising is carried out in different forms based on the interactions between the two participants. The advertisers can either sponsor the programs or they could place advertisements within programs or in between programs [1, 2].

In this paper, we consider placement of advertisements within the programs. The time periods of commercial videos located within the TV are called "advertising breaks." An advertisement to be placed in a break is usually between 10 and 60 s.

1.1 Phases of Advertising in Television

While considering advertising from the perspective of the television, essentially the process of advertising in television can be divided into four main phases: (1) selling of slots, (2) selection of advertiser, (3) optimal scheduling of slots, and (4) evaluating achievement of objectives [1, 2].

1.1.1 Selling of Slots

The television managers plan the television scheduling by considering the basic current year schedules. After the completion of scheduling, the TV network makes

---

[1]http://www.ibef.org/industry/media-entertainment-india.aspx.

the announcement of the programs; schedule for the next broadcasting year, immediately forecasts the ratings, and estimates the market demand for setting the rate card of advertising breaks [3]. The TV network divides these breaks into day parts (slots) and conducts an auction for selling these slots so as to choose that advertiser that yields maximum revenue. It is not possible to choose all the advertisers for placement of advertisements as there is a fixed number of advertising breaks. These breaks are brought to the advertiser mainly in two ways: up-front market and scatter market.

The Up-front Market: The TV network sells its advertising time at a particular rate during a brief period after announcing the programs' schedule to advertisers. This period of selling is known as the up-front market. It starts with the identification and planning of the television programs scheduled by the TV managers as presented in the [3].

The Scatter Market: During the up-front market, only 60–80% of slots are sold and the remaining slots are sold throughout the broadcast season, which is called scatter market. The inventories saved are used to attract the additional customers who are ready to pay higher price per unit time than the up-front market. During this period, no penalties are to be paid by the advertisers for canceling orders at the last minute. The overall process summary is shown in [3].

1.1.2 Selection of Advertiser

The advertisers raise bids to purchase the time for advertising from the TV network. A typical advertiser bid comprises of the package cost, the target group, the program mix, weights for preferences, a negotiated cost per 1,000 per viewers, etc. The TV network decides and accepts those bids that maximize its revenue. As presented in [4], the TV network lists the requests of advertisers which have been accepted in order of their priority and develops a sales plan that consists of the complete schedule of advertising break times. If the sales plan meets the requirements of both the advertiser and the network, a sales contract is signed between the two parties. The contract includes a list of commercials to be aired according to the show and as per the date of airing for a prescribed planning period, although the exact positioning of these advertisements breaks is decided closer to the date of airing of the show.

1.1.3 Optimal Scheduling

The optimal scheduling of these purchased advertising breaks is carried out in consultation with both parties, keeping in mind different objectives.

By Advertiser: An advertiser has certain objectives in mind while considering advertising for the product(s) in advertisement breaks of television. The advertiser decides a certain amount they are willing to pay for space in the advertising breaks. This forms the basis of negotiation of the advertiser with the television network. The aim of the advertiser is two-pronged wherein they would want to maximize the reach of advertising for their product and minimize the cost of advertising. These objectives could be considered one at a time or could be considered simultaneously as in the case of the multi-objective model [1].

By TV Network: During the broadcast season, each advertiser sends their orders to the TV network with some general rules. The TV network decides the selection of the advertisers' order and then schedules their advertisements in the advertisement breaks that best satisfy the advertiser's request and also increase the revenues and productivity of TV network [5, 6]. The optimal arrangement of advertiser's advertisement is one of the most popular areas for the researchers, and limited research has been carried out in this field.

1.1.4 Evaluating the Achievements of Objectives

The objective of the TV network is to maximize the revenue and/or minimize the penalties incurred for not meeting the requirements of the advertisers. The main purpose of optimal scheduling advertisements is to avoid the penalty incurred by the TV network. The penalty could be incurred due to the reasons, namely not reaching the specified audience rating points in the entire advertising campaign, irregularity in multiple airing of the advertisements, or not meeting the position goals. Advertisers check for their requirements on a weekly basis and puts a penalty is on the TV network for not meeting any of their requirements.

As per certain rules, advertisers are also allowed to cancel orders for advertising in the advertising breaks. The cancellation can be done up to 48 h before the airing of the program. Depending on the rules, the TV network may or may not charge a cancellation fee for canceling the order and may also refund the entire amount or part which was going to be spent by the advertiser.

Moreover, there are several changes that occur before the starting of the broadcast week such as show format alterations, request of advertisers for revision in the schedule of the commercials, product conflicts owing to some cancellation of orders by advertisers. There are many improved processes that have provided the benefits and have resolved the infeasibility occurring in the problem by taking care of the conflicts and shifting certain advertisements to other TV shows in consultation with advertisers.

The paper is presented in four sections. Section 2 describes the methodology used in reviewing the research in the field of scheduling of advertisements in television. Section 3 describes the review of extant literature. Further, Sect. 4 concludes the paper.

2 Methodology

This is a short review paper based on some of the literature in the field of scheduling of advertisements in breaks of programs aired on television channels. Initially, we provide an overview of the variety of research that has been carried out in this field. Further, we also concentrate on certain specific models in the literature, which could be broadly classified as follows: (i) minimum space requirements models, (ii) penalty minimization models, (iii) multi-objective optimization models, and (iv) revenue maximization models. Through this review, we try to bring forward the research that is carried out in this field as well as develop an understanding of the future scope in the field.

3   Review of Existing Literature

While considering the research in scheduling, there are some researchers which have considered the "lead-in" strategy. In this strategy, it is considered that preceding programs boost the ratings of subsequent programs (existing or new) [7–12]. These papers mainly focused on the scheduling of programs (not on commercials) on the television from the perspective of the television network such that the scheduling objectives are met. In [13], the author provided an explicit two-stage model with multiple, half-hour program segments to develop program schedule. A linear regression model was used to predict the ratings for prime-time slot, and a mathematical programming model is proposed to determine an effective program schedule such that the audience ratings are maximized. [14–16] developed a heuristic algorithm for scheduling of programs in a television network for improving the audience viewership as well as audience share. [17] developed an optimization model to schedule the television programs with the objective to maximize the weekly audience rating by taking several assumptions regarding the competitive environments. [18] proposed an analytical approach for optimizing the television scheduling during the prime time.

The strategies for scheduling the slots on TV have also been studied by [19, 20]. In these papers, the objective was to maximize the advertising effectiveness. [21] discussed the various difficulties in manual scheduling the slots in the TV channel "Thames Television." Furthermore, an algorithm was developed for scheduling such that a systematic exchange of slots (different lengths) between breaks was possible and additional space for the others slots was created. [22] presented a simulation study concerned with the impact of advertising schedule on the slots. [23] also proved that the problem defined by the [21] is NP-complete. [3, 22, 24] developed a mathematical programming model and proposed a heuristic algorithm for generating optimal sales plan that satisfies the requirements of advertisers. A sales plan includes the complete schedule details of the commercials by show and air dates during the broadcast year. The algorithm in [3] is currently implemented by the NBC television network for improving its revenue. [25, 26] discussed the scheduling of multiple commercials such that schedule developed must allow for appropriate spacing of multiple airings of the same advertisements. Further, [27] considered the differential weighting of conflicts between the pairs of commercials. [4] found the best combination of the advertisements within a specified budget limit in order to gain the maximum rating. They proposed a hybrid genetic algorithm for generating optimal advertising campaigns for multiple brands. [28] have studied advertising scheduling problem faced by the Portuguese TV station with the objective of maximizing the audience ratings. [29] focused on the scheduling of advertisements by considering the combinatorial auction. In this, a mathematical programming model is developed to maximize the revenue of television network by placement of advertisements. A genetic algorithm methodology was also proposed.

3.1 Minimum Space Requirements Models

This section discusses mathematical models related to minimum space requirements between multiple airings of advertisements. Each advertiser wants that the multiple airing of the same commercials to be as much evenly spaced as possible. [25, 26, 30] have discussed optimization models in this area. The paper determines the schedule of advertisements of the advertisers, and minimize the difference of the actual distance and the ideal distance between the two-successive airing of the same advertisements. In [25], mixed integer programming (MIP) model is formulated and is solved with several heuristics and algorithms. In [26], a loss function is considered and simulated annealing heuristic is developed for the larger problems to obtain a solution. In [30], two mixed integer linear programming (MILP) models are presented as a new network flow-based model. Local search procedures and simulated annealing (SA) approaches are presented for solving the medium and large problems.

3.2 Penalty Minimization Models

The television networks promise advertisers that they would try and meet their preferences such as that of location preferences (first or last position of the advertisement breaks) as these are likely to have higher audiences, avoiding product conflicts by placing their advertisements close to the advertisements of their competitors, etc. Although it is not always possible to meet all the preferences of the advertisers, the television channel tries to meet the preferences as much as possible. Advertisers set a certain percentage of the audience that they would want to reach; if the television network is able to meet that requirement, the advertisers are satisfied to a greater extent. The schedule for placement of advertisements is made such that the penalty incurred for not meeting the requirements of the advertisers is minimizes. In [31], the advertisement arrangement problem is presented as a commercial juggling problem and an algorithm is designed for automatically scheduling the commercials that minimizes the penalty cost of the television network while incorporating the costs of sales personnel, meeting the promised positions considering product conflict [32]. Similar problem is solved in [32] by adopting ant colony optimization (ACO) heuristic.

3.3 Multi-objective Optimization Models

In this section, we consider models with multiple objectives while scheduling of advertisements in television medium. Here, the multi-objective nature of the models could either be considered from the perspective of the television network or from that of the advertiser. In [33], a multi objective approach for the television scheduling is discussed from the perspective of the television network. An integer mathematical programming model is formulated with two conflicting objectives of maximization of audience and minimization of cost. The values of ratings and costs are first forecasted by the regression model and are then used further in the optimization model as inputs. The ε-constraint method and the Pareto frontiers are applied to obtain a suitable trade-off among the two conflicting objectives. [34] considered a multi-objective model for an NP-hard combinatorial optimization problem with numerous constraints. A modified multi-objective genetic algorithm (MOGA) was developed to optimally generate a TV advertising campaign for multiple brands. [35] considered a multi-objective model from the perspective of the advertiser. The model considers conflicting objectives of maximization of reach and minimization of cost of advertising for the product of an advertiser. The advertising of the product is carried out by placing advertisements in advertising breaks (slots) of the multiple programs of different channels over a planning horizon. The model is solved under constraints of diversification over channels, limits on frequency of advertisements, and product conflict constraints. A goal programming approach is proposed to achieve a trade-off between the two conflicting objectives.

3.4 Revenue Maximization Models

The TV network generates major part of its revenue by selling its advertising time in breaks of various programs to advertisers at a particular rate [1]. While maximizing the revenue, the television network must consider increasing the audience ratings and at the same time meeting advertiser preferences. [5] formulated the integer programming model in which the TV station accepts some advertisements and then schedules these advertisements with the objective of maximizing the revenue. [6] developed an interdependent problem and five heuristics to solve the problem of accepting orders and scheduling of accepted advertisements at the same time. [36] proposed an integer programming model for allocating personal TV advertisements to viewers for the multiple periods considering the advertisements and viewers as the main entities of the problem's input. Two different versions of this problem have been adapted to handle the uncertainty of viewer's viewing capacity: (i) deterministic single period, wherein the viewer's viewing capacity is known in advance and (ii) stochastic multi-period, wherein the uncertainty in the viewing capacity is considered. A sequential solution procedure is presented and used it to develop several heuristic algorithms to solve both the versions of the problem.

4   Conclusion and Future Scope

The following are the observations from the extant literature reviewed:

1. The scheduling of advertisements in television can be considered from different perspectives, namely (i) television network and (ii) advertiser.
2. The optimization models considered for scheduling of advertisements are either single-objective or multi-objective.
3. While considering the model from the perspective of the television network, the objectives could be to maximize the audience ratings, minimize the cost, or obtain a trade-off between the two.
4. On the other hand, when the model is considered from the perspective of the advertiser, the objective is to maximize the reach and minimize the cost of advertising.
5. There are a large number of decision-making tools and methods for automatic optimal scheduling of advertisements in the advertisement breaks.
6. There has been more research that has been carried out from the perspective of the television network as compared to that from the perspective of the advertiser.
7. Apart from the mathematical programming models, several heuristics and algorithms have been developed and tested to solve the large real-world problems individually and sequentially.
8. Some papers have applied generic population-based meta-heuristic optimization algorithms for obtaining a good efficient solution for an optimization problem.
9. Allocation of personal TV advertisements has also been considered by researchers. Work in this direction could be considered in the future as limited research has been carried out till now.
10. The research discussed in this review paper is a base for future research on dynamic and stochastic approaches to these problems.

References

1. Belch, G., Belch, M., Purani, K.: Advertising and Promotion: An Integrated Marketing Communications Perspective. Tata McGraw Hill, India (2010)
2. Sissors, J., Baron, R.B.: Advertising Media Planning. McGraw Hill Education Pvt. Ltd., India (2010)
3. Bollapragada, S., Cheng, H., Phillips, M., Garbiras, M., Scholes, M., Gibbs, T., Humphreville, M.: NBC's optimization systems increase revenues and productivity. Interfaces 32(1), 47–60 (2002)
4. Pashkevich, M.A., Kharin, Y.S.: Statistical forecasting and optimization of television advertising efficiency. In: Proceedings of the 3rd International Conference Modeling Simulation, pp. 25–27 (2001)

5. Bai, R., Xie, J.: Heuristic algorithms for simultaneously accepting and scheduling advertisements on broadcast television. J. Inf. Comput. Sci. 1(4), 245–251 (2006)
6. Kimms, A., Muller-Bungart, M.: Revenue management for broadcasting commercials: the channel's problem of selecting and scheduling the advertisements to be aired. Int. J. Revenue Manag. 1(1), 28–44 (2007)
7. Goodhardt, G.J., Ehrenberg, A.S.C., Collins, M.A.: The Television Audience: Patterns of Viewing. Saxon House, UK (1975)
8. Goodhardt, G.J., Ehrenberg, A.S., Collins, M.A.: The Television Audience: Patterns of Viewing. An update. Gower Publishing Co. Ltd., UK (1987)
9. Headen, R.S., Klompmaker, J.E., Rust, R.T.: The duplication of viewing law and television media schedule evaluation. J. Mark. Res. 16(3), 333–340 (1979)
10. Henry, M.D., Rinne, H.J.: Predicting program shares in new time slots. J. Advert. Res. 24(2), 9–17 (1984)
11. Tiedge, J.T., Ksobiech, K.J.: Counterprogramming primetime network television. J. Broadcast. Electron. Media 31(1), 41–55 (1987)
12. Webster, J.G.: Program audience duplication: a study of television inheritance effects. J. Broadcast. Electron. Media 29(2), 121–133 (1985)
13. Horen, J.H.: Scheduling of network television programs. Manag. Sci. 26(4), 354–370 (1980)
14. Rust, R.T.: Advertising Media Models: A Practical Approach. Free Press, USA (1986)
15. Rust, R.T., Alpert, M.I.: An audience flow model of television viewing choice. Mark. Sci. 3 (2), 113–124 (1984)
16. Rust, R.T., Eechambadi, N.V.: Scheduling network television programs: a heuristic audience flow approach to maximizing audience share. J. Advert. 18(2), 11–18 (1989)
17. Kelton, C.M., Stone, L.G.S.: Optimal television schedules in alternative competitive environments. Eur. J. Oper. Res. 104(3), 451–473 (1998)
18. Reddy, S.K., Aronson, J.E., Stam, A.: SPOT: scheduling programs optimally for television. Manag. Sci. 44(1), 83–102 (1998)
19. Simon, H.: ADPULS: an advertising model with wearout and pulsation. J. Mark. Res. 19(3), 352–363 (1982)
20. Mahajan, V., Muller, E.: Advertising pulsing policies for generating awareness for new products. Mark. Sci. 5(2), 89–106 (1986)
21. Brown, A.R.: Selling television time: an optimisation problem. Comput. J. 12(3), 201–207 (1969)
22. Balachandra, R.: A simulation model for predicting the effect of advertisement schedules. In: Proceedings of the 9th conference on Winter simulation, vol. 2, pp. 580–589. Jan 1. Winter Simulation Conference (1977)
23. Hägele, K., Dunlaing, C.O., Riis, S.: The complexity of scheduling TV commercials. Electron. Notes Theor. Comput. Sci. 40, 162–185 (2001)
24. Bollapragada, S., Ghattas, O., Hooker, J.N.: Optimal design of truss structures by logic-based branch and cut. Oper. Res. 49(1), 42–51 (2001)
25. Bollapragada, S., Bussieck, M.R., Mallik, S.: Scheduling commercial videotapes in broadcast television. Oper. Res. 52(5), 679–689 (2004)
26. Brusco, M.J.: Scheduling advertising slots for television. J. Oper. Soc. 59(10), 1363–1372 (2008)
27. Gaur, D.R., Krishnamurti, R., Kohli, R.: Conflict resolution in the scheduling of television commercials. Oper. Res. 57(5), 1098–1105 (2009)
28. Pereira, P.A., Fontes, F.A., Fontes, D.B.: A decision support system for planning promotion time slots. In: Operations Research Proceedings 2007. Springer, Berlin, pp. 147–152 (2008)
29. Ghassemi Tari, F., Alaei, R.: Scheduling TV commercials using genetic algorithms. Int. J. Prod. Res. 51(16), 4921–4929 (2013)
30. García-Villoria, A., Salhi, S.: Scheduling commercial advertisements for television. Int. J. Prod. Res. 53(4), 1198–1215 (2015)
31. Bollapragada, S., Garbiras, M.: Scheduling commercials on broadcast television. Oper. Res. 52(3), 337–34 (2004)

32. Wuang, M.S., Yang, C.L., Huang, R.H., Chuang, S.P.: Scheduling of television commercials. In: IEEE International Conference on Industrial Engineering and Engineering Management (IEEM), Dec 7. IEEE, pp. 803–807 (2010)
33. Lupo, T.: Non-dominated "trade-off" solutions in television scheduling optimization. Int. Trans. Oper. Res. **22**(3), 563–584 (2015)
34. Fleming, P.J., Pashkevich, M.A.: Optimal advertising campaign generation for multiple brands using MOGA. IEEE Trans. Syst. Man Cybern. C (Appl Rev) **37**(6), 1190–1201 (2007)
35. Kaul, A., Aggarwal, S., Jha, P.C., Gupta, A.: Optimal advertisement allocation for product promotion on television channels. In: Sinha, A.K., Rajesh, R., Ranjan, P., Singh, R.P. (eds.) Recent Advances in Mathematics, Statistics and Computer Science, pp. 173–184. World Scientific, Singapore (2016)
36. Adany, R., Kraus, S., Ordonez, F.: Allocation algorithms for personal TV advertisements. Multimed. Syst. **19**(2), 79–93 (2013)

Preliminary Study of E-commerce Adoption in Indian Handicraft SME: A Case Study

Rohit Yadav and Tripti Mahara

Abstract Small and medium enterprises (SME) play a vital role in Indian economy. It is crucial for a SME to innovate if it wants to thrive in the competitive environment. Adopting e-commerce as a sales channel is one of the ways to increase customer base for an organization. This paper investigates the reasons why small and medium size enterprises employed in handicraft sector move from traditional commerce to e-commerce through case study based approach. The research methodology involved interviews with owner and employee/s. The findings were assessed using technology-organization and environment (TOE) framework. The results reveal that the owner had a crucial and important role in e-commerce adoption by a SME. It is the willingness of the owner to innovate that opens the new sales channel for a SME. The importance of good infrastructure becomes a driver for e-commerce adoption.

Keywords E-commerce · Small and medium enterprise · Handicraft
Innovation · Technology-organization and environment

1 Introduction

Small and medium enterprises (SME) are backbone of nation's economy, particularly in developing countries. They constitute the bulk of the industrial base and also contribute significantly to their exports. India has nearly three million SMEs, which account for almost 50% of industrial output and 42% of India's total exports [1].

R. Yadav (✉)
Department of Applied Science and Engineering, Indian Institute of Technology Roorkee,
Roorkee, India
e-mail: rohit13885@gmail.com

T. Mahara
Department of Polymer & Process Engineering, Indian Institute of Technology Roorkee,
Roorkee, India
e-mail: triptimahara@gmail.com

© Springer Nature Singapore Pte Ltd. 2018
M. Pant et al. (eds.), *Soft Computing: Theories and Applications*,
Advances in Intelligent Systems and Computing 584,
https://doi.org/10.1007/978-981-10-5699-4_48

With the advent of planned economy from 1951 and subsequent industrial policy by the government, SMEs started playing a vital role in Indian economy through contributions in terms of employment, export earnings, and innovation [2].

Innovation is defined as the implementation of a new significantly improved product, service or process, a new marketing method or a new organizational method in business practices, workplace organization, or external relations [3]. It consists of changes in the business model which means the way the product or the service is offered to the market. There are two prominent approaches to innovation. First, firm's internal capabilities is a primary driver for innovation and in the second approach the effect of demanding environment becomes the primary driver for innovation. With globalization, SMEs are facing a fierce competition to remain competitive in market and find new sales channel. Thus, the effect of demanding environment is the leading driver for innovation. Adoption of e-commerce can be regarded as one form of innovation [4] to meet the rising need of alternative sales channel.

Benefits of e-commerce for SMEs have been studied by various authors and agencies due to its potential and effects on economy [5, 6]. As per [7] few benefits of engagement of e-commerce for firms and individuals include:

- Geographical Limits: No limits for either party, i.e., customer or seller as both can do business from anywhere anytime through global communications.
- Speed: Interactions and transactions are fast and reliable and occur within a short span.
- Productivity: As Internet provides fast communications and transactions speed, e-commerce participants saves time and labor that can be devoted to different activities of interest, i.e., ability to pursue higher output than anticipated.
- Share Information: Easy and quick transfer of different types of information like textual, audio, video, graphical or animation to networked/connected users.
- Innovation in features: Potential to introduce new features of product and service and test them is added benefit for users of e-commerce.
- Lower Cost: Online transactions prove cheaper than customary or traditional ways. Cost is also reduced as most businesses' work on the platform to remove middlemen and gain profits directly from customers.
- Competitive Advantage: As argued in most studies that e-commerce offers distinctive advantage to firms over others as competitors cannot provide same services or product and operations capabilities.

In reality, utilization and adoption of e-commerce are mostly practiced in developed countries and large companies [8, 9]. Authors have confirmed with results, large firms easily gains from e-commerce adoption by disposal of resources at their usage [10, 11]. SME in developing nations lacks various resources with lack in wisdom on business and political environment of their countries gives little support to increase e-commerce adoption [5]. Few studies have shown the effect of e-commerce adoption on SMEs in developing countries [12, 13], but not much work has been done in Indian context. Handicraft is one of the important segments where many SMEs are employed. The artisans in this segment need to find new markets for their products.

Hence, the aim of this research is to investigate the factors that play a vital role in e-commerce adoption of handicraft segment. Case study approach is used in a SME engaged in manufacturing wooden handicraft items in Saharanpur (SRE), India. SRE, an important industrial city and municipality lies in the state of Uttar Pradesh in northern India (29°58 N 77°33 E/29°97, 77°55) and is internationally famous for its wooden handicraft industry. There are approximately 3500 unregistered and 1500 registered SMEs that work in this industry and provide employment to 150,000 artisans [14]. A variety of other agro-based industrial enterprises such as textile, sugar, paper, and cigarette factories are also located in the city. The export revenue share of the wooden-based handicraft is close to 40% of the total handicraft industry (US$2 billion) in India [15].

2 Research Methodology

Case based approach is selected to gain in-depth understanding of various factors affecting e-commerce adoption in wooden handicraft SME of SRE. This approach has been used by various researchers as an exploration tool [16–18]. Data were gathered through interviews with few representatives from companies which were SME owner and the IT employee. The interviewers were selected by an approach best described as purposive sampling. As per [19], the aim of purposive sampling is to strategically select interviewers that are relevant to the research topics. This was done in order to get understanding within an industry.

TOE is one of the widely used organizational level frameworks [20] to categorize the various factors of e-commerce adoption. TOE is principled on three technology innovation dimensions: technology, organization, and external environment. Organizational context of TOE includes formal and informal systems, communication processes, size, and slack as basic elements. Environmental context of TOE includes structure of market and its characteristics, technology–support infrastructure and regulations by government; finally, technology context includes availability of technology and its characteristic [20]. TOE framework under SME environment is extended to comprise of the CEO of organization, i.e., principal decision-maker [21]. Various studies have used TOE framework to assess adoption decisions on numerous technologies which are shown in Table 1.

Table 1 Various studies conducted through TOE framework

Technology	References
E-business	[23–27]
Cloud computing	[25, 28]
E-commerce	[29]
Electronic data interchange (EDI)	[30]

3 Case Study

3.1 Business Model

AB corp.(alias) was established in the year 1992 by the founder and is currently managed by his son (owner). It is a small shop that sells variety of wooden handicraft gift and furniture items. AB corp. owns a karkhana (manufacturing unit) where the wooden handicraft carving items are manufactured. The owner has employed few artisans and workers on daily wage and contract basis for production. Owner manages the karkhana from the shop where he sits. He has three full-time employees working under him in his shop to manage the customers. One of the employees is an IT employee who handles their online business. It includes posting new items on various Websites and tracking the orders placed.

Retail customers and distributors were the sales channel in 1992 when the organization began its operations. The distributors were located in many places in North India. The owner joined the organization in year 2000. In 2004, owner decided to expand their customer base by selling the products online. Hence, AB corp. became a seller on Baazi.com (now Ebay.in) to sell wooden handicraft gift and furniture items online. Low adoption of e-commerce by customers due to low quality and high prices of Internet services was a major hurdle for the seller. Also, initially there were very few delivery options available in Saharanpur due to which it was difficult to fulfill the orders on time. Indian post lacked reliability and the private courier were very costly leading to delayed fulfillment of orders.

Most of the online business diminished in great depression of economy from 2009 to 2011. Once, the economy started blooming, in 2012 AB corp. began its online business again on Ebay. As the understanding of the owner grew, he had tied up with new courier companies to maintain on time delivery of products. Due to good quality of products and on time delivery of goods, he enjoyed all the special perks and provisions of a powerful brand on EBay. He also set up his own static Website to disburse information to clients.

In the meanwhile, a major initiative was taken by the owner by shedding production of heavy wooden handicraft items, especially furniture from the shelf. The main reason attributed was the huge shelf space required to store furniture and less demand of these items. He started focusing more towards the client base online, because they were leaders in the offline sales. With more and more retail sites coming up in the last couple of years, AB corp. grew it's online business by becoming a seller in majority of these sites including Amazon, Flipkart, Pepperfry and Shopclues. In many of these Websites, he is listed as a power seller. Thus, with the awareness and literacy level of the owner the overall business/sales multifold.

3.2   Factor Affecting Online Adoption

The factors affecting e-commerce adoption at AB. Corp were found out through detailed discussion with the owner and the IT employee. In one of the interview session, the owner mentioned the lack of availability of basic infrastructure in SRE region. According to him

Power cuts are usual and power supply is very low here, Back up services is not enough.

It implies that the acute shortage of electricity that hinders the online business. Internet speed is also owner's concern. The other thing that really worries the owner is the limited availability of delivery services in Saharanpur and its connectivity to other cities as quoted by [22]. The Website on which AB corp. sell products has different delivery options. For instance Ebay deploys two kinds of shipping services, where first one is shipping by seller and Ebay.in shipping (Power ship). Shopclues.com offers three shipping services like velocity premium, basic and merchant direct fulfillment. But due to unavailability of premium services in the region, AB corp. has chosen basic and direct fulfillment methods. This to some extent has affected their online sales. The unavailability also affects the product prices offered to the customer as the price includes all the cost components.

The owner quoted:

We have hired one extra employee who looks after all online business and also ensures timely delivery of order to designated area. His being on leave creates ruckus but at his arrival he handles the chaos well and ensures all orders are delivered correctly.

The factors affecting e-commerce adoption by this SME are mapped to TOE framework to understand most significant factors as in Table 2.

3.3   Analysis

Innovation in an organization can be because of the internal or external drivers. In this case, the organization innovation started with the external driver of advent safe net/e-commerce and its utilization to increase sales. But without the initiative of the owner this would not have been possible Thus, it is evident from the case study that organization factor is most important for e-commerce adoption in SME followed by technological and environmental factors.

Here the role of support services in terms of better delivery options can help gain business profit. Also, success would not have been possible without innovation in the business model. For instance, adoption of new sales channels (e-commerce) and shedding of heavy items from shelf. Understanding of customer preferences and knowledge of their taste in handicraft items leads to choice of items to be displayed online. The long tail problem also finds a solution when products are online. Thus, adoption of e-commerce was a radical change to the old business model of AB corp.

Table 2 Factors affecting e-commerce adoption

Factors	Constructs	Effects of constructs	AB corp.	
			Type	Effect level
Technology	Internet speed	Affects real-time update of information	Barrier	Moderate
	Power supply	Interrupted power supply hinders online transactions and other support services	Barrier	Moderate
	ICT availability	Better availability leads to higher customer services	Driver	High
Organization	E-mail and internet services	Assists in communications between firm and customers	Driver	High
	Owner/manager educational background	Education of owner paws way for strategic approach to adoption	Driver	High
	Maintenance and setup cost	E-commerce adoption requires extra costs incurred for device and services	Barrier	Low
	Trained staff/employees	Supports in proper distribution of work and timely delivery of services	Driver	High
	Managerial skills of owner	Ability to lead his/her staff/employees and make decisions for company	Driver	High
	Government as business partner and regulator of policies	Affects procurement, tax, and distribution policies	Driver	Moderate
Environment	Customer product preferences	Impacts e-commerce adoption and strategic direction of firm. Understanding leads to better offerings and reduction in shelf space	Barrier	Moderate
	Support from suppliers	Impacts on e-commerce adoption and diversification decision (manufacture or not)	Barrier	Moderate
	Lack of support from local industry firms (courier and IT support)	Disrupts consistency in delivery of services by firm. Firm may outsource various IT and technical support work in abundance of support	Barrier	Moderate to high

3.4 Post Adoption Benefits of E-commerce to AB Corp.

- AB corp. can now deliver and showcase their offerings globally, and it has added a new sales channel to its business model successfully. New revenue stream offers extra profits to kitty of organization.

- With support from delivery services and e-commerce companies AB corp. provides 24*7 services to customers.
- AB corp. uses its product information on Website to showcase product offerings to B2B clients too.
- As branded seller on various Websites AB corp. has set up a brand for itself on various portals. This phenomenon is rare in handicrafts items business in India. AB corp. now enjoys all features of a power brand on few portals.

4 Conclusion

Combining the observations from the study, interesting patterns and remarks originate as follows:

- AB corp. owner has an experimental and open minded approach, though they are in a rural market and a capital intensive industry. They have learnt important lessons with their venture into e-marketplace and understood the market and its dramatics.
- Knowledge and guidance of owner plays an important part for flourishing future of organization.
- Careful and well understood moves are the need to monetize existing resources. Therefore, ICT paws way for rural Indian organizations to expand globally.

The company's strong desire to flourish and expand leads to venture into online business. This was well executed by the wisdom of the owner. Thus, AB corp. has become a brand in Saharanpur market with straightforward attitude to innovation. Other SMEs in this region has taken the experience of AB corp. and started to venture online for business growth.

Acknowledgements We thank all the participants of this study (AB corp. owners and employees) toward their contributions in the research work. We are immensely grateful to them for their remarks, although any errors in manuscript are our own and should not affect reputations of participants.

References

1. Taylor, R.: 12. Epilogue: Labour Mobility and Human Resources. Emerging Asian Economies and MNCs Strategies, p. 235 (2016)
2. BalaSubrahmanya, M. H.: Technological innovation in Indian SMEs: need, status and policy imperatives. Curr. Opin. Creativity, Innov. Entrepreneurship 1(2), (2012)
3. OECD: Innovation in Firms. A Microeconomic Perspective. OECD Publishing (2009). Available on http://www.oecd-ilibrary.org/science-and-technology/innovation-in-firms_9789264056213-en. Accessed 27 Sept 2016

4. Lertwongsatien, C., Wongpinunwatana, N.: Empirical study of e-commerce adoption SMEs in Thailand, pp. 1040–1044 (2005)

5. OECD: Promoting Entrepreneurship and Innovative SMEs in a Global Economy: Towards a More Responsible and Inclusive Globalization—ICT, e-business and SMEs", OECD Digital Economy Papers (2004). Available on http://www.oecd.org/cfe/smes/31919590.pdf. Accessed on 27 Sept 2016

6. Van Akkeren, J., Cavaye, A.L.: Factors affecting entry-level internet technology adoption by small business in Australia: an empirical study. In: Proceedings 10th Australasian Conference on Information Systems (1999)

7. Senn, J.A.: Information Technology: Principles, Practices, and Opportunities. Pearson Prentice Hall, NJ (2004)

8. Martinsons, Maris G.: Relationship-based e-commerce: theory and evidence from China. Inform. Syst. J. **18**(4), 331–356 (2008)

9. Torsten Eriksson, L., Hultman, J., Naldi, L.: Small business e-commerce development in Sweden—an empirical survey. J. Small Bus. Enterp. Develop. **15**(3), 555–570 (2008)

10. Kartiwi, M., MacGregor, R.C.: Electronic commerce adoption barriers in small to medium-sized enterprises (SMEs) in developed and developing countries: a cross-country comparison. J. Electron. Commer. Organ. **5**(3), 35–51 (2007)

11. Thatcher, S.M., Foster, W., Zhu, L.: B2B e-commerce adoption decisions in Taiwan: the interaction of cultural and other institutional factors. Electron. Comm. Res. Appl. **5**(2), 92–104 (2006)

12. Daniel, E., Wilson, H., Myers, A.: Adoption of e-commerce by SMEs in the UK towards a stage model. Int. Small Bus. J. **20**(3), 253–270 (2002)

13. Grandon, E.E., Pearson, J.M.: Electronic commerce adoption: an empirical study of small and medium US businesses. Inform. Manag. **42**(1), 197–216 (2004)

14. Virendra Singh R: Wood carving industry in Saharanpur hit by riots. Business Standard. August 4, (2014). http://www.business-standard.com/article/sme/wood-carving-industry-in-saharanpur-hit-by-riots-114080401359_1.html. Accessed 27 Sept 2016

15. Planning Commission. Uttar Pradesh state development report, vol. I and II, Government of India, New Delhi (2005). Available on http://www.Planningcommission.nic.in/plans/stateplan/upsdr/vol-2/Chap_b4.pdf. Accessed 27 Sept 2016

16. Cavaye, A.L.M.: Case study research: a multi-faceted research approach for IS. Inform. Syst. J. **6**(3), 227–242 (1996)

17. Peng, Shao, Ping, Hu: A case study of business model innovation and evolution of the e-commerce platform. Sci. Res. Manag. **7**, 010 (2016)

18. Yang, N., Ouyang, T.H., Hu, J.B., Zeng, D.L.: The strategy of e-commerce platform based on the perspective of resource orchestration: a case study of Haier. In: Proceedings of the 22nd International Conference on Industrial Engineering and Engineering Management 2015, pp. 629–637. Atlantis Press (2016)

19. Bryman, A.: The End of the Paradigm Wars. The Sage Handbook of Social Research Methods, pp. 13–25 (2008)

20. Tornatzky, L.G., Fleischer, M., Chakrabarti, A.K.: Processes of Technological Innovation. Lexington Books (1990)

21. Thong, J.Y.L.: An integrated model of information systems adoption in small businesses. J. Manag. Inform. Syst. **15**(4), 187–214 (1999)

22. Yadav, R., Mahara, T.: An exploratory study to investigate value chain of Saharanpur wooden carving handicraft cluster. Int. J. Syst. Assur. Eng. Manag. 1–8 (2016)

23. Ifinedo, P.: An empirical analysis of factors influencing Internet/e-business technologies adoption by SMEs in Canada. Int. J. Inform. Technol. Decis. Making **10**(04), 731–766 (2011)

24. Lin, Hsiu-Fen, Lin, Szu-Mei: Determinants of e-business diffusion: a test of the technology diffusion perspective. Technovation **28**(3), 135–145 (2008)

25. Oliveira, T., Martins, M.F.: Understanding e-business adoption across industries in European countries. Ind. Manag. Data Syst. **110**(9), 1337–1354 (2010)

26. Zhu, K., Kraemer, K.L.: Post-adoption variations in usage and value of e-business by organizations: cross-country evidence from the retail industry. Inform. Syst. Res. **16**(1), 61–84 (2005)
27. Zhu, K., Kraemer, K.L., Sean, Xu: The process of innovation assimilation by firms in different countries: a technology diffusion perspective on e-business. Manag. Sci. **52**(10), 1557–1576 (2006)
28. Abdollahzadegan, A., Hussin, C., Razak, A., Moshfegh Gohary, M., Amini, M.: The organizational critical success factors for adopting cloud computing in SMEs. J. Inform. Syst. Res. Innov. **4**(1), 67–74 (2013)
29. Ghobakhloo, M., Arias-Aranda, D., Benitez-Amado, J.: Adoption of e-commerce applications in SMEs. Ind. Manag. Data Syst. **111**(8), 1238–1269 (2011)
30. Kuan, K.K., Chau, P.Y.: A perception-based model for EDI adoption in small businesses using a technology–organization–environment framework. Inform. Manag. **38**(8), 507–521 (2001)

Optimization of End Milling Process for Al2024-T4 Aluminum by Combined Taguchi and Artificial Neural Network Process

Shilpa B. Sahare, Sachin P. Untawale, Sharad S. Chaudhari, R.L. Shrivastava and Prashant D. Kamble

Abstract In this research paper, the attempt is made to optimize the end milling process for Al2024-T4 workpiece material. The input process parameters used are cutting speed, feed per tooth, depth of cut, and the cutting fluid flow rate, and the response parameters used are surface roughness, cutting force, and MRR. Taguchi L_9 orthogonal array is used for the experimentation. To make the experimental design robust, a noise factor-hardness of the workpiece material is considered. To match the current scenario of manufacturing process, minimum quantity lubrication (MQL) is used. The experimental setup is designed and fabricated for minimum quantity lubrication. Regression analysis is perfumed by the Artificial Neural Network in MATLAB software. The optimized setting of end milling process is obtained by ANN. The obtained results revealed that the ANN combined with Taguchi method is suitable for optimization.

Keywords ANN · End milling · Optimization · Taguchi method · S/N ratio

S.B. Sahare (✉) · S.S. Chaudhari · R.L. Shrivastava · P.D. Kamble
Department of Mechnical Engineering, Yeshwantrao Chavan College of Engineering,
Nagpur, Maharashtra, India
e-mail: ssahare83@yahoo.com

S.S. Chaudhari
e-mail: sschaudharipatil@rediffmail.com

R.L. Shrivastava
e-mail: rlshrivastava@yahoo.com

P.D. Kamble
e-mail: pdk121180@yahoo.com

S.P. Untawale
Department of Mechanical Engineering, DMIETR, Sawangi, Wardha, Maharashtra, India
e-mail: untawale@gmail.com

© Springer Nature Singapore Pte Ltd. 2018
M. Pant et al. (eds.), *Soft Computing: Theories and Applications*,
Advances in Intelligent Systems and Computing 584,
https://doi.org/10.1007/978-981-10-5699-4_49

1 Introduction

Metal cutting processes are manufacturing industrial processes in which the required shapes in the metals parts are obtained by removing unwanted materials. Materials were removed by traditional chip forming processes such as turning, milling, boring, and drilling. In these operations, metal is removed as a plastically deformed chip of appreciable dimension. Nowadays numerical and artificial neural networks (ANN) methods are widely used for both modeling and optimizing the performance of the manufacturing technologies. Optimum machining parameters are of great concern in manufacturing environments, where economy of machining operation plays a key role in competitiveness in the market. Therefore, the present research is aimed at finding the optimal process parameters for end milling process. The end milling process is a widely used machining process in aerospace industries and many other industries ranging from large manufacturers to a small tool and die shops, because of its versatility and efficiency. The reason for being widely used is that it may be used for the rough and finish machining of such features as slots, pockets, peripheries, and faces of components [1]. In metal cutting process, especially turning, milling process, besides the basic cutting process parameters like cutting speed, feed rate, depth of cut, tool geometry, environment of cutting, and the type of tool play an important role to decide the performance of quality characteristics. Proper selection of machining conditions (spindle speed, feed rate, and depth of cut,) and cutting tool geometries (helix angle, rake angle, relief angle, and nose radius) impact severely machining performance. An improperly selected machining parameters and poorly designed cutting tool not only deteriorate surface finish but also increase cutting forces, affecting stability and causing thermal deformation due to variation of temperature rise at the cutting zone and rapid tool wear.

Lou et al. [1] studied surface roughness prediction technique for CNC end milling. Input factors were cutting speed, feed, depth of cut and coolant flow to examine surface roughness and metal removal rate. Pare et al. [2] used particle swarm optimization to optimize cutting conditions in end milling process. Oktem et al. [3] predicted minimum surface roughness in end milling process. Hybrid neural network and genetic algorithm is used to optimize the response. It is observed that the application of ANN with optimization technique is helpful to give better results. Oktema et al. [3] used neural network and genetic algorithm to minimize surface roughness in the end milling. It is observed that the experimental and computed values are within the range given by confidence interval. Zaina et al. [4] developed a model using artificial neural network to optimize the surface roughness. In ANN, higher value of regression (R) can be achieved by retraining the data. Ighravwe et al. [5], Krenker et al. [6], Lakshmi and Subbaiah [7] and Li et al. [8] used the artificial neural network to develop a model to predict the different responses.

2 Design of Experiment

The experiments are planned and designed with the help of Taguchi technique. Taguchi method L27 OA has been used in order to explore the process interrelationships within the experimental frame. Taguchi recommends analyzing the mean response for each run in the inner array and also suggests to analyze variation using an appropriately selected signal-to-noise ratio (S/N). There are three signal-to-noise ratios of common interest for optimization of static problem is given by Eqs. 1, 2, and 3 respectively (Figs. 1 and 2; Table 1).

- *Nominal the better characteristics*

$$S/N = -10 \log \left(\frac{\hat{y}}{s^2} y \right) \tag{1}$$

- *Lower-the-Better characteristics*

$$S/N = -10 \log \left(\left(\frac{1}{n} \right) \sum y^2 \right) \tag{2}$$

- *Higher-the-Better characteristics*

$$S/N = -10 \log \left(\left(\frac{1}{n} \right) \sum \frac{1}{y^2} \right) \tag{3}$$

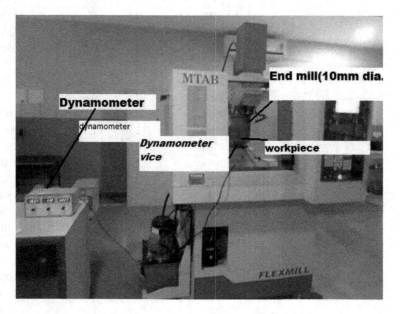

Fig. 1 Experimental setup

Fig. 2 Workpieces before and after machining

Table 1 Selection of cutting parameters

Cutting parameters	Units	1	2	3
Spindle speed	rpm	1910	2866	3821
Feed	mm/rev	1.5	2	2.5
Depth of cut	mm	0.5	1	1.5

Table 2 Taguchi L27 OA (input for ANN)

Run	Cutting speed	Feed	Depth of cut	Flow
1	1910	1.5	0.5	500
2	1910	1.5	1	500
3	1910	1.5	1.5	500
4	1910	2	0.5	1000
5	1910	2	1	1000
6	1910	2	1.5	1000
7	1910	2.5	0.5	1500
8	1910	2.5	1	1500
9	1910	2.5	1.5	1500
10	2866	1.5	0.5	500
11	2866	1.5	1	500
12	2866	1.5	1.5	500
13	2866	2	0.5	1000
14	2866	2	1	1000
15	2866	2	1.5	1000
16	2866	2.5	0.5	1500
17	2866	2.5	1	1500
18	2866	2.5	1.5	1500
19	3821	1.5	0.5	500
20	3821	1.5	1	500
21	3821	1.5	1.5	500
22	3821	2	0.5	1000
23	3821	2	1	1000
24	3821	2	1.5	1000
25	3821	2.5	0.5	1500
26	3821	2.5	1	1500
27	3821	2.5	1.5	1500

3 Artificial Neural Network (ANN)

Artificial neural networks are composed on simple elements operating in parallel. These elements are inspired by biological nervous systems shown in Fig. 3.

Three different neural network models are used for the prediction of surface roughness, cutting force, and MRR. The networks were trained with Levenberg–Marquardt algorithm. This training algorithm was chosen due to its high accuracy in similar function approximation [5]. In order to improve the generalization of the network, a regularization scheme was used in conjunction with the Levenberg–Marquardt algorithm. The automatic Bayesian regularization was used. For training with Levenberg–Marquardt combined with Bayesian regularization, the input/output dataset was divided randomly into two categories: training dataset and test dataset. The ANN modeling is carried in two steps: The first step is to train the network, whereas the second is to test the network with data, which were not used for training. In this research, a two-layer back propagation network was employed as a tool for mapping the complex and highly interactive process parameters such as cutting speed, feed, and depth of cut. The input data, target, and testing data in ANN are shown in Tables 2, 3 and 4 (Fig. 4).

The material samples are prepared, and hardness of each sample is tested. Based on the hardness of the material, the samples are categorized in three categories, that is, low hardness samples, medium hardness samples, and high hardness samples. Every run of the experiments can be performed thrice, i.e., for low, medium, and high hardness of the samples.

Fig. 3 Biological nervous system

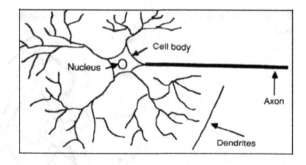

Table 3 Testing data

Run	MRR (m³/min)	Cutting forces (N)	Surface roughness (μm)
5	723	348.5	0.75
10	524	332.1	0.41
15	995	395.1	1.37
20	797	374.1	0.78
25	899	382	1.55

Table 4 Output measured

Run	MRR	Cutting forces	Surface rough
1	419	312.4	0.3
2	587	333.5	0.45
3	756	361	0.6
4	554	327.8	0.5
6	890	376.8	1
7	690	343.1	0.7
8	857	366.2	1.06
9	1025	386	1.41
11	693	353.4	0.62
12	861	373.5	0.82
13	660	347.9	0.69
14	827	366	1.03
16	795	361.1	0.96
17	963	381	1.44
18	1130	410.9	1.92
19	629	348.3	0.52
21	966	394.4	1.04
22	765	361	0.87
23	932	389.9	1.3
24	1100	406	1.74
26	1067	402.1	2.05
27	1236	429.3	2.55

Fig. 4 Layers in neural networks

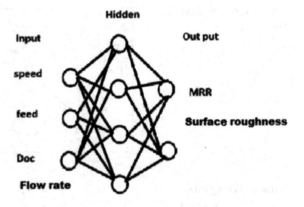

3.1 ANN Model Design for Prediction of Output Parameters

The steps involved in the model design for MRR and cutting forces, and surface roughness area as follows:

(a) *Assigning the number of nodes in input and output layers.*

The number of input parameters will be number of nodes; similarly, the number of output parameters would be number of outer layers, i.e., one in this case.

(b) *Normalizing Input and output.*

The neural network works well if the input and output would be in the range of [0, 1]. Therefore, it is necessary to convert the inputs and outputs into required form, and then the following equations are developed.

$$Mi = Ai/Maxi; \quad Mo = Ao/Maxi$$

(c) *Fixing of Number of hidden layers.*

Generally, single and double hidden layers networks are used to solve the most of the problems. Here, also single hidden layer is used. Double hidden layer has given the minimum error than the single hidden layer.

3.1.1 Network Model-Surface Roughness, MRR Cutting Force

See Tables 5, 6 and Fig. 5.

Table 5 Network model

Particular	Surface roughness	MRR	Cutting force
Network type	Feed forward back propagation (FFBP)	FFBP	FFBP
Training	Levenberg–Maquardtl algorithm (LMA)	LMA	LMA
No.. of layers	2	2	2
Output layer	1	1	1
No. of neurons	0–10	0–20	0–20
Performance	Mean square error (MSE)	MSE	MSE
Training function	TRAINLM	TRAINLM	TRAINLM
Hidden layer transfer function	Tran sigmoid	Tran sigmoid	Tran sigmoid
Output layer transfer function	Pure linear	Pure linear	Pure linear
Adaption of learning rate	LEARNGDM	LEARNGDM	LEARNGDM

Table 6 Best performance error with hidden nodes for CF

	Surface roughness	MRR		Cutting force	
Hidden Layer	Best performance error (BPE)	Hidden layer	BPE	Hidden layer	BPE
3	0.010	15	0.0262	14	0.0251
4	0.092	17	0.0123	16	0.0113
11	0.0017	18	0.0041	17	0.0031
16	0.031	19	0.079	18	0.069
20	0.061	20	0.0977	20	0.0957

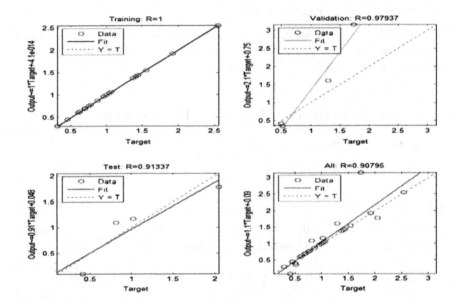

Fig. 5 Best regression plot for surface roughness

4 Results

This paper presents an experimental investigation on surface finish and material cutting force and MRR during the high speed end milling of Al2024-T4 aluminum in order to develop an appropriate roughness prediction model and optimize the cutting parameters using Taguchi and ANN. The predicted values and experimental values of MRR, CF, and SR training and testing are shown in the Figs. 6, 7 and 8, and it is observed that the testing data also work well with the network (Figs. 9 and 10).

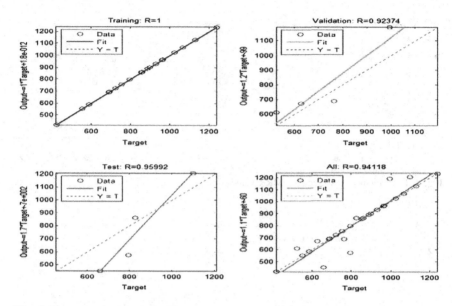

Fig. 6 Best regression plot for MRR

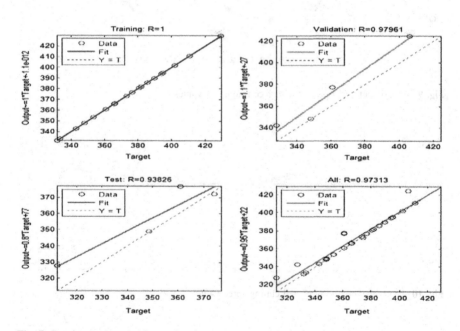

Fig. 7 Best regression plot for cutting force

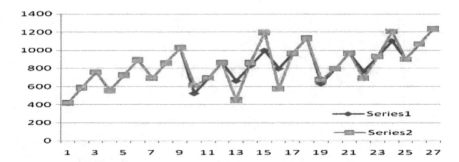

Fig. 8 Experimental and predicted MRR values

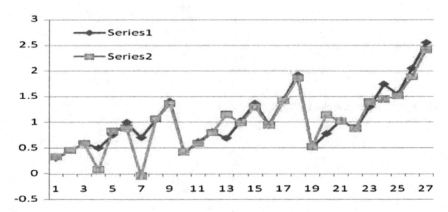

Fig. 9 Experimental and predicted surface roughness values

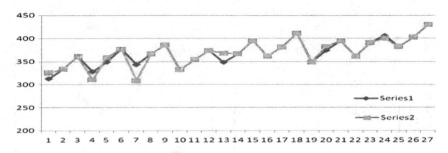

Fig. 10 Experimental and predicted cutting force values

5 Conclusion

The fuzzy logic with Taguchi method is successfully applied in this study to optimize end milling process for Al2024-T4 aluminum. The results are discussed as follows:

1. The Taguchi method is successfully applied with ANN.
2. The models were developed to predict the surface roughness, cutting force, and MRR through ANN techniques.
3. The best model is selected based on the best performance error for different network configurations.
4. The models have been evaluated by means of the percentage deviation between the predicted values and actual values.
5. It is shown that the ANN predicted results show good agreement with the experimental results. Hence, it proved its efficiency in optimizing the end milling process.

References

1. Lou, M.S., Chen, J.C., Li, C.M.: Surface roughness prediction tech-nique for CNC end milling. J. Indus. Technol. 17(2–7), 2–6 (1999)
2. Pare, V., Agnihotri, G., Krishna, C.M.: Krishna optimization of cutting conditions in end milling process with the approach of particle swarm optimization. Mechanical Engineering Department, Maulana Azad National Institute of Technology, Bhopal, India
3. Oktema, H., Erzurumlu, T., Erzincanli, F.: Prediction of minimum surface roughness in end milling mold parts using neural network and genetic algorithm. Mater. Des. 27(9), 735–744 (2006)
4. Zain, A.M., Haron, H., Sharif, S.: Prediction of surface roughness in the end milling machining using artificial neural network. Expert Syst. Appl. 37(2), 1755–1768 (2010)
5. Ighravwe, D.E., Oke, S.A.: Machining performance analysis in end milling: predicting using ANN and a comparative optimisation study of ANN/BB-BC and ANN/PSO. Eng. J. 19(5), (2015)
6. Krenker, A., Bester, J., Kos, A.: Introduction to the artificial neural networks. In: Suzuki, K. (ed.) Artificial Neural Networks: Methodological Advances and Biomedical Applications, pp. 3–18. Croatia: InTech (2011)
7. Lakshmi, V.V.K., Subbaiah, K.V.: Modelling and optimisation of process parameters during end milling of hardened steel. Int. J. Eng. Res. Appl. 2(2), 674–679 (2012)
8. Li, B., Hu, Y., Wang, X., Li, C., Li, X.: An analytical model for oblique cutting with application to end milling. Mach. Sci. Technol. 15(4), 453–484 (2011)

Dynamic Classification Mining Techniques for Predicting Phishing URL

Surbhi Gupta and Abhishek Singhal

Abstract The online community faces the security challenge due to Phishing attack because many numbers of transactions are performed online to save time of users. It is a continual threat, and the risk of this attack is mostly on social networking sites such as Facebook, Twitter, LinkedIn, and Google+. In this paper, a dynamic approach based on minimum time to detect Phishing URL by using classification technique is described. This approach is based on different parameters such as high accuracy, Recall, Precision, Specification, and many more. The analysis result of these algorithms shows that Random tree is a good classification technique with comparison to others; it takes very less training build data time.

Keywords Phishing · Classification · Precision · Recall · Accuracy

1 Introduction

In today's world, Internet is a global system for individual users to link billion of peoples through social networking sites, to do business, for academic purpose, etc. But through Internet people are facing many difficulties like social engineering attack such as Phishing, Hacking, etc. Phishing attack is an online threat that may cause loss of personal data, financial loss through disclosing bank details and credit card details. Phishing is a way of impersonating a Web site of honest site aiming online people to disclose private data such as username, password of social networking sites, and others. These types of fake Web sites are created by deceptive person to imitate legal Web sites. These Web sites have the same level of visual as legitimate Web sites to manipulate the end users.

S. Gupta (✉) · A. Singhal
Department of Computer Science and Engineering, Amity University Uttar Pradesh,
Noida, India
e-mail: surbhiamity15@gmail.com

A. Singhal
e-mail: asinghal1@amity.edu

© Springer Nature Singapore Pte Ltd. 2018   537
M. Pant et al. (eds.), *Soft Computing: Theories and Applications*,
Advances in Intelligent Systems and Computing 584,
https://doi.org/10.1007/978-981-10-5699-4_50

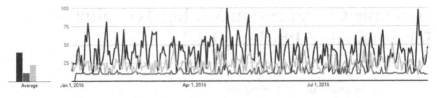

Fig. 1 Represent the attack types spear, cloning and whaling ratio in 2016

Phishing attack is increasing rapidly day by day by the report of Wikipedia [1] of Phishing attack, March 2016 have highest number Phishing attack attempt. Another method of Phishing attack is email spoofing. An ordinary Phishing attack is planted on various techniques such as exploiting vulnerabilities of browsers or man-in-the-middle is based on proxy by sending an email that seems from an authenticate organization to target user. These types of emails ask them to update their personal information and bank account information by clicking on a link within an email. Phishing is a typical issue of classification that the given URL is Phishing or legitimate. Phishing attack has many types such as Spear Phishing, Phone Phishing, Whaling, cloning, and many more. Through the Web search, it is shown in Fig. 1 that spear phishing attack is more common from the above-mentioned attacks.

In this paper, the problem of Phishing URL detection is inspected using classification technique using data mining tool. We primarily test on a developed classification algorithm and compare them with each other. The Phishing data is collected from UCI repository archive [2] that includes data from Phish tank [3], APWG group. These are the Web sites to check if the given URL is Phishing or not. The evaluation is based on different parameters such as Recall, Precision [4], and Accuracy of various Algorithms [5].

This Paper Organization includes Introduction in Sects. 1 and 2 provides the literature survey in this area, and Sect. 3 highlights the proposed work, which describes Phishing attack cycle and dataset. Section 4 represents the classification and result, and Sect. 5 concludes the study.

2 Literature Survey

Pradeepthi and Kannanl [6] has researched on the classification technique. The technique used for detecting fake URL based on different measures such as Recall, Precision, Accuracy, and many others. In this paper pattern recognition capabilities are used to detect domain of URL and to achieve the improvement in performance. He compares many classification algorithms and identified best classifier as tree-based classifier to detect Phishing URL. We can add more attribute for better result.

Lew et al. [7] has proposed a technique, in this they hybrid the features of emails or attribute to detect Phishing emails. The hybrid features contain behavior, content based, IP address URL based, and others. They select the two resources to select the common features that are used by attacker in Phishing Web sites. This technique used support vector machine for classification. The result of this classifier is good with the hybrid of features. But there are some restrictions to use black-list keyword features. Future work can make it better by using different features from graphical elements.

Debars et al. [8] proposed a methodology which is based on traffic behavior, clustering, and random forest classification algorithm [8]. In this paper, the first step is applying filtering technique such as LDA (latent Dinchelt Allocation) that helps to identify that type of Phishing email that is tough. Then they apply spectral cluster analysis on the link between URL; after this he applies Random Forest classification algorithm to calculate Precision, Recall, Accuracy of that type of dataset. It provides significant improvement in result in the entire list of metrics and compared to filtering technique (LDA). For the spear Phishing emails detection, spectral clustering is a good technique.

Pati et al. [9] have proposed a model to determine the Phishing sites to safeguard the online users from the attacker. In this paper, they determine the features based on URL & HTML tags. They used clustering & Naïve Bayes Classifier to Predict the class of given URL that is Phishing or not. They used clustering technique (k-means) that generates the output fast with compromising result, and Bayesian approach gave result with high accuracy but it is time-consuming process so they implement the combination of both algorithm which is more efficient and reliable compared with them.

Li et al. [10] describe a novel approach which is based on ball support vector machine (BVM) to Phishing Web site detection. This approach is used to enhance the integrity of feature vector. The first step of this approach is performing an analysis to the structure of Web site with reference to DOM tree. Then BVM classifier detects the feature vector. This method has high precision of detecting and complements. The result shows that the BVM has the better performance in compare than SVM. This method makes efficient to detection Phishing attack.

3 Proposed Work

3.1 Phishing Attack Cycle

The infrastructure of phishing attack cycle is shown in Fig. 2. There are basic steps to execute the phishing attack which is used by the attacker.

- First step is Attacker Plans the scenario of the attack and find the victims through the mails and social networking sites. Then he sends the test message through the emails with an attached link.

Fig. 2 Phishing attack cycle

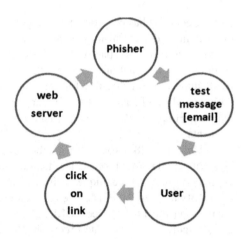

- The message is delivered in a way that the victims disclose their personal information example id and Password to verify, bank account information, credit card details, etc.
- Third step, when the user clicks on that link and enter the information which is asked by the attacker's Web page. All information has gone to the attacker.
- Sometimes, this type of link has a malicious code, which is run automatically in your system and the victims cannot see it.

3.2 Dataset

A dataset with 11055 URLs was collation for the purpose to implement classification. Of these 7262 are Legitimate URLs, and 3793 are Phishing or Suspicious URLs. This dataset is collected from uci repository archive [2]. This dataset contains 31 feature attributes [11] which is help to classified URL that Phishing or not. It is a collection from Phish tank [3] and APWG. Phish tank is a center site of Phishing where user can submit the suspicious URL link and others user vote to confirm that it is Phish site or not. These URLs are the mixture of Online shopping, Bank emails, Games, and education fake Web page, etc.

3.3 Features List

In this paper, we used URLs, which is combination of Phishing and legitimate URLs. Dataset is collected from UCI repository achieve [2] which is shown in Fig. 3. It contains 31 attributes. We discuss some features, which is based on Address bar.

```
@relation phishing

@attribute having_IP_Address  { -1,1 }
@attribute URL_Length    { 1,0,-1 }
@attribute Shortining_Service { 1,-1 }
@attribute having_At_Symbol   { 1,-1 }
@attribute double_slash_redirecting { -1,1 }
@attribute Prefix_Suffix  { -1,1 }
@attribute having_Sub_Domain  { -1,0,1 }
@attribute SSLfinal_State  { -1,1,0 }
@attribute Domain_registeration_length { -1,1 }
@attribute Favicon { 1,-1 }
@attribute port { 1,-1 }
@attribute HTTPS_token { -1,1 }
@attribute Request_URL   { 1,-1 }
@attribute URL_of_Anchor  { -1,0,1 }
@attribute Links_in_tags { 1,-1,0 }
@attribute SFH   { -1,1,0 }
@attribute Submitting_to_email { -1,1 }
@attribute Abnormal_URL { -1,1 }
@attribute Redirect   { 0,1 }
@attribute on_mouseover  { 1,-1 }
@attribute RightClick  { 1,-1 }
@attribute popUpWidnow  { 1,-1 }
@attribute Iframe { 1,-1 }
@attribute age_of_domain  { -1,1 }
@attribute DNSRecord   { -1,1 }
@attribute web_traffic  { -1,0,1 }
@attribute Page_Rank { -1,1 }
@attribute Google_Index { 1,-1 }
@attribute Links_pointing_to_page { 1,0,-1 }
@attribute Statistical_report { -1,1 }
@attribute Result  { -1,1 }
```

Fig. 3 AList of features

3.3.1 URL Contains IP Address

URL Contains IP Address: when URL contains IP address, it is the alternate form of URL. Example: http//123.11.12.25/facebook.html, this type of URL is considered in Phishing email.

If URL contains Domain Part as IP Address → Phishing
Otherwise → Legitimate.

3.3.2 Hide the Suspicious Fake Part with Lengthy URL

This type of URL is in the category of Phishing attack, No one can catch it that what is hiding within the suspicious fake link. Example: http://1a2b3c4d123-0876757t?fake12785ab7ediueyeyjdie9ywjkdhdoifye8r35hnj9t6be8h4/fake.html

$$(\text{If URL length} < 54 \;\rightarrow\; \text{feature} = \text{Legitimate URL}$$
$$\text{else if URL length} \geqslant 54 \;\rightarrow\; \text{feature} = \text{Suspicious}$$
$$\text{otherwise} \;\rightarrow\; \text{feature} = \text{Phishing URL})$$

So, there are many features attribute, which is used in dataset. The list of attribute is given below.

4 Classifications and Result

The classification is based on the feature attributes with the help of different algorithm Random tree, Random forest, Naive bayes, J48, and LMT. A good classification is based on some parameters such as Precision, Recall [12, 4], Accuracy, Training build data time, Specification, etc. We use data mining tool Weka 3.8.0 [13] to calculate these values. The result shows the classification using various algorithms. This tool is mostly used in machine learning techniques such as supervised and unsupervised, associative and others.

Parameterized value of various algorithms is shown in Table 1 which is used to select the most suitable among them. From the above table, it is clearly shown that Random Tree is a good algorithm for Phishing URL detection.

4.1 Classification Evaluation Metrics

We get a confusion matrix after applying different classification techniques as shown in Fig. 4.

Table 1 Represents some values of different classification algorithm

	Random tree	Random forest [8]	J48	NB	LMT	SMO
Kappa statistic	0.9787	0.9791	0.9445	0.8582	0.9784	0.87775
Mean absolute error	0.0133	0.025	0.0461	0.0892	0.015	0.0603
Root mean squared error	0.0816	0.0893	0.1518	0.2301	0.0853	0.2456
Relative absolute error	2.6969%	5.075%	9.3349	18.07%	3.0482	12.2255%
Root Relative square error	16.4224%	17.9771%	30.5533	46.319%	17.171%	49.448%

Fig. 4 Shows the confusion matrix. *TP* true positive, *FN* false negative, *FP* false positive, *TN* true negative

	Predicted 1	Predicted 0
True 1	TP	FN
True 0	FP	TN

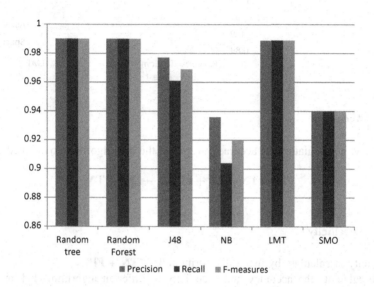

Fig. 5 Shows precision, recall, *F*-measures

PRECISION: It defines by using this formula **TP/(TP + FP)**
RECALL: **TP/(TP + FN)**
F-MEASURES: **(2 * Precision * Recall)/(Precision + Recall)**.

Figure 5 depicts the Precision, Recall, *F*-measures of various algorithms.

4.2 Parameters of Classification Algorithms

By using graph, we display the result of accuracy and specificity in Fig. 6.

Fig. 6 Represents the
accuracy and specificity of
various algorithms

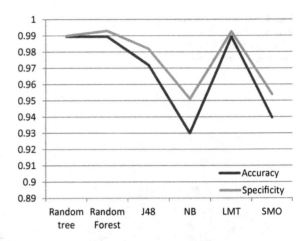

4.2.1 Accuracy

Accuracy by calculating the accuracy, we select the best algorithm to the prediction.

$$(\mathbf{TP} + \mathbf{TN})/(\mathbf{TP} + \mathbf{FP} \;+\; +\; \mathbf{FN} + \mathbf{TN})$$

4.2.2 Specificity

Specificity is calculate by using this formula **TN/(TN + FP)**.

We calculate the accuracy and specificity of different algorithm [14], and the high accuracy of both algorithms such as Random tree and Random forest [8]. We now show the difference between both algorithm by using the time measurement which is time taken to build model and time taken to test model on training data. Table 2 and Fig. 7 represent the time taken to build model of an algorithm and also show that random tree algorithm is taking less time as compared to random forest algorithm.

Table 2 Represents the time taken to build model of algorithms

Algorithm used	Time taken to build model	Time taken to test model on training data
Random tree	0.05	0.03
Random forest	1.94	0.62

Fig. 7 Represents the time taken to build of algorithms

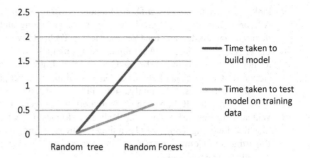

5 Conclusion

In this paper, we have used six different data mining classification algorithms such as Random forest, Random tree, LMT, J48, NB, SMO to analysis the dataset of URLs. This analysis shows the prediction based on URL classification by using data mining tool weka. The result of this analysis is that Random tree and Random forest both are the best classification algorithm. This result is based on the accuracy and specificity of the algorithm. We have shown other parameters like Precision, Recall, and F-measures which show the better algorithm out of them. But the Random tree algorithm is better than Random forest in terms of time taken to build model and time taken to test model on training data. With the above comparison of different algorithm, we can say that tree-based algorithms are best classifier.

References

1. Phishing Information, https://en.wikipedia.org/wiki/Phishing
2. UCI Repository, http://archive.ics.uci.edu/ml/
3. Phishtank, http://www.phishtank.com
4. Usmani, T.A., Bhatt, A.S., Pant, D.: A comparative study of google and bing search engines in context of precision and relative recall parameter, 2012, pp. 21–34
5. Lee, J., Kim, D., Hoon, C., Lee.: Heuristic-based approach for phishing site detection using URL features. In: 3rd International Conference on Advances in Computing, Electronics and Electrical Technology—CEET 2015, pp. 131–135 (2015)
6. Pradeepthi, K.V., Kannan, A.: Performance study of classification techniques for phishing URL detection. In: 6th IEEE International Conference on Advanced Computing (ICoAC), pp. 135–139 (2014)
7. Form, L.M., Chiewy, K.L., Szez, S.N., Tiongx, W.K.: Phishing email detection technique by using hybrid features. In: 9th IEEE International Conference on IT in Asia (CITA), pp. 1–5 (2015)
8. DeBarr, D., Ramanathan, V., Wechsler, H.: Phishing detection using traffic behavior, spectral clustering, and random forests. In: IEEE International Conference on Intelligence and Security Informatics (ISI), pp. 67–72 (2013)

9. Patil, R., Dhamdhere, B.D., Dhonde, K.S., Chinchwade, R.G., Mehetre, S.B.: A hybrid model to detect phishing sites using clustering and bayesian approach. In: IEEE International Conference for Convergence of Technology (I2CT), pp. 1–5 (2014)

10. Li, Y., Yang, L., Ding, J.: A minimum enclosing ball-based support vector machine approach for detection of phishing websites. Optik—Int. J. Light Electron Opt. 127(1), 345–351 (2015)

11. Gupta, R.: Comparison of classification algorithms to detect phishing web pages using feature selection. Int. J Res. Grayanthlya 4(8), 118–135 (2016)

12. Phishing and Recall, https://en.wikipedia.org/wiki/Precision_and_recall

13. Witten, l.H., Frank, E., Trigg, L., Hall, M., Holmes, G., Cunningham, S.J.: Weka: Practival Machine Learning Tools and Techniques with Java Implementations. Working Paper 99/11, Department of Computer Science, The University of Waikato, Hamilton (1999). (http://www. cs.waikato.ac.nz/ml/weka/)

14. Va, S.L., MSb, V.: Efficient prediction of phishing websites using supervised learning algorithms. Procedia Eng. 30, 798–805 (2011)

15. Shekokar, N.M., Shah, C., Mahajan, M., Rachh, S.: An ideal approach for detection and prevention of phishing attacks. Procedia Comput. Sci. 49, 82–91 (2015)

16. Abdelhamid, N., Ayesh, A., Thabtah, F.: Phishing detection based associative classification data mining. Exp. Syst. Appl. 41(13), 5948–5959 (2014)

17. Aldeen1, Y.A.S., Salleh, M., Razzaque, M.A.: A comprehensive review on privacy preserving data mining. Springer Plus, Nov 12 (2015)

18. Jeeva, S.C., Rajsingh, E.B.: Intelligent phishing URL detection using association rule mining. Human-centric Computing and Information Sciences, July 10 (2016)

Protection from Spoofing Attacks Using Honeypot in Wireless Ad Hoc Environment

Palak Khurana, Anshika Sharma, Sushil Kumar
and Shailendra Narayan Singh

Abstract Wireless ad hoc network or mobile network is the class of networks that have allowed user the freedom of movement from one geographical location to another. Their ease and use are unquestioned and wide enough. But apart from the boons that wireless networks give us, the bane comes handy. Mobile networks are prone to large number of attacks. Since the wireless medium is built upon a shared medium of network, they are highly prone to spoofing attacks. Typically speaking spoofing attacks are the class of attacks that cause harms to a network or a node by pretending to be someone else. The attacker steals important information or injects harmful information in a network. Our paper explains the detection and preventive measures that should be taken in order to avoid spoofing attacks, and finally our main aim in this paper is to apply the technique of honeypot in order to detect and prevent the spoofing attacks.

Keywords Honey pot · Spoofing attack · Wireless network · Authentication
Mobile node

P. Khurana (✉) · A. Sharma · S. Kumar · S.N. Singh
Department of Computer Science and Engineering, Amity University
Uttar Pradesh, Noida, India
e-mail: pkhurana187@gmail.com

A. Sharma
e-mail: anshikaintegral@gmail.com

S. Kumar
e-mail: kumarsushiliitr@gmail.com

S.N. Singh
e-mail: Snsingh36@amity.edu

© Springer Nature Singapore Pte Ltd. 2018   547
M. Pant et al. (eds.), *Soft Computing: Theories and Applications*,
Advances in Intelligent Systems and Computing 584,
https://doi.org/10.1007/978-981-10-5699-4_51

1 Introduction

A mobile network is the collection of various nodes that are physically disconnected. These nodes align themselves to form a network that is temporary, this
network then shares all the information needed. But, there are various methods that
can alter the ease of this mode of communication. This causes great risk to the
security. Attackers with various means can read our private information or put some
harmful information in our mobile systems.

Spoofing can be undoubtedly considered as one of the most dangerous tool in
order to harm our data and devices and hamper working of our network. In laymen
language, spoofing is an attack when a user or a program disguises itself to be
someone else, this impersonification by the attacker causes our data to leak.
Spoofing attacks can be the root cause for different advanced attacks, hampering
network's performance [1–5]. The harms are concentrated in mobile networks,
since users today carry their social, professional and personal life in their devices.

The earlier trends of detecting a spoofing attack were done by cryptographic
authentications in the network, but this required an additional amount of overhead
and infrastructure in order to maintain cryptographic keys. Power and resources are
the major concern of today's scenario and this technique, however, utilized them to
a great extent. Hence, in this paper we would like to propose the method of
honeypot to detect and prevent the spoofing attacks.

Honeypot is a technique that is used to aloof the masquerades in the current
network. In the mobile environment, honeypots act as an essential node of the
wireless network. This deviates the attention of the attackers towards them. As soon
as the alien network or person attacks these nodes, they capture all the information
from them like their strategy of attacking, reason behind it and its techniques. In this
way, the mobile network gets to know about the external malicious link.

The paper is arranged in five sections, Sect. 1 enclosed the introduction to the
paper, Sect. 2 consists of the literature review, Sect. 3 consists of analysis of various techniques discussed in the literature review, Sect. 4 introduces our honeypot
technique with spoofing attacks and finally in Sect. 5 we give our conclusion and
future scope.

2 Literature Review

In order to complete the analysis of our paper, we referred to some 300 papers from
various digital libraries namely: Science Direct, ACM, IEEE Xplore, Google
Scholar and a lot more. Out of which we have selected few papers for our analysis.

Yang et al. [6] proposed a method for detection of spoofing attacks while
the nodes were dynamic by introducing a DEMOTE system which worked on the
concept of received signal strength (RSS). The experimental results showed that the

error disclosure time was reduced to 190 from 220 s while gaining the detection rate of 100% and false positive as 0%.

Chen et al. [7] presented techniques that detect spoofing attack in wireless network. They also explained the integration of their detector with the localized system in indoor environment. In order to produce the test statistics, they used k-means algorithm for cluster analysis. Further their results showed that the detection rate of 95% and above and false positive of less than 5% have been achieved from their algorithm.

Sheng et al. [8] discussed an approach that was dependent upon mixture models given by Gaussian, by constructing their RSS profiles. The outcomes of the proposed work show that their method is fit for the technique antenna diversity. The detection rate of their algorithm was calculated to be 97.8% at a false positive rate of 3%.

Vijaykumar et al. [9] proposed a clustering technique based on mediod which they have used to identify spoofing attacks. Along with this another approach named enhanced partitioning around medioid (EPAM) has been used in integration with average in order to generate better accuracy. They used MAC address allocation technique along with MD5 hashing in order to provide higher detection rates of spoofing. The results generated showed a detection rate of 98.9% with the false positive as 1% only, provided the threshold value was assumed to be 6.5 db.

Lee et al. [10] addressed problems that were related to location spoofing. They examined error relative to the location and not its absolute value. They named the technique as relative error detection (RED) and topological residual fingerprint matching (TRFM). The technique proposed does not require any a priori statistical information, already established map or any database type.

Xiao et al. [11] proposed a technique based on the authentication and investigation over the physical layer. It worked on various parameters such as strength of the signals received, in order to reveal spoofing attack. The attack detection technique was based on reinforcement learning, in order to achieve the appropriate threshold value using Q-learning approach. They finally implemented it on USRP (universal s/w radio peripherals). The experimental results proved the fact that if the bandwidth of the system was around 200 MHz, the error rate of around 0.1% was achieved.

Zeng et al. [12] discussed spoofing attacks in GPS, in mobile network, on the synchronization of time. The deterioration of performance is calculated by symbol error rate (SER) from FH-CDMA in wireless environment. The paper presented a technique which was quick in detecting the attacks and was named as CUMSUM algorithm for testing. In future, the paper will discuss remedies for the spoofing attack in GPS.

Wang et al. [13] explained the fact that although authentication schemes such as cryptography can be used in a network, but they are not feasible in environments such as ad hoc network. The paper works on the concept of fusion and extraction of features in order detect a spoofing attacker. The paper selected two nodes and tested correlation amongst them. The results identified the conditions of attacking.

In future, the scheme will attempt to prolong the sparse portrayal of channel in other mediums of wireless security.

Dharani et al. [14] proposed an approach that was used for: (a) detection of attacks (spoofing), (b) finding out the total no. of attackers when multiple invaders masquerade same node, (c) localization of numerous intruders. The system used the concept of received signal strength (RSS) with spatial correlation. The concept of K-nearest neighbour classifier (KNN) was used in order to cluster the class of attackers.

Yadav et al. [15] presented a technique that detects and prevents the spoofing attacks at the MAC layer by using an intermediary dummy or fake node. The node was supposed to lie between user and the server and it served two purposes: (a) over passed the intruders request to network, (b) helped decrease network's traffic. The intruder detection was done by using relative signal strength (RSS), by disuniting the no. of users into different clusters using K-means classifier. Generalized attack detection model (GADE) provided the total invaders and integrated and detected localization system (IDOL) identified intruders itself.

Raguvaran et al. [16] discussed an algorithm that was profitable in terms of diagnosing the masquerader or attacker. The proposed algorithm was performed in two steps: (a) identify unique client address, (b) modify ACL list. The concept of Hash gathering was also used. The technique solves the problem of attack by spoofer when the user once authenticates it to exchange data by projecting re-authentication mechanism. If any unwanted activity appears, the user (attacker) gets disconnected.

3 Analysis on Various Software Security Techniques

To do the analysis of our technique, namely spoofing attacks, we reviewed all the technologies and techniques that have so far been developed and presented. We realized that the techniques basically used for detection of spoofing attacks had fallen under two categories techniques that used relative signal strength (RSS) and techniques that did not use RSS.

All these techniques that can be used to detect and/or prevent masquerades are listed as under in Table 1.

4 Reducing Spoofing Attacks with Honeypots

Honeypots as discussed above are the security tactics that are used in order to lure cyber attackers. They act as temptation to spoofers by imitating as if they are very important part of the network. This leads the attackers to throw their attention towards them. The main advantage of honeypots is that they are very simple in deploying and maintenance. The best part of using a honeypot is that they

Table 1 Techniques used for detection and/or prevention of spoofing attacks

Various detection/prevention techniques developed so far					
S. No.	Technique's name	Ref. No.	S. No.	Technique's name	Ref. No.
1	DEMOTE system	[5]	6	Reinforcement-based attack detection	[10]
2	Integration of detector with localized system	[6]	7	CUMSUM	[11]
3	Gaussian mixture models	[7]	8	Extraction of features and fusion	[12]
4	Mediod-based technique	[8]	9	RSS with spatial co-relation	[13]
5	RED and TRMF	[9]	10	GADE and IDOL	[14]
			11	Hash gathering	[15]

incorporate minimum danger since the criminal never reaches the actual nodes for attacking purpose.

Honeypot ploy the attackers in such a way that the attacker assumes honeypot node to be a very important node in the mobile ad hoc network and at the same time in the back end, it collects all the required data such as criminals identity, their reason for attacking, attacking strategy and their technique. As soon as all the information is collected by this node, it sends all this information to the base station for further actions.

4.1 Architecture

The basic architecture of honeypots is shown in Fig. 1. The architecture comprises of various components, namely:

(a) Honey nodes,
(b) Mobile/wireless nodes,
(c) Base stations (BS).

Fig. 1 Architecture for honey nodes

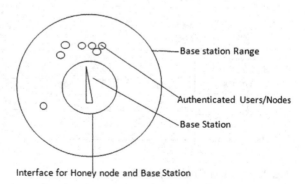

Base station Range

Authenticated Users/Nodes

Base Station

Interface for Honey node and Base Station

Honey nodes can be explained as the nodes where the honeypots are implemented. It is this honey nodes that fake a signal frequency to which attackers fall prey. The attackers assume these to be important nodes in the wireless network and start behaving their steps. In the backend, all the data are collected and in the mean time it gives immense time to honeypots to notify with a signal message to the base station (BS).

The Base station on the other hand sends the signal to all other legitimate users/nodes to change their frequency for operating.

4.2 Implementation Algorithm

As shown in Fig. 2, the algorithm for implementation of honeypots works by first of all testing whether the attack is detected or not. If the attack is detected then the node on which the attack occurs is measured. If the attacked node is the honey node then BS is informed about the attack, and new frequency is chosen for communication by informing every node in the mobile network about the new frequency. In the mean time, the honey node keeps its regular communication with the criminal node in order to deceive the attacker.

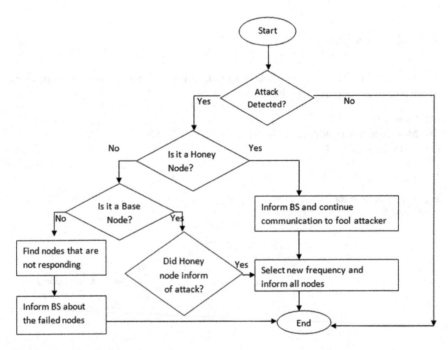

Fig. 2 Algorithm for honeypots

If the node attacked is not a honey node, then we are supposed to check whether it was initially notified by the honeypot or not. If it was notified in advance then it can easily change the frequency and inform other nodes. If it is neither honey node nor a base node, then we can consider that spoofer has attacked one of our legitimate node and will inform BS about is as soon as possible.

5 Conclusion

In this paper, we revised one of the most common and dangerous attack prevalent in ad hoc environment. The literature survey showed one of the most common techniques prevalent in detection or/and prevention of Spoofing attacks, with their results and future scope.

Techniques such as DEMOTE gave a detection rate of 100% with false positive of 0%, while technique such as location identifier using K-means gave a detection rate of 95% and false positive of less than 5%. Similarly, EPAM produced the detection rate of 98.9 and 1% of false positive rate and so on.

All the techniques discussed produced results with advantages, disadvantages and future scope as discussed in the literature survey.

Finally, we proposed the technique of honeypots to be used in the spoofing environment. Honeypot is one of the most common and widely used techniques in networking environment that is used to fool the attackers. The algorithm for honeypots in the spoofing environment is also presented which can be used to detect and prevent the criminals and attackers at the same time.

References

1. Ferreri, F., Bernaschi, M., Valcamonici, L.: Access points vulnerabilities to dos attacks in 802.11 networks. In: IEEE Wireless Communications and Networking Conference, pp. 634–638 (2004)
2. Bellardo, J., Savage, S.: 802.11 denial-of-service attacks: real vulnerabilities and practical solutions. In: USENIX Security Symposium, pp. 15–28 (2003)
3. Faria, D., Cheriton, D.: Detection of identity-based attacks in wireless networks using signal prints. In: IEEE Green Computing and Communications (GreenCom), IEEE/ACM International Conference on and International Conference on Cyber, Physical and Social Computing (CPSCom), pp. 35–41 (2010)
4. Li, Q., Trappe, W.: Relationship-based detection of spoofing-related anomalous traffic in ad hoc networks. In: IEEE Third Annual IEEE Communications Society Conference on Sensor, Mesh and Ad Hoc Communications and Networks (SECON), vol. 1 (2006)
5. Khurana, P., Sharma, A., Singh, P.K.: A systematic Analysis on mobile application software vulnerabilities: issues and challenges. Indian J. Sci. Technol. 9(32), 1–6 (2016)
6. Yang, J., Chen, Y., Trappe, W.: Detecting spoofing attacks in mobile wireless environment. In: IEEE Sixth Annual Society Conference on Sensor, Mesh, and Ad Hoc Communications and Netwoks, pp. 1–8 (2009)

7. Chen, Y., Trappe, W., Martin, R.P.: Detecting and localizing wireless spoofing attacks. In: IEEE Fourth Annual Communications Society Conference on Sensors, Mesh and Ad Hoc Communications and Networks, pp. 193–202 (2007)

8. Sheng, Y., Tan, K., Chen, G., Kotz, D., Campbell, A.: Detecting 802.11 MAC layer spoofing using received signal strength. In: IEEE Twenty Seventh Conference on Computer Communications (INFOCOM), pp. 1–9 (2008)

9. Vijaykumar, R., Selvakumar, K., Kulothungan, K., Kannan, A.: Prevention of multile spoofing attacks with dynamic MAC address allocation for wireless network. In: IEEE International Conference on Signals and Communication Processing (ICCSP), pp. 1635–1639 (2014)

10. Lee, J.H., Buehrer, R.M.: Location spoofing attack detection in wireless network. In: IEEE Global Telecommunication Conference (GLOBECOM 2010), pp. 1–6 (2010)

11. Xiao, L., Li, Y., Liu, G., Li, Q., Zhuang, W.: Spoofing detection with reinforcement learning in wireless networks. In: IEEE Global Telecommunication Conference (GLOBECOM 2010), pp. 1–5 (2015)

12. Zeng, Q., Li, H., Quian, L.: GPS spoofing attack on time Synchronization in wireless networks and detection scheme design. In: IEEE Millitary Communications Conference (MILCOM), pp. 1–5 (2012)

13. Wang, N., Lv, S., Jiang, T., Zhou, G.: A novel physical layer spoofing detection based on sparse signal processing. In: IEEE Global Conference on Signal and Information Processing (GlobalSIP), pp. 582–585 (2015)

14. Dharani, S.B., Kumar, P.: Spoofing attack detection and localization of multiple adversaries in wireless network. Int. J. Recent Trends Electr. Electron. Eng. 3(2), pp. 133–138 (2014)

15. Yadav, A.S., Natu, P.M., Sethia, D.M., Mundkar, A.B., Sambare, S.S.: Prevention of spoofing attack in wireless network. In: IEEE International Conference on Computing Communication Control and Automation (ICCUBEA) pp. 164–171 (2015)

16. Raguvaran, S.: Spoofing attack: prevention in wireless network. In: IEEE International Conference on Communications and Signal Processing (ICCSP), pp. 117–121 (2014)

Threat Detection for Software Vulnerability Using DE-Based Adaptive Approach

Anshika Sharma, Palak Khurana, Sushil Kumar
and Shailendra Narayan Singh

Abstract Cloud computing has become a view in mind's eye of the upcoming entrepreneur architecture of the IT enterprise but still their exits security issues of information and communication technology (ICT), which has become primary mission since decades. In today's world, several blending vulnerabilities are making ICT infrastructure more prone to the worldly attacks and left behind disastrous impacts. Presently, the organizations are taking cloud computing, valuable option for outsourcing the important IT services so that it could cut the cost. In this research paper, we have proposed a security infrastructure for cloud and also a productive path to reduce the vulnerability threats and optimize the cost by using the differential evolutionary algorithm. We wind up our result by ensuring that the cloud computing providers and the business man can have an efficient control on the security services of the IT companies and reducing the cost amount by using our submitted work.

Keywords Differential evolutionary algorithm · Cloud computing
Optimization · Cloud security · Vulnerabilities · Information
and communication technology (ICT)

A. Sharma (✉) · P. Khurana · S. Kumar · S.N. Singh
Department of Computer Science and Engineering, Amity University, Noida,
Uttar Pradesh, India
e-mail: anshikaintegral@gmail.com

P. Khurana
e-mail: pkhurana187@gmail.com

S. Kumar
e-mail: kumarsushiliitr@gmail.com

S.N. Singh
e-mail: snsingh36@amity.edu

© Springer Nature Singapore Pte Ltd. 2018
M. Pant et al. (eds.), *Soft Computing: Theories and Applications*,
Advances in Intelligent Systems and Computing 584,
https://doi.org/10.1007/978-981-10-5699-4_52

1 Introduction

Cloud computing, a vast paradigm, is a service model that, on the one hand, provides customers on-demand network accesses to an enormous pool of shared computing resources (the cloud) and, on the other hand, provides us the security against the various software vulnerabilities. In the upcoming trend, globally all the businesses are constantly striving to cut the cost; among them, the most targeted areas are under information and communication technology (ICT). Cloud computing seems to be a very promising choice to carry through its objective by granting the entrepreneur to select the perfect business applications and information technology (ICT) framework, which is easily accessible in a fraction of amount. Computing on to a cloud is gaining a lot of importance among the businesses as a result of varied flexible opportunities. Fundamental characteristics that are symbolizing cloud computing include (a) service-based, (b) scalable and elastic, (c) shared, (d) metered by use, and (e) Internet technology dependable [1].

In today's scenario, adapting cloud computing sounds to be a very tempting on-demand service due to distinct profits such as quick implementation, greater computing power, pay-per-use, scalability, accessibility, availability, reduced costs, speedy elasticity, ubiquitary network access, higher flexibility, safety against network attacks, on-demand security controls, lesser cost from restoration against the disaster, and solutions for data storage. Diverse opportunities provided by cloud computing became primary driving force behind the commercial conception financed by virtualizations, increased performance, and physical server's capacity.

In spite of multitudinous benefits, large companies are still hesitant in joining the hands with the world of cloud computing until and unless the most prominent threats are pinpointed, clearly understood, and handled adequately, and ICT security has a leading controversy with respect to computing, along with the performance, availability, and inadequacy of standards of interoperable networks [2]. Nonetheless, the advancement of computing has been moreover apart from our predictions. From smaller organizations to the larger organizations, all are translocating their IT framework to the cloud. According to the analysis carried out by the International Data Corporation (IDC), some important funds have been kept aside for the use of cloud computing [2]. The prime fundamental of our paper is to present an optimized, analytical and sequenced technique to determine an optimal solution to discover the various software vulnerabilities and optimized by using the differential evolutionary (DE) algorithm.

The odds and ends of this paper are organized into various sections. Section 1 presents the introductory part followed by the literature survey in Sect. 2. Sections 3 and 4 deal with security framework and mapping on to different vulnerabilities. Optimization of security in cloud computing using differential evolutionary algorithm is described in Sect. 4.1. Section 5 presents the usage of algorithms in cloud computing optimization of security. Analysis, weaknesses, conclusion, and future work are discussed in the Sect. 6. The last section concludes the paper [3].

2　Literature Survey

The upcoming corporate world from all sizes includes small, medium, or large organizations are looking to grab the opportunities in order to fulfill the set standards and regulations to enhance the services provided without much expenditure. The banks are an interesting example in terms of huge investments to meet the financial service standards. The cloud computing in the beginning came into view as a metaphor for the Internet. But anyhow, in present days, it becomes the reflection of hidden framework [4]. It is expressed in terms of a way of technology where tremendously robust IT-enabled services are transported to the external customers with the help of Internet technologies.

2.1　Delivery Models of Cloud

The delivery model of cloud is indicated as cloud service delivery model because each model is distributed into various types of cloud computing services, such as software as a service (SaaS); in this, the applications of software may be used by the users using the cloud [5], and platform as a service (PaaS) that approves the customers to set up their own applications, which is created using the various programming languages, services, tools and libraries [6]. Here, users of infrastructure as a service (IaaS) access the needed storage, processing power, bandwidth, and the varied other accessories to install and get their software run [6], and data storage as a service (dSaaS) comprises of the primary memory effectiveness in Web (Fig. 1).

2.2　Cloud Computing Security Challenges

Security is the matter of interest when talking about data, framework, and virtualization. Private cloud is considered as a resistant deployment model from the other models of cloud. However, the safety in our model is dependent upon IT staff and the protection mechanisms of the place. Security issues become a major concern in the use of public, hybrid, and community models. Computing encourages in outsourcing data and the applications to the third party. Since, the availability, integrity, confidentiality, authenticity, and accountability are the essential objectives of the security services offered and the information systems [7]. When locating the IT assets on to the remote locations, they are being exposed to distinct security issues like the Web application vulnerabilities, database vulnerabilities, and the accessibility vulnerabilities to name a few. Some distinguishable hacking techniques that make the IT assets to station on to the cloud include cross-site scripting (XSS), domain name system (DNS) poisoning, man in the middle attacks,

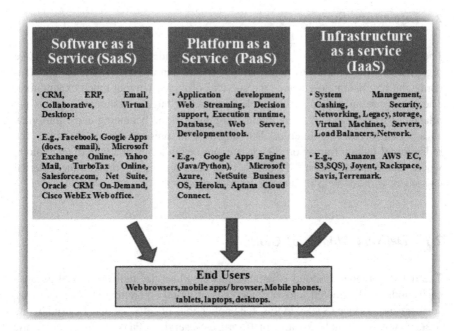

Fig. 1 Cloud computing delivery models and service offerings

structured query language (SQL) injection to name a few [7]. Furthermore, the issues related to the virtual memory traffic, virtual memory live migration, and virtual memory forensic can be migrated by the smart management of MAC addresses [8]. Digital forensic and pinpointing the attack vectors are a challenging task when we talk about cloud. In extension to susceptibility in security, cloud setup provides a platform to shoot the economical and disastrous attacks and also provide access on the powerful computing resources [9]. Gartner Inc. [10] determined that the security-related matters that are seven in number has to be forwarded by the cloud provider before consideration: (a) privileged user access, (b) data location, (c) segregation of data, (d) recovery, (e) investigation support team, (f) long-term viability, and (g) regulatory consent.

2.3 Cloud Computing Security Benefits

Cloud computing has multidiscipline environment where each discipline can use varied privacy, security, and interfaces. In spite of the various benefits brought by the cloud computing, still it faces numerous security-related challenges than other computing approaches. The security of the IT industry can be outsourced using monitoring as a service (MaaS) model, which is becoming popular since last decade [4]. Different cloud models help us in reducing the varied attacks of malwares by safeguarding our Internet applications and traffic. Moreover, this can be believed that "most cloud apps can be an effective tool in investigations of forensic and incident response" [11]. Logging is considered as a major activity for monitoring and analyzing the security issues. The susceptible applications and the software are eye-catching, so cloud computing provides power of computing in order to test password's strength. Password cracking time can be minimized, and the password assurance testing can become more enhance using the dedicated virtual machine (VM_s). They are basically designed and used for testing the functionalities and changes in security prior to the utilization for an instances [12], which ultimately reduces the cost of security testing.

2.4 Hacking Cases of Cloud Computing

2.4.1 Case Study on Amazon

The hackers made an attempt to break through the cloud's infrastructure of Amazon by attacking over a weak Web site for an unauthentic access to EC2 service of Amazon and quietly download password of zeus for occupying botnet. When the security researchers spotted botnet running, they quickly removed it [13].

2.4.2 Case Study on Sony

The incident that took place in Sony affected the 24.6 millions gamers on PC, caused major damage to the organization and to the idea of entertainment using cloud. The hackers earned a key to the protected and authenticated information by which they can easily get into the Sony's PC game network. This attack had bad idea of trusting among the users on the Sony's PC game's Web and computing technologies [14].

2.4.3 Case Study on Dropbox

The information of the accounts that are saved on cloud is weak. In the upcoming generations, users tend to use similar passkeys for various accounts. Dropbox user's logging credentials are used by the hackers for obtaining password from other Web sites and then use the same for other malicious attacks. A two-step authentication was implemented, in order to test the presence of virus and auditing them [15]. Still, the trust of the users was badly shattered and they have now shifted to parallel services that are now safer.

3   Security Framework for Cloud Computing

In the various studies done, we have adopted the IT security infrastructure from Fig. 2 [15].

Nonetheless, vulnerabilities' identification and remediation infrastructure (Fig. 3) have been developed to lead this research. The proposed framework has the following seven steps, which are listed below:

Step 1: **Finding the source**: Firstly, we have to find the source from where we can gather the relevant information and the data for our approach.

Step 2: **Gathering the information**: This is a very critical step for securing the IT infrastructure. Secondly, we have to gather the data and information, which are relevant to the systems and networks. Data include logs, configuration files to name a few.

Step 3: **Mapping of the Network and System/Classification of services**: Distinct systems require distinct IT security mechanisms. Therefore, we have to analyze and process the information and data collected in the step 1 in order to map the network and classify the various systems and services offered.

Step 4: **Exploring the Vulnerabilities/Penetrating Testing**: For this step, there are certain requisitely conditions, which includes mapping on to the network, classifying the different systems and services offered to them. Suitable techniques include auditing and penetration testing for this step.

Step 5: **Categorization of the Vulnerabilities**: Vulnerabilities that are identified have to be prioritized according to the severity of the risk and the impact on the

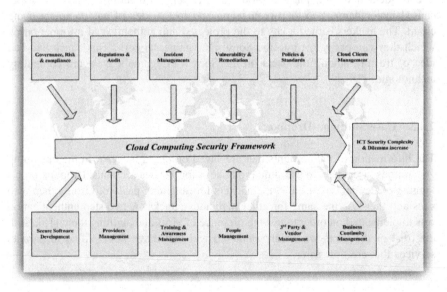

Fig. 2 Cloud computing security framework

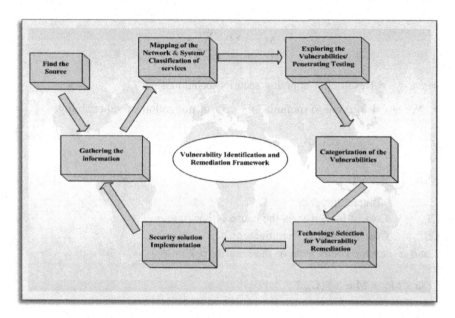

Fig. 3 Vulnerability identification and remediation framework

software. The technologies that are used to mitigate these vulnerabilities are also identified in this step.

Step 6: **Technology Selection for Vulnerability Remediation**: The various technologies claim to mitigate the similar types of vulnerabilities. Nonetheless, the intercommunication between these solutions can develop new vulnerabilities in addition to residual security issues.

Step 7: **Security solution**: The security solution is proposed in this paper.

4 Cloud Computing: Optimizing the Security Using the Differential Evolution (DE)

The main objective of optimizing the cloud computing security is to search the security-related technologies, in order to effactually reduce the vulnerabilities and the various security-related risks and threats to our important data onto a cloud. The aim of our paper is to.

- We are minimizing the residual vulnerabilities, and the coverage vulnerability has been maximized.

$$Mv = \text{Max} \sum_{pq=1}^{lm} \tau_{pq}$$

where τ_{pq} = Result after applying security techniques.

- We are also trying to minimize the cost of mitigation of vulnerability.

$$C_{v_q} = \lambda CL_{v_q} - \mu CS_{v_{pq}}; \lambda + \mu = 1$$

where

v_p	Vulnerability,
CL_{v_q}	Cost of loss, if v_p is the cause of hacking,
$CS_{v_{pq}}$	Cost of solution to mitigate v_p,
λ and μ	are the coefficients of the security risk.

i.e., $Mc = \text{Min} \sum_{q=1}^{m} (C_{vq})$

$$Mc = -\text{Max} \sum_{q=1}^{m} (-C_{vq})$$

- In the last, we are also minimizing the risk of proxy attacks, which is only possible if we maximize the distance between the victim node and vulnerable node.

$$\varepsilon = \text{Max} \sum_{q=1}^{l} (A_q + B_q)$$

where

A_q Host distance,
B_q Network distance.

Fitness function of a cloud network, to maintain the security level, represented as:

$$F = \alpha \sum_{pq=1}^{lm} \tau_{pq} + \beta \sum_{q=1}^{m} (-C_{vq}) + \gamma \sum_{q=1}^{l} (A_q + B_q)$$

where $\alpha\beta\gamma$ = positive coefficient for balancing security level and the cost and $\alpha + \beta + \gamma = 1$.

4.1 Differential Evolutionary Algorithm

Differential evolution (DE) is a parallel direct search method that uses NP D-dimensional parameter vector.

$X_{a,G}$, $a = 1, 2, 3,..., $ NP

For each generation, G. At the minimization process, NP did not alter. It is also a meta-heuristic algorithm that optimizes the problem.

Begin:

Initialize the DE population and find the vector population
Generate the perturbation: $X_{r1} - X_{r2}$
DE mutates and combines the population to produce a population of NP trial vector
Selection process begins
New population vector is mutated with a randomly generated perturbation
Again selection process

End.

5 Application of Algorithm to Cloud Computing Security Optimization

Case 1: A company conducted the survey on IT security and found a certain set of vulnerabilities, $V = \{V_1, V_3, V_5, V_7, V_9, V_{11}\}$, they are categorized. Since the fund was limited, the team decided to reduce the cost while covering all possible vulnerability. Therefore, they decided that β should be greater than $\gamma + \alpha$, where $\beta = 0.755$, $\alpha = 0.155$, and $\gamma = 0.105$. Result is found according to the company choice as shown in Table 1 and Fig. 4. The outstanding value in this case is negative, since the solutions are selected according to the cost.

Case 2: In this case, the funds are available and the team decided in covering all the security vulnerabilities. Therefore, they decided that β should be less than $\gamma + \alpha$, where $\beta = 0.105$, $\alpha = 0.805$, and $\gamma = 0.105$. Result is found to cover all the security vulnerability as shown in Table 2 and Fig. 5. The outstanding value in this case is

Table 1 Solutions and outstanding values of the fitness function

	V_1	V_3	V_5	V_7	V_9	V_{11}
Outstanding values	−0.2050	−0.3550	−0.3050	−0.1550	−0.1550	−0.0550
Outstanding solutions	P (22, 1)	P (8, 3)	P (9, 5)	P (3, 7)	P (23, 8)	P (16, 10)

Fig. 4 Fitness values are
plotted according to the
solutions found in Case 1

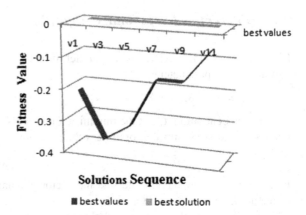

Table 2 Solutions and outstanding values of the fitness function

	V_1	V_3	V_5	V_7	V_9	V_{11}
Outstanding values	1.5050	1.6050	1.5050	1.8050	1.7050	1.9050
Outstanding solutions	P (3, 2)	P (9, 4)	P (13, 4)	P (2, 6)	P (11, 8)	P (16, 10)

Fig. 5 Fitness values are
plotted according to the
solutions found in Case 2

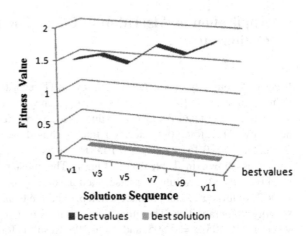

positive, which implies that all the detected vulnerabilities are mitigated using the
outstanding solution with very less residual security risk.

In Fig. 6, we can see the difference between the solutions selected in both the
cases. Coefficients α, β, and γ are considered successful, as a parameter to proportionate the need of security and the financial situation of the company.

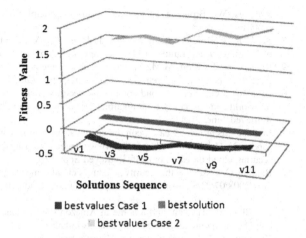

Fig. 6 Fitness values are plotted according to the solutions found in both Case 1 and Case 2

6 Conclusion and Future Scope

The methodology, which was proposed, is developed for cloud computing. On the other hand, by slightly changing the fitness function, this methodology can be used for other type of computing as well. These days, cloud computing is being followed by all the companies at a very high pace. A huge amount of security solutions are easily available, among them not all are effective. In this paper, we have used the DE algorithm to identify the optimal solution for the identified vulnerabilities as shown in the experimental results.

References

1. Plummer, D.C., Smith, D.M., Bittman, T.J., Cearley, D.W., Cappuccio, D.J., Scott, D.: Five refining attributes of public and private cloud computing. Gartner Res. (2009). http://www.gartner.com/newsroom/id/103501. Retrieved March 2012
2. Mullins, R.: IDC survey: risk in the cloud (2010). http://www.networkcomputing.com/cloud-computing/idc-survey-risk-in-the-cloud/229501529. Retrieved Sept 2016
3. Zineddine, M.: Vulnerabilities and mitigation techniques toning in the cloud. A cost and vulnerabilities coverage optimization approach using Cuckoo search algorithm with Levy flights. Comput. Secur. **48**, 1–18 (2015)
4. Rittinghouse, W.J., Ransome, F.J.: Cloud computing implementation, management, and security. CRC Press. Taylor & FrancisGroup (2010)
5. Zhu, J.: Cloud computing technologies and applications. In: Furht, B., Escalante, A. (eds) Hand of cloud computing (2010). http://www.springer.com/cda/content/document/cda_downloaddocument/9781441965233c1.pdf?SGWID¼0-0-45-989760-p173998774. Retrieved March 2012
6. Mell, P., Grance, T.: The NIST definition of cloud computing. National Institute of Science and Technology (2011). http://csrc.nist.gov/publications/nistpubs/800-145/SP800-145.pdf. Retrieved Aug 2016

7. Stallings, W., Brown, L.: Computer security: principles and practice. USA, Prentice Hall (2008)
8. Salah, K., Calero, M.J.A., Bernabe, B.J., Perez, J.M.M., Zeadally, S.: Analyzing the security of Windows 7 and Linux for cloud computing. Comput. Secur. **34,**113–22 (2013)
9. Ramgovind, S., Eloff, M.M., Smith, E.: The management of security in cloud computing. In: IEEE International Conference on Cloud Computing (2010)
10. Brodkin, J.G.: Seven cloud-computing security risks. Network World (2008). www.infoworld.com/d/security-central/gartner-seven-cloudcomputing-security-risks-853. Retrieved June 2016
11. McAfee. Secure cloud-based communications: manage risk while embracing cloud services. Jones, Q., California (2011). http://www.mcafee.com/us/resources/technology-blueprints/tb-securing-cloud-based-comm.pdf. Retrieved Sept 2016
12. Balding, C.: Assessing the security benefits of cloud computing (2008). http://cloudsecurity.org/blog/2008/07/21/assessing-the-security-benefits-of-cloud-computing.html. Retrieved August 2016
13. McMillan, R.: Hackers find a home in Amazon's EC2 cloud (2009). http://www.infoworld.com/d/cloud-computing/hackers-find-home-in-amazons-ec2-cloud-742. Retrieved Sept 2016
14. Kennedy, J.: Latest cyber attack on Sony a breach too far for entire cloud industry (2011). http://siliconrepublic.com/cloud/item/21613-latest-cyber-attack-onsony. Retrieved Aug 2016
15. Barron, C., Yu, H., Zhan, J.: Cloud computing security case studies and research. In: Proceedings of the World Congress on Engineering, vol. 2.. London, UK (2013)
16. Patil, J.: Information security framework: case study of a manufacturing organization (Master dissertation). Mercy College, New York (2008)

An Implementation Case Study on Ant-based Energy Efficient Routing in WSNs

Kavitha Kadarla, S.C. Sharma and K. Uday Kanth Reddy

Abstract The earlier era of Wireless Sensor Networks (WSNs) is extending its hands towards Internet-of-Things (IoT) and Cyber Physical Systems (CPS). The emergence of new technologies using WSNs leads to the dire need of real-time implementations to be utilized in typical applications like health care, agriculture, industries, and much more. These sensors are generally characterized by several limitations such as limited energy availability, low memory, and reduced processing power. Routing in WSNs is a well-studied research domain leaving several holes for further research. A large number of network routing algorithms have been developed to effectively pass the sensor data from individual nodes to base station. This paper presents an implementation case study using Ant Colony Optimization (ACO) heuristic for an energy efficient routing in WSNs. The real-time experiments were carried out using iSense proprietary hardware and software platform for WSNs. Many practical challenges like neighbor detection failure, packet loss, interference from external objects were taken into consideration and ACO algorithm was modified accordingly to successfully solve those problems. The results achieved are reported in this paper.

Keywords Wireless Sensor Networks · Ant Colony Optimization
Energy efficiency · Routing · Implementation · Internet-of-Things
Cyber Physical Systems

K. Kadarla (✉) · S.C. Sharma
Indian Institute of Technology Roorkee, Saharanpur, UP 247001, India
e-mail: satkavi.23@gmail.com

S.C. Sharma
e-mail: scs60fpt@gmail.com

K. Uday Kanth Reddy
Indian Institute of Technology Roorkee, Roorkee, Uttarakhand 247667, India
e-mail: kudaykanth96@gmail.com

© Springer Nature Singapore Pte Ltd. 2018   567
M. Pant et al. (eds.), *Soft Computing: Theories and Applications*,
Advances in Intelligent Systems and Computing 584,
https://doi.org/10.1007/978-981-10-5699-4_53

1 Introduction

Wireless Sensor Networks (WSN) [1] have captured the attention of researchers over the past decade because of the diverse applications they support and the flexibility of network deployment options they provide. These advantages, along with the remarkable advances in sensor technology towards Internet-of-Things (IoT) and Cyber Physical Systems (CPS), make them a favorite option for many tracking and monitoring applications. In such applications, sensors nodes can collaboratively monitor the network environment and report real-time information about the monitored phenomenon. With the advent of miniaturization and improved capabilities in microprocessors, sensors and transceivers, several new sets of applications appeared around WSNs. These WSNs are highly useful in typical applications like battlefield, natural calamity monitoring, volcano monitoring, forest fire monitoring, agriculture, health care, etc. By nature, the WSNs are hypersensitive and vulnerable to energy. The limited energy is the key issue influencing WSNs. So, routing algorithms have to be designed to maximize the life of WSNs. A family of ant colony optimization (ACO) [2] algorithms have been tested in the past for checking their efficiency in routing data in WSNs. The efficiency found was comparatively high to other routing algorithms. Through this paper, ACO for energy efficient routing has been implemented using special parameters to maximize the life time of WSNs.

This paper presents a case study on the implementation of energy efficient routing in WSNs using ACO. The rest of the paper is organized as follows. The recent literature survey is covered in Sects. 2 and 3 introduces the utilization of ACO for energy efficient routing in WSNs. Section 4, presents the experimental setup, the parameters used for experiments, the methodology of experiments, and results achieved during the implementation of the work. Section 5, concludes the paper.

2 Related Work

Several techniques are proposed in the literature to address the routing problem in sensor networks in an energy efficient [3–8] manner. Some of the recent studies are presented in this paper to understand the way with which routing problem is addressed using ACO.

Schurgers et al. [9] presented an approach in twofold to improve the life time of the sensor nodes. First, it aims at aggregating the streams of packets by optimal traffic scheduling and the second, to achieve the uniform resource utilization with the help of network shaping. The approach used is gradient-based routing with spreading for balancing the traffic flow and data combining entities that act like cluster heads to aggregate the data received from the sensors. Several simulation studies are done to showcase the work presented by the authors. As rightly indicated by the authors the approach may fail if the nodes are critical in network connectivity. Yet another method to improve the life time of the sensor nodes is developed by Kalantari et al. [10]. This paper presented a method based on partial differential equations to find shortest paths to cluster head from the source node. Gui et al. [11] have made an extensive study of the literature survey on application of swarm intelligence for routing in WSNs. Farzana et al. [12] presented a method to assess the link quality and link delay parameters based on packet reception rate, received signal strength, and link quality index. The link quality and link delay parameters are used for choosing a reliable path between source and destination. This research is based on several simulations. But some of the results indicated a weakness of poor delivery of the packets. Ghosh et al. [13] presented a dominant set based modified LEACH protocol with the application of ACO to construct a cluster and select a cluster head based on remaining energy level at the current node. However, the entire study is based on static and homogeneous nodes. A similar effort is made by the same authors in [14] by using PEGASIS to improve the life time of the sensors and reduce the redundancy and latency in the communications. Both these methods are simulation-based approaches. Gupta et al. [15] considered different packet sizes as a parameter to enhance the energy component of the sensor nodes within the cluster and outside the cluster. Several studies are made to assess the energy for both cluster and non-cluster environments. Yao et al. [16] developed an energy efficient, delay-aware model based on ACO to improve the life time and balance the data collection in WSNs. The model uses status gossiping and route construction methods for optimization. In this approach, ants will choose the nodes with highest remaining energy for routing the packets and further route construction process. Unlike many other researchers, the authors considered the heterogeneous nodes instead of homogeneous nodes for the study. This paper has done several simulation studies apart from the implementation of the system with a test bed of several motes and a workstation to study the performance analysis of the model presented in their work. Zhong et al. [17] focused on issue of depleting energy at mobile sinks due to large traffic load. To address the problem, they used ACO to mobile sink to optimize the traffic movement and maximize the life time of the nodes. The paper is based on simulations considering the factors like forbidden regions and the maximum movement distance of the sink in the optimization procedure. Liu et al. [18] developed a model that allows the network to be energy balanced and energy efficient by limiting the local transmission distance to be optimal using ant colony algorithm.

Though considerable research is reported in WSNs, there is little research reported with real-time implementations and performance analysis while applying ACO for energy efficient routing. Keeping this fact in mind, our paper attempts to implement an energy efficient routing for WSNs by applying ACO. Several experiments are carried out to analyze the established test bed. The results are reported.

3 Ant Colony Optimization (ACO)

ACO [2] is based on the behavior of ants that hunt for food by depositing pheromone on their path. Among a large number of random paths available, the path that leads to a food source will have high probability because of collective behavior of ants that deposit large amounts of pheromone along those paths. In this process, ACO denotes query packets as forwarding ants and response packets as backward ants. The concentration of pheromone is in inverse proportion to the distance to the destination node and directly proportional to the average of the remaining amount of energy level of the nodes on the path. In route maintenance process, the sensor nodes send a certain number of forwarding ants to the destination node periodically to monitor the quality. In this paper, we have adapted primary principles of ACO from [19]. In this approach, each ant attempts to find the optimal path in terms of shortest and energy efficient path. In their journey from source node to destination node, ants move from node i to a neighboring node j with a transition probability [19]

$$P_{ij}[k] = \frac{\tau_{ij}^{\alpha}\eta_{ij}^{\beta}}{\sum_{l \in N_i[k]} \tau_{il}^{\alpha}\eta_{il}^{\beta}}, \; j \in N_i[k], \tag{1}$$

where $P_{ij}[k]$ is the probability for ant k to move from node i to node j, τ_{ij} is the pheromone value deposited on link (i, j), η_{ij} a heuristic value assigned to the link, and (α, β) are weights used to control the importance given to the pheromone and the heuristic values, respectively. Here, $N_{ij}[k]$ is the list of neighbors of node i visited by ant k. The pheromone value $\tau_{ij}[k]$ on a given link depends on the likelihood that the ants pass by the link while constructing the solutions. The heuristic value η_{ij}, on the other hand, depends on the calculated cost of the link. The initial pheromone value is usually set to be equal for all links. The heuristic value of the link (i, j) is the inverse of the link cost, C_{ij}. The heuristic variable is calculated as follows.

$$\eta_{ij}(t) = \frac{e_j(t)}{\sum_{S_l \in C(i)} e_l(t)}, \tag{2}$$

where $e_j(t)$ is the remaining amount of energy of neighbor j for node i. The denominator is the sum of remaining energy levels of all neighbors for node i. When the forward ant gets to the destination node, the destination node generates a backward ant, and sends it back along the reverse path. The backward ant in each visited node release a certain amount of pheromone $\Delta\tau$ and it is given by

$$\Delta\tau = c.(Hop_{max} - Hop_{count}).Eavg_n \tag{3}$$

where c is the variable parameter, Hop_{max} represents the maximum allowed number of hops for forward ants; Hop_{count} represents the number of hops forward ant has taken to reach the destination node.

Therefore, when node receives the backward ant from the destination node by the nth neighbor node, the node will update the pheromone concentration $\tau_{n,d}$ given by

$$\tau_{n,d} = (1 - \rho).\tau_{n,d} + \Delta\tau \tag{4}$$

where ρ is the pheromone evaporation coefficient, $1 - \rho$ is the pheromone residue factor. The range of pheromone is from $[0, 1]$. The proposed algorithm considers both energy and distance. In order to take into account the movement of destination node, the above formula is slightly modified as

$$\tau_{n,d} = (1 - \rho).\tau_{n,d} + \frac{\Delta\tau}{\omega.Hop_{count}} \tag{5}$$

where ω is a control factor for controlling the influence of a number of hops over pheromone update rule and Hop_{count} is the number of hops traveled by the backward ant. In order to limit the maximum and minimum values of pheromone values to the range between 0 and 1, the minimum amount of pheromone for a particular neighbor is 0.005 and the maximum value is 0.9.

In this paper, temperature data have been used as the phenomenon that is being monitored by WSNs. The data and forward ants have been sent at specific intervals and the entire network is made to operate at a single synchronized time for coordination between forwarding ants and data packets by using time synchronization protocol provided in iSense SDK.[1]

4 Implementation and Results

In this paper, we have attempted to implement the ACO algorithm applied to WSNs to optimize the route in an energy efficiency manner. Several experiments are conducted on a practical test bed setup in the lab with actual motes, base station, and laptop. The following sections present the details of the experimental setup and results achieved during the experiments.

---

[1]http://www.coalesenses.de/index.php/products/solutions/isense-devices/. Last access May 2016.

Fig. 1 Wireless network
experimental setup

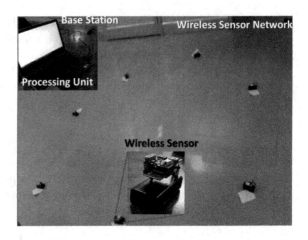

4.1 Experimental Setup

Figure 1 shows the experimental test bed setup at out laboratory which includes a
maximum of 10 motes connected wireless to a base station. The base station is connected
to a laptop used for collecting the routing information and processing it for performance
analysis. The sensors kit available in the lab is procured from Coalesenses,[2] a
German-based company. These modules are prefixed with a C++ based proprietary
iSense software. All motes were coded with the energy efficient routing protocol using
ACO. For successful implementation of ACO in WSNs, a time frame was proposed as
shown in Table 1 for route discovery process and route maintenance process using
ACO. The variation of energy of the network with time was observed in real time. Also,
the variation of a total number of forwarding ants that are required for the entire network
to get converged to their optimal paths from nodes to the destination was observed. The
network was subjected to various disturbances like switching off nodes in the network
dynamically (in order to create a case of node fault), hindrance from other objects which
lead to variation in the signal strengths between nodes. Table 2 lists the several important
parameters used during the experiments in the implementation phase.

4.2 Results

In this work, several experiments were conducted to observe the impact of ACO on
the experimental test bed setup as shown in Fig. 1 by varying the number of nodes
in the network over a number of ants and the behavior of network with the depleting
battery power over a period of time. Figure 2 shows a screenshot of the data

---

[2]http://www.coalesenses.de/-last. Access May 2016.

Table 1 Time line of the events defined for implementation

Step	Time tick	Event occurrence
1	0	Boot and Wait for time synchronization
2	6	It takes one second for getting synchronized. But time has been given for six seconds
3	66	Start the route discovery process by flooding sample data to neighboring nodes. Neighboring nodes can then identify its neighbors
4	76	Send flooding data again in order to compensate for the lost flooded data initially
5	86	Send flooding data for one last time and then enter into the energy sending process
6	93	Send energy level of the node to the top four neighbors in terms of signal strength
7	99	Send energy levels again in order to compensate for the lost energy level data packet sent previously because of unpredictable environment conditions
8	110	Check whether the given node is in the immediate neighbor of base station. If it is, then no need to send the forward ant. Or else, send a forward ant
9	116	Go to Step 3

Table 2 Parameters used for implementation

Parameter		Value
α		1.0
β		2.0–3.0
ρ		0.5
ω		0.6
Default pheromone		0.5
Number of motes		1–10
Sensors type		Temperature sensors
Number of base stations		1
Number of processing units		1
Radio type		ZigBee
Transmission range of each node		600 m
Internet protocol applied		IPv6 and 6LoWPAN
Transport protocol		UDP
Port number		8080
Identifiers used	Energy data	200
	Forward ant	202
	Backward ant	203

collection of the routing information and energy values of the individual nodes at base station. Figure 3 provides results of the assessment on the impact of a total number of ants as the number of nodes increase. Since the growth is almost linear,

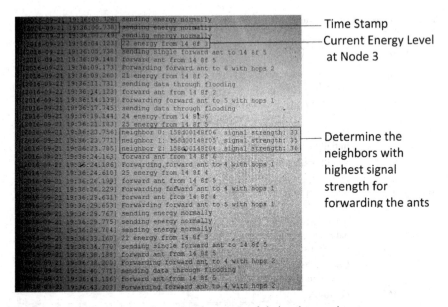

Time Stamp
Current Energy Level
at Node 3

Determine the
neighbors with
highest signal
strength for
forwarding the ants

Fig. 2 Routing information and energy values captured during the experiments

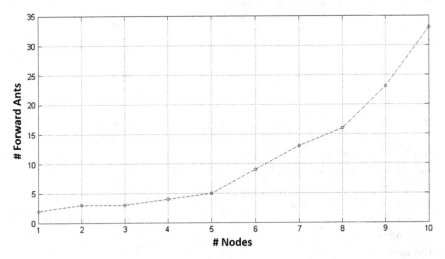

Fig. 3 Number of forward ants growing in the network as the number of nodes in the entire network increases

we believe that the ACO model used for experiments is scalable to a larger network. Further analysis is done to verify the impact of ACO on energy factor of the sensors over a period of time as shown in Fig. 4. We can observe from the figure that the depletion rate is also almost linear. However, further optimization is required to enhance the sustainability of the energy for a longer period. During the study, it is

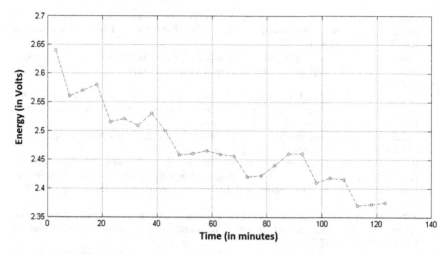

Fig. 4 Depletion of energy over a period of time

observed that there is an impact of a total number of forwarding ants that are required for the entire network to get converged to their optimal paths from nodes to destination. It is also observed that several disturbances on the network lead to variation in the signal strengths between nodes and still the network got dynamically adjusted to the variations and optimal paths were recorded.

Overall, the experimental studies indicate that ACO is a suitable algorithm for WSNs for energy efficient routing, even in real-time scenarios.

5 Conclusions

Through this work, we have implemented ACO for energy efficient routing and also minimum path distance in WSNs. We have proposed a new time frame for a better implementation of the above protocol in WSNs because of which the network could sustain many disturbances converge to its optimal path. The future work could be to investigate ways in which the total number of forwarding ants required to converge to optimal path for the entire network can be reduced.

References

1. Estrin, D., Govindan, R., Heidemann, J., Kumar, S.: Next century challenges: scalable coordination in sensor networks. In: Paper Presented at the Proceedings of the 5th Annual ACM/IEEE International Conference on Mobile Computing and Networking, Seattle, Washington, USA (1999)
2. Dorigo, M., Stutzle, T.: Ant Colony Optimization. Bradford Co., Scituate, MA, USA (2004)

3. Ganesan, D., Govindan, R., Shenker, S., Estrin, D.: Highly-resilient, energy efficient multipath routing in wireless sensor networks. ACM SIGMOBILE Mob. Comput. Commun. Rev. **5**(4), 11–25 (2001)
4. Mann, P.S., Singh, S.: Energy efficient hierarchical routing for wireless sensor networks: a swarm intelligence approach. Wireless Pers. Commun. (2016). doi:10.1007/s11277-016-3577-1
5. Muruganathan, S.D., Ma, D.C., Bhasin, R.I., Fapojuwo, A.O.: A centralized energy efficient routing protocol for wireless sensor networks. IEEE Commun. Mag. **43**(3), S8–S13 (2005)
6. Razaque, A., Abdulgader, M., Joshi, C., Amsaad, F., Chauhan, M.: P-LEACH: energy efficient routing protocol for wireless sensor networks. In: Long Island Systems, Applications and Technology Conference (LISAT), pp 1–5. IEEE (2016)
7. Razaque, A., Mudigulam, S., Gavini, K., Amsaad, F., Abdulgader, M., Krishna, G.S.: H-LEACH: Hybrid-low energy adaptive clustering hierarchy for wireless sensor networks. In: Long Island Systems, Applications and Technology Conference (LISAT), pp 1-4. IEEE (2016)
8. Tabus, V., Moltchanov, D., Koucheryavy, Y., Tabus, I., Astola, J.: Energy efficient wireless sensor networks using linear-programming optimization of the communication schedule. J. Commun. Netw. **17**(2), 184–197 (2015)
9. Schurgers, C., Srivastava, M.B.: Energy efficient routing in wireless sensor networks. In: Military Communications Conference, 2001. MILCOM 2001. Communications for Network-Centric Operations: Creating the Information Force, pp. 357–361. IEEE (2001)
10. Kalantari, M., Shayman, M.: Energy efficient routing in wireless sensor networks. In: Proceedings of Conference on Information Sciences and Systems (2004)
11. Gui, T., Ma, C., Wang, F., Wilkins, DE.: Survey on swarm intelligence based routing protocols for wireless sensor networks: an extensive study. In: 2016 IEEE International Conference on Industrial Technology (ICIT), pp. 1944–1949. IEEE (2016)
12. Farzana, A.F., Neduncheliyan, S.: Ant based mobility aided routing in mobile Wireless Sensor Networks. In: 2016 International Conference on Information Communication and Embedded Systems (ICICES), pp. 1–5. IEEE (2016)
13. Ghosh, S., Mondal., S, Biswas, U.: A dominating set based modified LEACH using ant colony optimization for data gathering in WSN. In: 2016 2nd International Conference on Advances in Electrical, Electronics, Information, Communication and Bio-Informatics (AEEICB), pp. 390–396. IEEE (2016)
14. Ghosh, S., Mondal, S., Biswas, U.: Enhanced PEGASIS using ant colony optimization for data gathering in WSN. In: 2016 International Conference on Information Communication and Embedded Systems (ICICES), pp. 1–6. IEEE (2016)
15. Gupta, A.L., Shekokar, N.A.: Novel approach to improve network lifetime in WSN by energy efficient packet optimization. In: 2016 IEEE International Conference on Engineering and Technology (ICETECH), pp. 117–122. IEEE (2016)
16. Yao, Y., Cao, Q., Vasilakos, A.V.: EDAL: An energy efficient, delay-aware, and lifetime-balancing data collection protocol for heterogeneous wireless sensor networks. IEEE/ACM Trans. Networking **23**(3), 810–823 (2015)
17. Zhong, J.-H., Zhang, J.: Ant colony optimization algorithm for lifetime maximization in wireless sensor network with mobile sink. In: Proceedings of the 14th Annual Conference on Genetic and Evolutionary Computation, Philadelphia, Pennsylvania, USA (2012)
18. Liu, X.: An optimal-distance-based transmission strategy for lifetime maximization of wireless sensor networks. IEEE Sens. J. **15**(6), 3484–3491 (2015)
19. Mohajerani, A., Gharavian, D.: An ant colony optimization based routing algorithm for extending network lifetime in wireless sensor networks. Wireless Netw. **1**(11) (2015)

Comparison Between Gaming Consoles and Their Effects on Children

Devansh Chopra, Ekta Sharma, Anchal Garg and Sushil Kumar

Abstract This study basically includes the understanding and definitions of Gaming Consoles, the comparison between three basic consoles of this generation —PS2, XBOX, and GameCube. Nowadays people play games to relief the tension from workload, workplace, and school and also from family problems like misunderstandings between siblings, etc. This paper discusses the effects or consequences of these gaming consoles on children As research is always intended to support, as they will not have to search three different papers in order to gather information of these three battling consoles instead they can find all the content related to gaming consoles here whether it is related to effects on children or to know more about best gaming console.

Keywords Gaming consoles · PS2 · XBOX · GameCube

1 Introduction

The study of gaming consoles is important for upcoming generations because these consoles are their source of entertainment, and they should know about the technology so that they can easily decide which console is best as per their requirements.

This research paper is quite different from other research papers because this paper includes different topics such as study of three different consoles and their important features so that this paper can help others to know about the best console and their technologies.

Nowadays people play games to relief the tension from job, home, and few additional stress-giving factors. Playing games help people to diverting their mind from stressful life to some stress-free environment. Games help in growth of mind

D. Chopra · E. Sharma · A. Garg · S. Kumar (✉)
Department of Computer Science and Engineering, Amity School
of Engineering and Technology, Amity University, Noida, UP, India
e-mail: kumarsushiliitr@gmail.com

© Springer Nature Singapore Pte Ltd. 2018
M. Pant et al. (eds.), *Soft Computing: Theories and Applications*,
Advances in Intelligent Systems and Computing 584,
https://doi.org/10.1007/978-981-10-5699-4_54

skills and alertness of mind. Games are of many types: console-based games, games for PCs, board-games, etc. Console games are most common, renowned games. Here, we are discussing three different gaming consoles:

XBOX: In 2001, gaming console XBOX was launched, and it is the only certified gaming console to run workable retained by Microsoft. Gaming console XBOX is truly a low-end personal computer.

PS2: Company Sony, launched PS2 in year 2001. This was the Sony's second launch in field of PlayStation Series. Then, PS2 became the best selling gaming console in the history. Game titles of more than 3,870 varieties have been introduced for PlayStation 2 since the time it was launched, and more than 1.5 million copies have been sold till date.

GameCube: GameCube was released in year 2001 by the company Nintendo. Optical disks were firstly used by the GameCube among all Nintendo consoles as a primary storage medium.

Disks are analogous to the mini DVD setup. Initially, system was not compatible to work with standard DVD or audio CD because of very small size of disks. A number of connectivity choices are provided to the GameCube by Nintendo.

Problems faced by people: Problems faced due to gaming are health problems like overweight, dry eyes, headaches, sleep disturbances, carpel tunnel syndrome, muscular problems, and skeletal problems and problems in postures of kids. Gaming also builds aggressive nature of gaming by playing violent games.

2 Literature Review

Amy Voida and Saul Greenberg inspected the cross-generational gaming practices of four different groups of gamers who play console games, age ranging from 3 to 83 mainly, the functions shot by distinctive groups when they play collected in groups. Their records highlighted the degree to which present gaming technologies help relations inside the cross-generational generation and also reveal generationally flexible suite of roles in these computer-mediated interactions and also for cross-generational generations, and they offered specifications to the creators of games who wanted to build console games.

Travis [1] presented differences in designing personal computers and gaming consoles, and he also examined the differences and similarities of processors and systems for past three generations of consoles as compared to their desktop counterparts of time.

Prof. Hamid Jahankhani stated and proved that human imagination is unlimited. Millions of people are entertained by advanced and amazing games. Prof. also said that these gadgets have also captured the interests of hackers, crackers, and cyber criminals. He also researched on forensic challenges and security of XBOX.

3 Research Methodology

3.1 Comparison Between Gaming Consoles [1]

In 1995, play station was launched by Sony (Japanese electronics company). In 2002, Sony achieves selling above $ 90 million only because of play station. A significant information was extracted from the world of gaming console, certainly called "PERIOD OF EXCITEMENT [2]," i.e., people will get bored after sometime and will demand something else. This is why NES released its new console every 5–6 years. GAME CUBE (2000) is the newest updated Nintendo console subsequent the publication of PlayStation 2 (PS2) in the same year as of GameCube by the company named Sony. The CEO of Microsoft, Bill Gates, was thrilled by the huge acceptance of these gaming consoles, and then, he participated in the battle of consoles and launched XBOX as their first gaming console.

Comparison between consoles on the behalf of main categories:

(a) Game Play (availability of different type of games, friendly to the user and age alignment according to the user)
(b) Specification
(c) Cost

According to me, Game Play is the best option among all the categories because if user can't play well with the console so there is no need of using it, no matter how advanced is the console.

3.1.1 Price

There is no precise price of a gaming console. Each console has two prices—one is buying a new console and another a used one. In the case of these three gaming consoles, GameCube seems to be the best, and it has a price of Rs. 14,000 for new one, and Rs. 7000 for old one, whereas the PS2 has a price of 23,000 for new one and Rs. 10, 000 for old one, and at last, the price for XBOX is Rs. 55,000 for new one and Rs. 15,000 for old one.

3.1.2 Specifications

XBOX has the best specifications till date and is also just similar to personal computer, which has a 733 MHz Intel Pentium III Processor and 6.4 gb/s as the highest transfer rate of bus, and XBOX becomes the gaming console with fastest processing speed.

Compared to XBOX, PS2 has only 300 megahertz processor with 3.2 gigabytes per second bus speed, whereas GameCube has 485 megahertz IBM processor [3].

Xbox and GameCube have a L2 caching system (256 kb) that allows the users to save the impermanent files and data for faster access (resulting in faster processing) [4].

Thus, PS2 has the lowest speed among the others but also has a feature that it has an "emotion engine" which uses the original PS1 I/O processors. Thus, PS1 games can also be played in PS2.

PS2 has the 150 MHz graphic synthesizer which can produce 75 million polygons/s. XBOX and GameCube have 250 and 162 MHz graphic processors that produce 125 million and 12 million polygons/s, respectively.

Xbox has a 8 gb inbuilt hard drive to store media whereas others can have a hard drive but only externally but external hard drive is very much costlier.

3.1.3   Game Play [5]

Game Play is the strongest point of PS2. PS2 have a biggest gaming collection. And the PS2 is the only console, which provides backward compatibility that is if you buy PS2, this does not means that you will dump PS1 games; PS1 games are also compatible in PS2. Thus, PS2 becomes the best valued console in playing games. Thus, PS2 has the highest resale value compared to others. The PS2 always had the greater demand, and it gained its high value in the market.

The other issue under the Game Play is the biasness issue. For example, GameCube games are meant for young users only. For adults, please do not think of buying this console unless you have a child taste in gaming. As GameCube is meant for younger generation, XBOX is meant for older generation who can afford costlier games. PS2 has a upper hand in this category as well as it has more game selection for every type of generation whether a younger one or an older one.

3.2   Effect of Gaming Consoles on Children [6]

There has been a lot of studies conducted to study the effect of video games on children. There are both good as well as bad effects on kids according to the studies. It depends on two factors. Firstly, if the child is playing age-appropriate games, and secondly, on how much time they spend in playing. We will see certain advantages and disadvantages of gaming and look on the various effects video gaming has on children.

3.2.1   Desirable Effects [6]

1. When a child plays video games, it helps in developing the child's perceptual and cognitive abilities and motor skills.
2. They also learn how to solve problems effectively, thus acquiring problem solving skills. They learn how to play and perform in teams and become aware.

It helps them memorize better. They learn to make decisions faster, make more correct decisions, and hence, they develop many skills.

3. The more interactive the video game is, the more it attracts and encourages the child to give in their best. Playing these games and winning them gives them motivation and builds up their confidence.

4. "Gamer Performs Better In Tests That involves focus, speed, multitasking, vision and accuracy than others." Thus, when you play video games, it actually teaches the child to concentrate, plan, and come up with different strategies to reach the goal. It makes the child a multitasking individual which is very crucial and thus handling difficult scenarios.

5. The fact is that this virtual environment is more or less the future of education. This is because of its interactive way for students to learn and succeed in this competitive world.

3.2.2 Negative Effects [7]

Physical health has been greatly affected by playing video games and also video games can increase the possibility of obesity, sleeping problems, headaches, dry eyes, carpel tunnel disease, muscular diseases, as well as skeletal, postural disorders in kids. Other adverse health effects involve anxiety, enlarged heart rate, and blood pressure.

Few researchers have showed a connection among violent conduct and playing video games that have viciousness. According to the researchers, children who are more associated with games which involve more shooting and killing rise anger in them and also lead to belligerent behavior.

These video games can hook children as playing and winning boosts their morale. A survey on children aged 9–15 years who play mature and violent games have a direct relation with the tendencies like bullying, becoming abusive, beating or clobbering siblings or peers.

A direct connection has been shown among those who spend time on playing video games and others who ignore their educations and resulting into bad performance in studies.

Playing certain intense video games can also lead to death of the people as shown by some records.

4 Survey

Now, let us conduct a survey to know the other people response about which gaming console is the best…why let us…?

Survey was conducted between different age groups that are 3–10 years, 11–18 years, 19–25 years, and 26 and above people. Every person who filled the survey was given a consent form as well, and people below 18 years of age were asked to get the consent form signed by their parents. The consent form for both minor and adults was different. No one was pressurized to fill up the survey.

LINK: https://docs.google.com/forms/d/1j8kivtB05uaDpOlWcDytr7P22207Srd-N5eHMJCdBRU/viewform?c=0&w=1&usp=mail_form_link.

4.1 Survey Conclusion

After receiving the opinion of 50 people, I gathered many different views about gaming consoles by people of different age group.

About more than 50% people play "play station", an amazing gaming console developed by SONY. And about 35% people like to play Xbox and rest play GameCube and pc games. To be more accurate, we can see the pie chart below which says that 58% people like to play "play station," 26% people like to play XBOX, 8% people like to play GameCube, and rest 8% people like to play pc games.

Majorly, the video games are played by 11–18 years of age group as seen in the response spreadsheet. Maximum people search the games by visiting online catalog of the games, and mainly, they check about the graphics, technology, and user friendliness of the games. After playing the video games, almost every person encounters the problem of dry eyes, and there are some people who are also suffering from carpel tunnel syndrome and obesity, and there are other diseases like headaches, muscular pain, etc., and reducing the time spent on video games and increasing outdoor games can prevent these problems.

Acknowledgements This study is a survey paper from the literature, and references are included. The information is analyzed only for research purpose not for any commercial work.

References

1. Travis, P.: Game Consoles Vs. Personal Computers Design, Purpose, and Marketability. University of Alaska Fairbanks
2. Nintendo GameCube Technical Specs. Retrieved from http://en.wikipedia.org/wiki/Nintendo_GameCube. 23 Oct 2012
3. Amy, V., Saul, G.: (2012) Console Gaming Across Generations: Exploring Intergenerational Interactions in Collocated Console Gaming
4. Mark, G., Christoph, B., Andreas, R.: Keeping the Game Alive: Evaluating Strategies for the Preservation of Console Video Games, pp. 64–90
5. Research Paper: The International Journal of Digital Curation 5(1) (2010)
6. Ghadrdanizadi, A., Hamid, J.: Games Consoles Security and Forensics Challenges

7. Ralph, B.: Recovering the History of the Video-Game. Smithsonian Lemelson Center (2012)
8. Do Consumers Still Need Game Consoles? The Question Arises as PCs and Mobile Platforms Draw Crowds of Game Players (2015)
9. Liu, H.: Dynamics of pricing in the video game console market: skimming or penetration? J. Mark. Res. **47**(3), 428–443 (2010)

Design of Chamfered H-bend in Rectangular Substrate Integrated Waveguide for *K*-band Applications

Anamika Banwari, Shailza Gotra, Zain Hashim and Sanjeev Saxena

Abstract This study concerns the designing and analysis of right-angled chamfered H-bend using rectangular substrate integrated waveguide technology. The proposed design combines the advantages of planer profile, such as high performance, low fabrication cost, easy integration with the low loss inherent to the waveguide solution. The waveguides get easily connected into any positions required, in miniature circuit by using proposed bend. The bend is optimized in *K*-band 18–26.5 GHz by HFSS code. Unlike others, this proposed bend is designed on modified epoxy substrate and the measured results demonstrate that the return loss is −44.89 dB and VSWR is 1.01 at the bandwidth of 23 GHz. Moreover, the return loss and VSWR remain below the value of −21 dB and 1.3, respectively, across the entire *K*-band. Thus, through the study of results, the optimal positions of metallic poles have been confirmed and the bend is very effective for signal transmission with low losses.

Keywords Waveguide · Rectangular substrate integrated waveguide
Waveguide bend · S parameter and HFSS

A. Banwari (✉) · S. Gotra · Z. Hashim · S. Saxena
Department of Electronics and Communication Engineering,
Amity University, Noida, Uttar Pradesh, India
e-mail: anamika.banwari@gmail.com

S. Gotra
e-mail: shailzagotra25@gmail.com

Z. Hashim
e-mail: zainhashim64@gmail.com

S. Saxena
e-mail: ssaxena@amity.edu

© Springer Nature Singapore Pte Ltd. 2018
M. Pant et al. (eds.), *Soft Computing: Theories and Applications*,
Advances in Intelligent Systems and Computing 584,
https://doi.org/10.1007/978-981-10-5699-4_55

1 Introduction

A large variety of microwave components using SIW (substrate integrated waveguide), such as power dividers, filters [1], transitions, circulator [2], phase shifter [3], antennas, and couplers [4], are designed and studied at high quality, high power, low cost, and easy to integrate with other components within the same dielectric substrate [5–7]. To acquire the advantages of rectangular waveguide while being in planar profile, the SIW technology is beneficial [8, 9]. The microwave bends are very important structures as it arrange the waveguide into the positions required. It is obligatory to ensure the bending is accomplished in the accurate manner else the electric and magnetic field will be distorted and the signal will not propagate in the required direction and causes loss and reflection. To overcome this, the waveguide bends are manufactured to alter the wave propagation without destroying the field patterns and instigate losses. In this paper, K-band 90° chamfered H-bend is proposed and optimized using RSIW with modified epoxy substrate RSIW. Bends are building blocks for many millimeter and microwave wave integrated circuits and telecommunication systems such as radar, satellite, astronomical observations, and automotive radar.

Previous works show that Rahali and Feham [10] designed V-band components using RSIW. Bhowmik et al. [11] designed multiple beamforming antenna system by using substrate integrated folded waveguide (SIFW) technology. Rhbanou et al. [12] designed dual-mode band-pass filters by using substrate integrated waveguide. Fojlaley et al. [13] introduced antenna array with SIW planar feed network.

The main contribution of the presented paper is that unlike others existing, the presented design used modified epoxy dielectric substrate, which has good temperature and chemical resistant. Thus, the bend has good survivability in harsh environment. Moreover, it has quite good VSWR and return loss values, which means that when operating at K-band losses are very less. This method is quite simple and quick to design microwave bends, found on RSIW technology.

This paper contains section as follow: Sect. 2 describes the design parameters, equations used to calculate parameters, and design of RSIW. Section 3 illustrates the design of right-angled chamfered H-bend using defined RSIW. Section 4 describes the results of presented design and last section consists of conclusion and future work of this paper.

2 Design of RSIW

The RSIW is built in a dielectric substrate of height h, by assembling two parallel lines of metallic poles on sidewalls, which connects the ground and upper plane of dielectric substrate, preserving the benefits of conventional metallic rectangular waveguide. The electric field dissemination in rectangular substrate integrated waveguide is identical to the classical waveguide without any leakage loss. The permittivity ε_r of substrate is mediate between two metal poles that are placed on upper side and lower side to allow propagation of TE_{10} mode. Smith [14] presented equations for propagating modes in SIW. The cutoff frequency (f_c) for TE_{10} made has been defined as formula (1).

$$f_c(TE_{10}) = \frac{c}{\sqrt[2]{\varepsilon_r}} \times \left(W_{eq} - \frac{d^2}{0.95p} \right)^{-1} \tag{1}$$

where

d is diameter of metallic pole
ε_r is relative permittivity
c is speed of light
W_{eq} is width of waveguide

The distance between two consecutive poles is p which should be kept proper so that leakage should be less, the diameter of the pole is d and the height of pole is same as the height of dielectric substrate h, the space between two parallel lines of metallic poles is W_{SIW}. The basic building structure for designing bends that is RSIW (rectangular substrate integrated waveguide) is demonstrated in Fig. 1. As depicts in Fig. 1, the rectangular SIW is of length 43.43 mm is conceived in the K-band (18–26.5 GHz), with a dielectric substrate modified epoxy of permittivity value 4.2, of height 0.408 mm. By using SIW technology, two parallel lines of metallic poles of diameter 1.524 mm and of height 0.408 mm are fabricated in dielectric substrate. The spacing between two metallic poles is 2.54 mm.

If the parameters are elected properly, the energy leakage through sidewalls is less and can get effective electric and magnetic field distribution in a waveguide [8]. In previous work, Cassivi et al. presented empirical equations, which are derived to

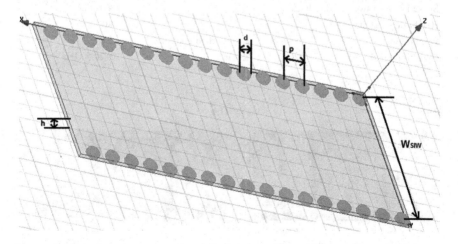

Fig. 1 Structure of rectangular substrate integrated waveguide

calculate the width of the rectangular wave guide, which gives the equivalent characteristics of the fundamental mode propagating in the RSIW having the same height and the dielectric constant. Equation (2) describes those formulas:

$$W_{eq} = W_{SIW} - \frac{d^2}{0.95p} \tag{2}$$

where, p is:

$$p < \frac{\lambda_0}{2}\sqrt{\varepsilon_r} \quad \text{or} \quad p < 4d$$

λ_0 is space wavelength, $\lambda_0 = \frac{c}{f}$.

The numerical values of design parameters are described in Table 1. First, the basic RSIW structure has been designed by using this parameter and further 90° chamfered H-bend has been designed. The distance between consecutive poles must be kept low to reduce the sidewall leakage losses, and in support of this statement the electric field distribution for equivalent rectangular waveguide and RSIW has been studied.

Figure 2 proves that the sidewall leakage of electric field in rectangular substrate integrated waveguide is negligible as compared to classic equivalent waveguide.

Table 1 Design parameters of the proposed RSIW

Parameters	Abbreviations	Numerical values
Length	L	43.43 mm
Pole diameter	D	1.524 mm
Pole spacing	P	2.54 mm
Space between two metallic lines	W_{SIW}	7.416 mm
Height of substrate	h	0.408 mm
Permittivity	ε_r (modified epoxy)	4.2
Operating band	K-band	18–26.5 GHz

(a)   **(b)**

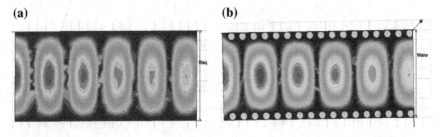

Fig. 2 Electric field distribution for **a** equivalent rectangular waveguide and **b** rectangular substrate integrated waveguide

3 Design of 90° Chamfered H-Bend

The direction of propagation is not always straight. To change the direction of propagation in stiff waveguides, it requires the use of bends. Therefore, it is an important factor to design and simulate these bends in an accurate and speedy manner. The most common methodology to compensate the disjointedness of right-angled bend is to chamfrain the corners in order to minimize the reflection. Through this paper, the SIW chamfered right-angled bends in the K-band have been designed. Thus, we have examined the impact of chamfrain position on the return loss and VSWR as these are important parameters for analyzing the performance of the device. Through this study, we have examined various structures such as the rectangular waveguide (Fig. 1). Figure 3 shows electric field distribution of a right-angled bend without SIW structure. The bend has decreasingly E field density in the direction of propagation because of sidewall losses.

Figure 4 depicts that when bend (Fig. 3) has assembly of SIW in 90°, parallel to sidewalls then the E field is propagates longer without any side leakage. But the electric field reflected back when hits the metallic rods at the end. To overcome this, bend's corner has been chamfered using SIW in circular manner.

If the positions of circular chamfrains are made properly, the energy leaking through the bend will be minimum and can get effective distribution of magnetic and electric field in a waveguide [15]. Figure 5 shows the right-angled H-bend with circular chamfered at corner. As depicts from the diagram, there is less reflection and wave propagates to the other end with low loss as compared with Figs. 3 and 4. The supportive results are discussed in result section.

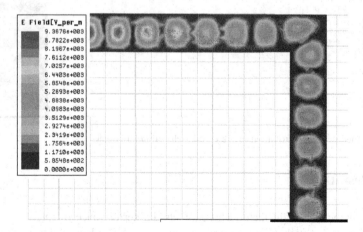

Fig. 3 Electric field distribution in the equivalent rectangular bend

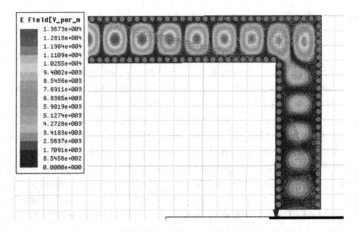

Fig. 4 Electric field distribution in the bend with chamfrains at angle of 90°

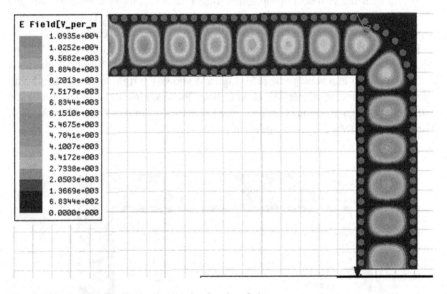

Fig. 5 Electric field distribution in the circular chamfrain at corner

4   Results

As discussed, the main factor which affects the performance of microwave components is losses such as return loss, component mismatch loss, and bends loss. The return loss and VSWR simulation for different geometries are illustrated under this section which has been selected as performance parameters.

4.1 Return Loss Plots (Frequency Vs. S11)

Figure 6a, b depicts the return loss in the bends geometries not having circular chamfrain (Figs. 3 and 4). On the study of these plots (Frequency GHz vs. reflection coefficient dB), the two chamfered bends without circular chamfrain at the corner have return loss values −21 dB and −37.40 dB, respectively, at K-band frequency 23 GHz. −37.40 dB is quite acceptable, but it can be improve by using circular chamfrain.

The bend having chamfered SIW at the corner in circular manner gives −44.89 dB return loss, which proves that design of Fig. 5 has less losses as compared to other two designs when it used for wave transmission (Fig. 7).

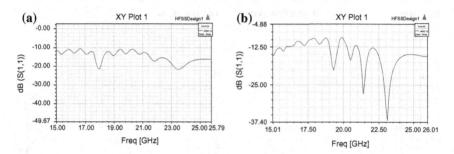

Fig. 6 Return loss of the **a** equivalent rectangular bend and **b** bend with a chamfrain angle of 90° at K-band frequencies

Fig. 7 Return loss of the bend with circular chamfrain corner at K-band frequencies

4.2 VSWR Plots (Frequency Vs. VSWR)

The ideal value of VSWR is 1 which means that there is no reflection and mismatch loss as this value is increases, losses also increase and affect the performance. From the curves presented in Fig. 8a, b, it is concluded that the bends without circular chamfrain at corners have VSWR value 1.96 and 1.28, respectively, at K-band frequency 23 GHz. These values are quite high and increase the losses as well as decrease the performance of bends. VSWR can be decrease by using SIW circular chamfrain at those areas where losses are occurs.

The bend having SIW at the corner in circular manner gives VSWR value of 1.02, which is quite close to the 1. It proves that design of Fig. 5 has less losses as compared to other two designs when it used for wave transmission (Fig. 9).

These results confirm that the circular chamfered bend is very effective for transmission of signal in microwave components with minimum losses.

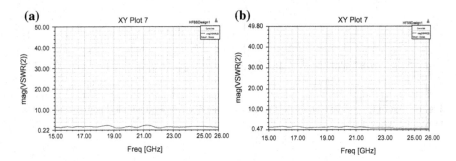

Fig. 8 VSWR of the **a** equivalent rectangular bend and **b** bend with a chamfrain angle of 90° at K-band frequencies

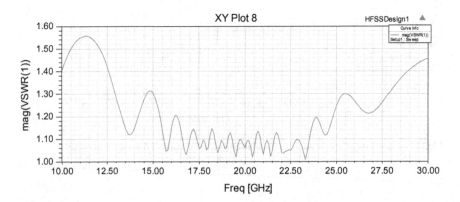

Fig. 9 VSWR of H-bend having circular chamfrain at K-band frequencies

5 Conclusion

In this paper, the design of a microwave 90° chamfered H-bend using substrate integrated waveguide for K-band applications is presented. The bend is designed on modified epoxy substrate. The presented method is quite simple and quick method to design microwave bends, built on RSIW technology using Ansoft HFSS code. The impact of different geometry of SIW bend on results is also studied and compared. Different performance parameters such as return loss and VSWR have been plotted for different bend geometry.. The measured results demonstrate that the return loss is −44.89 dB and VSWR is 1.01 at the bandwidth of 23 GHz. The bend holds good survivability in harsh environment and very effective for telecommunication. The applications of presented designs can be very useful for microwave systems such as radars, diplexers and multiplexers and satellite. This proposed design achieves significant results; future work may be efficient in terms of fabrication and designing other microwave components using proposed rectangular substrate integrated waveguide.

References

1. Adabi, A., Tayarani, M.: Substrate integration of dual inductive post waveguide filter. Electromagnet. Res. B 7, 321–329 (2008). doi:10.2528/PIERB08051002
2. Che, W., Ji, X.J., Yung, E.K.N.: Miniaturized planar ferrite junction circulator in the form of substrate-integrated waveguide. Int. J. RF Microwave Comput. Aided Eng. 18, 8–13 (2007)
3. Rahali, B., Feham, M.: Design of Ku-band substrate integrated waveguide phase shifter. Int. J. Inform. Electron. Eng. 4(3), 225–229 (2014)
4. Tan, K.-J., Luan, X.-Z.: Compact directional coupler based on substrate integrated waveguide. IEEE Microw. Wirel. Compon. Lett. 1–4 (2009). doi:10.1109/IWAT.2009.4906898
5. Bochra, R., Mohammed, F., Tao, J.: Design of optimal chamfered bends in rectangular substrate integrated waveguide. Int. J. Comput. Sci. Issues (IJCSI) 8(4(2)), 1694–0814 (2011)
6. Deslandes, D., Wu, K.: Accurate modeling wave mechanisms, and design considerations of a substrate integrated waveguide. IEEE Trans. Microw. Theory Tech. 54, 2516–2526 (2006). doi:10.1109/TMTT.2006.875807
7. Che, W., Wang, D., Deng, K., Xu, L, Chow, Y.L.: Characteristics of H-plane rectangular waveguide bends integrated into thin substrate. In: Antennas and Propagation Society International Symposium IEEE, pp. 1969–1972 (2006). doi:10.1109/APS.2006.1710962
8. Cassivi, Y., Perregrini, L., Arcioni, P., Bressan, M., Wu, K., Conciauro, G.: Dispersion characteristics of substrate integrated rectangular waveguide. IEEE Microw. Wirel. Compon. Lett. 12(9), 333–335 (2002). doi:10.1109/LMWC.2002.803188
9. Rahali, B., Feham, M., Tao, J.: Analysis of S-band substrate integrated waveguide power divider, circulator and coupler. Int. J. Comput. Sci. Eng. Appl. (IJCSEA) 4(2) (2014)
10. Rahali, B, Feham, M.: Design of V-band substrate integrated waveguide power divider, circulator and coupler. Comput. Sci. Inf. Technol. (CS & IT) 35–44 (2013)
11. Bhowmik, W., Srivastava, S., Prasad, L.: Design of multiple beam forming antenna system using substrate integrated folded waveguide (SIFW) technology. Electromagnet. Res. B 60, 15–34 (2014)

12. Rhbanou, A., Sabbane, M., Bri, S.: Design of dual-mode substrate integrated waveguide band-pass filters. Circ. Syst. **6**(12), 257–267 (2015). doi:10.4236/cs.2015.612026

13. Fojlaley, M., Amirkabiri Razian, S., Mohammadi, B., Pourahmad Azar, J., Nourinia, J.: 10-GHz antenna array with substrate integrated waveguide planar feed network. Int. J. Tech. Res. Appl. **2**, 77–78 (2014). e-ISSN: 2320-8163

14. Smith, N.A.: Substrate integrated waveguide circuits and systems. Thesis for the degree of Master of Engineering, Department of Electrical & Computer Engineering McGill University Montréal, Québec, Canada, May (2010)

15. Che, W., Deng, K., Chow, Y.L.: Equivalence between Waveguides with side walls of cylinders (SIRW) and of regular solid sheets. In: IEEE Microwave, APMC, Asia-Pacific Conference, vol. 2, INSPEC Accession Number: 8874441 (2005). doi:10.1109/APMC.2005. 1606377

Journal Article

16. Bochra, R., Mohammed, F., Tao, J.: Design of optimal chamfered bends in rectangular substrate integrated waveguide. Int. J. Comput. Sci. Issues (IJCSI) **8**(2)(4), 1694–0814 (2011). ISSN (Online): 1694-0814

17. Rahali, B., Feham, M., Tao, J.: Analysis of S-band substrate integrated waveguide power divider, circulator and coupler. Int. J. Comput. Sci. Eng. Appl. (IJCSEA) **4**(2) (2014)

18. Fojlaley, M., Amirkabiri Razian, S., Mohammadi, B., Pourahmad Azar, J., Nourinia, J.: 10-GHz antenna array with substrate integrated waveguide planar feed network. Int. J. Tech. Res. Appl. **2**, 77–78 (2014). e-ISSN: 2320-8163

Journal Article Only by DOI

19. Che, W., Ji, X.J., Yung, E.K.N.: Miniaturized planar ferrite junction circulator in the form of substrate-integrated waveguide. Int. J. RF Microwave Comput. Aided Eng. **18**, 8–13 (2007). doi:10.1002/mmce.20260

20. Rahali, B., Feham, M.: Design of Ku-band substrate integrated waveguide phase shifter. Int. J. Inf. Electron. Eng. **4**(3), 225–229 (2014). ISSN: 2010–3719. doi:10.7763/IJIEE.2014.V4. 438

21. Rahali, B, Feham, M.: Design of V-band substrate integrated waveguide power divider, circulator and coupler. Comput. Sci. Inf. Technol. (CS & IT) 35–44 (2013). doi:10.5121/csit. 2013.3804

22. Bhowmik, W., Srivastava, S., Prasad, L.: Design of multiple beams forming antenna system using substrate integrated folded waveguide (SIFW) technology. Electromagnet. Res. B **60**, 15–34 (2014). doi:10.2528/PIERB14022603

23. Rhbanou, A., Sabbane, M., Bri, S.: Design of dual-mode substrate integrated waveguide band-pass filters. Circ. Syst. **6**(12), 257–267 (2015). doi:10.4236/cs.2015.612026

An Approach to Vendor Selection on Usability Basis by AHP and Fuzzy Topsis Method

Kirti Sharawat and Sanjay Kumar Dubey

Abstract The exceptionally quick change in innovation and development in the field of portable devices puts major test for engineers to find out the best features which users want in their devices and strategically plan accordingly to serve best to their users. Mobile technology is the only technology which owns 87% of the world's total population of portable device users. Mobile technology also reduces the dependency of peoples on fixed physical places and allowed them to do business from a far place. However, this decision becomes complicated in case of multiple mobile companies vendors and multiple conflicting criteria. Therefore, for the selection of mobile companies, AHP method is used as an approach which is a multi-criteria decision-making method. This paper aims to find out the best mobile company and find out the one which serves the best applications.

Keywords Usability · AHP · Fuzzy · Topsis · Model

1 Introduction

This century has marked rapid growth in the smartphone market. Mobile applications are now used for almost all areas of service as many business houses have deployed mobile applications due to competitive environment. The mobile application market has become competitive due to their rapid growth and availability of different type of mobile applications provided by these companies. It increases the growth in the different types of mobile phones as well as vendors. The amazing journey of mobile devices has been proved that there is a lot more than to do than just only making calls and used for communication purpose only. This rapid

K. Sharawat (✉) · S.K. Dubey
Department of Computer Science and Engineering, A.S.E.T, Amity University
Uttar Pradesh, Sec.-125, Noida, UP, India
e-mail: kirtisharawat21@gmail.com

S.K. Dubey
e-mail: skdubey1@amity.edu

© Springer Nature Singapore Pte Ltd. 2018
M. Pant et al. (eds.), *Soft Computing: Theories and Applications*,
Advances in Intelligent Systems and Computing 584,
https://doi.org/10.1007/978-981-10-5699-4_56

development in mobile applications makes our life faster, comfortable, and easier. The latest mobile phones have lot of features that we could have never think of. For developer, it is difficult to develop an accurate, useful, and adoptable application. To ensure that the mobile application is accurate and useful one, we need to evaluate the usability of mobile applications. Therefore, evaluating usability means to measure usability of mobile applications. There are lots of categories available in mobile phones. Our lives are transformed truly by features offered by these mobile devices like MMS, Bluetooth, Internet access, and video recording. The main objective behind the paper is to take review from others and on the basis of the review to out the usability metrics for different mobile companies and find out the best company by using AHP theory and validate the result by using Fuzzy Topsis method. The analysis of current techniques and review will result in a set of selected usability guidelines for mobile applications. In the next section, a review of several usability evaluation techniques is described and model is presented and highlights the various factors on the basis of which we evaluate the usability of different mobile companies. Finally, the conclusion takes place.

2 Model Development

A model is proposed to develop an approach for finding the best mobile companies which serve the best applications to their users. These mobile applications are compared in terms of various factors, and these are simplicity, efficiency, functionality, and attractiveness. To find the mobile company which serves the best mobile applications on the basis of the above-mentioned factors, the data is collected. Hence, they are evaluated by using MCDM (multi-criteria decision making) approach; in this, we calculate each criterion weight using AHP and then ranking is done by using Fuzzy Topsis method.

2.1 Defining Factors for Mobile Company Selection

Firstly, we establish the factors which are used to evaluate the mobile companies. In order to meet the factors and their importance for the selection process of mobile companies, the data is collected by taking reviews from the peoples itself. Defining factors for mobile company selection contains the step by step procedural collection of factors such as simplicity, efficiency, effectiveness, and error tolerance, and based on these features each feature is passed through and given the rank on the scale of four points.

In order to select and decide which factor will be useful to the research model, each factor is given numbers rating (1–9); if a factor rates in between the numbers (1–7), then it may not be considerable for the model; but if a criterion falls under

(7–9), then it is selected to be processed further. In order to choose the most important factor, it is mandatory for the factor to have average above seven.

2.2 Decision Parameters

(i) *Simplicity*: Simplicity is the state or quality of being something simple, which is easy to understand or explain seems simple, in contrast to something being complicated. Sometimes, we found some problems like a mobile functionality is very good but it is very difficult to operate.

(ii) *Efficiency*: Productivity is the capacity to abstain from squandering materials, vitality, endeavors, time, and cash in accomplishing something useful or in delivering a craved result. Effectiveness is measured as the proportion of valuable yield to the aggregate info.

(iii) *Effectiveness*: It is the ability to produce the expected outcome. When something seems to be effective, then it means it has an intended and desired result. Effectiveness sometimes used as a quantitative term.

(iv) *Error Tolerance*: It is the evaluation of the error correlated to the size of each solution component. It allows any system or application to work properly even in case of event failure. An error tolerant system enables an application to continue its function may be at a lower level but prevents them from a complete failure.

2.3 Approach

A relationship is established between the mobile companies (V1, V2, and V3) and the selection factors (C1, C2, C3, and C4) as shown in Fig. 1. After creating the decision hierarchy for the concerning problem, the weights of the criteria are used for evaluation and AHP method is used for the calculation in the second phase. The assessment is done by the AHP theory as shown in Sect. 3 (Fig. 2).

3 The AHP

AHP is due to Saaty [2], and this method is very popular and concisely used especially in military for analysis. It is a highly outstanding management tool for complex multi-criteria decision-making problems.

Analytical hierarchy process has its own ability to make rank of the choices, in order to make an effective result out of conflicting objects. If the judgments between two or more objects are relatively important and equal than one another, using AHP

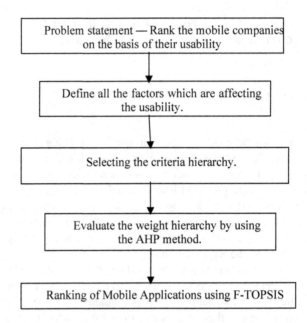

Fig. 1 Steps of the proposed model

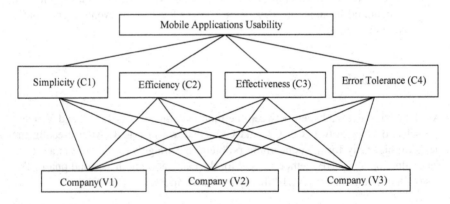

Fig. 2 Proposed hierarchical model

it can be resolved out. The logical consequences for those objects may be look quite hard but these cannot be impossible to fix the problem. Limitation of the AHP is that it only works with the positive matrices which all are of same mathematical form. AHP is very useful technique for discriminating between two or more objects competing with one another [2].

For making decisions in an organized manner, priorities are generated and decision is decomposed into the given following steps.

Step 1: Firstly, the issue is characterized and every one of the criteria is resolved.

Step 2: At that point by assessing the objective of choice, a choice progressive system is organized.

Step 3: Then, an examination lattice is created where we perform pairwise correlations.

Step 4: There are all out $n(n - 1)/2$ examinations are required to build up an arrangement of grids in step 3.

Step 5: At that point, the total of all the weighted eigenvectors are figured.

Step 6: In the wake of playing out all the pair, savvy correlations are finished and the consistency of the examinations is evaluated by utilizing the eigen esteem, λ, to ascertain the consistency record,

$$CI: (\max - n)/(n - 1)$$

where n is the framework size.

Step 7: The last consistency proportion (CR) is computed as the proportion of the CI and the irregular file (RI), as demonstrated.

$$CR = CI/RI$$

where RI remains for random consistency index.

4 Fuzzy Topsis

This is a procedure utilized for finding the request of inclination called Fuzzy Topsis which is a basic leadership strategy, created by Hwang and Yoon in 1981. Later with couple of more improvements, it was given by Yoon in 1987. In this strategy, we picked the right option by finding the geometric mean from the positive perfect arrangement (PIS) and the biggest geometric separation from the negative perfect arrangement (NIS).

The calculation of this strategy can be portrayed as follows:

Step 1: Firstly, create all the plausible components, set up a gathering of chiefs, and decide the assessment. Suppose there are k leaders, m choices and n assessment criteria.

Step 2: Then, pick the phonetic variables for finding the significance weight of the criteria and the phonetic evaluations for options as for these criteria.

Step 3: Then, to locate the collected fluffy weight of foundation C_j, we total the heaviness of the criteria.

Step 4: Then, a fluffy choice lattice is built.

Step 5: Then, we standardize the fluffy choice lattice.

Step 6: Then, a standardized weighted fluffy choice lattice is figured.

Step 7: Later on positive perfect arrangements (PIS) and negative perfect arrangements (NIS) are discovered.

Step 8: Then, we ascertain every option separation from A* and A−.
Step 9: Closeness coefficient is ascertained for every one of the choices.
Step 10: Finally, on the premise of the estimation of closeness coefficient for every, we rank every one of the options in plunging request.

5 Experimental Work

By using AHP method, we rank the mobile companies on the basis of priority vector out of three alternatives. These alternatives are taken out from experts' judgement. First, we compared all three mobile companies w.r.t each factor (C1 to C4). Then, these factors are compared with each other to show their relative importance. These are as follows.

The priorities of factors are calculated and shown in the above matrices. These matrices are used to carry out the final priorities. Value of CR (Consistency Ratio) for each matrix is calculated, and it should be less than or equal to 0.1 for the result to be acceptable. This is true for all above matrices, i.e., weight and judgments about all factors are acceptable.

Now the weight of each factor and the respective vector is combined to get the reliability index. C1, C2, C3 and C4 are the factors in the proposed model Simplicity, Efficiency, Effectiveness and Error Tolerance respectively. W_i are the eigenvectors, calculated from the pairwise judgments taken for the factor only (Tables 1, 2, 3, and 4). V1, V2, and V3 are the mobile companies, and their eigenvectors are taken from each of the matrix of factors (Table 5). To calculate the rank of the mobile companies, we calculate the reliability index as follows.

Reliability index of mobile companies = $\sum$weight of C_iV_i* weight of C_iW_i ($i = 1$ to 4) for the sample tables.

Table 1 Matrix for simplicity (C1)

	V1	V2	V3	Nth	EV
V1	1	1/3	1/5	0.4	0.103
V2	3	1	1/3	1	0.259
V3	5	3	1	2.449	0.636
Total				3.849	0.998

$\lambda_{max} = 3.033$, CI = 0.016, CR = 0.027

Table 2 Matrix for efficiency (C2)

	V1	V2	V3	Nth	EV
V1	1	5	1/5	1	0.266
V2	1/5	1	1/7	0.307	0.081
V3	5	3	1	2.444	0.651
Total				3.751	0.998

$\lambda_{max} = 3.076$, CI = 0.038, CR = 0.065

Table 3 Matrix for effectiveness (C3)

	V1	V2	V3	Nth	EV
V1	1	3	5	2.444	0.642
V2	1/5	1	1/9	0.606	0.159
V3	1/7	3	1	0.755	0.198
Total				3.805	0.999

$\lambda_{max} = 3.03$, CI = 0.015, CR = 0.025

Table 4 Matrix for error tolerance (C4)

	V1	V2	V3	Nth	EV
V1	1	5	1/5	1	0.220
V2	1/5	1	1/7	0.307	0.067
V3	5	7	1	3.232	0.712
Total				4.539	0.999

$\lambda_{max} = 3.177$, CI = 0.08, CR = 0.137

Table 5 Matrix for criteria

	C1	C2	C3	C4	Nth	EV
C1	1	3	7	5	3.2010	0.5452
C2	1/3	1	9	3	1.7320	0.2950
C3	1/7	1/9	1	1/3	0.2685	0.0457
C4	1/5	1/3	3	1	0.6687	0.1139
Total					5.8702	0.9998

$\lambda_{max} = 4.173$, CI = 0.057, CR = 0.0633

Table 6 Reliability index

	C1	C2	C3	C4	Reliability index
W_{Ci}	0.5452	0.2950	0.0457	0.1139	
W_{V1}	0.103	0.266	0.642	0.220	1.082
W_{V2}	0.259	0.081	0.159	0.067	0.441
W_{V3}	0.636	0.651	0.198	0.712	1.744

Mobile company having the highest reliability index will be ranked on the top, and others will be followed by it. From the above formula, we calculated the respective rank of mobile companies and it is shown in Table 6.

For the taken samples, mobile company V3 has the highest reliability index; therefore, vendor V3 is ranked I, V1 is ranked II, and V2 is ranked III.

5.1 Validation

After using the application of AHP (analytical hierarchy process), now we validate the result in the form of priorities for the various mobile companies. For this purpose, Fuzzy Topsis method is used.

Table 7 Decision matrix (X)

0.1029	0.3748	0.6228	0.3018
0.2913	0.0811	0.0906	0.0653
0.6056	0.5441	0.2867	0.6328

Table 8 Normalized performance matrix (Z)

0.0464	0.1409	0.0277	0.0385
0.1314	0.0305	0.00	0.0083
0.2733	0.2046	0.0127	0.0808

The positive ideal solutions are
(0.2733, 0.2046, 0.0277, 0.0808)
The negative ideal solutions are
(0.0464, 0.0305, 0.0040, 0.0083)

Table 9 Distance of each alternative from FPIS and FNIS

	d+	d−
C1	0.2389	0.1161
C2	0.2368	0.0848
C3	0.0148	0.2947

Table 10 Closeness coefficient

Mobile company	CCi
C1	0.3270
C2	0.2636
C3	0.9521

In this section we, use the Fuzzy Topsis method for the selection of mobile companies. Three decision makers defined as D1, D2, D3 are used to determine the importance and weight of the criteria (Tables 7, 8, 9, and 10).

On the basis of the closeness coefficient of three alternatives, the ranking order of three alternatives is determined as C3 > C1 > C2.

Therefore, $R^3 I_1 = 0.9521$. This is the rank for first mobile company.

Similarly, for the second company $R^3 I_2 = 0.3270$.

And for the third company, $R^3 I_3 = 0.2636$.

6 Results and Analysis

From the results of Fuzzy Topsis method, we can observe that the mobile company V3 has the rank I among all other available mobile companies. This ranking is also same as the rank found by AHP. So, our result is validated and thus mobile company V3 is considered as the best alternative among all the available alternatives. The analysis graph between mobile company's factors and weight is shown in Fig. 3.

Fig. 3 Analysis and ranking chart

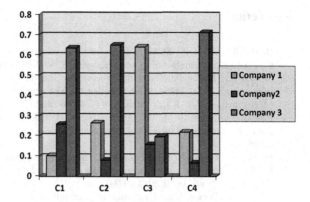

7 Conclusion

As we know mobile devices are developing so rapidly, so it become necessary to find out their usability and this plays a critical role in the economical functionality of any organization. In this paper, presented mobile company selection factors using AHP approach and results showed the best optimum selection among the available list of mobile companies based on some important factors. Result is also validated by using Fuzzy Topsis method. Decision parameters are very important and are defined to find the ranking among all the alternatives on the basis of subjective data. The present approach will help to select best mobile company on the basis of defined factors. In future, more number of decision parameters will be used for the research purpose and with different methodologies to get more effective and optimum results.

8 Future Scope

This term usability evaluation has a lot of scope of improvement for any software application by finding some subcriteria which help us to improve the result of selection of usable window devices or any other kind of devices. The methods which we used in any decision making problem where we have to select the best option from all the available options like finding out the most secured devices or systems so that people can choose the best choice from all available versions of these mobile phones. We can also take more decision parameters to get more clear results. These above two methods can be used anywhere for finding out the best option among all the available options.

References

1. Bindu, R., More, P.: SCM in India—a perspective. In: Proceedings of the International Conference on Manufacturing and Management, Vellore, India (2004)
2. Saaty, T.L.: The analytic hierarchy process. McGraw-Hill, New York (1980)
3. Rouyendegh, B.D., Erkan, T.E.: Selecting the best supplier using analytic hierarchy process (AHP) method. Afr. J. Bus. Manag. 6(4), 1455–1462 (2012)
4. Farzad, T., Osman, R.M., Ali, A., Rosnah, Y., Esfandyari, M.: AHP approach for supplier evaluation and selection in steel manufacturing company. J. Ind. Eng. Manag. 1(2):54–76 (2008)
5. Farzad, T., Dabbagh, M., Ebrahim, N.: Supplier assessment and selection using analytic hierarchy process in a steel manufacturing company. J. Sci. Res. Reports 3(10), 1319–1338 (2014). Article no. JSRR.2014.10.003
6. Mishra, A., Dubey, S.K.: Evaluation of reliability of object oriented software system using fuzzy approach. In: 2014 5th International Conference on Confluence the Next Generation Information Technology Summit (Confluence), 5–6 Sept 2014, pp. 806–809 (2014)
7. Mishra, A., Dubey, S.K.: Fuzzy qualitative evaluation of reliability of object oriented software system. In: International Conference on Advances in Engineering & Technology Research (ICAETR—2014), 01–02 Aug 2014, pp. 685–690
8. Yuichi, T., Marks, T.K., Hershey, J.R.: Entropy-based motion selection for touch-based registration using Rao-Blackwellized particle filtering. In: 2011 IEEE/RSJ International Conference on Intelligent Robots and Systems (IROS), pp. 4690–4697. IEEE (2011)
9. Pandey, G.K., Dukkipati, A.: Minimum description length principle for maximum entropy model selection. In: 2013 IEEE International Symposium on Information Theory Proceedings (ISIT), pp. 1521–1525. IEEE (2013)
10. Yang, G., Wenjie, H.: Application of the TOPSIS based on entropy-AHP weight in nuclear power plant nuclear-grade equipment supplier selection. In: ESIAT 2009. International Conference on Environmental Science and Information Application Technology, 2009, vol. 3, pp. 633–636. IEEE (2009)
11. Panda, B.N., Biswal, B.B., Deepak, B.L.V.: Integrated AHP and Fuzzy Topsis approach for the selection of a rapid prototyping process under multi-criteria perspective. In: 5th International & 26th All India Manufacturing Technology, Design and Research Conference (AIMTDR 2014)
12. Gupta, S., Singh, S.K.., Agrawal, K., Nagaraju, D.: Comparison of fuzzy AHP and Topsis evaluations of luggage bag design alternatives. ARPN J. Eng. Appl. Sci.
13. Ayhan, M.B.: A fuzzy Ahp approach for supplier selection problem: a case study in a gearmotor company. Int. J. Manag Value Supply Chains (IJMVSC) 4(3) (2013)
14. Sun, C.-C.: A performance evaluation model by integrating fuzzy AHP and Fuzzy Topsis methods. Expert Syst. Appl. 37, 7745–7754 (2010)
15. Sharma, C., Dubey, S.K.: Reliability evaluation of software system using AHP and Fuzzy Topsis approach. In: Proceedings of Fifth International Conference on Soft Computing for Problem Solving (SocProS 2015). Advances in Intelligent Systems and Computing, Springer, IIT, Roorkee, India, pp. 81–92, 18–20 Dec 2015
16. Mukhtar, E., Alexis, C., Pete, F.: Land evaluation techniques comparing fuzzy AHP with TOPSIS methods. In: 13th AGILE International Conference on Geographic Information Science 2010

Analysis and Comparative Exploration of Elastic Search, MongoDB and Hadoop Big Data Processing

Praveen Kumar, Parveen Kumar, Nabeel Zaidi
and Vijay Singh Rathore

Abstract The word "Big Data" describes innovatory tools and techniques to store, share, capture, manage and examine very large data sets with different structures. A Big Data may be unstructured, semi-structured or structured which results in incapacity of storing these data using any conventional data management methods. In order to use these data in an efficient and costless way, parallelism is used. Hadoop is open-source software and is the main platform for making Big Data structural and making it useful for different purpose. Furthermore, to solve different types of problem, different types of DBMS are being developed along with their application program interface. One of them is a MongoDB which is a very famous NoSQL database and is free from schema. It is oriented toward document whose performance for query processing is very high. Moreover, Elastic Search is a search engine which provides a way to organize data, so that it can be easily accessed. It is a tool for querying the word written. Hence, the term Elastic Search, MongoDB and Hbase are closely related. In this paper, we provide a comparative study of each one of them.

Keywords Elastic search · Hbase · MongoDB · Big Data · Hadoop
Node · Cluster

P. Kumar (✉) · P. Kumar
Department of Computer Science, NIMS University, Jaipur, India
e-mail: praveenvashisht07@gmail.com

N. Zaidi
Department of Computer Science and Engineering, Amity University, Noida,
Uttar Pradesh, India
e-mail: nabeelzaidi@ymail.com

V.S. Rathore
Shri Karni College, Jaipur, Rajasthan, India
e-mail: vijaydiamond@gmail.com

© Springer Nature Singapore Pte Ltd. 2018   605
M. Pant et al. (eds.), *Soft Computing: Theories and Applications*,
Advances in Intelligent Systems and Computing 584,
https://doi.org/10.1007/978-981-10-5699-4_57

1 Introduction

Big Data refers to data or a combination of data which have very large size, complexity and growth rate which makes it difficult to be stored, processed and managed by old technology. There is no proper size limit of Big Data [1]; for example, we may consider 50–60 TB (terabytes) to multiple PB (Petabytes) as Big Data as shown in Fig. 1. We may decompose Big Data into different layers as infrastructure layer, computing layer and application layer.

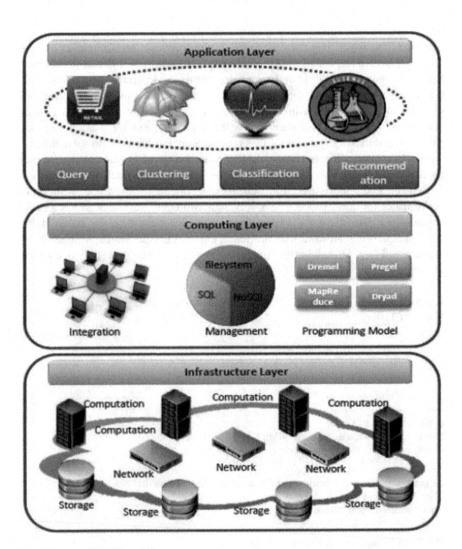

Fig. 1 Layers of Big Data [2]

There are some difficulties in processing Big Data as given below:

A. Incompleteness and heterogeneity: Whenever humans take information, a great heterogeneity is in fact tolerated. But the machine language algorithm cannot understand heterogeneous data. Hence, they need homogeneous data. It would be efficient if we can store multiple data together with same size and structure.

B. Size [2]: It is the first thing one thinks about Big Data when it comes to mind. Rapidly increasing data makes it a challenging issue for storing, processing and retrieving with minimal resources. In the past, Moore's law was followed, but now data size is rapidly increasing in terms of computer resources.

C. Time limit: The major factors that get affected due to rapidly increasing size are the time to analyze this Big Data. The larger the size, the longer it will take to process data. When we speak of velocity in context of Big Data, we also mean the rate challenge to be viewed.

D. Privacy: It is one of the huge concerns when one talks about Big Data. For medical health [3] records, there are laws managing for the things that can be done and not done. However, privacy has always been a concern in both technical and social areas.

2 Hadoop

Hadoop is a software platform where very large data sets are processed in a distributed computing environment. It was developed by using the concept of Google's MapReduce technique in which we break data into various parts and then process. Current Apache Hadoop structure as shown in Fig. 2 contains MapReduce, Hadoop Kernel, Hadoop distributed file system (HDFS) and other various components such as Base, Apache Hive and Zookeeper.

Fig. 2 Hadoop structure [2]

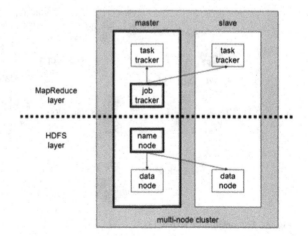

A. *Architecture of HDFS*

Hadoop consists of a fault-tolerant system of storage called Hadoop distributed file system (HDFS) [2]. Hadoop can store lots of data and even be able to survive the failure of important parts of storage without the loss of data. It consists of clusters of machines which coordinates among themselves. Suppose if one of the machines fails, Hadoop continues to do the work without the data loss by just shifting the work to remaining machine which is working in a cluster. Hadoop distributed file system (HDFS) works by breaking the files into small pieces called blocks and then storing these blocks across the server's pool. Generally, HDFS stores three copies of a single file by breaking each file into block and storing them at different servers. Hadoop architecture is shown in Fig. 3.

B. *Architecture of MapReduce*

The important pillar of Hadoop structure is the MapReduce framework as shown in Fig. 4. It works by applying the operation on a huge data set, dividing the data as well as problem and then running them in parallel. For example, we have very large data and we need to process it. If we apply a traditional way, then we might need to apply an ETL [2] (external, transfer and load) [4] operation to produce something that can be used by an analyst. But in Hadoop using high-level languages such as Hive and Pig, we can easily write the above operation. Then, the output can be written back to the HDFS system or stored in a traditional warehouse. The two functions which we use in MapReduce [1] are as follows:

1. Map—This function takes key/value pairs as input and then generates an intermediate set of key/value pairs.
2. Reduce—This function will merge all the intermediate values which have the same key.

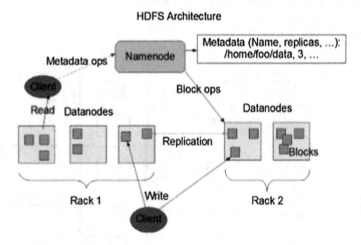

Fig. 3 Architecture of HDFS [3]

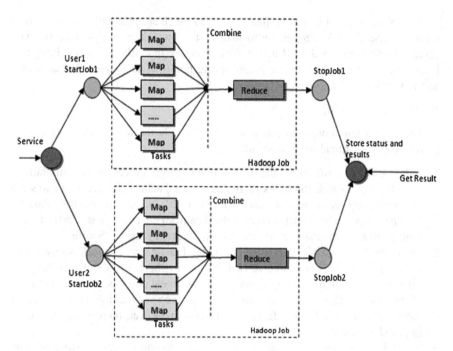

Fig. 4 MapReduce structure

3 Elastic Search

The concept of Elastic search's intelligent search engine is basically another software project called Lucene. It is easy to understand that this elastic search as a part of infrastructure is built nearby Lucene's Java libraries [5]. In elastic search everything present, there is related to the set of actual algorithms which can be used for matching text and storing optimized indexes of query in terms of execution by Lucene. This itself provides both functional and compact API, scalability and moreover operational tools overhead Lucene's search implementation. Lucene is ancient in terms of internet years, if we seek back to 1999. It is widespread and established. Lucene is used by so many companies that the number is inexpressible; they run in scope from huge corporations such as Twitter [6], to even smaller start-ups. Lucene is also demonstrated, is tested, and is one which is widely considered one of the best-of-breed in open-source search software. Most of the effort that users of elastic search allocate to the task of search is related to nothing but use of the Lucene APIs elastic search exposes.

A. The value adds

While Lucene is one of the eccentric tools, which is unwieldy to use directly, it provides with few features for scaling previous single machine. Elastic search has the capability to provide a more instinctive and much simpler API as compared to basic Lucene API. Moreover, substructure story which makes scaling on data centers is also provided by Lucene. Few features that Elastic search brings to Lucene are a simple API, interoperation using non-java languages, ease of access and replication and clustering.

B. Basic concepts [5]

There are a few more concepts that come with Elastic Search, and how Elastic search works is critical to understand.

1. Index: Elastic search stores the data in one or more indices. It uses similarities from the SQL world, and indexing is similar to a database. It is also used to store the documents and is used to read from them. Elastic search also uses Apache Lucene library to both write and read the data from the index. Their index is also built of more than a single Apache Lucene Index by using "Shards."
2. Document: It is one of the main entities in the Elastic search world. All use cases of using Elastic search can be carried to a point where it is about searching for documents and moreover analyzing them. Document consists of fields, and each of them is identified by its name and can contain one or multiple values. Each of the documents may have different set of fields; moreover, there is no schema or imposed structure.
3. Type: Each document has its type defined. This allows us to store various document types and allow different mapping for different documents types.
4. Mapping: All the documents are further analyzed before being indexed. Moreover, the input text is divided into tokens, which tokens should be filtered out, or what processing, such as removing HTML tags, is needed. This is the point where mapping comes into play; it also holds all set of information about the analysis chain.
5. Node: The single instance of this server is called a node. A single node in deployment can be sufficient for many other use cases. Elastic search is also designed to index and to search our data, so the first type of node needed is the data node. Such types of nodes hold the data and also search on them. The second type which comes into play is the master node, a node that works as supervisor of the cluster controlling other nodes' work. The third type is the tribe node, which is new and was introduced in Elastic search [6]. This tribe node can join multiple clusters and can also act as a bridge between them.
6. Cluster: Cluster is a set of nodes that works together. This distributed nature of it allows us to easily handle data which are too large for a single node to handle.
7. Shard: Clustering allows storing information in volumes that exceed abilities of a single server. To achieve this requirement, it spreads data to several physical Lucene indices and these indices are called Shards, and all parts of this index are called sharding. Elastic search can do this automatically, and all the parts of it are visible to the user as one big index.

8. Replica: Sharding allows pushing more data into Elastic search that are handled by single node. The idea is simple and creates an additional copy of shard, which can be used for queries just as we use original, primary shard.
9. Key concepts behind Elastic search architecture [6]: Elastic search was built with concepts in mind. The development team wanted easy-to-use and highly scalable concepts. This feature was visible in every corner of Elastic search. The main features are as follows:

   (a) Reasonable default values which allow users to start using Elastic search just after they have installed. This also includes built-in discovery and auto-configuration.
   (b) Working of this in distributed mode by default. Node of this assumes that they are or will be a part of the cluster.
   (c) Moreover, peer-to-peer architecture without single point of failure (SPOF). Nodes can automatically connect to other machines in the cluster for the data interchange and also for mutual monitoring. This covers our automatic replication of shards.

C. It is easily scalable in terms of both capacity and the amount of data by adding new nodes to the cluster.
D. Elastic search does not impose any restrictions on the data organization in the index. This also allows user to adjust to their existing data model near-real-time (NRT) [5] searching and versioning. Because of this distributed nature of the elastic search, it is very impossible to avoid the delays and the temporary differences between the data located on the different nodes present. Elastic search also tries to reduce this issue and provide additional mechanisms as versioning.

4 MongoDB

It is a scheme with less document-oriented database. Its name came from "humongous." This database is scalable and is written in C++. The main reason for moving away from relational model helps to make scaling easier. The basic concept is for replacing the concept of "row" with much more flexible model. By using embedded document and arrays, the approach makes it possible for representing complex hierarchical relationship with single record. MongoDB is schema-free [7]: Documents key is not pre-defined or fixed.

A. *Features*

MongoDB supports BSON [8] data structures for storing complex data types, supports powerful and complex query language and high access to mass data, stores as well as distributes large binary files like images and videos instead of storing procedure developers that can use or store Java Script, functions and values on the server side, it supports an easy-to-use protocol to store large file and file metadata, and it will give fast serial performance for single client and use memory mapped file for fast performance. Because of this characteristic of MongoDB, many projects have started considering using MongoDB instead of relational database.

B. *Data Design*

MongoDB holds collection of set. The collection does not contain pre-defined schema like table and store data as document. BSON [9] (binary-encoded JSON-like objects) are currently being used to store documents. The document is considered as a set of fields which can be though as a row in collection. It contains complex structures like lists, or even documents. Each and every document will have its ID field, which can be used as primary key, each collection contains any kind of document, and the queries and index can be made against a collection. MongoDB instead of embedded objects will support indexing and arrays and thus has special array features called "multi-keys." This feature will allow using array as index, which will be used to search documents by associated tags. The figure shows a typical structure of MongoDB.

C. *Architecture*

Basically its cluster is made up of three components [8] as shown in Fig. 5. They are:

1. Shard nodes: Cluster of MongoDB is made up of shard nodes in which each and every shard node stores data. It has the advantage of failure that if primary server fails, secondary automatically takes its places. Hence, the server keeps on working.

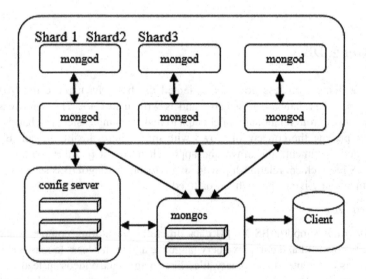

Fig. 5 Architecture of MongoDB [6]

2. Configuration servers: Cluster consists of group of servers which is called configuration servers and is used to store huge amount of data as well as the routing information which indicates that which shard contains particular data.
3. Routing services or Mongos: They are those servers which take care of the task performed by the client. Data are received in the form of queries and then processed to the appropriate shards. Thus, that particular requested data are sent back from the shard to the client.

5 Comparison of Elastic Search, Hbase, MongoDB

After discussing the basic concept of each and every topic listed above, we are now in a position to compare them. Elastic Search is a latest search and its engine is based on Apache Lucene, while Hadoop database is a wide column [3] which is based on Apache Hadoop and also on some concept of Big Table. Furthermore, MongoDB is one of the most popular document stores. Now we can divide Elastic Search, Hadoop Base and MongoDB in categories, so it will become easy to compare:

1. Database model: Elastic Search is considered as a search engine, while Hadoop is a wide-column store and MongoDB is a document store.
2. Developer: Elastic Search was developed by elastic with its initial release on 2010, Hbase by Apache Software Foundation with its initial release on 2008 and MongoDB by MongoDB, Inc. with its initial release on 2009.
3. Current release and license: Elastic Search has an open-source license with its 2.4.1, September 2016 version currently going on, Hbase is also an open source with its 1.2.3, September 2016 version going on, while MongoDB is also an open source with its 3.2.9, August 2016 version currently going on.
4. Database as a service: All the three of them do not support database as a service.
5. Server operating system: Elastic Search server can have all the operating system but with a java virtual machine. Hbase server works on Linux, UNIX or Windows, while MongoDB server works on Linux, OS X, Solaris or Windows.
6. Data schema: All of them are schema-free with typing (editing) except for Hbase in which we cannot do typing.
7. APIs and other access methods: Elastic Search has Java API and RESTful HTTP/JSON API, while Hbase has Java API, RESTful HTTP API and Thrift. Furthermore, MongoDB has proprietary protocol using JSON.
8. Programming languages support: There are great varieties of programming languages supported by them. They are as follows:
(a) Elastic Search- .Net, Scala, Clojure, Ruby, Erlang, Python, PHP, Lus, Perl, Haskell, JavaScipt, Java, Groony, Go.

(b) Hbase-Scala, Java, Python, C++, PHP, Groony, C, C#.

(c) MongoDB—Clojure, C, C++, C#, Scala, Python, Ruby, Prolog, PHP, Power Shell, Perl, Matlab, Liep, Lua, Java, Haskell, JavaScript, Go, Groony, Delphi, Erlang, D, Dart, ColdFusion, ActionScript, Smalltalk, Scala, R.

 9. Server-side scripts and triggers: Elastic Search and Hbase support server-side scripts, while MongoDB supports JavaScript with no triggers.

10. Replication method: We can use replication method in Elastic Search, whereas Hbase supports selectable replication factor and MongoDB supports Master–slave replication.

11. MapReduce: No MapReduce technique is used in Elastic Search, while in Hbase and in MongoDB it is used.

12. Consistency concept: Elastic Search obtains eventual consistency, while Hbase obtains immediate consistency, whereas MongoDB can obtain consistency in both ways depending on data.

13. Foreign key and transaction concepts: None of them have foreign key and do not support transaction concept.

14. User concepts: MongoDB has certain rights for users and roles, while Hbase has access control list (ACL) for users.

6 Conclusion

We have discussed the basic concept of Elastic Search, Hadoop and MongoDB in the beginning of the paper and then comparative study between them. We have found that all the three are equally important and are irreplaceable with each other. They may support different language but are equally useful when it comes to functionality. Furthermore, we have also provided the architecture of all three of them.

References

1. Vidya Sagar, S.D.: A study on role of Hadoop in information technology era. 2(2), (2013). ISSN No 2277
2. Bhosale, H.S., Gadekar, D.P.: Review paper on big data and Hadoop. Int. J. Sci. Res. Publ. 4(10), (2014)
3. Research on MongoDB design and query optimization in vehicle management information system, pp. 12–13 (2012)
4. Kirkpatrick, K.: Software-defined networking. Commun. ACM 56(9), 16–19 (2013)
5. Kononenko, O., Baysal, O., Holmes, R., Godfrey, M.W.: Mining modern repositories with elastic search (2013)
6. Kumar, P., Rathore, V.S.: Improvising and optimizing resource utilization in big data processing. In: Proceedings of Fifth International Conference on Soft Computing for Problem Solving: SocProS 2015, vol. 2. Advances in Intelligent Systems and Computing (2015)

7. Aghi, R., Mehta, S., Chauhan, R., Chaudhary, S., Bohra, N.: A comprehensive comparison of SQL and MongoDB databases. Int. J. Sci. Res. Publ. **5**(2), (2015)
8. Khan, S., Mane, V.: SQL support over MongoDB using metadata. Int. J. Sci. Res. Publ. **3**(10) (2013)
9. Pragya, G., Sreeja N.: Elastic search. Int. J. ISSN (Online) 2319–7064

Methods to Choose the 'Best-Fit' Patch in Patch-Based Texture Synthesis Algorithm

Arti Tiwari, Kamanasish Bhattacharjee, Sushil Kumar and Millie Pant

Abstract This paper proposes the comparative analysis of accuracy and efficiency in patch-based texture synthesis among different color models and SSD (sum of squared distance) method. Here we have taken the basic texture quilting method of Efros and Freeman as a synthesizer for analysis. To find the "best-fit" patch for synthesizing the image, this paper shows these different color models and SSD method on randomly fetched patch. Finally comparison is done on these different methods to analyze the result and synthesized images have shown for better visualization.

Keywords Patch-based texture synthesis · Image quilting · Color models

1 Introduction

Texture synthesis is the process of creating large texture image from the small source texture by using its structural components which can be defined as the small blocks (patch). Patch-based is one of the methods (pixel based, patch based, chemistry based) for synthesizing texture which gives best result in comparison to other. In this method the patches (small blocks of the given image) are put on the given output sized image in the overlapping manner.

A. Tiwari (✉) · K. Bhattacharjee · S. Kumar
Department of Computer Science and Engineering, Amity School of Engineering
and Technology, Amity University, Noida, Uttar Pradesh, India
e-mail: arti.tiwari94@gmail.com

K. Bhattacharjee
e-mail: kamanasish_b@live.com

S. Kumar
e-mail: kumarsushiliitr@gmail.com

M. Pant
Department of Applied Science and Engineering, IIT Roorkee, Roorkee, India
e-mail: millidma@iitr.ac.in

© Springer Nature Singapore Pte Ltd. 2018
M. Pant et al. (eds.), *Soft Computing: Theories and Applications*,
Advances in Intelligent Systems and Computing 584,
https://doi.org/10.1007/978-981-10-5699-4_58

617

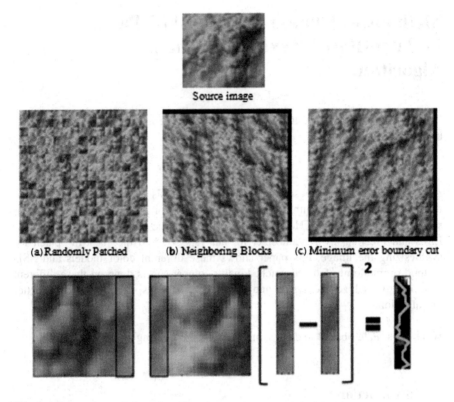

Fig. 1 Minimum error path-cut boundary

Overlapping of patches sometimes leads to the irregularity in the output image. To overcome this problem we use minimum cost path i.e. minimum error boundary cut, this technique finds the pixel wise minimum path between two overlapping patches. For plotting the patches on the output image. Figure 1 shows three methods, a. Random Patches, b. Neighboring patches and c. Minimum error boundary cut.

This paper has focused on choosing the "best-fit" patches using different color spaces and SSD (sum of squared difference) method.

Several techniques have been used for texture synthesis; using pixel-based method [1, 2, 3] and using patch-based method [4, 5, 6, 7] to choose the "best-fit" patch, [8] used wang-tiles method, [9] used graph-cut method for synthesizing the image and video.

Here we used the three color spaces, first RGB standard luminance weight for R, G and B channels, second YIQ color space, third YCbCr color space and SSD (sum of squared distance) method. Figures 2 and 3 shows the comparative result of the used methods.

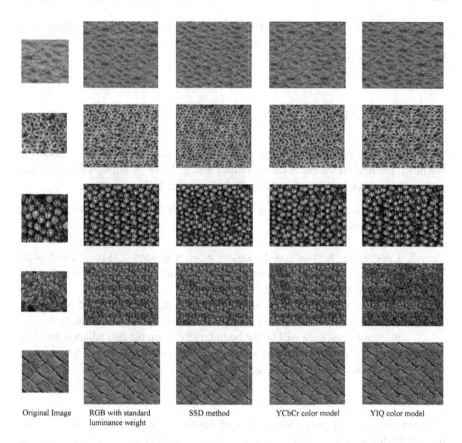

| Original Image | RGB with standard luminance weight | SSD method | YCbCr color model | YIQ color model |

Fig. 2 Resultant images

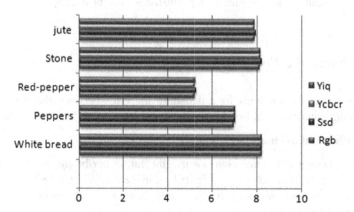

Fig. 3 Comparison of PSNR values of images, *red line* indicates the SSD method; it shows better result in comparison to other methods

Further, Introduction is followed by Sect. 2, illustrates our contribution as Proposed Work, Sect. 3 will demonstrate the Result and Analysis and finally Sect. 4 is Conclusion.

2 Proposed Work

In this paper analysis is done on the significance of using different color models to lay down the best suited patch on the output image. If we go through the algorithm proposed by Efros and Freeman [10], it performs the following steps:

Step1. Get a random patch of a given size from input image and paste it on the output image.
Repeat
Step2. Choose the "best-fit" patch (we described it in Sect. 2.1) that "satisfies the overlap constraints within some error tolerance" to lay down on the output image in the raster scan order.
Step3. Find the overlapping error surface between the old patch and new patch, then find the minimum error boundary-cut on the error surface. This minimum cost path will be the boundary between the two adjacent patches.
Step4. Paste this patch on the output image (if left overlap then paste the patch on right side and if there is top overlap then paste the patch below the old patch).
End.

Patch size, overlap size and number of patches in a row and column (n) are defined by the user then resultant image size is evaluated as:

$$\text{Output\_image\_size} = n^*\text{patchsize} - (n-1)^*\text{overlapsize} \qquad (1)$$

2.1 Choose the Best Patch

Here we define the methods for calculating the error between the two adjacent overlapping patches. Let's first know the significance of all the used methods.

RGB with standard luminance weight

For better human eye perception we use the luminance/brightness parameter to choose the patch. There are three weighted parameters as 0.299, 0.587 and 0.114 for R, G and B channels respectively. Error tolerance can be computed between two overlapping patches as:

$$Eov = \sum sqrt(0.299^*(P1_r^2) + 0.587^*(P1_g^2) + 0.114^*(P1_b^2)$$
$$- sqrt(0.299^*(P2_r^2) + 0.587^*(P2_g^2) + 0.114^*(P2_b^2)) \tag{2}$$

$P1$ and $P2$ are two overlapping patches and r, g, b are the red, green and blue components of the image.

YIQ color space

YIQ (luminance-inphase-quadrature), model is prominently used for image processing and is generally used for color television. This color model is defined by NTSC, in this grayscale is separated from the color data. Error computation is formed as:

$$Eov = \sum (P1_y - P2_y) + (P1_i - P2_i) + (P1_q - P2_q) \tag{3}$$

y, i and q are the Y, I and Q components respectively of patches $P1$ and $P2$.

YCbCr color space

YCbCr color model is used in digital photography and videos, here Y contains luminance, Cb and Cr are the blue and red difference chrominance components. The components of this color model can be obtained mathematically by coordinate transformation of related RGB image. Error computation by this model is done using:

$$Eov = \sum (P1_y - P2_y) + (P1_{cb} - P2_{cb}) + (P1_{cr} - P2_{cr}) \tag{4}$$

y, cb and cr are the Y, Cb and Cr components of patches $P1$ and $P2$ respectively.

Described color methods that can be used to choose the best patch, there is one more method (SSD) which is used by Efros and Freeman [10] in their work.

SSD (Sum of Squared Distance)

SSD method is equivalent to squared L2-norm; this method is much efficient to deal analytically because of the absence of absolute function. In patch selection, SSD is used to calculate the minimum distance error value between the patch $P1$ and $P2$.

3 Result and Analysis

Analysis of resultant images (Fig. 2) by applying all the methods has been done on the basis of PSNR value, shown in Table 1 and comparison is shown in Fig. 3. In Fig. 2 first column is for original image and then synthesized images by applying

Table 1 PSNR value of images

Image	Illumination weight (RGB)	SSD	YCbCr	YIQ
White bread	8.1944	8.2087	8.2111	8.2071
Peppers	6.9552	7.0232	7.0353	7.0109
Red-pepper	5.2186	5.2854	5.2163	5.2361
Stone	8.1134	8.2184	8.1415	8.1416
Jute	7.8621	7.9429	7.8873	7.8834

luminance weight method, SSD, YCbCr and YIQ in column second, third and fourth respectively.

The result shown above is analyzed on the basis of PSNR values of all the image with respect to original image to recognize the quality of the image. The PSNR values are shown in Table 1.

Comparison is done by the bar chart is shown in Fig. 3.

4 Conclusion

This paper shows the different methods to choose the 'best-fit' patch. Analysis is given over all the used methods by evaluating the PSNR values. One can observe that the method used by Efros and Freeman, SSD (L2-norm) shows (Fig. 3) the better result in comparison to other methods and not much difference can be obtained by all the color model methods. One can choose any of these methods for choosing the 'best-fit' patch in image synthesis and texture transfer.

References

1. Ashikhmin, M.: Synthesizing natural textures. In: Proceedings of ACM Symposium Interactive 3D Graphics, pp. 217–226 (2001)
2. Zelinka, S., Garland, M.: Jump map-based interactive texture synthesis. ACM Trans. Graph. **23**, 930–962 (2004)
3. Ashikhim, M.: Fast texture transfer. IEEE J. Comput. Graph. Appl. **23**, 0272–1716 (2003)
4. Attea, B.A., Rashid, L.M.: A genetic algorithm for texture synthesis and transfer. In: Texture Proceedings of the 4th International Workshop on Texture Analysis and Synthesis, pp. 59–64 (2005)
5. Cui, H.F., Zheng, X., Ruan, T.: An efficient texture synthesis algorithm based on WT. Int. Conf. Mach. Learn. Cybermetics. **6**, 3472–3477 (2008)
6. Varadarajan, S., Karam, L.J.: Adaptive texture synthesis based on perceived texture regularity. In: International Workshop on Quality of Multimedia Experience (QoMEX), pp. 76–80 (2014)
7. Celaya-Padilla, J.M., Galvan T., C.E Delgado C., J.R., Galvan-Tejada, I., Sandoval, E.I.: Multi-seed texture synthesis to fast image patching. In: International Meeting of Electrical Engineering Research ENIINVIE, vol. 35, pp. 210–216 (2012)

8. Zhang, X., Kim, Y.J.: Efficient texture synthesis using strict Wang Tiles. Graph. Models **70**, 43–56 (2008)
9. Lefebvre, S., Hornus, S., Lasram, A.: By-example synthesis of architectural textures. ACM Trans. Graph. **29**, 1–15 (2010)
10. Efros, A.A., Freeman, W.T.: Image quilting for texture synthesis and transfer. ACM SIGGRAPH, 1–6 (2001)

Block Matching Algorithm Based on Hybridization of Harmony Search and Differential Evolution for Motion Estimation in Video Compression

Kamanasish Bhattacharjee, Arti Tiwari and Sushil Kumar

Abstract In video compression, the most efficient technique for motion estimation is Block Matching and there are many algorithms to implement it. In this paper, two such algorithms are discussed, which are based on Differential Evolution (DE) and Harmony Search (HS), and a new algorithm is proposed by hybridizing these two algorithms to get better results. In the proposed algorithm, pitch adjustment operation of HS is replaced by mutation and crossover operations of DE.

Keywords Block Matching · Differential Evolution (DE) · Harmony Search (HS)

1 Introduction

Block Matching (BM) is a very important technique for motion estimation in video compression, where frames of a video sequence are divided into macro blocks. For each block in the current frame, the best matching block is identified in the search space of the previous frame to minimize the mean absolute difference (MAD) or mean squared error (MSE) or sum of absolute differences (SAD) between two blocks.

This MAD/MSE/SAD calculation is the most time-consuming operation in Block Matching. So, it can be treated as an optimization problem, where the objective is to identify the best matching block in a search space. The Full Search Algorithm (FSA) or Exhaustive Search (ES) gives the most accurate motion vector by checking every block in the search space, which makes this method slow. Hence, to speed up the Block Matching, many fast algorithms are proposed which

K. Bhattacharjee (✉) · A. Tiwari · S. Kumar
Department of Computer Science, Amity School of Engineering and Technology,
Amity University, Noida, Uttar Pradesh, India
e-mail: kamanasish_b@live.com

A. Tiwari
e-mail: arti.tiwari94@gmail.com

S. Kumar
e-mail: kumarsushiliitr@gmail.com

© Springer Nature Singapore Pte Ltd. 2018
M. Pant et al. (eds.), *Soft Computing: Theories and Applications*,
Advances in Intelligent Systems and Computing 584,
https://doi.org/10.1007/978-981-10-5699-4_59

calculate only a fixed subset of blocks at the price of poor accuracy. This accuracy-speed trade-off is a very important aspect of Block Matching.

The evolutionary and swarm intelligence techniques have shown good optimization of this trade-off. Several evolutionary algorithms, swarm intelligence and population-based metaheuristic methods like Genetic Algorithm (GA) [1, 2], Harmony Search (HS) [3], Artificial Bee Colony (ABC) [4], Particle Swarm Optimization (PSO) [5], Differential Evolution (DE) [6] have been used by researchers for Block Matching. Empirical studies have shown that hybridization of these evolutionary algorithms provides better result than the individual algorithms. Therefore, in this paper we propose a new algorithm for Block Matching through hybridization of DE and HS.

The paper is divided into six sections. The basic concept of Block Matching is described in Sect. 2. Block Matching algorithms based on Differential Evolution (DE) and Harmony Search (HS) are discussed in Sects. 3 and 4, respectively. The proposed algorithm is discussed in Sect. 5. Section 6 presents the conclusion and future work.

2 Block Matching

Block Matching is the method to identify the matching blocks in a sequence of video frames. It utilizes the temporal redundancy for video compression.

In Fig. 1, (a) represents the previous frame and (b) represents the current frame. The current block is marked with green border in the current frame. Its counterpart in the previous frame is also marked same. Search space is specified by the maximum allowed displacement or the displacement parameter (p). Here, $p = 2$. The search space is $(2p + 1) \times (2p + 1)$ dimensional, i.e., 5×5 in this case, which is marked with blue border in the previous frame. The blocks are selected from this search space of 25 blocks in case of HS, DE, and the proposed algorithm.

Here, the objective is to find the best matching block, i.e., the minimum difference between the two blocks. Figure 2 represents the method. Sum of absolute difference (SAD) is used as the difference measure.

Fig. 1 a Previous frame.
b Current frame

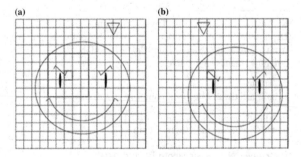

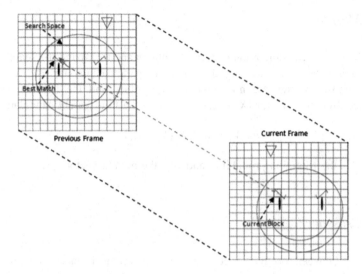

Fig. 2 Block Matching

$$\text{SAD}(h, v) = \sum_{j=0}^{N-1} \sum_{i=0}^{N-1} |I_t(x+i, y+j) - I_{t-1}(x+h+i, y+v+j)| \qquad (1)$$

where $I_t(.)$ and $I_{t-1}(.)$ are the intensities of a pixel in the current frame and previous frame respectively. The motion vector in (h, v) is defined as

$$(h, v) = \min \text{SAD}(h, v)$$

where $(h, v) \in$ Search Space and $-p \leq h, v \leq p$.

3 Block Matching Using Differential Evolution (DE)

In [6], Cuevas et al. proposed the DE algorithm for Block Matching. It is based on Differential Evolution, a heuristic optimization technique originally proposed by Storn et al. [7]. It reduces the search locations in Block Matching. The following steps are implemented to optimize the problem through DE.

3.1 Population Generation

In this step, a random population of D dimensional parent vectors X_i is generated up to NP candidates where X_i is the ith parent vector. In the proposed method, to generate the initial group of particles, motion vector distribution analysis is done.

3.2 Mutation

To exchange information among several solutions and to expand the search space, mutation is used. For mutation, three random parent candidates X_p, X_q, X_r are chosen whose indexes are not equal to each other and the iteration number. Weighted difference vector $(X_q - X_r)$ is added to the third parent candidate X_p:

$$V = X_p + F^*(X_q - X_r) \qquad (2)$$

where F is the Mutation probability and V is the mutant vector.

3.3 Crossover

Crossover between original vector and mutant vector is used to increase the diversity. Uniform Crossover between parent vector (X_i) and mutant vector (X_m) to generate utility vector or child vector (X_u) with CR as the Crossover probability. A random number is generated between 0 and 1. If the value is less than CR, then attribute value is chosen from mutant vector, otherwise from parent vector.

$$U = \begin{cases} V_{j,i} & \text{if } \mathrm{rand}(0,1) \leq \mathrm{CR} \quad \text{or} \quad j = j_{\mathrm{rand}}, \\ X_{j,i} & \text{otherwise} \end{cases} \qquad (3)$$

where $j, j_{r \text{ and}} \in (1, 2, \dots D)$.

3.4 Selection

Fitness value is calculated through objective function for each parent vector-utility vector pair. If utility vector is superior to corresponding parent vector, then it replaces the parent in the population otherwise parent remains same. Through this operation, population is generated for next generation.

$$X_i = \begin{cases} V_{j,i} & \text{if } f(u_i) \leq f(x_i), \\ X_{j,i} & \text{otherwise} \end{cases} \qquad (4)$$

DE based Block Matching Algorithm

Step 1: Set the parameters. F = 0.5, CR = 0.5, D = 2.

Step 2: Randomly select the initial population of 5 blocks from the search space. NP = 5.

Step 3: Calculate SAD values between the current block and each block from NP using Eq. 1.

Step 4: While stopping criteria is not satisfied, do

    for i = 1 to NP

        Select 3 random blocks B_p, B_q, B_r from NP where $p \neq q \neq r \neq i$

        Calculate mutant $V = B_p + F*(B_q - B_r)$

        for j = 1 to D

            if (rand(0,1) ≤CR)

                trial block $U_{j,i} = V_{j,i}$

            else

                trial block $U_{j,i} = B_{j,i}$

            end if

        end for

        Calculate SAD value between the current block and each trial block using

Eq. 1.

    if (SAD(U_i) < SAD(B_i))

        $B_i = U_i$

    end if

    end for

Step 5: Select the block with the minimum SAD value for Motion Vector calculation.

---

4 Block Matching Using Harmony Search (HS)

In [3], Cuevas proposed the HS algorithm for Block Matching. It is based on Harmony Search, a heuristic optimization technique originally proposed by Geem et al. [8]. Number of search locations is reduced in this algorithm. The following steps are implemented to optimize the problem through HS.

4.1 Initialization of the Problem and the Parameters

The global optimization problem can be generally summarized as

$$\min f(\mathbf{x_i})$$

where $f(.)$ is the objective function; $\mathbf{x_i} = (x_i(1), x_i(2), ..., x_i(n))$; $x_i(j) \in [l(j), u(j)]$, $j = 1, 2, ..., n$; n is the number of dimensions; $l(j)$ and $u(j)$ are lower and upper bounds, respectively, for $x_i(j)$. The algorithm parameters are

- Harmony Memory Size (HMS)
- Harmony Memory Consideration Rate (HMCR) [$0 \leq \text{HMCR} \leq 1$]
- Pitch Adjustment Rate (PAR) [$0 \leq \text{PAR} \leq 1$]
- Distance bandwidth (BW)
- Number of improvizations (NI)

4.2 Initialization of Harmony Memory

In this step, Harmony Memory (HM) is initialized with randomly generated HMS vectors with D dimensions.

$$\text{HM} = \begin{bmatrix} x_1 \\ x_2 \\ \vdots \\ x_{\text{HMS}} \end{bmatrix}$$

$x_i = \{x_i(1), x_i(2), \ldots, x_i(n)\}$ is the ith randomly generated harmony vector.

$$x_i(j) = l(j) + \text{rand}(0, 1) \cdot (u(j) - l(j)) \tag{5}$$

where rand(0, 1) is a uniform random number between 0 and 1; $j = 1, 2, \ldots, D$; $i = 1, 2, \ldots$, HMS.

4.3 Improvizing New Harmony

In this step, improvization is performed by generating a new harmony vector x_{new} as follows

$$x_{\text{new}}(j) = \begin{cases} x_i(j) \in \{x_1(j), x_2(j), \ldots, x_{\text{HMS}}(j)\} & \text{if rand}(0, 1) < \text{HMCR} \\ l(j) + \text{rand}(0, 1) \cdot (u(j) - l(j)) & \text{otherwise} \end{cases} \tag{6}$$

Every component generated through the above step is pitch-adjusted as follows

$$x_{\text{new}}(j) = \begin{cases} x_{\text{new}}(j) \pm \text{rand}(0, 1) \cdot \text{BW} & \text{if rand}(0, 1) < \text{PAR} \\ x_{\text{new}}(j) & \text{otherwise} \end{cases} \tag{7}$$

PAR assigns the frequency of the adjustment and BW controls the local search around the selected elements of HM. Pitch adjustment generates new potential harmonies by modifying the original variable positions, which is similar to the mutation operation in evolutionary algorithms. Hence, each dimension of the vector is either perturbed by a random number between 0 and BW or left unchanged.

4.4 Updating Harmony Memory

In this step, the decision is taken whether the new harmony vector x_{new} will replace the worst harmony vector x_{worst} in HM or not through the following criteria

$$x_{worst} = \begin{cases} x_{new} & \text{if } f(x_{new}) < f(x_{worst}) \\ x_{worst} & \text{otherwise} \end{cases} \tag{8}$$

HS based Block Matching Algorithm

Step 1: Set the parameters. HMS = 5, HMCR = 0.5, PAR = 0.5, BW = 8, NI = 50, D = 2.

Step 2: Randomly select 5 blocks B_i (i = 1, 2,..., HMS) from the search space.

Step 3: Calculate SAD values between the current block and each of these 5 blocks using Eq. 1.

Step 4: Determine the block B_{worst} with worst SAD value i.e. the highest SAD value.

Step 5: Improvise new block B_{new}:

    for j = 1 to D, do
        if (r1<HMCR)
$$B_{new}(j) = B_i(j) \text{ where } i = 1, 2, ..., HMS$$
            if(r2<PAR)
$$B_{new}(j) = B_{new}(j) \pm r3 \cdot BW \text{ where r1,r2,r3} \in (0,1)$$
            end if
            if B$_{new}$(j) < l(j)
                B$_{new}$(j) = l(j)
            end if
            if B$_{new}$(j) > u(j)
                B$_{new}$(j) = u(j)
            end if
        else
$$B_{new}(j) = l(j) + r \cdot (u(j) - l(j)) \text{ where r} \in \text{rand}(0, 1)$$
        end if
    end for

Step 6: $B_{worst} = B_{new}$ if $SAD(B_{new}) < SAD(B_{worst})$

Step 7: If NI is completed, Select the block with the minimum SAD value for Motion Vector calculation; otherwise go back to Step 4.

---

5 Proposed Algorithm for Block Matching

Hybridization of DE and HS has already been proposed [9, 10]. In [10], Chakraborty et al. used mutation operator of DE to perturb the target vector instead of pith adjustment. In this paper, we have proposed a version of HS-DE hybrid algorithm where we have used the crossover operator of DE also to increase the diversity of the perturbed vector. The hybrid method is presented below.

Hybrid HS-DE based Block Matching Algorithm

Step 1: Set the parameters. HMS = 5, HMCR = 0.5, PAR = 0.5, BW = 8, NI = 50, F = 0.5, CR = 0.5, D = 2.

Step 2: Randomly select 5 blocks B_i (i = 1, 2,..., HMS) from the search space.

Step 3: Calculate SAD values between the current block and each of these 5 blocks using Eq. 1.

Step 3: Determine the block B_{worst} with worst SAD value i.e. the highest SAD value.

Step 4: Improvise new block B_{new} :

    for j = 1 to D, do

        if (rand(0,1)<HMCR)

$$B_{new}(j) = B_i(j) \text{ where } i = 1, 2, \ldots, HMS$$

        else

$$B_{new}(j) = l(j) + r \cdot (u(j) - l(j)) \text{ where } r \in rand(0, 1)$$

        end if

        Select 2 random blocks B_p, B_q from initially selected blocks where p $\neq$ q

        Calculate mutant $V = B_{new} + F * (B_q - B_r)$

        if (rand(0,1)$\leq$CR)

            trial block $U_j = V_j$

        else

            trial block $U_j = B_{j,new}$

        end if

    end for

    Calculate SAD value between the current block and trial block using Eq. 1.

    if (SAD(U_i) < SAD(B_{new}))

        $B_{new} = U$

    end if

Step 5: $B_{worst} = B_{new}$ if $SAD(B_{new}) < SAD(B_{worst})$

Step 6: If NI is completed, Select the block with the minimum SAD value for Motion Vector calculation; otherwise go back to Step 3.

5.1 Justification

In [10], Chakraborty et al. have proposed a hybrid algorithm of HS and DE and have done a comparative study between one most popular DE variant called DE/rand/1/bin, different variants of HS and their proposed hybrid algorithm using different standard benchmark minimization functions—Sphere (*f*1), Rosenbrock (*f*2), Rastrigin (*f*3), Griewank (*f*4), Ackley (*f*5). Graphs shown in Figs. 3, 4, 5, and 6 are formulated from the results of [10]. These graphs show the comparative performance analysis of hybrid HS-DE, HS, and DE algorithms on various dimensions. It can be seen that their proposed hybrid HS-DE algorithm performs better than both DE and HS.

A different variant of hybridization of HS and DE is proposed in [9]. The result of this paper also establishes the superiority of hybrid HS-DE algorithm.

After reviewing these proposed hybrid HS-DE algorithm, it is seen that the hybrid algorithm performs better for minimization problems. In this paper, the problem is also a minimization problem. Hence, it can be inferred that the hybridization of HS and DE will perform better than the individual algorithms for Block Matching.

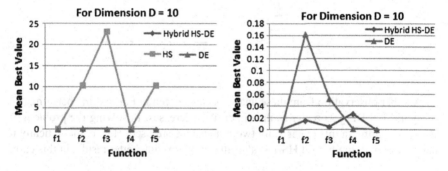

Fig. 3 Performance analysis of hybrid HS-DE, HS and DE using $D = 10$

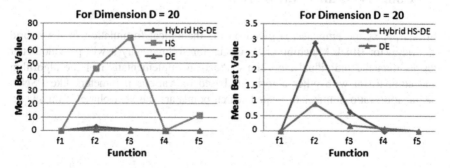

Fig. 4 Performance analysis of hybrid HS-DE, HS, and DE using $D = 20$

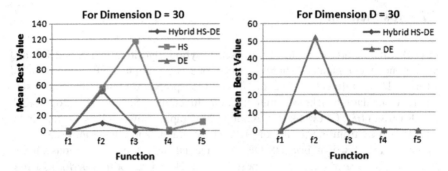

Fig. 5 Performance analysis of hybrid HS-DE, HS, and DE using $D = 30$

Fig. 6 Performance analysis of hybrid HS-DE w.r.t different dimensions

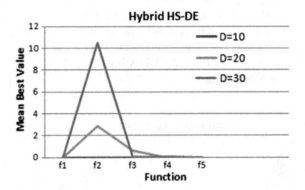

Another observation from these analyses can be seen in Fig. 6—the performance of hybrid HS-DE gets better as the dimension decreases. In solving the problem of Block Matching in this paper, only two dimensions are used. Hence, theoretically it can be inferred that hybrid HS-DE algorithm will provide better results in this case.

6 Conclusion and Future Work

In this paper, the emphasis is given on the theoretical formulation of the hybrid algorithm for Block Matching and a discussion follows why it will provide better results than the individual algorithms. Practical implementation of the algorithm and comparison with other algorithms based on various parameters like PSNR, degradation ratio, number of visited search points are the next phase of this work.

References

1. Chun-Hung, L., Ja-Ling, W.: A lightweight genetic block-matching algorithm for video coding. IEEE Trans. Circuits Syst. Video Technol. **8**(4), 386–392 (1998)
2. Wu, A., So, S.: VLSI implementation of genetic four-step search for block matching algorithm. IEEE Trans. Consum. Electron. **49**(4), 1474–1481 (2003)
3. Cuevas, E.: Block-matching algorithm based on harmony search optimization for motion estimation. Appl. Intell. **39**(1), 165–183 (2013)
4. Cuevas, E., Zaldívar, D., Pérez-Cisneros, M., Sossa, H., Osuna, V.: Block matching algorithm for motion estimation based on artificial bee colony (ABC). Appl. Soft Comput. J. **13**(6), 3047–3059 (2013)
5. Yuan, X., Shen, X.: Block matching algorithm based on particle swarm optimization. In: International Conference on Embedded Software and Systems (ICESS2008), 2008
6. Cuevas, E., Zaldívar, D., Pérez-Cisneros, M., Oliva, D.: Block-matching algorithm based on differential evolution for motion estimation. Eng. Appl. Artif. Intell. **26**(1), 488–498 (2013)
7. Storn, R., Price, K.: Differential evolution-a simple and efficient heuristic for global optimization over continuous spaces. J. Global Optim. **11**, 341–359 (1997)
8. Geem, Z.W., Kim, J.H., Loganathan, G.V.: A new heuristic optimization algorithm: harmony search. Simulation **76**(2), 60–68 (2001)
9. Ammar, M., Bouaziz, S., Alimi, A.M., Abraham, A.: Hybrid harmony search algorithm for global optimization. In: Nature and Biologically Inspired Computing (NaBIC) (2013)
10. Chakraborty, P., Ghosh Roy, G., Das, S., Jain, D., Abraham, A.: An improved harmony search algorithm with differential mutation operator. Fundamenta Informaticae **95**, 1–26 (2009)

A Rigorous Investigation on Big Data Analytics

Kajal Rani and Raj Kumar Sagar

Abstract Nowadays Big Data becomes a new trend in science, technology, business, and marketing. Traditional data analytics techniques are not able to be applied straightforward toward big data. There is a requirement to developed high-performance platform that analyzes big data more efficiently. It is challenging for the organizations to unlock the patterns of information actionable value in massive volume of data, enabling great improvements in business and technical processes, customer analytics. Datasets are heterogeneous in granularity and accessibility. There are many issues and challenges that company faces while storing and handling Big Data. The skill to automatically store, organize, review, and analyze the data is essential. This paper will tell why there is need of big data? Why big data is such a big hype? What is the need of analytics? This paper describes need of Big Data and its analysis. Paper discusses a brief investigation on big data analytics. The use of tools like HADOOP, HIVE PIG, and SPARK in summarizing the data.

Keywords Big data analytics · Big data issues and challenges
Analytics techniques · Apache hadoop · Apache drill · Project storm

1 Introduction

Big Data is one of the most hyped business issues today. Big data refers to large volume of datasets produced by millions of users. Big Data analytics process uncovered unseen samples that help to make better decision. Big Data sets originated from multiple data sources in different forms. These datasets are varied of

K. Rani (✉)
Department of CSE, Amity University, Noida, Uttar Pradesh, India
e-mail: Er.kajalchauhan6apr@gmail.com

R.K. Sagar
Amity University, Noida, Uttar Pradesh, India
e-mail: rksagar@amity.edu

© Springer Nature Singapore Pte Ltd. 2018
M. Pant et al. (eds.), *Soft Computing: Theories and Applications*,
Advances in Intelligent Systems and Computing 584,
https://doi.org/10.1007/978-981-10-5699-4_60

nature. Big Data comes in different formats. Traditional processing tools and technologies cannot cope with large datasets. Big Data Analytics defines the analysis of large collection of data that may be social media data, log files machine data, and enterprise data. New technologies are evolved to address large quantity data HealthCare, web traffic, enterprise data, sensor data, social data, business data, machine data, and global positioning system data. Researchers and traders purposed solutions to big data systems. Big Data is a proactive approach. Big data sets a point at which traditional technologies and tools are not sufficient for uncovered value or insights in cost-effective manner. Big data analytics model required to be re-evaluated. Big data has become very popular. Big Data analytics helps to gain more profit and productivity and improve efficiency of public and private sectors. Big data may be incomplete, inaccurate, and duplicate noisy. Here, we need to analyze these data. Big Data analytics indicates the starting of new form of technologies. Big data contains various features. Big data storage and management is very complicated task. Big data management handles large amount of data. New approaches, tools, and techniques need to be developed to analyze Big Data. Big Data is related to all aspect of human activity.

The definition of Big Data gives tools, set of methods, and technology to compare traditional data with Big data. Comparison is presented in Table 1. Big data is constantly update

It is semi-structured and unstructured data whereas traditional data is structured data. Data integration is easy in traditional database but difficult in big data.

This paper is organized as follows: Sect. 2 explains the features, characteristics of Big Data, discusses the motivation for adapting Big Data Analytics, and briefly highlights on Big Data Analytics Techniques; Sect. 3 focuses on Issues and challenges related with Big Data; Sect. 4 explains three open-source Big Data Analytics Framework and comparisons; and Section 5 is a conclusion to the study.

Table 1 Comparison between traditional data and big data

	Traditional data	Big data
Volume	GB	Constantly updated
Generated rate	Per hour, day	More rapid
Structure	Structured	Semi-structured or unstructured
Data source	Centralized	Fully distributed
Data integration	Easy	Difficult
Data store	RDBMS	HDFS, NoSQL
Access	Interactive	Batch

2 Big Data Analytics Characteristics

In Big Data input, data could babble from web server logs, broadcast audios, social networking sites, mp3s of music web server logs and documents traffic flow sensors, scan of government documents, satellite imagery, the details of web pages, banking transactions PS trails, financial market data, telemetry from automobiles, etc. For the identification of information from Big data, it has to be analyzed through different aspects.

2.1 Big Data Characteristics

To simplify the characteristics of Big Data, the five V's of volume, velocity, veracity, value, and variety are mainly used to organize varied aspects of big data. They are different aspects which work as the lens that picture the behavior of data and the software platforms available to develop them. Most possibly you will compete with the every V's of Big Data characteristics with one another for the better analysis of input data.

Volume—The main area of interest of analytics is processing complex and large volume of data. Volume depicts the astounding challenge to traditional IT structures. The distributed approach of querying and measurable storage is performed using this feature. Large numbers of archived data, which can be in the structure of logs, are becoming difficult for the companies to process it as they lack that ability.

Velocity—The significance of the velocity of the data is the pace at which the more and more data is generated. Immediately, data is provided to user when required. In the 8 years of 2005–2013, the digital universe spreads out from 130 million to 40 trillion. The use of smartphones has increased the rate of streamed data flow. The velocity is not just the problem but also the storage of streaming fast data inflow for later batch processing. There are two main reasons for the consideration of processing of streaming. First one is when the storage of input data in their loyalty is fast enough this implies that some level of assessment should appear as the streams of data for the storage requirements to become practical. The second reason for the consideration of streaming is the range of data originated from various resources is batch to real time.

Variety—It is least expected that data portrays itself in a much organized manner which is ready for processing. The big data of source data is diverse, and it does not lay itself into the appropriate relational structures [1]. The structures can be in a form of text from social networking sites, a raw feed directly from a sensor source and an image data from the websites [2]. This data does not come prepared for the integration into an application varied browsers send a number of data to users having information perhaps using varying software versions to exchange information with you, and there will be inaccuracy if human involvements is there. Flaws and inconsistency will also immerge.

Veracity—Due to the increasing rate of data, there is no time to spend in the cleaning of the massive data before using it. Analyzing of the data for the business process to enhance the productivity and decision making the organization requires a mechanism that should deal with the imprecise data. The processed and unprocessed, the clean and unclean, and the precise and imprecise combine to form big data.

Value—By processing and analyzing high-volume data with increased velocity, variety and veracity, here comes the need to unlock the hidden patterns to uncover the useful information or data for the business organizations to understand the habits of consumers and to facilitate the requirements of them. This unlocked data represents the actionable value of big data. The actionable value of data is that factor that can transform their business processes from the strategic view in revenue (Fig. 1).

2.2 Motivation for Big Data Analytics

Traditional technologies and tools are not satisfactory to reserve and collect large volume of datasets. Statistics [8] shows data grows rapidly. Analytics of Big data is the process that is used to collecting, storing, and analyzing high-volume high-velocity heterogeneous data. Big data Analytics process helps to extract hidden useful patterns and meaningful insights. Big data analytics allows to extract valuable insights it provide better understanding of customer behavior and understanding. Further, the result is used for decision making which helps in business grow, online browsing, online business, social media, network weather forecast, and telecom. Big data Analytics can help to make better decision it helps in future

Fig. 1 5 V's of big data [8]

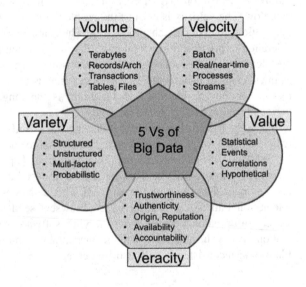

prediction of sales, in predicting customer demands. Due to heterogeneity of data, analytics become difficult. Analytics can be divided into three categories predictive analytics that uses statistical data, descriptive analytics that contain historical data which is related with business intelligence, prescriptive analytics is used to find out the optimized solutions for the concerned problem.

Big Data Analytics in Healthcare—Big data analysis gives important insight into medical field. Hospital and medical researchers store large- and high-volume data about patient's historical state, current state data, medication, and other details. Drug manufacturing firms store large volume of data. Analysis of Big data helps in both public and private medical services. In health sector, Big Data analytics help to detect which department needs to reorganize. The analysis of Big Data supports medical decision makers. Analysis of Big data helps to monitor and assess quality of medical services. Medical Data analytics helps to provide new life to patients gives valuable insights and discovery of new solutions to the doctors. It helps in measure the performance of medical units. It offers decision support tools and reduces high cost of medical sector. Analysis of Big data helps doctors and practitioners to find new solutions and ways to genetic and hereditary attacks.

Big Data Analytics in Intelligence Services—Many intelligence firms collects high-volume data from heterogeneous sources such as web sources, sensor data, publicly available sources, social networking websites, signal intercepts they make analysis based on these gathered data. Linking and connecting all the data and information makes the available information discovered. Thefts can be detected. Sound Analytics technique can handle high volume of unstructured data.

Environments and Big Data Analytics—Perception of Environment conditions in a better way needs large size of data from various sources as water being monitored using sensors and air quality system, mixture of gasses present in environment. These examples shows that adaptation of technologies, methods, tools, and frameworks to provide better valuable insights hidden from data sources.

Big Data Analytics and Marketing Campaigns—By analyzing Big Data Organizations help to discover new insights and valuable information that is present in large amount datasets. Insights can be used for predicting Customer behavior. Customer gives their reviews and ratings to the product, so Organizations can analyze this big amount of data to increase profits and values. Organizations can forecast the behavior of customers. Big Data Analytics lead to targeted market analysis. Organizations can developed new strategy and provide high-level satisfaction to the customer.

2.3 Big Data Analytics and Techniques

Analytics of Big Data uses advanced technologies. Big Data analytics process analyze high amount datasets and discover hidden patterns, meaningful information, insights, market tendency, recognized relationships. Analytics process is useful to make better decisions. It provides business benefits, business values, new strategies, and enhanced efficiency. Big Data Analytics uses wide variety of modern techniques (Table 2).

Blackett et al. Categorized Big Data Analytics into three categories: predictive analytics, descriptive analytics, and prescriptive analytics. All categories give valuable insights and profits to the organizations. In Big Data System, the first step is to capture lots of data and information from various sources.

Predictive Analytics—Predictive analytics predicting future trends based on statistical techniques. The nature of predictive analytics is probabilistic. Predictive analytics can only predict and forecast future probabilities. Predictive analytics uses logistic and linear regression to predict the future trends and outcomes. It can only forecast what might happen in the future. Regression and logistics techniques are used to extract patterns from big datasets. Predictive analytics uses data mining, statistical methods, and machine learning to make predictions about the future trends. Predictive analytics gives answer what will happen.

Descriptive Analytics—Descriptive method is the simplest method of analytics. The purpose of descriptive analytics summarizes historical data to what come about. This analytics answers the questions what happened? Descriptive analytics is connected with business intelligence. Descriptive analytics looks historical data and understands the reason behind success or failure. Marketing operations and sales uses descriptive analysis.

Prescriptive Analytics—Prescriptive analytics is the third business analytics. It is useful for decision making. Prescriptive analytics summarizes the big data, business rules, computational science, and then make predictions to business analytics. Business analytics adopts these predictions to make better profits and satisfying the customers. Prescriptive analytics take new current data every time. Prescriptive analytics improve accuracy and provide better decisions. Prescriptive analytics takes hybrid datasets as input and prescribe how to take advantages of this to predict future.

Table 2 Big data analytics techniques

Big data analytics	Techniques
Sql analytics	Count, mean, OLAP
Descriptive analytics	Univariate distribution, central tendency, dispersion
Data mining	Associations rules, clustering, feature extraction
Predictive analytics	*Classification, regression, forecasting, spatial, machine learning, text analysis*
Simulation	Monte Carlo, agent-based modeling
Optimization	Linear optimization, non-linear optimization

3 Issues and Challenges Related with Big Data Analytics

Big Data is diverse in nature, Organizations face challenges companies and organizations, and traditional systems are facing problems to store and analyzing hybrid data to make useful decision. This is a challenging task to companies and organizations. Big data analysis needs to store efficient data and requires queries to large datasets. Big data contain heterogeneity and incompleteness. Data comes from different data sources while machine language algorithm works on homogenous data. Big data challenges will be difficult to resolve as data is generated continuously. There is always a need of efficient, cost effective, appropriate tools and technologies. Some challenges are given below.

Privacy, Security—Big data is new technology trend: Privacy and security are two important aspects of any organization. Mostly, information present in such datasets is important. So there is a need of security mechanisms like encryption to secure the data from unauthorized external sources. Traditional resources are not sufficient to dealing with Big Data to ensure privacy and security. The analysis of Big data would make the system safer. Advanced techniques need to develop in terms of infrastructure, application, and data.

Big Data Management and Sharing—Big Data Management is important for any organization and companies to build new business model. Big Data management helps organization to organize the data and uses this data for future purposes. Analytics of Big Data extract meaningful insights from the use of Big Data management. This managed data is used by the companies to gaining more profits in business and market. Big Data Analytics contain large dataset. The datasets should be detectable, approachable, and usable. The agencies must persist to privacy laws, thigh having these urges. The present tendency toward the open-source datasets has noticed an importance on the development of such datasets ready for access of the public. The agencies should pay some more concern on making data conventional, standardize. It always permits them to use it and for the collaboration of privacy laws to the maximum extent possible.

Analytical Skills and Technology—Big data Analytics has given a maximum effort on ICT provider to develop new technology and tools to handle more complex datasets. The current technologies and tools are not able to process, store, and analyze huge volume of discrete datasets. Developers and vendors of such a huge data system suggest some solutions for making reliable effective, useful tools to reduce the complexity of huge datasets. Big data analytics helps to discover tools for Big Data integration and tools for manage resources.

Data Representation—Big Data sets comes from heterogeneous sources. Big data is varied in nature. Big Data represents huge amount data that is found in structured and unstructured forms. Datasets are heterogeneous in granularity and accessibility. An Efficient Data Representation should be designed to contemplate the hierarchy, collection of such huge amount of data. Integration technique should be designed in such a manner that performs operations efficiently.

Big Data Quality—As Big Data grows, data quality is important and challenging. Currently, quality assessment standards and methods are lacking. Data quality is an important aspect. Data quality defines as a set of quality attributes. Many factors affect data quality at different levels of processing. It is very difficult to find out manually generated data quality errors. There is a need to maintain data integrity. This includes the following:

How can be data preprocessed in order to improve quality of the data and result?
How can be data preprocessed so as to improve efficiency?
How to confirm the integrity of data?
How to calculate the worth of information in large datasets?

Complexity—Big Data deals with large records of structured and unstructured data, and this large datasets contains complex relationship between them it increase complexity of Data. Data records contain complicated relationships. As data grows rapidly, more data increases more complexity. Incompleteness is also a big challenge to deal with Big Data Analytics.

4 Analytics Frameworks for Big Data

Big Data required several types of frameworks to run several types of data analytics. A large number of different frameworks available to processing enterprise data.

Batch Analytics for Historical Data—Map Reduce and associate technology are very useful for batch analytics on Big data. Map Reduce is a framework using which we can write application to process huge amounts of data. Map Reduce is also a programming model. In this every data processing is divided into Map Reduce step. Map Reduce works in parallel fashion. Map Reduce framework breaks large data set into smaller ones and performs parallel processing which are executed on slave nodes. Map Reduce computes nodes independently. Map reduce contains master node (job tracker) and slave node (task tracker). Map Reduce includes several phases of process such as input phase, map phase, shuffle, and reduce phase. The result of map phase is intermediate key value pair. Map Reduce is not for random data. It works on sequential data. Map Reduce is fault tolerance framework. Map Reduce framework plays an important role in Big Data Analytics. Batch processing is widely adopted, and it gives faster response to real-time applications [3].

Stream Processing for Current Online Data—Stream processing takes input in the form of stream data with storm being a representative framework [4]. It is used for online analytics. It takes infinite data size as input. Data quality is an important aspect. Data comes in streams. It requires only few passes over streams to find approximation results. It provides result as quickly, and it takes milliseconds. Data stream processing is highly active technique.

- Interactive ad hoc queries and analysis with apache drill.

4.1 Apache Hadoop

Big Data deals with Apache Hadoop. It is available publically and open-source framework with no licenses fees that enables distributed processing of large volume of data. Apache Hadoop is basically based on Map Reduce programming model. Apache Hadoop supports java language. Apache Hadoop is highly scalable. Apache Hadoop supports distributed file system. IT companies and Organizations need to analyze huge volume data set they need more advanced technologies apache hadoop provide solution to the organizations. As Hadoop has become popular platform, it assures high availability at application layer [5]. Apache Hadoop includes listed sections.

Apache Hadoop includes listed sections:

(a) HDFS: It provides high throughput. HDFS supports large number of datasets. HDFS supports Master Slave Design.
(b) Hadoop core: It contains common utilities that support other modules.
(c) Hadoop YARN: Yet Another Resource Negotiator: framework for scheduling the job and cluster management. Hadoop Map Reduce model: programming model for massive volume of datasets.

Figure 2 represents how to manage data with Apache Hadoop. Here, user submits a query. Apache Hadoop framework is fault tolerant and robust. Hadoop is based on master (name node)/slave (data node) architecture. Master node manages file system policies and namespace and provides access to files. Job trackers

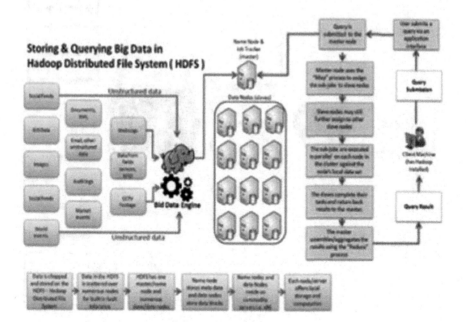

Fig. 2 Data store and retrieval operation in Apache Hadoop [9]

assigning the jobs to data nodes, data nodes are responsible for read and write operation from client sites. The data nodes also execute creation, deletion, and replication operations. The name node performs open, close, and rename operation. Data node stores each file, which is split into a sequence of blocks.

4.2 Storm Project

Hadoop can process and store large amount of data that previously unthinkable by the use of its related technologies have made it possible to process and store huge amount of data. Storm provides real-time analytics. Storm is fault tolerant and very simple to use. Storm processes streams of data. Processing of real-time data at such an immense scale going to be big requirements for business. Storm defines rules for real-time computation like how Map Reduce can make it more ease in writing of parallel batch processing. Parallel real-time computation can become easy using storm's rule [4].

Figure 3 shows storm cluster architecture. A storm cluster is pretty likely to Hadoop cluster, but on Hadoop cluster, you can run map reduce jobs on storm, and you run topologies, jobs, or topologies that are very different [4].

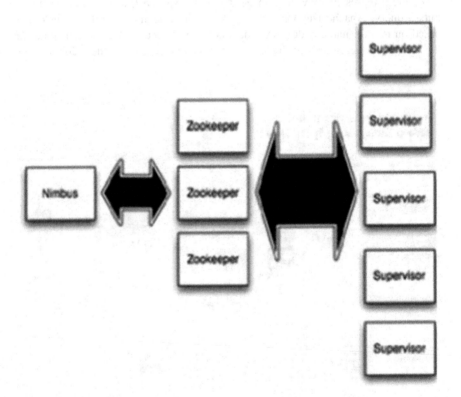

Fig. 3 Architecture of storm cluster [10]

Basically, nodes are divided into two types in storm cluster: (1) master node and (2) worker nodes. The master node runs a demon thread, i.e., "Nimbus," which is similar to Hadoop's "job tracker." Nimbus is mainly responsible for distributing code across the cluster and assigns tasks to different machines and monitor for failures occurring. Each worker node executes a daemon known as the supervisor. The supervisor takes account for work given to its machine and performs start and stop worker processes as required based on what Nimbus has assigned to it. Each and every process runs a part of a topology, a topology that is working and has many worker processes spread around many machines. All coordination between Nimbus and the supervisor is done through Zookeeper [6]. The Nimbus daemon and supervisor daemons are failing fast and stateless. All state is kept in zookeeper or on local disk. The design leads to storm clusters being incredibly stable.

4.3 Apache Drill

Apache Drill is a user- and developer-friendly software. Drill is an important project of Apache. Apache Drill is open-source software with no licenses fees. Apache Drill is available publicly. Apache Drill is a query engine. The Apache Drill is simple to use and scale to petabytes of data. Apache Drill provides connectivity to many data stores.

Figure 4 shows the architecture of Apache drill. It contains following parts.

User—it provides command line interfaces, and JDBC and ODBC and RESET interfaces for human.
Processing—allowing for pluggable query languages.
Data sources—pluggable data sources either local or in cluster setup, providing data sources.

Apache Drill is mainly focused on non-relational database and ad hoc queries. Apache Drill supports RDBMS. It can deal with multiple databases and file formats. Drill is capable to process millions of records in seconds. Apache Drill supports NoSQL databases. Apache Drill supports ANSI SQL standards. ANSI SQL can be used to get better result quickly. There is no requirement to define any schema. Apache drill is a query layer that works with underlying multiple data sources. Apache drill provides a flexible query execution framework that enables quick aggregation of statistics to explore data analytics. Big data analytics has become more accessible to a big range of users. Many times user need to run ad hoc queries against business application. Apache drill provides the solutions for that kind of issues [7].

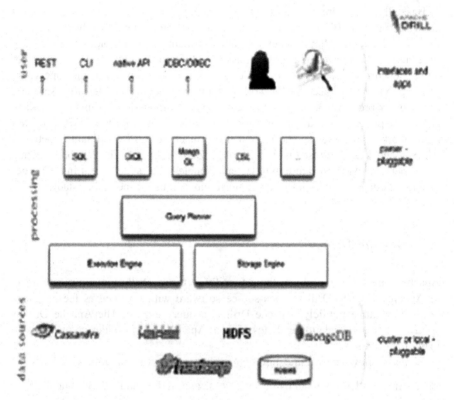

Fig. 4 Architecture of Apache drill [7]

5 Conclusion and Future Scope

Detailed study and analytics of Big Data has been accomplished, and comparisons between different frameworks are described below.

As Table 3 represents that Apache Hadoop is suitable for batch processing. Apache Hadoop has high latency. Hadoop can absorb any size of data. Apache Hadoop is more scalable and cheaper. Hadoop works on Map Reduce model. Project storm works with real-time computation and stream analysis. Project storm is easy to handle and use. It provides fast processing and process unbounded stream of data with high velocity. Storm is not suitable for batch processing of high-volume data. Apache drill is complex. It works well for interactive and ad hoc applications. Apache drill has low latency.

Table 3 Comparisons between big data analytics frameworks

Features	Apache Hadoop	Project storm	Apache drill
Owner	Community	Community	Community
Workload	Batch processing	Real-time computation, steam analysis	Interactive and ad hoc analysis
Source code	Open	Open	Open
Low latency	No	Yes	Yes
Complexity	Easy	Easy	Complex

5.1 Future Scope

Cost-effective and efficient tools are needed to analyze Big Data sets in real time. Data is generated continually. Here is a need to develop Big Data analytics frameworks for IT companies and Organizations. Data Analytics is hastily changing area. Traditional techniques need to revolution that can cope with Big Data. As data grows continuously, it is required to revisited current policies and system to manage large, hybrid datasets. Different frameworks need to analyze that practices with Big Data.

A new set of integration techniques should be designed. Traditional approaches are sufficient to extract value from huge amount of Data. Modified paradigms are required to develop.

Many organizations and companies face many challenges to cope with diverse Big Data. They need better tools and technologies to make better business decision.

For batch processing, it is not easy to adapt hastily growing data volume and real-time requirements. A big amount of time is wasted during batch processing transmission. Data Analytics provides new opportunities for real-time applications.

Security and privacy is also an important challenge for organizations dealing with Big Data. Big Data Analytics should find effective data control and quality mechanism. Big Data Analytics required facilitating new data processing techniques.

Big Data Analytics research needs more attention. Research on Big data can create more benefits for Business and organization. Big Data Analytics build a robust model that is able to analyze all size of data.

References

1. StructuredData: http://www.webopedia.com/TERM/S/structured_data.html
2. Semi structured Data: http://en.wikipedia.org/wiki/Semistructured_data
3. Apache-Hadoop: http://hadoop.apache.org/#What+Is+Apache+Hadoop%3F
4. Project Storm: http://storm-project.net/

5. Dean, J., Ghemawat, S.: MapReduce: Simplified data processing on large clusters. In: Sixth Symposium on Operating System Design and Implementation, San Francisco, CA, December 2004
6. Apache Zookeeper: http://zookeeper.apache.org/Big%20Data%20statistics-wikibon.org/blog/big-data-statistics
7. Hausenblas, M., Nadeau, J.: Apache Drill ad-hoc interactive analysis at scale, June 2013
8. Characteristics of Big Data: http://www.datatechnocrats.com/tag/bigdata/
9. Storing and querying data Big Data in HDFS: http://ecomcanada.wordpress.com/2012/11/14/storing-and-querying-bigdata-in-hadoop-hdfs/
10. Storm cluster: https://github.com/nathanmarz/storm/wiki/Tutorial
11. Katal, A., Wazid, M., Goudar, R.H.: Big data: Issues, challenges, tools and good practices. In: Sixth International Conference on Contemporary Computing (IC3) (2013)
12. Stephen, K., Frank, A.J., Alberto, E., William, M.: Big data: Issues and challenges moving forward. In: IEEE, 46th Hawaii International Conference on System Sciences (2013)
13. Conference on Communication, Information & Computing Technology (ICCICT), 19–20 Oct 2012
14. Michael, K., Miller, K.W.: Big data: New opportunities and new challenges. IEEE Technol. Soc. Mag. **13**
15. Sergey, M., Andrey, G., Jing Jing, L., Geoffrey, R., Shiva, S., Matt, T., Theo, V.: Dremel: Interactive analysis of web-scale datasets. Google (2013)
16. Apache-Dri: https://cwiki.apache.org/confluence/display/DRILL/Apache+Drill+Wiki
17. Apache HBase: http://hbase.apache.org/

Proposed Algorithm for Identification of Vulnerabilities and Associated Misuse Cases Using CVSS, CVE Standards During Security Requirements Elicitation Phase

C. Banerjee, Arpita Banerjee, Ajeet Singh Poonia and S.K. Sharma

Abstract Measurement of security during the development of a software application is a vital step in determining its future course of action. Security is defined as non-functional requirement which is often difficult to gather. Although many methods and approaches exist to provide definition of security much before the software application is developed. Among those methods and approaches, misuse case modeling may be used to define security as non-functional requirements during the software requirements elicitation phase. This paper proposes an algorithm for identification of vulnerabilities based on a specific software application and associated misuse cases using the industry-accepted standards like common vulnerability scoring system (CVSS) and common vulnerability enumeration (CVE). This paper also advocates the creation of software application-specific database so that the concept of data repository and reusability could be used for defining the security requirements with modification of such software applications to be made in future. The paper also highlights the areas where further research work can be carried out to further strengthen the entire system.

Keywords Vulnerability · Misuse cases · CVSS · CVE
Security requirements elicitation phase

C. Banerjee (✉)
Department of CS, Pacific University, Udaipur, India
e-mail: chitreshh@yahoo.com

A. Banerjee
Department of CSE, St. Xavier's College, Jaipur, India
e-mail: arpitaa.banerji@gmail.com

A.S. Poonia
Department of CSE, Govt. College of Engg. & Tech., Bikaner, India
e-mail: pooniaji@gmail.com

S.K. Sharma
Pacific University, Udaipur, India
e-mail: sharmasatyendra_03@rediffmail.com

© Springer Nature Singapore Pte Ltd. 2018
M. Pant et al. (eds.), *Soft Computing: Theories and Applications*,
Advances in Intelligent Systems and Computing 584,
https://doi.org/10.1007/978-981-10-5699-4_61

651

1 Introduction

Current practices for incorporating security during the software development process practically address security only in the late stages of development. Misuse cases does not have a precise application process and are too general in nature, and due to this fact they have been misunderstood and a general conception has been developed about its under definition and misinterpretation [1].

Security is one of the non-functional requirements, which is very difficult to identify. Even when the security requirements are identified, they are difficult to quantify. Identification of vulnerability in a software application during the initial stages of software development lifecycle can assist in the quantification of security requirements. The resultant software application would then be able to withstand malicious attacks thereby reducing cost and its rework on security aspects [2].

Practicality of misuse case model for the analysis of security requirements is a proven fact, but there should be proper knowledge for the alignment of security assets with goals [3]. Misuse cases can prove to be a way of specifying negative requirements which should not in any case occur in the system, either externally or internally. They are a convenient way of modeling any form of attacks being made to the system and their subsequent countermeasures if any [6].

Every day more and more vulnerabilities are being found and are continuously to be alarming high. This is attributed to the lack of interest of software development team to address security during the initial stages of software development. Moreover, the security requirements team is also not equipped with the knowledge about how to tackle and specify security as a non-functional requirement [7].

In this research paper, we propose an algorithm for identification of vulnerabilities based on a specific software application and associated misuse cases using the industry standards like common vulnerability scoring system (CVSS) and common vulnerability enumeration (CVE) which is properly aligned with the business and security goals of the system. Creation of a central repository is also proposed so that it may be used in reference to the similar future projects. Apart from the introduction, the remainder of the paper is organized as follows: Sect. 2 describes related work in this field and Sect. 3 presents the misuse case ontological classification with a brief description. Section 4 showcases the proposed algorithm, whereas results and discussions are covered in Sect. 5, conclusion and future work is given in Sect. 6.

2 Related Work

Matulevicius et al. have presented an analysis of the alignment of misuse case model with the information system security risk management (ISSRM) reference model and have also suggested improvements for the conceptual appropriateness of misuse case for the security risk domain [1]. Abdulrazeg et al. have proposed security metrics model whose basis is goal question metrics (GQM) approach. The proposed work focuses on the design of misuse case model and provides the security team to identify defects and vulnerabilities so that it can be fixed on time [2].

Takao Okubo et al. have proposed an extension for misuse case model, and they have included some new elements in the model like assets and security goal for specifying the security requirements of a computing system [3]. Malone and Siraj proposed an approach tracking of requirements and threats (TREAT) which employs several other modeling techniques to conceptualize and design a secure software system [4].

Hartong et al. have proposed a meta-model that covers graphical, textual, and OCL models for misuse cases that augments the existing UML 2.0 Use Case meta-models [5]. Whittle et al. proposed misuse case modeling language which is executable in nature through which a modeler may be able to specify misuse case scenarios in correspondence with the use case model. They further advocated that the proposed model is useful for analyzing potential attacks and their counter-measures [6].

Tøndel et al. have proposed a method for getting a high level view of threats using the concept of misuse cases and attack trees by linking them and describing security activities for each identified threat [7]. Banerjee et al. have proposed misuse and abuse case-oriented quality requirements (MACOQR) metrics from defensive perspective [8] and attacker's perspective [10] and also showed to measure the predicted and observed ratio of flaw and flawlessness in misuse case modeling, but the ontological classification and validation lacks depth.

El-Attar has proposed a model which may assist the modelers of misuse case modeling to improve the quality of their current and future models. As per their claim, even a person having lack of knowledge about the misuse case modeling can use it to their advantage [9].

3　Misuse Case Ontology

This paper also proposes an improvised version of ontological classification of misuse case as shown in Fig. 1.

- Database Vulnerability: Those security loopholes which can exploit the database like sql injection, etc.
- Application Vulnerability: Those security loopholes which can exploit the application like code injection, etc.
- Internal Threat: Those security loopholes which can be exploited by an insider.
- External Threat: Those security loopholes which can be exploited by an outsider.
- Mitigated Cases: Those security loopholes for which some kind of counter-measure is available.

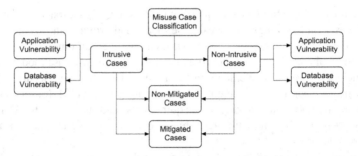

Fig. 1 Misuse case classification

- Non-mitigated Cases: Those security loopholes for which no countermeasure exist till now.
- Intrusive Cases: Those security loopholes which are direct and destructive and dangerous in nature.
- Non-intrusive Cases: Those security loopholes which are indirect and doesn't pose direct impact on the system in question.

4 Proposed Algorithm for Vulnerability Identification and Misuse Case Classification

Following is a step-by-step proposed algorithm which is a part process of misuse case-oriented quality requirements (MCOQR) framework which aims at identification of vulnerabilities and classification of misuse cases using the industry standards like CVSS and CVE.

Step 1	Identify software business goals
Step 2	Identify related assets
Step 3	Identify related security risk and goals
Step 4	Identify associated vulnerabilities
Step 5	For each of the identified associated vulnerability repeat Step 6 to Step 27
Step 6	Open and read CVE_DATABASE*
Step 7	Correlate the identified vulnerability in CVE_DATABASE for vulnerability categorization
Step 8	Associate the categorized vulnerability with vulnerability type**
Step 9	Open and read CVE_DATABASE
Step 10	If a match of vulnerability type is not found in CVE_DATABASE, assign a unique Vulnerability_Type_ID to the newly found vulnerability type and append in CVE_DATABASE
Step 11	Open CVSS_Document***
Step 12	Gather data according to CVSS base metrics datasheet (access vector, access complexity, authentication, confidentiality impact, integrity impact, availability impact)

(continued)

(continued)

Step 13	Using the data gathered using access vector, categorize the newly identified misuse case into Internal_Threat or External_Threat (i.e., if the access vector has a value as 'local' or 'adjacent network,' it is Internal_Threat and if it has a value as 'network,' it is External_Threat)
Step 14	Apply CVSS base metrics on data gathered
Step 15	Calculate the CVSS base scoring according to CVSS base metrics **CVSS Base Metrics** *Exploitability = 20** *AccessVector*AccessComplexity*Authentication* *Impact = 10.41*(1-(1-ConfImpact)*(1-IntegImpact)** *(1-AvailImpact))* *f(impact) = 0 if Impact = 0, 1.176 otherwise* *BaseScore = round_to_1_decimal* *(((0.6*Impact) + (0.4*Exploitability)-1.5)*f(Impact))*
Step 16	Gather data according to CVSS temporal metrics datasheet (exploitability, remediation level, report confidence)
Step 17	Using the data predicted and gathered using remediation level categorize the newly identified misuse case into Mitigated_Cases or Unmitigated_Cases (i.e., if the remediation level has a value as 'official fix,' 'temporary fix,' or 'workaround,' it is 'Mitigated_Case' and if it has a value as 'unavailable' or 'not defined,' it is 'Unmitigated_Case')
Step 18	Apply CVSS temporal metrics on data gathered
Step 19	Calculate the CVSS scoring using CVSS temporal metrics (it also includes CVSS base metrics scoring) **CVSS Temporal Metrics** *TemporalScore = round_to_1_decimal* *(BaseScore*Exploitability*RemediationLevel*ReportConfidence)*
Step 20	Gather data according to CVSS environmental metrics datasheet (collateral damage, target distribution, security requirements)
Step 21	Apply CVSS environmental metrics on data gathered
Step 22	Calculate the final CVSS scoring using environmental metrics (it also includes CVSS temporal metrics scoring) **CVSS Environmental Metrics** *AdjustedImpact = min(10,10.41*(1-(1-ConfImpact*ConfReq)*(1-IntegImpact*IntegReq)*(1-AvailImpact*AvailReq)))* *AdjustedTemporal = TemporalScore recomputed with the BaseScores Impact sub-equation replaced with the AdjustedImpact equation* *EnvironmentalScore = round_to_1_decimal* *((AdjustedTemporal + (10-AdjustedTemporal)** *CollateralDamagePotential)*TargetDistribution)*
Step 23	Using the final CVSS scoring and ranking (according to ranking table of CVSS document) categorize newly identified misuse case into Intrusive_Cases or Non_Intrusive_Cases (i.e., if the final CVSS scoring is '0-3.9,' it is Non_Intrusive_Case and if it is '4.0-6.9' or '7.0-10.0,' it is Intrusive_Case

CVSS Scoring	Severity of Vulnerability
0-3.9	Minor
4.0-6.9	Major
7.0-10.0	Critical

Step 24	Open and read MISUSE_CASE_DATABASE

(continued)

(continued)

Step 25	Match the data gathered so far with records in MISUSE_CASE_DATABASE
Step 26	If a match is not found, assign a unique MISUSE_CASE_ID and write the associated data gathered and calculated so far as a new record in MISUSE_CASE_DATABASE
Step 27	Write the data gathered and calculated as new record into APPLICATION_SPECIFIC_MISUSE_CASE_DATABASE

5   Validation and Results

The proposed algorithm was applied to a real-life project from industry (on the request of the company, identity is concealed). The resultant database was created with two versions, i.e., a central repository consisting of CVSS metric, CVE, misuse case databases with revised CVSS document and software-specific repository consisting of application-specific misuse cases database. Since the proposed algorithm helps identify vulnerabilities and associated misuse cases using CVSS and CVE and is a part process of misuse case-oriented quality requirements framework, when the MCOQR framework was applied in totality, the level of security implementation was compared with the other project's security implementation in which this algorithm was not applied.

The security indicators and estimators thus obtained using the MCOQR framework assisted the security engineering team to identify vulnerabilities, their associated misuse cases, and any mitigation strategy if applicable. This framework provided them with all the necessary analysis to plan in advance and to eliminate any defects that may occur during the modeling of misuse cases in the requirements elicitation phase of the similar projects. The study shows that the level of risk is minimized up to a significant level. Due to the page limit constraint, we are not providing the details of validation results in this paper; we will discuss in our next paper.

6   Conclusion and Futurework

Research and studies have proved that misuse case modeling may prove to be a potential source of defining security requirements during the software requirements elicitation phase. Researchers have also advocated that the misuse case modeling should be collaborated and synchronized with industry-accepted standards like CVSS, CVE, etc., so that, the model/framework/metrics thus proposed should be acceptable to the industry. Although researchers have contributed significantly to the removal of defects in the misuse case modeling so that security requirements could be taken care of in the initial stages of software development, so that, a more secure software could be built. But, most of the research work carried out does not comprehensively implement the industry-accepted standards like CVSS, CVE, etc.

In our paper, we have proposed an algorithm for identification of vulnerabilities and associated misuse cases through the tunneling of CVSS and CVE. The basis of the various vulnerabilities identified is closely associated with the security goals and assets of the system to be built. If proper process is adopted to identify vulnerabilities and associated misuse cases, then it may be appropriate to say that the further process of implementation of misuse case-oriented quality requirements (MCOQR) framework could be achieved. Further, proper security indicators and estimators could be found for elimination of defects in the misuse case modeling thereby resulting in a more secured software.

There are few points which needs special mention here like the CVE_DATABASE created during the implementation of the proposed algorithm should be updated using CVE external repository from time to time. In our research, we have categorized the newly identified misuse case as Application_Vulnerability or Database_Vulnerability depending upon the correlationship between the Source_Threat and Vulnerability_Type. Also the CVSS_DOCUMENT from time to time should be updated using CVSS external document repository.

Future work may include subcategorization of application vulnerability and database vulnerability for more in-depth analysis. One of the future works may include further categorization of vulnerability type other than application vulnerability and database vulnerability. Future work may also include automatic synchronization of CVE_DATABASE with CVE external repository. Another future work may include automatic synchronization of CVSS_DOCUMENT with CVSS external document repository.

References

1. Matulevicius, R., Mayer, N., Heymans, P.: Alignment of misuse cases with security risk management. In: Availability, Reliability and Security, 2008. ARES 08. Third International Conference on IEEE, pp. 1397–1404 (2008, March)
2. Abdulrazeg, A.A., Norwawi, N.M., Basir, N.: Security metrics to improve misuse case model. In: Cyber Security, Cyber Warfare and Digital Forensic (CyberSec), 2012 International Conference on IEEE, pp. 94–99 (2012, June)
3. Okubo, T., Taguchi, K., Kaiya, H., Yoshioka, N.: Masg: Advanced misuse case analysis model with assets and security goals. J. Inf. Process. 22(3), 536–546 (2014)
4. Malone, B., Siraj, A.: Tracking requirements and threats for secure software development. In: Proceedings of the 46th Annual Southeast Regional Conference on XX, pp. 278–281, ACM (2008, March)
5. Hartong, M., Goel, R., Wijesekera, D.: Meta-models for misuse cases. In: Proceedings of the 5th Annual Workshop on Cyber Security and Information Intelligence Research: Cyber Security and Information Intelligence Challenges and Strategies, p. 33, ACM (2009, April)
6. Whittle, J., Wijesekera, D., Hartong, M.: Executable misuse cases for modeling security concerns. In: 2008 ACM/IEEE 30th International Conference on Software Engineering IEEE, pp. 121–130 (2008, May)

7. Tøndel, I. A., Jensen, J., Røstad, L.: Combining misuse cases with attack trees and security activity models. In Availability, Reliability, and Security, 2010. ARES'10 International Conference on IEEE, pp. 438–445 (2010, February)
8. Banerjee, C., Banerjee, A., Murarka, P.D.: Measuring software security using MACOQR (misuse and abuse case oriented quality requirement) metrics: defensive perspective. Int. J. Comput. Appl. **93**(18) (2014)
9. El-Attar, M.: A framework for improving quality in misuse case models. Bus. Process Manage. J. **18**(2), 168–196 (2012)
10. Banerjee, C., Banerjee, A., Murarka, P.D.: Measuring software security using MACOQR (misuse and abuse case oriented quality requirement) metrics: attackers perspective. Int. J. Emerg. Trends Technol. Comput. Sci. **3**(2), 245–250 (2014)

Vulnerability Identification and Misuse Case Classification Framework

Ajeet Singh Poonia, C. Banerjee, Arpita Banerjee and S.K. Sharma

Abstract Specification of security as a non-functional requirement has always been a matter of difficulty as far as its identification and implementation are concerned. There are a lot number of methods and techniques for the implementation of security during the development of software application. The main concern is the identification of vulnerabilities and its proper treatment. Also, one more major concern in the alignment of business goals and its associated assets and related risk with the identified vulnerabilities. Further, assessing these vulnerabilities for finding a better countermeasure has also been a matter of a question. In this paper, we propose a novel framework for the identification of vulnerabilities properly aligned with the business goals and associated assets and its risk of the software application in development. Further, for assessing these vulnerabilities, we also present an ontological classification of the misuse case through a system model. The paper also showcases the implementation mechanism of the proposed framework in detail. Industry-accepted standards like CVSS and CVE have been applied in the proposed framework for authentication and validation of the work proposed.

Keywords Vulnerability identification · Misuse cases · CVSS
CVE · Misuse case classification

A.S. Poonia (✉)
Department of CSE, Govt. College of Engg. & Tech., Bikaner, India
e-mail: pooniaji@gmail.com

C. Banerjee
Department of CS, Pacific University, Udaipur, India
e-mail: chitreshh@yahoo.com

A. Banerjee
Department of CSE, St. Xavier's College, Jaipur, India
e-mail: arpitaa.banerji@gmail.com

S.K. Sharma
Pacific University, Udaipur, India
e-mail: sharmasatyendra_03@rediffmail.com

© Springer Nature Singapore Pte Ltd. 2018   659
M. Pant et al. (eds.), *Soft Computing: Theories and Applications*,
Advances in Intelligent Systems and Computing 584,
https://doi.org/10.1007/978-981-10-5699-4_62

1 Introduction

Vulnerability identification is the first step toward realization of any security loophole in a software application while it is being developed [1]. To identify correct and related vulnerability, the business goals should be properly aligned with the assets and its associated risk so that a clear picture of the various threats can be thought off. As it seems, it is not an easy step and an expertise is needed in this matter. This work should be done from the beginning of the software development process and a dedicated security team should be employed for the same [2].

After the vulnerabilities of the software applications that need to be developed are identified, the misuse cases could be drawn properly synchronized with the various predicted use cases of that software application [3]. As it is evident from the research work carried out so far, use cases are a potential contender for defining the functional requirements of a software application during the requirements engineering phase. Similarly, for defining the security requirements of a software application during its development, misuse cases are ideal contender [4].

When the vulnerabilities are identified and properly aligned to associated misuse cases, a system model, called misuse case modeling [5] should be developed so that the software application system in development could represent that various threats to its assets. Further, this model could be analyzed by the security team to develop countermeasures of these threats so that the security of the software application could be improved. This also results in the improvisation of misuse case model [6].

Since misuse cases are a representation of unified modeling language (UML), hence misuse case modeling may be applied to latest and future technological software application for specification of security requirements during the requirements engineering phase. The researchers have shown that using misuse case modeling there is a significant improvement in productivity and a secure software application could be developed [7].

In this research paper, we have proposed a framework for vulnerability identification of software application in development phase properly aligned with the system's business goals and its associated assets and risk factors. Using the identified vulnerability, the associated misuse cases are identified and model as a whole system. Industry-specific standards like common vulnerability scoring system (CVSS) and common vulnerability enumeration (CVE) are applied in the framework to give an ontological classification of the misuse cases identified, and a repository is also proposed. Apart from the introduction, the remainder of the paper is organized as follows: Sect. 2 describes related work in this field. Sect. 3 presents the proposed framework. Section 4 explains the implementation mechanism of the proposed framework, whereas results and discussions are covered in Sect. 5, and conclusion and future work is given in Sect. 6.

2 Related Work

Zech et al. have proposed a novel approach for risk-driven model-based security testing of cloud computing environment using misuse case modeling [8]. Loucopoulos et al. have advocated for the inclusion of non-functional requirements for early requirements and for a domain-specific non-functional approach. Through their finding, they have shown that rather than focusing on the late non-functional requirements, the software application development team should focus on the early non-functional requirements [9].

Raspotnig et al. have suggested a combined method covering the harm identified and analysis part of the assessment process using UML-based model [10]. El-Attar have proposed a framework for improving the quality of misuse case model using the concept of antipatterns to help remedy defective misuse case models and poor modeling practices [11].

Ficco et al. have proposed a set of stereotypes for the definition of a vocabulary for annotating UML-based models with information which is relevant and evident for the integration of the security requirements specification of cloud-based software applications [7]. Karpati et al. have shown the evaluation of misuse case maps, and the analysis shows better understanding of the intrusions and good ability for suggesting countermeasures [12]. El-Attar has proposed using structured misuse case descriptions (SMCD) to reduce inconsistencies in misuse case modeling [13].

Stålhane and Sindre through an experimental comparison of system diagrams and textual use cases have shown the potential for the identification of safety hazards [14]. Abdulrazeg et al. have proposed a security metrics with focus on the design of misuse case modeling using goal question metric approach. They suggested that the proposed metric may assist the development team in examining the misuse case model to identify vulnerabilities and to fix the defects before moving to the next stages of software development process. The proposed security metrics model has been based on the OQASP top 10-2010 in addition to misuse case modeling antipatterns [15].

3 Proposed Framework

A framework is proposed for the identification of vulnerability and subsequent misuse case classification. Industry-accepted standards like CVSS and CVE are used and synchronized in this approach to validate our work (Fig. 1).

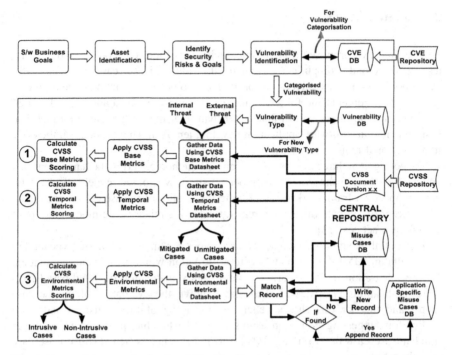

Fig. 1 Proposed framework of vulnerability identification and misuse case classification

4 Implementation Mechanism

In this section, the implementation mechanism identification of vulnerability and subsequent misuse case classification is explained in detail. First of all, the business goals of the software which needs to be built are to be identified and listed. Then, the process of associated asset identification needs to be carried out which should be properly aligned with the identified business goals.

Once the business goals are identified and properly aligned with the identified asset, the next step is to identify related security risk and goals. Using the identified security risk and goal, the various loopholes are studied and identified and become the step for vulnerability identification. This step is very crucial for further processing and needs expertise because if proper vulnerability identification is not done then the whole process will become futile.

There is a central repository consisting of common vulnerability enumeration (CVE) database, vulnerability database, common vulnerability document, misuse case database, and a repository which contains only application specific misuse case database. It is the responsibility of the security team to update the common vulnerability enumeration (CVE) database with the external CVE repository and CVSS document with external CVSS repository from time to time.

Once the identification of vulnerability is done, the CVE database is used to categorize it. For our research purpose, we have taken two categories of main vulnerability, i.e., database vulnerability and application vulnerability. The framework has been designed in such a way that more categories of vulnerability could be accommodated. Once the vulnerability is classified into database vulnerability or application vulnerability the associated record is written/updated into the vulnerability database. In this way, all possible vulnerability for that software application to be built should be identified and stored in the vulnerability database.

Now, one by one the identified vulnerability will be taken from the vulnerability database and will pass through the following process for its subclassification and score counting (for this the CVSS document may be implemented to further subclassify the identified vulnerability):

Process 1: The CVSS base metrics datasheet would be used to subclassify the identified vulnerability and associated misuse case into internal threat and external threat which forms the source of threat. The CVSS base metrics datasheet is shown in Fig. 2. Further, CVSS base metrics would be applied to calculate CVSS base metrics scoring.

CVSS Base Metrics

$$Exploitability = 20 * AccessVector * AccessComplexity * Authentication$$
$$Impact = 10.41 * (1 - (1 - ConfImpact) * (1 - IntegImpact) * (1 - AvailImpact))$$
$$f(impact) = 0 \text{ if } Impact = 0, 1.176 \text{ otherwise}$$
$$BaseScore = round\_to\_1\_decimal(((0.6 * Impact) + (0.4 * Exploitability) - 1.5) * f(Impact)).$$

Process 2: The CVSS temporal metrics datasheet would be used to subclassify the identified vulnerability and associated misuse case into mitigated cases and unmitigated cases. The CVSS temporal metrics datasheet is shown in Fig. 3. Further, CVSS temporal metrics would be applied to calculate CVSS temporal metrics scoring.

CVSS Temporal Metrics

$$TemporalScore = round\_to\_1\_decimal$$
$$(BaseScore * Exploitability * RemediationLevel * ReportConfidence).$$

Process 3: The CVSS environmental metrics datasheet would be used to subclassify the identified vulnerability and associated misuse case into intrusive cases and non-intrusive cases. The CVSS environmental metrics datasheet is shown in Fig. 4. Further, CVSS environmental metrics would be applied to calculate CVSS environmental metrics scoring.

CVSS Environmental Metrics

$AdjustedImpact = min(10, 10.41 * (1 - (1 - ConfImpact * ConfReq) * (1 - IntegImpact * IntegReq) * (1 - AvailImpact * AvailReq)))$

$AdjustedTemporal = TemporalScore$ recomputed with the $BaseScores$ Impact sub $-$ equation replaced with the $AdjustedImpact$ equation

$EnvironmentalScore = round\_to\_1\_decimal((AdjustedTemporal + (10 - AdjustedTemporal) * CollateralDamagePotential) * TargetDistribution).$

After going through all the subclassification, the associated vulnerability is appended/updated in the misuse case database residing in the central repository and a copy will be appended to the application specific misuse case database. This data may be further used to create a misuse case modeling datasheet and the security team may use the created datasheet to analyze the defects in the misuse case modeling. This modeling may be used to specify security requirements during the requirements engineering phase and may also aid in the further improvement of security of the system to be built.

5 Validation and Experimental Results

The proposed framework was applied to a real-life project from industry (on the request of the company, identification is concealed), and the identification of vulnerabilities and associated misuse cases with their classification was done as per the prescribed implementation mechanism. This studies done while using and implementing this framework resulted in increased security of the software application being built, because the security team were able to remove many defects in the misuse case modeling with detailed precision which otherwise would have got unnoticed had this framework being not implemented. In this paper, the details of the validation results are not been provided due to the page limit constraint and which we will discuss in our next paper.

6 Conclusion and Future Work

Proper vulnerability identification aligned with the business goals and related assets and its risk involved for the software application in development has been a matter of concern. Also, choice of applicability of some correct method or technique for assessing and analysis of identified vulnerability has also been a cause of concern. Because if these above-mentioned steps are not performed wisely and in a synchronized manner may result in the wrong specification of security requirements for a software application in development.

In this paper, we have proposed a novel framework for identification of vulnerabilities of software application in development phase right from the requirements elicitation phase. Through the proposed framework, a proper alignment of the system's business goals and its associated assets and risk factors has been shown. Once the vulnerabilities are identified, using the industry-specific standards like common vulnerability scoring system (CVSS) and common vulnerability enumeration (CVE), the associated misuse cases are identified and further subclassified for in-depth analysis of the misuse case model so that the defects if found may be corrected and appropriate countermeasures could be developed to eliminate any threat to the system.

Future work may include the development of an algorithm for identification of vulnerabilities and associated misuse cases using industry-accepted standards like CVSS and CVE. Another future work may include automation of the entire process of vulnerability identification and misuse case classification. Some software security metrics may also be developed to provide in-depth analysis of the misuse case model for its improvisation.

References

1. Banerjee, C., Banerjee, A., Murarka, P.D.: Algorithmic approach for development of misuse case modeling framework for iMACOQR metrics. In: Proceedings of the 2015 Fifth International Conference on Advanced Computing and Communication Technologies IEEE Computer Society, pp. 564–568. (2015, February)
2. Sindre, G., Opdahl, A.L.: Eliciting security requirements with misuse cases. Requirements Eng. 10(1), 34–44 (2005)
3. Banerjee, C., Banerjee, A., Murarka, P.D.: Measuring software security using MACOQR (misuse and abuse case oriented quality requirement) metrics: defensive perspective. Int. J. Comput. Appl. 93(18) (2014)
4. Alexander, I.: Misuse cases help to elicit non-functional requirements. Comput. Control Eng. J. 14(1), 40–45 (2003)
5. Banerjee, C., Banerjee, A., Pandey, S.K.: MCOQR (Misuse Case-Oriented Quality Requirements) Metrics Framework. Problem Solving and Uncertainty Modeling through Optimization and Soft Computing Applications, 184 (2016)
6. Alexander, I.: Misuse cases: use cases with hostile intent. IEEE Softw. 20(1), 58–66 (2003)
7. Ficco, M., Palmieri, F., Castiglione, A.: Modeling security requirements for cloud-based system development. Concurrency Comput.: Pract. Exper. 27(8), 2107–2124 (2015)
8. Zech, P., Felderer, M., Breu, R.: Towards a model based security testing approach of cloud computing environments. In: Software Security and Reliability Companion (SERE-C), 2012 IEEE Sixth International Conference on IEEE, pp. 47–56. (2012, June)
9. Loucopoulos, P., Sun, J., Zhao, L., Heidari, F.: A systematic classification and analysis of NFRs (2013)
10. Raspotnig, C., Karpati, P., Katta, V.: A combined process for elicitation and analysis of safety and security requirements. In: Enterprise, business-process and information systems modeling (pp. 347–361). Springer Berlin Heidelberg (2012)
11. El-Attar, M.: A framework for improving quality in misuse case models. Bus. Process Manage. J. 18(2), 168–196 (2012)
12. Karpati, P., Opdahl, A.L., Sindre, G.: Investigating security threats in architectural context: experimental evaluations of misuse case maps. J. Syst. Softw. 104, 90–111 (2015)

13. El-Attar, M.: Using SMCD to reduce inconsistencies in misuse case models: a subject-based empirical evaluation. J. Syst. Softw. **87**, 104–118 (2014)
14. Stålhane, T., Sindre, G.: An experimental comparison of system diagrams and textual use cases for the identification of safety hazards. Int. J. Inf. Syst. Model. Design (IJISMD) **5**(1), 1–24 (2014)
15. Abdulrazeg, A. A., Norwawi, N. M., Basir, N.: Security metrics to improve misuse case model. In: Cyber Security, Cyber Warfare and Digital Forensic (CyberSec), 2012 International Conference on pp. 94–99. IEEE (2012, June)

Revisiting Requirement Analysis Techniques and Challenges

Shreta Sharma and S.K. Pandey

Abstract Requirement analysis is one of the major activities of requirement engineering (RE) process, which is an iterative process of discovering and analysing feasible features to produce a contracted set of complete and consistent requirements. Revolution and developing business environments demand well-defined and complete requirements for successful software development. In spite of various trend-setting requirement analysis methods/tools/framework/ techniques, it has been observed that few techniques are useful with certain systems, but at the same time, few of them may not be useful for others. Accordingly, identification/selection of correct requirement analysis (RA) techniques becomes more important before moving further towards other activity in RE. Accordingly, an attempt has been made in this paper to illustrate the requirement analysis process, various techniques and their related challenges. It is expected that the same would facilitate the concerned stakeholders to understand and choose the most suitable RA technique to be used in their project/s.

Keywords Software engineering · Requirement engineering (RE)
Requirement analysis (RA) · Requirement analysis techniques
Issues and challenges in requirement analysis

1 Introduction

A key goal of software engineering (SE) is to consistently produce high-grade software within the project plan and budget limit. Contrary to this, research studies indicate that more than 85% of software are rejected by the customers after delivery

S. Sharma (✉)
Jagannath University, Jaipur, India
e-mail: shretasharma@hotmail.com

S.K. Pandey
Ministry of Communications & IT, Govt. of India, New Delhi 110003, India
e-mail: santo.panday@yahoo.co.in

© Springer Nature Singapore Pte Ltd. 2018
M. Pant et al. (eds.), *Soft Computing: Theories and Applications*,
Advances in Intelligent Systems and Computing 584,
https://doi.org/10.1007/978-981-10-5699-4_63

because the software does not meet the customer requirements, and accordingly, more than 50% of all money is spent on maintenance of the product after delivery [1, 2]. Moreover, software procurement as compared to that of hardware procurement has been showing a disquieting trend over the years [3]. Researchers have identified the primary cause behind the same, and accordingly, they found that majority of software fail due to 'incomplete requirements and specification' [2, 3].

The discipline of RE is mainly concerned with gathering, analysing, specifying and validating user requirements and tries to overcome related issues *right from the beginning* in the SDLC [2, 4]. Many researchers emphasized that RE process is an efficient approach through which engineers can gather requirements from various sources and implement them into software development procedure and help to steer the improvement towards producing the efficient software. RE process is seen as a group of well-defined activities, transformation, techniques that people used to develop system requirements and maintain the requirement specification and related artefacts. It encloses a group of phases for eliciting, analysing, recording, validating and maintaining the group of requirements [5, 6].

RA is a term, which is used to define all the activities that go into the initiation and scoping of the project. The major aim of the analysis phase is to alter the needs and advanced requirements specified in earlier phases into consistent, traceable, complete, unambiguous requirements. RA is the essential step towards creating a specification and design. In the literature, there are numerous techniques and tools for the same, which provide a boundless sustenance towards refining the efficiency and quality of software [7, 8].

This paper provides a broad review of state-of-the-art methods, and process of requirement analysis has been accomplished by emphasizing their key features along with the related challenges [2]. The major objective of this paper was to highlight the challenges of the analysis phase. This paper is designed in five segments. This segment, being an outline, offers a brief overview of requirement analysis, and the stages of the RA process have been summarized in Sect. 2, whereas various analysis techniques are used for analysis phase and their challenges are provided in Sect. 3. The Conclusion and Future Work have been articulated in Sect. 4.

2 RA Process

This includes documenting and knowledge about users' present work, as well as focus on future works from the user's perspective. Complete analysis process can be divided into nine stages, which are shown in Fig. 1. And the details are given as follows:

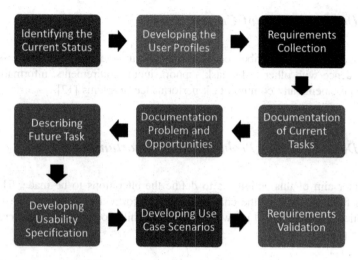

Fig. 1 Requirement analysis(RA) process

2.1 Identifying the Current Status

It is essential in the beginning of the process to explore the area/s of the problem and the position profoundly in which system will participate. This task includes various sub-activities such as the current stage/s of the task and significant information about concern project, defining interface design constraints, the scope of the project documentation and management [2, 9, 10].

2.2 Developing User Profiles

Developing user profile is an essential task in requirement analysis. The major objective of this phase is to find out user characteristics that will be more significant to interface design trade-offs. These include user experience with hardware environment/software environment and frequency of use and so on [11, 12].

2.3 Requirement Collection

During this stage, requirements are gathered from the users, which include verified user profile as well as their present and forthcoming task flows. There are various methods to collect the same such as bringing user in, role playing and field studies [12].

2.4 Documentation of Current Tasks

The documentation describes each activity during the task, such as task flow, independence with other tasks, task support, input requirements, information display requirements and common task performance problems [12].

2.5 Documentation Problems and Opportunities

The major aim of this activity is to define the alterations to be made. There are various fundamentals for the changes such as information from possible work or difficulties or prospects, which were uncovered while documenting the present task analysis [12].

2.6 Describing Future Tasks

Earlier activities have been designed for understanding and documenting the users and their current work. However, this stage focuses on future tasks, which produces a new task line that describes the future plan [11, 12].

2.7 Developing Usability Specifications

After analysing current and future tasks, this task is to define usability description. This summarizes main user, training and document assumption in a set-up that is independent, assessable and testable [12].

2.8 Developing Use Case Scenarios

It is a framework of tasks and subtasks that define 'how the users will proceed further'. It produces a conceptual design [12].

2.9 Requirement Validation

It is essential to verify the decisions during the analysis phase. There are various assessment events that go on throughout analysis phase such as validation of user profile, use case scenario and user specification [12].

3 Major RA Techniques and Challenges

RA is one of the most crucial processes in RE, which includes many activities with a variety of available techniques [12, 13]. In spite of good research contributions towards RA techniques to provide efficiency and documentation to analysis process, these techniques are still facing a few challenges, which are shown in Fig. 2. And the details are given in subsequent subsections:

3.1 Stakeholder's Interviews

Stakeholder's interviews are one of the simple methods used in requirement analysis. However, in any big system, a number of people need to be interviewed, which increases the time and cost. But these interviews may reveal major

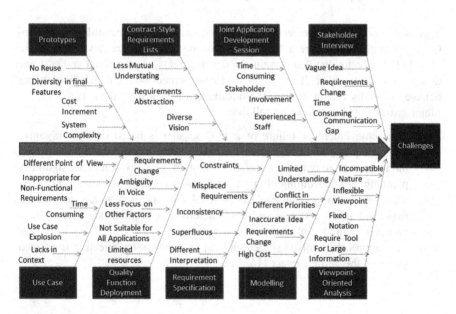

Fig. 2 Requirement analysis techniques

differences to know the existing business process and how this process should work in future. Additionally, different users will have different ideas of their expectation, which is useful in the identification of barriers and roadblocks during the process [11, 13]. Major issues are given as follows:

(i) Defining precise user requirements are the source of successful software growth. The major issue in requirement analysis is that users may not have a correct idea of their need and they may provide a vague idea. Developers need to perform the analysis to turn this amorphous vision to fully documented specification [11, 12].

(ii) Informally, user frequently changes their requirements even when the system or product process has been started, which can result in the unrealistic schedule and budget expectation [13, 14].

(iii) Various activities performed in interviews such as analysing, interviewing, setting up, transcribing, reporting, feedback can be a time-consuming process [14].

(iv) Often, project stakeholders and developers fail to communicate clearly with each other because stakeholders are geographically dispersed. They do not dynamically contribute in the development. This problem is generally faced with tasks within the large organization, where projects have been subcontracted to an exterior organization [13].

3.2 Joint Application Development (JAD) Session

This is a process that accelerates the plan of information technology solutions. Gathering requirements are a difficult task during the software development process. JAD is a technique that permits the developments, management and customer groups to work together. It is a controlled environment wherein stakeholders participate in the discussion to elicit the information and analyse their details [2, 15]. Major issues stated are given as follows:

(i) The achievement or failure of a JAD session is based on how organizer manages plans and the session efficiently. The major issue with this technique is that it requires a huge amount of time to complete significant planning and scheduling efforts to maintain well the JAD session [15, 16].

(ii) The contribution of stakeholders in the JAD session is widely accepted as important to its ultimate success. Therefore, it requires significant stakeholder's commitment to the time and efforts [17].

(iii) This approach involves skilled and expert workers for effective implementation of the entire project [16, 17].

3.3 Contract-Style Requirement Lists

Contract-style requirement lists provide a specification of requirements and agreement between the project designers and sponsor/s. It also provides a high-level description of the large system [18]. Major issues identified by various researchers regarding usage of this technique are given as follows:

(i) These lists create an unreliable logic of contract between the stakeholders and developers [19].
(ii) Such requirement list is non-representational, which makes difficult to identify which requirement is most important and by what procedure the requirements suit together [20].
(iii) Many people who read such requirements propose diverse visions of the system as per their understanding [20].

3.4 Prototypes

Prototyping provides a concrete representation of all parts of the system. Prototypes are a commonly renowned fundamental stone of discovering and articulating designs. It helps users to get an idea of system design decision. Early blueprint of application focused on minor changes in next stages and hence reduced total costs of software [21, 22]. Major issues identified with the above technique are given as below:

(i) Prototype model may be of no use for further projects in the same way after showing it to the client [22].
(ii) Sometime features which appear in a prototype are dissimilar from the finishing system so users can become confused [22].
(iii) Building a prototype during development can increase cost factor also [22].
(iv) The scope of the system does not meet the original plans, which increase the complexity of the system [23].

3.5 Use Cases

Use cases are the widely used mechanisms to discover and record functional requirements. Use cases include a text document by which a user can easily understand and describe the details using the system. Informally, these are sections of using a system to accomplish goals. Each use case offers one or more states that express the interaction with end users to achieve system goals [24–26]. Major issues reported in the literature related to use cases are given as below:

  (i)   Use cases are not appropriate to capture non-functional requirements such as
        security, timing and performance critical aspects and their related aspects
        [24].
 (ii)   Generally, a use case is not written from the actors' point of view but from
        the system's view. The use case names define the flow of the system, rather
        than the goals, the actor wants to accomplish [24].
(iii)   It takes a lot of time to prepare a complete set of a use case for an application
        [25].
(iv)   When a set of use case has too many cases, they usually do not describe user
        goals, rather insignificant interactions or minor actions of the users. In a large
        project, there are too many use cases, which are involved in the use case set
        that creates the use case explosion problem [26].
 (v)   Use case specifications are confusing, too long and unsystematic, and they
        lack in context [27].

3.6   Quality Function Deployment (QFD)

It is a technique that explains the needs of clients into technical requirements for
software. QFD identifies objectives and goals (e.g., 'detailed system functions,
types of graphic displays') and estimated requirements [26–28]. Quality function
deployment is an enormously convenient approach to aid planning, communication
and decision-making within a product development team. It is not an administrative
exercise or additional records that must be concluded in demand to continue to the
next expansion milestone. It does not only fetches the new product closer to the
planned target but diminishes development cycle time and budget in the process
[28]. Major issues stated regarding QFD are given as follows:

  (i)   Major disadvantage of the QFD is that customer needs change quickly and
        unpredictably, which require assumptions [29].
 (ii)   Sometimes customer's responses are tough to organize as requirements due
        to the ambiguity in the customer's voice. [28–30].
(iii)   QFD focuses especially on what a business essential to do to fulfil its cus-
        tomers. The major disadvantage of QFD is that it does not focus on other
        factors such as the duration of product life cycle, cost, durable strategy and
        progressive ideas and available resources [31].
(iv)   QFD does not support to all the applications such as in the automotive
        industry; there are only a few possible customers; the customer categorizes
        their requirements; and the supplier acts to gratify them [32].
 (v)   QFD methodology executes the need to deal with the large amount of data
        gathered from customers, competitors and cross-functional teams, etc. [31].

3.7 Requirement Specification

Requirement specification is a broad elucidation of the desired actions of the system. Requirement specification contains non-functional and functional requirements and also describes an environment in which system operates. It contains set of use cases, which describe the interaction between users and software [33, 34]. The major issues observed with requirement specification are given as follows:

(i)   Some requirements cannot be executed due to system restrictions, budget and human resource limitation [32].
(ii) Sometimes the necessary information related to the problem being solved by the software may be misplaced from requirement document or may not be complete [32].
(iii) One part of the requirements document is inconsistent with another part/s or with the problem that the specification explains [32].
(iv) Some specification documents are superfluous, and some part of the information may not be relevant to the problem being solved [33].
(v)   Another major challenge is the information, or requirements written in a document may have more than one clarification; sometimes the meaning of the report may not be clear in the document [33, 34].

3.8 Modelling

It supports the expertise in understanding the function, information and behaviour of a system. The model develops the central point for analysis in the phases of comprehensiveness, consistency, and correctness of the requirements. It becomes the groundwork for design, providing a needed representation of software to designer [35]. It deals with the problem of size by portioning the system. Major issues stated in modelling method are given as follows:

(i)   The key issue of modelling is lacking or limited understanding of 'Software development'. Basically, the developers need the lively provision of project stakeholders to be successful [34].
(ii) The system must reflect the needs of numerous areas such as senior management, support staff. Each of these groups has different priorities and aims, which conflict during development, to be addressed [35].
(iii) Another general problem at the foundation of the project is that project stakeholder may not be clear about their needs [35].
(iv) Project stakeholders change their requirements after the plan, and the cost has been fixed during the progress of the system [35].
(v)   Modelling is incapable to cheaper projects because the cost of the modelling and automated code generation is very high [35].

3.9 Viewpoint-Oriented Analysis

All information about the system supplies cannot be shown by taking a particular perspective; rather, it is essential to collect and manage information from diverse viewpoints. Viewpoint analysis is an approach, which encapsulates the partial information about system's requirements [36]. The main aim of viewpoint analysis is to collect and classify different types of information such as application domain and system environment, which are needed for a system's development [26]. Major challenges stated by specialists and practitioners in viewpoint method are given as follows [2]:

(i) The foremost practical issue coming from the previous study is incompatible nature with other software engineering methods. Organizations generally have existing design models, and they might get profit from the use of viewpoint method; therefore, it should be compatible with existing methods [37].

(ii) Inflexible viewpoint models are also a concerning issue. If the viewpoint model is too limited in its definition of a viewpoint, it will not include all of the potential stakeholders and domain viewpoints, which may be essential. Many requirement challenges are organizational, human and social. Viewpoints need to be able to reproduce these points and not just technical terms of system requirements [37].

(iii) Some of the viewpoint methods have a fixed notation for stating requirements. Requirement sources often will not have time to precise requirements in any suitable notations [37].

(iv) Viewpoint-oriented methods tend to produce a huge amount of information, which must be managed, and this apparently requires tool support of some kind. This method is seeking for industrial—strength tool to support the management of large amounts of information [37].

4 Conclusion and Future Work

RA is an approach that demands a blend of hardware, software and human aspects, engineering expertise as well as skills in dealing with people either existing or future users to ensure the successful outcomes. The paper offered the procedure of requirement analysis by discussing the entire foremost activities preliminary from the identification of the current status of requirements till the requirement validation. Additionally, a broad description of various analysis techniques has also been presented. The literature reveals that there are various challenges and issues in requirement analysis phase, which outcome incomplete requirement part of the analysis. This research paper has attempted to showcase major challenges and issues in different types of requirement analysis techniques [2].

Future work may be to introduce innovative ideas/methods to overcome these issues concerns along with valid authenticate outcomes. It is apparent from the above-mentioned discussion that the main source of each issue is the extreme human involvement in this procedure. Integration of state-of-the-art artificial intelligence (AI) techniques may limit such interference up to some level and appears to deliver some successful results. Therefore, further investigation may be directed to provide the procedure for incorporating AI techniques with various activities of requirement analysis. This study may provide an important direction to the RE experts for developing a quality software with fewer efforts and cost [2].

References

1. Future of CIO, Five ponderings on Why IT projects Fail. Re-imagine future of IT. CIO. Leadership in 21st Century. http://futureofcio.blogspot.in/2012/01/five-pondering-why-it-projects-fail.html/20/2/2016 (2013)
2. Sharma, S., Pandey, S.K.: Requirements elicitation: issues and challenges. In: International Conference on Computing for Sustainable Global Development (INDIACom (2014). Doi:10.1109/IndiaCom.2014.6828119
3. Britto, R., Wohlin, C., Mendes, E.: An extended global software engineering taxonomy. J. Softw. Eng. Res. Dev (2016). Doi:10.1186/s40411-016-0029-2
4. Shilpi, S., Saikia, L.P.: Ambiguity in requirement engineering documents: importance approaches to measure and detect, challenges and future scope. Int. J. Adv. Res. Comput. Sci. Softw. Eng. 5(10), 791–798 (2015)
5. Kaur, H., Singh, D., Kaur, P.: Preview Analysis in Requirement Engineering. Int. J. Adv. Res. Comput. Commun. Eng. 5(7), 370–372 (2016)
6. Foster, E.C.: Three Innovative Software Engineering Methodologies. In: Annual Global Online Conference on Information and Computer Technology (GOCICT). pp. 1–20 (2015). ISBN 978-1-5090-2314-1
7. Wong, W.E., Gao, R., Li, Y.: A survey on software fault localization. IEEE Trans. Softw. Eng. 42(8), 707–740 (2016)
8. Abbas, Jalil: Quintessence of traditional and agile requirement engineering. J. Softw. Eng. Appl. 9(3), 63–70 (2016)
9. Sharma, S., Pandey, S.K.: Integrating AI techniques in requirements phase: a literature review. Int. J. Comput. Appl. 3(2), 21–25 (2014)
10. Sharma, S., Pandey, S.K.: Integrating AI techniques in SDLC: requirements phase perspective. Int. J. Comput. Technol. Appl. 5(3), 1362–136 (2014)
11. Batra, M., Bhatnagar, A.: Descriptive literature review of requirements engineering models. Int. J. Adv. Res. Comput. Sci. Softw. Eng. 5(2), 289–293 (2015)
12. Nuseibeh, B., Easterbrook, S.: Requirements engineering: a roadmap. In: Proceedings of Conference on the Future of Software Engineering. ACM. 1(1), 35–46 (2000)
13. Majumdar, S.I., Rahman., M.S., Rahman, M.M.: Thorny issues of stakeholder identification and prioritization in requirement engineering process. IOSR J. Comput. Eng. 15(5), 73–78 (2013)
14. Sharp, H., Galal, G.H., Finkelstein, A.: Stakeholder identification in the requirements engineering process. In: Proceedings of the 10th International Workshop on Database and Expert System Applications. 1(2), 387–391 (1999)
15. Jalote, P.: A Concise Introduction to Software Engineering. Undergraduate Top. Comput. Sci. Springer. ISSN: 1863-7310. 37–41 (2008)

S. Sharma and S.K. Pandey

16. McConnell, S.: Rapid development: taming wild software schedules. Edition 7. Microsoft Press. ISBN 1-55615-900-5. pp. 449–463 (1996)
17. Hathaway, T., Myers, D.: Performing effective requirements gathering JAD sessions. Requirements Solutions Group, LLC. 2(0), 1–39 (2008)
18. Huzooree, G., Ramdoo, V.D.: A systematic study on requirement engineering processes and practices in mauritius. Int. J. Adv. Res. Comput. Sci. Softw. Eng. 5(2). 40–45 (2015). ISSN 2277 128X
19. Pressman, R.S.: Software engineering: a practitioner's approach. Software Engineering. pp. 243–250 (2010). ISBN 978-0-07-337597-7
20. Reffat, R.: Utilization of artificial intelligence concepts and techniques for enriching the quality of architectural design artifacts. In: Proceedings of the 1st International Conference in Information Systems. 5(1), 1–13 (2002)
21. Christof. E., Kuhrmann, M., Prikladnicki, Rafael.: Global software engineering: evolution and trends. In: IEEE 11th International Conference on Global Software Engineering (ICGSE) (2016) Doi:10.1109/ICGSE.2016.19
22. Gordon, V.S., Bieman, J.M.: Reported effects of rapid prototyping on industrial software quality. Softw. Qual. J. 1(2), 1–11 (1993)
23. Maheshwari, M.S., Jain, D.: A comparative analysis of different types of models in software development life cycle. Int. J. Adv. Res. Comput. Sci. Softw. Eng. 2(5), 285–289 (2012)
24. Maguire, M., Bevan, N.: User requirements analysis. In: Proceedings of IFIP 17th World Computer Congress, Montreal, Canada. 1(2), 133–148 (2002)
25. Bruseberg, A., McDonagh-Philp, D.: New product development by eliciting user experience and aspirations. Int. J. Hum Comput Stud. 55(4), 435–452 (2001)
26. Rus., I., Lindvall. M.: Guest editors' introduction: knowledge management in software engineering. IEEE Softw. 19(3), 26–38 (2002)
27. Jaiswal, E.S.: A case study on quality function deployment (QFD). IOSR J. Mech. Civil Eng. (IOSR-JMCE). 3(6), 27–35 (2012)
28. Delgado, D.J., Aspinwall, E.M.: QFD methodology and practical applications—a review. In: Proceedings of the Ninth Annual Postgraduate Research Symposium. The University of Birmingham. 3(6), 1–5 (2003)
29. Shahin, Dr. Arash.: Quality Function Deployment: A Comprehensive Review. Department of Management, University of Isfahan.Citeseer. 1–25. (2000)
30. Vonderembse, M.A., Raghunathan, T.S.: Quality function deployment's impact on product development. Int. J. Qual. Sci. 2(4), 253–271 (1997)
31. Kathawala, Y., Motwani, J.: Implementing quality function deployment—a system approach. TQM Magazine. 6(1), 31–37 (1994)
32. Franceschini, F., Rossetto, S.: QFD: the problem of comparing technical/engineering design requirements. Res. Eng. Design 7(4), 270–278 (1995)
33. Faulk, R.S.: Software Requirements. A Tutorial.: Software Requirements Engineering 2nd Edn, pp. 1–3. IEEE Computer Society Press (1997)
34. Lang, M., Duggan, J.: A tool to support collaborative software requirements management. Requirements Engineering. Springer, 6(3), 161–172 (2001)
35. Gunter, C.A., Gunter, E.L., Zave, P.: AT&T Laboratories.: A Reference Model for Requirements and Specifications. IEEE Softw. Requir. Eng. 17(3), 37–42 (2000)
36. Kotony, G.: Practical experience with viewpoint-oriented requirements specification. J. Requir. Eng. Springer London. 4(3), 115–133 (1999)
37. Kotonya, G., Sommeiville, I.: Requirements engineering with viewpoints. BCS/IEE Softw. Eng. J. 11(1), 1–15 (1996)

Test Data Generation Using Optimization Algorithm: An Empirical Evaluation

Mukesh Mann, Pradeep Tomar and Om Prakash Sangwan

Abstract This paper aims to design an approach for making an efficient fitness function for tests case generation based on distance from the goals. The designed function is given as an input to the genetic algorithm, and the result of the search process using the formed fitness function is evaluated in terms of time and the average number of test cases generated. This paper also investigates the effect of parameter setting such as the size of the initial population on the performance of genetic algorithm using the proposed fitness function. The experimental result shows that the proposed approach is both time and cost efficient in comparison with manual and random testing. It is also found that initial larger population size gives better results in comparison with low initial population.

Keywords Automatic test case generation · Genetic algorithm · Random testing

1 Introduction

The customer satisfaction has been growing as a priority area while developing quality software. To deliver software system on time and with quality demands both the speed and extent of testing. However, the extent of testing is directly proportional to the time and the resources involvement. Such factor requires more than 50% of

M. Mann (✉) · P. Tomar
Department of Computer Science and Engineering, SOICT,
Gautam Buddha University, Greater Noida, India
e-mail: Mukesh.gbu@gmail.com

P. Tomar
e-mail: Parry.tomar@gmail.com

O.P. Sangwan
Department of Computer Science and Engineering, Guru Jambheshwar University
of Science and Technology, Hisar, Haryana, India
e-mail: Sangwan_op@rediffmail.com

© Springer Nature Singapore Pte Ltd. 2018
M. Pant et al. (eds.), *Soft Computing: Theories and Applications*,
Advances in Intelligent Systems and Computing 584,
https://doi.org/10.1007/978-981-10-5699-4_64

679

total development cost [1]. Thus, it becomes the necessity to cutoff such costs using some automation approach [2].

This paper deals with generation of test data using genetic algorithm. By generating test data automatically one aim to introduce an automation approach in testing and thereby reducing both times spent in manual creation, execution of test data and simultaneously increasing the testing efficiency by reducing the manual human involvement which by default is error prone.

Further, the performance of GA is evaluated for it applicability in automatic test case generation (ATCG) with parameters such as initial population size. The ATCG is a well studied undecidable problem [3] in software testing. Previous studies have shown that both static and dynamic methods have been developed and implemented successfully for ATCG. Methods such as symbolic execution [4] and domain reduction [5, 6] fall in the category of static methods, whereas random testing, chaining method [7], local search [8], and evolutionary approaches [9–15] are dynamic methods. Although static methods are vibrantly used in ATCG, but such methods inherently suffer from problems like reference pointer, array, procedural calls, and infinite loops [10]. Thus, it is advantageous to use dynamic methods in which input is generated by executing the program. Past studies reveal that the use of genetic algorithm has huge potential to solve a problem, which requires optimal points in search space and as ATCG can be considered as a search problem which aims to find best input data according to the requirements, thus it is advisable to consider GA as a good candidate for ATCG. Our biasing toward genetic algorithm is also due to the success of GA for in various optimization problems [10, 16–18].

2 Genetic Algorithm Approach for ATCG

The genetic algorithm is inspired by the theory of evolution and was first proposed [19] as an effective and efficient method for searching optimal points in a large search space. The basic parameters that govern the working of GA are crossover and mutation of individuals in a population that leads to a fit individual in each next generation. Each individual is selected for crossover and mutation on the basis of selection parameters (such as roulette wheel, stochastic uniform, uniform and remainder [20], and fitness function). The fitness function varies from problem to problem, i.e. it is problem dependent. More efficient the fitness function more is the efficiency of GA. Thus, genetic algorithm search for optimal points in a given search space using a well-defined fitness function such that the goal is either to minimize or maximize the function $f(x)$, subject to some constraints $c(x)$. In a similar manner, the ATCG can be modeled as an optimization problem in which the goal is to find best input data subject to program path constraints. Let a program P has n paths/decision nodes. In order to follow a particular decision and to reach the goal in P one must traverse the paths containing that decision node(s). Thus, such a decision to decision traversing in order to reach the goal state is possible if one find such input values on which execution of P makes the goal state reachable.

Table 1 Fitness functions for various predicates

Expression	Target false	Target true
x1==x2	$-\|x1-x2\|$	$\|x1-x2\|$
x1! = x2	$\|x1-x2\|$	$-\|x1-x2\|$
x1 < x2	x2−x1	x1−x2
x1 < = x2	x2−x1	x1−x2
x1 > x2	x1−x2	x2−x1
x1 > = x2	x1−x2	x2−x1
x1‖x2	f(constrain1) + f(constrain 2)	min(f(constrain 1), f(constrain 2))
x1 && x2	min(f(constrain 1), f(constrain 2))	f(constrain 1) + f(constrain 2)

To reach the goal state, one can break the goal into various sub-goals such their fitness's sum leads to a final goal. The fitness of each sub-goal can be formed using the Table 1 in which the absolute distance between the targeted goal and the branch predicate node is calculated. Table 1 [21] shows fitness function for different types of predicate conditions.

The proposed steps for generating ATCG are given in listing 1.

---

input: Program under Test

Output: Generated test cases

1. *Generate all independent paths that lead to targeted goal(s)*

2. *For each independent path; create a fitness function*

3. *Adjust GA parameters such as pop size, crossover, and mutation rate and input the formed fitness function in step 2 to GA*

4. *Validate output (generated test data) by executing the optimal Test Data on the ith independent path.*

5. *If validation is correct go to step 6 else go to step 3.*

6. *Stop*

Listing 1: Algorithm for ATCG using GA

Triangle classification (Tri_Typ) is used as a benchmark to generate and validate test data. The Tri_Typ is a famous benchmark [15, 22–26] used in several software engineering problems, its aim is to classify a Triangle into (1) Scalene, (2) Equilateral, (3) Isosceles, and (4) Invalid Triangle on the basis of given sides. The sample instrumented Tri_Typ is given in listing 2. To generate testing data for the Triangle classification problem, first identify the various independent paths to the goal of the program and for each independent path, using Table 1, fitness is calculated for each path. An instrumented program with fitness function for each independent path is shown in listing 2.

```
function [tri, Line]= Triangle_classification(x1,x2,x3)
```

% net Fitness t for Scalene is $f_{Scal}= \textbf{f0}_t+\textbf{f1}_t+ \textbf{f2}_t$

% net Fitness for equilateral is $f_{equi}= \textbf{f0}_t+\textbf{f1}_t+ \textbf{f2}_f+\textbf{f3}_t$

% net Fitness for isoceleous is $f_{iso}= \textbf{f0}_t+\textbf{f1}_t+ \textbf{f2}_f+ \textbf{f3}_f+ \textbf{f4}_t$

% net Fitness invalid triangle is $f_{inv}= \textbf{f0}_t+\textbf{f1}_t+ \textbf{f2}_f+ \textbf{f3}_f+ \textbf{f4}_f$

% $\textbf{f0}_t\textbf{=(0-x1)+(0-x2)+(0-x3)}$ → **Follow True Branch of line no 0**

% $\textbf{f0}_f\textbf{=min((min(x1-0),(x2-0),(x3-0))}$ → **Follow false Branch of line no 0**

    0. if(x1>0 && x2>0 && x3>0)

% $\textbf{f1}_t\textbf{=(c- (x1+x2))+ (a-(x2+x3)) + (b-(x3+x1))}$ → **Follow True Branch of line no 1**

% $\textbf{f1}_f\textbf{=((x1+x2)-c)+ ((x2+x3)-a) + ((x3+x1)-b)}$ → **Follow false Branch of line no 1**

    1. if (x1+x2>x3 && x2+x3>x1 && x3+x1>x2)

%$\textbf{f2}_t\textbf{=-abs(x1-x2)+-abs(x2-x3)+-abs(x3-x1));}$ → **Follow True Branch of line no 2**

% $\textbf{f2}_f\textbf{=min(min(abs(x1-x2),abs(x2-x3)), abs(x3-x1));}$ → **Follow False Branch of line no 2**

    2. if(x1~=x2&&x2~=x3 && x3~=x1)
        type= 1; % scalene

    L=4;

% $\textbf{f3}_f\textbf{=abs(x1-x2)+ abs(x2-x3)+ abs(x3-x1);}$ → **Follow True Branch of line no 3**

% $\textbf{f3}_t\textbf{=min(min(-abs(x1-x2),-abs(x2-x3),-abs(x3-x1));}$ → **Follow False Branch of line no 3**

    3. elseif ((x1==x2) && (x2==x3) && (x3==x1))
        type= 2; % Equilateral

    L=7;

% **p1=(abs(x1-x2)+-abs(x2-x3));**

%**p2= (abs(x2-x3)+-abs(x3-x1));**

%**p3= (abs(x3-x1)+-abs(x1-x2));**

%**p4=min(-abs(x1-x2),abs(x2-x3));**

%**p5= min(-abs(x2-x3),abs(x3-x1));**

%**p6= min(-abs(x3-x1),abs(x1-x2));**

%**f4**$_t$**= min(min(p1,p2),p3); -** → **Follow True Branch of line no 4**

%**f4**$_f$**=p4+p5+p6;** → **Follow False Branch of line no 4**

    4. elseif (((x1==x2) && (x1~=x3)) ||((x1 == x3) && (x1 ~= x2)) ||((x2 == x3) && (x2 ~= x1)))
        type= 3; % Isoceleous

    L=10;

```
5.  else
    type= 4; % not a valid triangle
    L=3;
    end
else
type= 5; % out of range
L=2;
end
Line=L;
tri=type;
end
```

Listing 2: An Instrumented Triangle classification Listing

After applying the proposed GA, the testing data are obtained as shown in Table 2.

With random testing, it takes 94 s to generate 4,863,962 test cases in order to cover full path to Tri_1, whereas it requires only 1.88 s to generate 39 test cases, which provide full coverage to path Tri_1. The data for the other paths are shown in Table 1. In this way, one can generate test data automatically in a highly efficient manner as a comparison to random testing.

Table 2 Results of applying GA and random testing

Range	Pop size	GA					Random testing	
		% of test cases in paths/time (in sec)			Average no.	Avg.	Number of	Avg.
		[a]Tr_1(scal)	Tr_2(equi)	Tr_3(ios)	of test case	time	test cases	time
1–100	100	[ad]42.5/1.66	44/2.2	71/1.8	39	1.88	4,863,962	94
1–100	1000	[bd]75/1.39	75.3/6.2	84.6/7.89	57	5.16	4,828,503	90
1–1000	100	[cd]31/7	22/5.8	49/1.83	25	4.87	4,863,930[c]	200[c]
1–1000	1000	[dd]56/5.81	82/6.7	87.4/7.6	62	6.7	4,826,418[c]	202[c]

[a]Final test cases are obtained by using auto-adjustment[b] in pop size, as $a_d = 7$; $b_d = 16$; $c_d = 16$; $d_d = 30$, where a_d, b_d, c_d, and d_d are pop size obtained after adjustment in the initial given pop size ($a = 100$, $b = 1000$, $c = 1000$, and $d = 1000$), respectively
[b]Auto-adjustment is a process in which the initial population is adjusted automatically until the output error is get minimized or zero, where output error = expected output-actual output by running GA
[c]The results are still not converged

Now in order to check the effect of initial population size on the performance of GA, the following experiment was performed.

Experiment 1: To study the effect to "Population Size" versus "Total Generation"

Set initial population size as 100, 1000 and run the experiment. With smaller population size for each path in Tri_Typ, GA generates 3, 44, and 71 test cases for Tri_1, Tri_2, and Tri_3, respectively, in the range of [1 1 1]–[100 100 100]; thus, the smaller pop size covers all paths in 39 average test cases in 1.88 average times. With a large population size of 1000, one gets improved test cases, i.e., on an average it requires 57 test cases in 5.16 average times to cover full paths in Tri_Typ. Although small population size finds quick solutions, but their premature convergence rate results in sub-optimal solutions. This reduced diversification due to population size strictly prohibits goal search. Large population size allows more search space with increased diversification that has fewer chances to be stuck in the sub-optimal area. This increased population size, although increases the processing cost as compared to random testing. A run with large population size = 1000 for Tri_Typ takes 5.16 s as compared to random testing (90 s) to cover full paths in Tri_Typ in range of [1 1 1]–[100 100 100]. However, it is interesting to note that when the range.

It was set to [1 1 1]–[1000 1000 1000], the total time in GA get increased from 1.82 s to 4.87 s, whereas in random testing, it was increased from 94 s to more than 200 s. This increasing range has a lesser effect on processing time as compared to random testing. Thus, GA outperforms random testing in both reduced test sets and processing time. Another big disadvantage associated with random testing is that it is rarely possible to cover all branches without a huge investment of time. Thus, few remaining branches need to cover manually.

3   Conclusion and Future Prospects

In this paper, an evolutionary-based method for automatic test case generation is proposed. The proposed approach is compared with random testing. It is investigated how an initial population effects the performance of the genetic algorithm for ATCG. The experimental study reveals that GA can be used for evolving test data which are not random in nature, but enough mature to hit the targets. The lesser time and reduced test sets make GA as a good choice for ATCG problem as compared to random testing. With increasing initial population size, one can allow more search space with increased diversification that has fewer chances to be stuck in the sub-optimal area.

As a future work, it is planned to study the effect of other parameters such as— crossover and mutation rates on the performance of GA for ATCG. It is also planned to test this study for mutation testing in which artificial mutants are introduced as faults in the program. Automatic generation of such mutant killing data is another future perspective of this work.

Acknowledgements This research work is supported by University Grant Commission, Govt. of India under Grant No.F/NFO201415OBCDEL16123

References

1. Myers, G.J.: Art of Software Testing. John Wiley & Sons Inc., New York, NY, USA (1979)
2. Bertolino, A.: Software testing research: achievements, challenges, dreams. Future Softw. Eng. **2007**, 85–103 (2007)
3. Weyuker, E.J.: The applicability of program schema results to programs. Int. J. Comput. Inform. Sci. **8**(5), 387–403 (1979)
4. King, J.C.: A new approach to program testing. ACM Sigplan Not. **10**(6), 228–233 (1975)
5. Chen, T.Y., Tse, T.H., Zhou, Z.: Semi-proving: an integrated method based on global symbolic evaluation and metamorphic testing. ACM sigsoft Softw. Eng. Not. **27**(4), 191–195 (2002)
6. Sy, N.T., Deville, Y.: Consistency techniques for interprocedural test data generation. ACM Sigsoft Softw. Eng. Not. **28**(5), 108–117 (2003)
7. Korel, B.: Automated test data generation for programs with procedures. ACM Sigsoft Softw. Eng. Not. **21**(3), 209–215 (1996)
8. Korel, B.: Automated software test data generation. Softw. Eng. IEEE Trans. **16**(8), 870–879 (1990)
9. Corana, A., Marchesi, M., Martini, C., Ridella, S.: Minimizing multimodal functions of continuous variables with the 'simulated annealing' algorithm Corrigenda for this article is available here. ACM Trans. Math. Softw. (TOMS) **13**(3), 262–280 (1987)
10. Michael, C.C., McGraw, G., Schatz, M.A.: Generating software test data by evolution. Softw. Eng. IEEE Trans. **27**(12), 1085–1110 (2001)
11. Xanthakis, S., Ellis, C., Skourlas, C., Le Gall, A., Katsikas, S., Karapoulios, K.: Application of genetic algorithms to software testing. In: Proceedings of the 5th International Conference on Software Engineering and its Applications. pp. 625–636 (1992)
12. Wegener, J., Buhr, K., Pohlheim, H.: Automatic test data generation for structural testing of embedded software systems by evolutionary testing. GECCO **2**, 1233–1240 (2002)
13. Frankl, P.G., Weyuker, E.J.: Testing software to detect and reduce risk. J. Syst. Softw. **53**(3), 275–286 (2000)
14. Wegener, J., Baresel, A., Sthamer, H.: Evolutionary test environment for automatic structural testing. Inf. Softw. Technol. **43**(14), 841–854 (2001)
15. Mann, M.: Generating and prioritizing optimal paths using ant colony optimization. Comput. Ecol. Softw. **5**(1), 1 (2015)
16. Levin, S., Yehudai, A.: Evolutionary testing: a case study. Springer (2006)
17. Mann, M., Sangwan, O.P., Tomar, P.: Hybrid test language processing based framework for test case optimization. CSI Trans. ICT, 1–11
18. Wegener, J., Baresel, A., Sthamer, H.: Suitability of evolutionary algorithms for evolutionary testing. In: Computer Software and Applications Conference, 2002. COMPSAC 2002. Proceedings of the 26th Annual International, pp. 287–289 (2002)
19. Holland, J.H.: Adaptation in natural and artificial systems: an introductory analysis with applications to biology, control, and artificial intelligence. U Michigan Press (1975)
20. Baker, J.E.: Adaptive selection methods for genetic algorithms. In: Proceedings of an International Conference on Genetic Algorithms and their applications, pp. 101–111 (1985)

21. Chen, Y., Zhong, Y.: Automatic path-oriented test data generation using a multi-population genetic algorithm. In: Natural Computation, 2008. ICNC'08. Fourth International Conference on, vol. 1, pp. 566–570 (2008)

22. Mann, M.: Test case prioritization using Cuscutta search. Netw. Biol. **4**(4), 179 (2014)

23. Sthamer, H.H.: The automatic generation of software test data using genetic algorithms. University of Glamorgan (1995)

24. Pargas, R.P., Harrold, M.J., Peck, R.R.: Test-data generation using genetic algorithms. Softw. Test. Verification Reliab. **9**(4), 263–282 (1999)

25. Berndt, D., Fisher, J., Johnson, L., Pinglikar, J., Watkins, A.: Breeding software test cases with genetic algorithms. In: System Sciences, 2003. Proceedings of the 36th Annual Hawaii International Conference on 10 p (2003)

26. Mohapatra, D.: GA based test case generation approach for formation of efficient set of dynamic slices. Int. J. Comput. Sci. Eng. (IJCSE), **3**(9) (2011)

Telugu Speech Recognition Using Combined MFCC, MODGDF Feature Extraction Techniques and MLP, TLRN Classifiers

Archek Praveen Kumar, Ratnadeep Roy, Sanyog Rawat,
Ashwani Kumar Yadav, Amit Chaurasia and Raj Kumar Gupta

Abstract Telugu is the standard language used to communicate mainly in Andhra Pradesh and Telangana states with approximately 100 million speakers. Every Telugu word ends with vowels. Automatic speech recognition has major applications in the smart world. Telugu speech recognition has a huge scope of research from past decade. This paper deals with Telugu speech recognition in a speaker-dependent format for 10 words which are numbers from one to ten uttered by 10 speakers which creates a data base of 100 samples. Numbers are very frequently spoken words where recognition of these words plays a major role in Telugu speech recognition. These spoken words are preprocessed using various techniques like de-noising, framing, sampling, transformations, and endpoint detection. Methods like mel frequency cepstral coefficients (MFCC) and combined MFCC with modified group delay functions (MODGDF) are used for extracting the features. Extracted features are used for training and testing phase. Multilayer perceptron (MLP) and time lagged recurrent neural network (TLRN) patterns are trained and tested. Comparison is done for MLP and TLRN with the feature extraction techniques MFCC and MODGDF-MFCC. Integrated MODGDF-MFCC gives best accuracy in training and testing phases.

A.P. Kumar (✉) · R. Roy · A.K. Yadav · A. Chaurasia · R.K. Gupta
Amity University, Jaipur, Rajasthan, India
e-mail: archekpraveen@gmail.com

R. Roy
e-mail: rroy@jpr.amity.edu

A.K. Yadav
e-mail: akyadav@jpr.amity.edu

A. Chaurasia
e-mail: achaurasia@jpr.amity.edu

R.K. Gupta
e-mail: rkgupta1@jpr.amity.edu

S. Rawat
Manipal University, Jaipur, Rajasthan, India
e-mail: sanyog.rawat@jaipur.manipal.edu

© Springer Nature Singapore Pte Ltd. 2018
M. Pant et al. (eds.), *Soft Computing: Theories and Applications*,
Advances in Intelligent Systems and Computing 584,
https://doi.org/10.1007/978-981-10-5699-4_65

687

Keywords MFCC · Speech recognition · MLP · TLRN · MODGDF
Endpoint detection · Telugu · ANN

1 Introduction

Speech is a proper systematic communication style. Living beings are relaxed and simple with the speech. Other styles of communication require more focus and cause humans more effort due to the uncontrollable area. Air is blown from the lungs through the vocal tract which generates the speech. Vocal speeches are produced with respect to the fundamental frequencies of vibration of the vocal tract. There are approximately 6500 languages spoken in the world. Across India, 22 official languages are spoken and 1652 mother tongue languages are spoken. Telugu is one of the famous languages spoken in south India. Experimentation and investigation in Telugu speech recognition are underdeveloped compared with the other Indian languages. Few researches published some papers regarding Telugu speech recognition in past few years. Most of the work is done on Hindi, Marti, Malayalam, and Tamil languages [1]. Telugu language has 60 letters and every letter ends with vowels so this means every word ends with vowels. The grammar part is more typical and the slang is different. Recognition of this specification needs some advanced techniques for accruing good accuracy. Authors extracted features, pitch, and energy to recognize the speaker in the work "Analysis and Design of speaker identification system using NNC" and proved accuracy with 78%. Scholars used DWT, WPD, and Naïve Bayes techniques and justified the recognition accuracy with 83.5, 80.7, and 86.20% irrespectively. Few researchers developed a speech recognition system for Telugu letter and analyzed promising results with 96 and 97.56% accuracy in testing phase. This paper demonstrates how the Telugu speech is recognized by numerous suitable methods and models. Research is done on feature extraction techniques, named MFCC and combined MODGDF-MFCC [2]. Features are classified into two models MLP and TLRN. Comparison of all these techniques is described with attractive results.

2 Method Used

The ambition of this paper is to recognize Telugu speech which has a flow of sound modules from the recorded speech signal so that the actual information can be decoded from the speech signal [3] and the process is shown in Fig. 1.

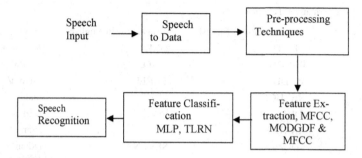

Fig. 1 Overview design

2.1 Telugu Language

Acoustic speech signal varies according to the fundamental frequencies exerted from vocal tract. Air is blown from lungs through vocal tract that produces the speech, which is dependent on vibration [4]. The speech exerted from the mouth is a physical signal where microphone converts this to an electrical signal. Later recognized speech exerted from the system is an electrical signal which is converted back to physical quantity through ears and transferred to brain. Speech is recorded by microphone in a silent environment since speech signals are easily affected by noise. This recorded speech is sampled at 16 kHz frequency [5]. Telugu is the property of the Dravidian family languages. Telugu has flourish script. Telugu language is spoken in two states Andhra Pradesh and Telangana. Telugu language consists of 60 symbols where 16 are vowels, 3 vowel modifiers, and 41 consonants. International Phonetic Alphabet (IPA) is used for processing the Telugu language [6].

2.2 Speech Database

The Telugu words in this research contain 10 words which are actual numbers from one to ten.

"OKATI, RENDU, MUDU, NALUGU, IDHU, AARU, EDU, ENIMIDHI, TOMIDHI, PADHI."

There are more than 44 letters in second stage, and similarly, IPA format is there for these letters. Table 1 represents the speaker words which are in Telugu script, pronunciation of each word in English, translation of that each word with its meaning, and IPA format of all the words.

The captured speech signal is the integration of information and noise. If a speaker speaks a word, there will be attached noise before word starts and after word ends. Endpoint detection and starting point detection mean finding out where

Table 1 IPA format for Telugu words

Telugu script	Pronunciation in English	Translation In English	IPA
ఒకటి	OKATI	ONE	/nəməsokaθi:/
రెండు	RENDU	TWO	/rəndə/
ముడు	MUDU	THREE	/mud:u/
నలుగు	NALUGHU	FOUR	/nalugu/
ఐదు	IDHU	FIVE	/ʌid:u/
ఆరు	AARU	SIX	/ʌarəu/
ఎడు	EDU	SEVEN	/əd:əu/ /gi:θʌm/
ఎనిమిది	ENIMIDHI	EIGHT	/a:ɳəmid:i/
తొమిది	THOMIDHI	NINE	/ʧme:di/
పది	PADHI	TEN	/pʰə:ɖə/

actual speech starts and ends. Endpoint detection is the major part for processing the recorded speech which extracts proper speech. The endpoint detected speech signals are segmented into a number of frames since total speech signal cannot be processed at a time. Features are extracted per each segment, and frame size considered is of 160 and 80 bits/frame. There are various techniques for smoothing, softening, and filtering. (Discrete wavelet transform (DWT) is used for these operations [7]. Framed discrete signals are processed under DWT to remove noise.

2.3 Feature Extraction

MFCC—The MFCC is mel frequency cepstral domain which is the best technique used by many speech recognition systems. MFCC results from discrete cosine transform (DCT) of the filter bank spectrum. Two processes are done in MFCC, namely cepstrum calculation and Mel scaling. MFCC method depends on the short-term analysis, and vectors are featured from each frame. DFT develops Mel filter bank. Mel frequencies are wrapped to produce coefficients. Lastly, IDFT is used for simplification. Around 12–14 first features are enough to extract the features.

MODGDF—The group delay function is designed to extract best parameters by applying negative derivative to the phase. The group delay function is defined in Eq. 1.

$$\tau(f) = -d(\theta(f)) \, df \tag{1}$$

The signal with minimum phase or poles of the transfer function is within the unit circle which is the condition to extract best features by GDF. To reduce the spike nature, the GDF are modified and called as modified group delay functions. Two new parameters are defined in modified GDF. The introduced new parameters are ranged from 0 to 1. To extract features from MODGDF, GDF are converted to cepstra by discrete cosine transform (DCT). The DCT acts as a linear de-correlator, which allows the use of diagonal covariances in modeling the speaker vector distribution.

The combined MODGDF and MFCC is the most apparent case with a group of feature set later used for speech recognition. Features of MFCC and MODGDF are mixed, which improves the efficiency of speech recognition. For a frame size of 160 bits, 80-dimensional MODGDF stream and 80-dimensional MFCC stream are combined by feature stream combination. Similarly, for 80 bits 40-dimensional stream for both MODGDF and MFCC is considered [8].

2.4　Feature Classification

Feature classification is the last method to be performed for recognition. There are many classifying techniques in different streams like pattern recognition, statistical, artificial neural networks, Naïve Bayes, and WPD. Classification is done by considering two phases: training phase and testing phase. Classifiers which are relevant with language pitch, and other parameters extracted is used. This research paper deals with the classifier based on artificial neural networks. Artificial neural network (ANN) is a supervised technology which reflects the parallel distributed processing of the human nervous system to deal with hi-tech complications in pattern recognition and functions. Hornik et al. [9] proved that ANN with tolerable problems is able to perform the approximation of any function with highest efficiency. ANN has the strength of proving correct output even if the input has some error. A recurrent neural network consists of one or more feedback connections [10]. Recurrent neural networks directs affair of inputs by preserving internal states that have a storage element. Passage of neurons is done between the layers through feedback connections.

Multilayer perceptron (MLP)—A multilayer perceptron is a subsection of ANN. MLP model is a feed-forward model. A group of input data is graphed to a group of output data. MLP name itself indicates multiple layers of nodes where every layer is interconnected and organized as feed-forward topology. Nodes other than input are represented as neurons. MLP promotes a managed algorithm called back-propagation algorithm for training. MLP is a conversion of the standard linear perceptron technique. MLP is a regular technique used by many researchers. There are different units where each unit is partisan by weighted sums and thresholds. Transfer functions are generated to get the output. MLP is shaped by functions with appropriate layers and number of units in every layer. The major advantage of using MLP is the number of hidden layers, and units in every hidden layer can be stated. The hidden layer number is specified according to the number of inputs and outputs. Generally, half of the input and output is considered for stating the number of hidden layers. The figure below shows an input layer with 3 neurons, one hidden layer with 3 neurons, and output layer with 3 neurons.

Time lagged recurrent neural network (TLRN)—Time is the major constraint in each frame; integrating time with neural world gives a technique called time lagged recurrent neural network. As declaration made in advance based on past input, TLRN is more accurate. TLRN tool is proposed to classify the extracted speech features. TLRN has motionless processing elements (PE). These PE have memory called gamma, the Laguarre or the tap delay line. There are layers in TLRN as similar to MLP, and each layer is affixed with memories, which produces practical neural topologies which are useful for time series prediction and system identification. The data to be stored or transferred are designed with time structure. TLRNs are the state of the art in nonlinear time series prediction, system identification, and temporal pattern classification.

3 Proposed Algorithm

The proposed process is shown in Fig. 2. In detail, how the signal is recorded to the recognition of speech is proposed with best techniques [11].

Analog speech signal cannot be processed at once so it is discretized by sampling at 16 kHz, and discrete speech signal is generated. These discrete signals are later framed and filtered [12, 13].

Fig. 2 Proposed algorithm

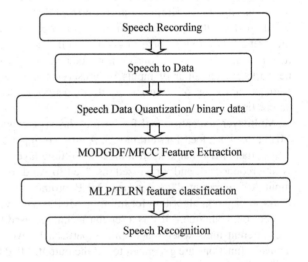

4   Experimental Results

Database is created for 100 words where 50% is used for training phase and 50% is used for testing phase. In all the ten words, only two words' sample graphs are shown in Fig. 3. These two words are OKATI and RENDU. These words after recording are preprocessed as mentioned before, and features are extracted.

After recording, feature extraction is done, firstly features are extracted by MFCC and MODGDF later combined MFCC and MODGDF are extracted. Comparative percentage of features extracted for work OKATI and RENDU is shown in Figs. 4 and 5. Joint feature for both words is around 94.28 and 94.37%, respectively, but individually the features extracted through MFCC and MODGDF for all the words are approximately 50–70%.

Fig. 3 Recorded speech for OKATI and RENDU

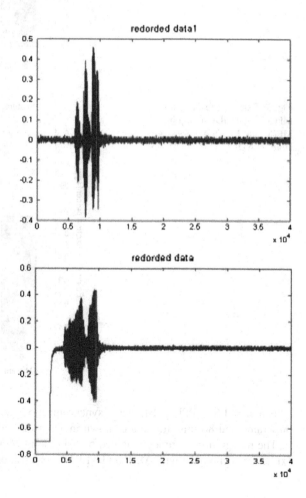

Fig. 4 Extracted features for
MFCC, MODGDF, and joint
MFCC and MODGDF for
word OKATI

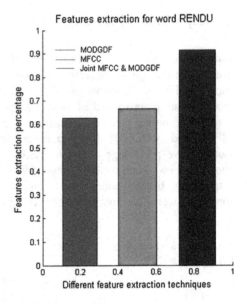

Fig. 5 Extracted features for
MFCC, MODGDF, and joint
MFCC and MODGDF for
word RENDU

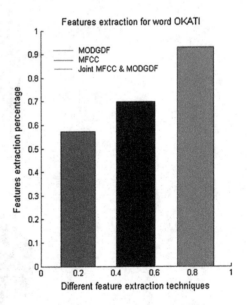

Features LSP, PPF, CBI, gain, synchronization, and FEC extracted for 160
bits/frame and 80 bits/ frame are shown in Tables 2 and 3.

The recognition accuracy obtained by using joint MFCC and MODGDF feature
extraction techniques with MLP and TLRN classification is 98.12%.

Table 2 Features extracted for word OKATI

Features	Bits of data 160 bits/frame	Bits of data 80 bits/frame
Line spectrum pairs (LSP)	40	22
Pitch prediction filter (PPF)	55	14
Code base indexes	23	16
Gain	32	25
Synchronization	5	2
FEC	5	1
Total	160	80

Table 3 Features extracted for word RENDU

Features	Bits of data 160 bits/frame	Bits of data 80 bits/frame
Line spectrum pairs (LSP)	38	21
Pitch prediction filter (PPF)	52	13
Code base indexes	25	18
Gain	35	26
Synchronization	5	1
FEC	5	1
Total	160	80

5 Conclusion

Speech recognition is done by feature extraction with the help of MODGDF and MFCC. Then, classification is done on extracted features by NLP and TLRN classifier. The techniques MFCC and MODGDF provide various parameters like LSP, pitch prediction filter, code base indexes, gain, synchronization, and FEC. Two frame sizes are taken: One is 160 bits/frames and other is 80 bits/frames. Recognition results are good. Further, the research can be extended from isolated words to continuous words and spontaneous words. The database can be extended with more number of speakers.

References

1. Dalmiya, C.P., Dharun, V.S., Rajesh, K.P.: An efficient method for Tamil speech recognition using MFCC and DTW for mobile applications. In: IEEE Conference on Information and Communication Technologies (2013)
2. Singh, S., Rajan, E.G.: Vector quantization approach for speaker recognition using MFCC and inverted MFCC. Int. J. Comput. Appl. 17(1), (2011)
3. Hossan, M.A., Memon, S., Gregory, M.A.: A novel approach for MFCC feature extraction. In: International Conference on Signal Processing and Communication Systems, pp. 1–5 (2010)
4. Ramesh Babu, P.: Digital Signal Processing, Fourth edition, SciTech Publications (2003)

5. Kumar, A.P., Kumar, N., Kumar, C.S., Yadav, A.K.: Speech compression by adaptive Huffman coding using Vitter algorithm. Int. J. Innov. Sci. **2**(5), (2015)
6. Kalyani, N., Sunitha, D.K.: Syllable analysis to build a dictation system in Telugu language. Int. J. Comput. Sci. Inform. Technol. **6**(3), (2009)
7. AYadav, A.K., Roy, R., Kumar, R., Kumar, C.S., Kumar, A.P.: Algorithm for de-noising of color images based on median filter. In: 2015 Third International Conference on Image Information Processing (ICIIP), Waknaghat, pp. 428–432 (2015)
8. MHegde, R.M., Murthy, H.A., Gadde, V.R.R.: Continuous speech recognition using joint features derived from modified group delay function and MFCC. Semant. Scholor J. (2012)
9. Hornik, K., Stinchcombe, M., White, H.: Multilayer feedforward networks are universal approximators. Neural Networks **2**, 359–366 (1989)
10. Suuny, S., Peter, S.D., Jacob, K.P.: Performance of different classifiers in speech recognition. Int. J. Res. Eng. Technol. **2**(4), (2013)
11. Kumar, A.P., Bansal, D.: Digital arithmetic coding with AES algorithm. In: IJCA Special Issue on International Conference on Electronic Design and Signal Processing, vol. 1, No. 2, pp. 15–18 (2013)
12. Kumar, A.P., Kumar, N., Kumar, C.S., Yadav, A.K., Sharma, A.: Speech recognition using arithmetic coding MFCC for Telugu language. In: 3rd International Conference on Computing for Sustainable Global Development, Proceeding IEEE digital library, BVICAm (2016)
13. Yadav, A.K., Roy, R., Kumar, A.P., Kumar, C.S., Dhakad, S.K.: De-noising of ultrasound image using discrete wavelet transform by symlet wavelet and filters. In: International Conference on Advances in Computing, Communications and Informatics (ICACCI), Kochi, pp. 1204–1208 (2015)

Speech Recognition with Combined MFCC, MODGDF and ZCPA Features Extraction Techniques Using NTN and MNTN Conventional Classifiers for Telugu Language

Archek Praveen Kumar, Ratnadeep Roy, Sanyog Rawat,
Rekha Chaturvedi, Abhay Sharma and Cheruku Sandesh Kumar

Abstract The automatic speech recognition systems are designed for human–computer interaction in trouble-free mode. Speech recognition has vast applications. Text to speech and speech to text transformations are mostly used segments in ASR. Perfect speech recognition is done by choosing proper extraction and classification techniques with respect to slang and pitch of the language. Telugu is a south Indian language which has around 120 million speakers. There are various feature extraction techniques such as LPC, MFCC, MODGDF, RASTA, DTW and ZCPA. This paper deals with the combined techniques and its comparison with individual techniques. The rate of features extracted in joint extraction techniques gives promising results comparatively with individual technique. Techniques MFCC, MODGDF and ZCPA are combined, and joint features are extracted. The next stage is to classify the features where selected technique using neural networks. Features are classified by NTN and MNTN classifiers for speaker-dependent recognition and presented by using closed and open sets. The MNTN is evaluated for several speaker recognition experiments. These include

A.P. Kumar (✉) · R. Roy · R. Chaturvedi · A. Sharma · C.S. Kumar
Amity University, Jaipur, Rajasthan, India
e-mail: archekpraveen@gmail.com

R. Roy
e-mail: rroy@jpr.amity.edu

R. Chaturvedi
e-mail: rchaturvedi@jpr.amity.edu

A. Sharma
e-mail: asharma2@jpr.amity.edu

C.S. Kumar
e-mail: cskumar@jpr.amity.edu

S. Rawat
Manipal University, Jaipur, Rajasthan, India
e-mail: sanyog.rawat@jaipur.manipal.edu

© Springer Nature Singapore Pte Ltd. 2018
M. Pant et al. (eds.), *Soft Computing: Theories and Applications*,
Advances in Intelligent Systems and Computing 584,
https://doi.org/10.1007/978-981-10-5699-4_66

closed- and open-set speaker identification and speaker verification. The MNTN is found to perform better than NTN classifier. Speech recognition rate for Telugu language by combine MFCC, MODGDF and ZCPA extraction techniques using NTN and MNTN classification techniques are compared with excellent results.

Keywords MFCC · MODGDF · ZCPA · NTN · MNTN · End point detection Telugu · ANN

1 Introduction

Communication is the major interaction between humans which shares concept, emotions and thoughts. Communication can be classified into 2 types: human to human and human to computer or machine. Human to machine interaction can be done perfectly if speech is recognized by the machine with great efficiency. Speech recognition is a branch of signal processing or communication. Speech recognition technology makes the machine to accept the duty assigned by humans where it can be technical or personal. Telugu is an Indian language. Language is a communication medium for human beings. There are totally 65,000 languages all over the world. According to the census of India, 122 major languages and 1599 other languages are used in India. Telugu is one of the major languages spoken by south Indians all over the world. There are 120 million speakers in India who speak Telugu. Telugu language consists of 60 symbols in which 16 are vowels, 3 vowel modifiers and 41 consonants [1]. Speech is classified into different classes depending on slang and utterance. Research on speech can be done on various formats such as isolated words, connected words, continuous speech and spontaneous speech [2]. This paper works on isolated words where individual words are recorded and recognized. Isolated words to be recorded and recognized are Telugu words which are very regularly used in daily life. These isolated words are recorded first in a silent environment since speech signals are easily affected by noise. The recorded signals are preprocessed or preemphasized by various techniques such as framing, segmentation, filtering, amplification, feature extraction and feature classification. Various techniques are used for preprocessing extraction and classification. Vectors are extracted and classified. Techniques used are given in Table 1.

Research on Telugu isolated words was done by few authors where the recognition rate is around 80–95% using different techniques and most of the papers

Table 1 Techniques used for speech recognition

Speech recognition stages	Techniques
Preprocessing	DWT
Feature extraction	MFCC, MODGDF, ZCPA, joint (MFCC, MODGDF and ZCPA)
Feature extraction	NTN, MNTN

researched on neural technology to improvise the accuracy and to increase the signal-to-noise ratio [3].

2 Design of Speech Recognition

This paper works on advanced techniques which work on cepstrum and vector technology. Speech input is recorded and database is created. Ten isolated Telugu words are recorded and preprocessed. The preprocessed signals have noise which is filtered and amplified. End point detection is a part of preprocessing. Zero crossing rate, short-time Fourier transform, etc. are considered with some esteemed useful techniques for preprocessing. Combining the feature extraction techniques is the advanced technology used by recent researchers which has many advantages compared to individual techniques. Vectors extracted with higher concentration are classified with NTN (neural tree network) and MNTN (modified neural tree network). The detailed explanation is shown in Fig. 1.

2.1 Telugu Language

Telugu is the Dravidian language which is exerted from Sanskrit. Telugu, Tamil, Kannada and Malayalam are Dravidian evolved languages in south India. Script of Kannada and Tamil is similar, but slang is different. Malayalam and Tamil slang is similar. Telugu language consists of 60 symbols. The interesting part of Telugu word is that every word ends with vowel. When air is pushed from lungs through vocal tract, the sound is produced. Features vary for speakers own speech every time [4], since due to this complexity speech recognition especially for Telugu language is difficult to generate 100% accuracy. International Phonetic Alphabet (IPA) is used for converting the speech to English format [5].

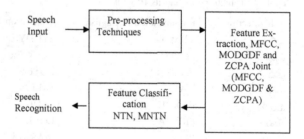

Fig. 1 Overview design

2.2 Speech Database

Speech database is created for 10 isolated words which are very regularly used in daily life. These 10 isolated words are recorded by 20 speakers. So 20 × 10 totally 200 speech database is created in which 60% is used for training phase and 40% is used for testing phase. The 10 Telugu words considered for speech recognition are given below. Table 2 represents the speaker words which are in Telugu script, pronunciation of each word in English, translation of that word with its meaning and IPA format of all the words.

"VIDYA, PUVU, THAMARA, NAVU, PANDU, PREMA, PELLI, PAISA, KUNDA, NEELLU".

Speech recorded is corrupted by noise, i.e., the signal is convoluted by noise. The noise may be additive white Gaussian noise or some thermal noise. This noise should be removed; there are various de-noising algorithms used in preprocessing. Wavelet de-noising is one of the best techniques used from the past to filter the noise. Soft thresholding and hard thresholding techniques can be used in DWT. This paper deals with soft thresholding. As the recorded signal has noise at starting point and ending point, the noise is removed by DWT soft thresholding. The next process is to detect the starting point and ending point of the speech signal. Zero crossing rate technique is used for detecting the points. Later the speech signal is to be segmented into number of frames where each frame is processed. Each frame is amplified, and features are extracted and further classified for recognition [6].

2.3 Feature Extraction

MFCC—MFCC is the most suitable technique for any language or any slang for proper vector features. Individual ear is mimed by MFCC, which helps audio

Table 2 Telugu words and its IPA format

Telugu script	Pronunciation in English	Translation In English	IPA
విద్య	VIDYA	STUDY	/vidjʰa/
పువ్వ	PUVU	FLOWER	/pu:vu/
తమర	TAMARA	LOTUS	/θa:mʌra/
నవ్వ	NAVU	SMILE	/na:vu/
పండు	PANDU	FRUIT	/pʌnd:u/
ప్రమ	PREMA	LOVE	/pr:eam/
పెళ్ళి	PELLI	MARRIAGE	/pəl:Li/
పైస	PAISA	MONEY	/pa:iəsa/
కుండ	KUNDA	POT	/kun:da/
నీళు	NEELLU	WATER	/neəl:Lu/

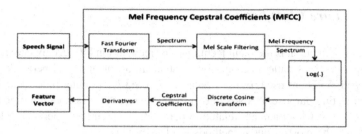

Fig. 2 MFCC block diagram

spectrum to be settled by nonlinear fundamental frequencies, so mel filters corrupt the frequency which agrees the spatial relationship of the hair cell distribution of the creature ear. MFCC purely depends on short-term analysis. Hamming window is used for windowing the speech signal. After windowing the signal, FFT is applied with respect to mel frequency wrapping, so log energy in critical band of frequencies is joined and added [7]. The vector features are extracted after wrapping, and finally, DCT is applied to generate cepstral features; the complete process is shown in Fig. 2.

MODGDF—Group delay functions are used in many applications, but now advanced techniques replace this technique, but the unique characteristics of this GDF can be used for extracting the features with a larger vector space. Similarly MODGDF which is modified group delay function has larger space to generate the vectors. Perfect sustained parameters are extracted by MODGDF [8]. The phase of the signal is considered, and negative derivative is applied to generate MODGDF coefficients as shown in Eq. 1.

$$\tau(f) = -\mathrm{d}(\theta(f))\mathrm{d}f \tag{1}$$

MODGDF technique needs the removal of partial and harmonics where prominent pitch is estimated. This is applied for flattened power spectrum.

ZCPA—ZCPA means zero crossing peak amplitude. This is also a feature extraction technique which deals with zero crossing. It consists of band-pass filter banks which approximate frequency responses. Frequency is detected with amplitude for all channels. The next zero occurrences are considered which detects the dominant frequency. Peak of amplitude is detected. For every channel, peak detection with respect to zero crossing provides coefficients [9]. At last, DCT is used to generate the ZCPA parameters. The total flow is shown in Fig. 3.

The joint features are extracted from MFCC, MODGDF and ZCPA. This group of features set is used for classification. The joint features prove that the vectors collected are in a larger set with higher rate. The jointly combined features also increase the efficiency or accuracy in the recognition. The joint features are characterized for a frame size of 180 bits per frame.

2.4 Feature Classification

Feature classification is the main technique for recognition [10]. Many classification techniques such as pattern recognition, statistical, artificial neural networks, naïve Bayes and WPD are used. 60% of database is used for training phase and 40% is used for testing purpose. ANN is one of the best areas for classification used for the past few years. Classifiers are related to parameters which are generated. NTN and MNTN are the most frequent methods to classify [11].

NTN—The neural tree network (NTN) is a combination of feed forward neural networks and decision trees. As shown in Fig. 4, single-layer perception (SLP) is there in tree. Each node of Fig. 3. SLP considers all the parameters for decision, so NTN does not have any borders compared to normal tree network. Architecture is self-designed in NTN. Training data are divided into subsets or smaller groups and assigned as a node. Feature parameters are separated by leaf nodes and form homogenous subsets [12]. The training is done by separating the tree with leaf nodes, where each leaf node has children and every children node has unique data. Similar vectors will be separated to their children nodes, so there will be 100% gain and operations in this NTN. Every speaker has his separate NTN. The feature vectors are simultaneously distributed by distrusted labeling scheme.

A NTN network is shown in Fig. 3 where spkr name represents as leaf with class 1 and antispkr represents class 0.

MNTN—The MNTN which is modified neural tree network is the advanced version of NTN. MNTN also provides greater accuracy in speech recognition. NTN and MNTN are similar but the NTN uses class leaf but MNTN uses class leaf and confidence measure at every leaf [13]. Let us consider a group of feature parameters; confidence measure is classified as speaker; similarly, antispeaker is the confidence measure of every speaker. Likelihood is evaluated as given in Eq. 2 [14, 15].

Fig. 3 ZCPA block diagram

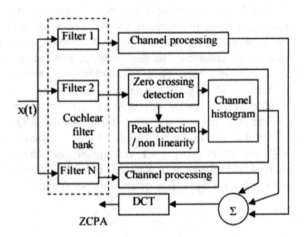

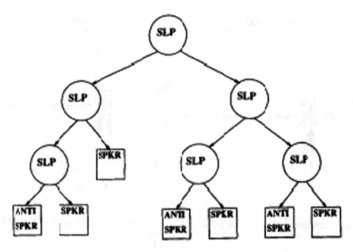

Fig. 4 NTN flow diagram

$$Likelihood = \frac{accumulated\ speaker\ confidence\ measure}{(accumulated\ speaker\ + antispeaker\ confidence\ measures)} \quad (2)$$

3 Experimental Results

Database is created for 200 speech signals which are recorded by a microphone in a silent environment. 60% of database is used for training phase and 40% is used for testing phase. VIDYA and PUVU are the two recorded sample words shown in Fig. 5a. The preprocessing of the recorded samples by DWT is also shown in Fig. 5b which represents the approximated coefficients of DWT.

After recording feature extraction is done, firstly features are extracted by MFCC MODGDF and ZCPA. Later combined MFCC, MODGDF and ZCPA are extracted. Comparative percentage of features extraction for words VIDYA and PUVU is shown in Fig. 6. Joint features for both the words are around 95.16 and 95.42%, but individually the features extracted through MFCC, MODGDF and ZCPA for all the words are approximately to 50–75%.

Features LSP, PPF, CBI, gain, synchronization and FEC extracted for 180 bits/frame are shown in Tables 2 and 3.

The recognition accuracy obtained by using Joint MFCC and MODGDF feature extraction techniques with MLP and TLRN classification is 98.37%.

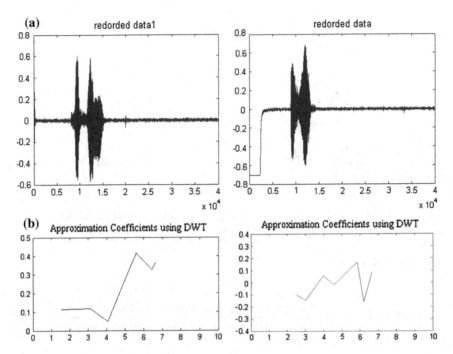

Fig. 5 a Recorded speech for VIDYA and PUVU. **b** Approximated coefficients of DWT for VIDYA and PUVU

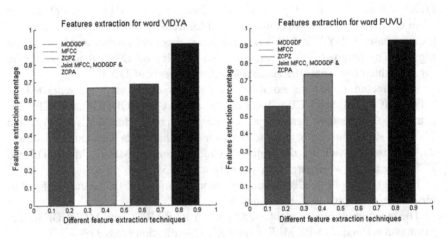

Fig. 6 Extracted features for MFCC, MODGDF, ZCPA and joint features for word VIDYA and PUVU

Table 3 Features extracted for word VIDYA and PUVU

Features	Bits of data 180 bits/frame for VIDYA	Bits of data 180 bits/frame for PUVU
Line spectrum pairs (LSP)	45	42
Pitch prediction filter (PPF)	52	50
Code base indexes	28	25
Gain	40	42
Synchronization	8	14
FEC	7	7
Total	180	180

4 Conclusion

Telugu isolated word speech recognition is done by feature extraction with the help of joint MFCC/MODGDF and ZCPA after some preprocessing. The extracted features or parameters are classified by considering training phase and testing phase by NTN and MNTN classifier. The technique MFCC/MODGDF and ZCPA provides various parameters such as LSP, pitch prediction filter, code base indexes, gain, synchronization and FEC. Frame size is considered as 180 bits/frames. Individually the features extraction rate is shown and combined rate is also given and proved that combined techniques have more extraction rate. Finally recognition results are comparatively proved good.

References

1. Kalyani, N., Sunitha, D.K.: Syllable analysis to build a dictation system in Telugu language. Int. J. Comput. Sci. Inform. Technol. **6**(3), (2009)
2. Ramesh Babu, P.: Digital Signal Processing, Fourth edition. SciTech Publications (2003)
3. Dalmiya, C.P., Dharun, V.S., Rajesh, K.P.: An efficient method for Tamil speech recognition using MFCC and DTW for mobile applications. In: IEEE Conference on Information and Communication Technologies (2013)
4. Kumar, A.P., Kumar, N., Kumar, C.S., Yadav, A.K., Sharma, A.: Speech recognition using arithmetic coding and MFCC for Telugu language. In: 2016 3rd International Conference on Computing for Sustainable Global Development (INDIACom), New Delhi, India, pp. 265–268 (2016)
5. Kumar, A.P., Bansal, D.: Digital arithmetic coding with AES algorithm. In: IJCA Special Issue on International Conference on Electronic Design and Signal Processing, vol. 1, No. 2, pp. 15–18 (2013)
6. Yadav, A.K., Roy, R., Kumar, R., Kumar, C.S., Kumar, A.P.: Algorithm for de-noising of color images based on median filter. In: Third International Conference on Image Information Processing (ICIIP), Waknaghat, pp. 428–432 (2015)

7. Hossan, M.A., Memon, S., Gregory, M.A.: A novel approach for MFCC feature extraction. In: International Conference on Signal Processing and Communication Systems, pp. 1–5 (2010)
8. Hegde, R.M., Murthy, H.A., Gadde, V.R.R.: Continuous speech recognition using joint features derived from modified group delay function and MFCC. Seman. Scholor J. (2012)
9. Kacur, J., Varga, M., Rozinaj, G.: ZCPA features for speech recognition. In: IX International Symposium on Telecommunications (BIHTEL), Sarajevo, pp. 1–4 (2012)
10. Kumar, A.P., Kumar, N., Kumar, C.S., Yadav, A.K.: Speech compression by adaptive Huffman coding using Vitter algorithm. Int. J. Innov. Sci. 2(5), (2015)
11. Suuny, S., Peter, S.D., Jacob, K.P.: Performance of different classifiers in speech recognition. Int. J. Res. Eng. Technol. 2(4), (2013)
12. Hornik, K., Stinchcombe, M., White, H.: Multilayer feedforward networks are universal approximators. Neural Networks 2, 359–366 (1989)
13. Farrell, K.R., Mammone, R.J., Assaleh, K.T.: Speaker recognition using neural networks and conventional classifiers. IEEE Trans. Speech Audio Process. 2(1), 194–205 (1994)
14. Singh, S., Rajan, E.G.: Vector quantization approach for speaker recognition using MFCC and inverted MFCC. Int. J. Comput. Appl. 17(1), (2011)
15. Kumar, C.S., Roy, R., Kumar, A.P., Yadav, A.K., Gupta, M.: Segmentation on moving shadow detection and removal by symlet transform for vehicle detection. In: 3rd International Conference on Computing for Sustainable Global Development (INDIACom), New Delhi, India, pp. 259–264 (2016)

Classification Model for Prediction of Heart Disease

Ritu Chauhan, Rajesh Jangade and Ruchita Rekapally

Abstract The heart disease is among the foremost problems faced by people around the globe. It is impossible for medical practioners to analyze this data without computation technology. Hence, there exists constant requirement to appropriately discover factors which can effectively and efficiently handle large scale data for future medical diagnosis. Adhering to needs of the healthcare practioners data mining technology is widely utilized as predictive analytics to discover unknown, hidden and useful information from complex and large scale databases. This paper intends to study varied classification techniques which can be utilized for prediction of heart related problems. In, proposed work the focus is to discover an appropriate technique which can assist and benefit future decision making.

Keywords Data mining · Classification · Accuracy · Classifier

1 Introduction

The expansion of data from varied application domains has inversely provided challenges for technology. The traditional tools utilized for such databases tends be insufficient to handle numerous data streams and complexity of data [1–3]. Hence, data mining is the approach utilized from past decade to effectively detect hidden patterns from large scale databases. We can say that it is among the most motivated areas for healthcare practitioners and scientist around the globe to derive information for future decision making. In, past data mining tools are vigorously applied to

R. Chauhan (✉) · R. Jangade · R. Rekapally
Amity University, Utta Pradesh Noida, India
e-mail: rituchauha@gmail.com

R. Jangade
e-mail: rkjangade@gmail.com

R. Rekapally
e-mail: ruchita.1106@gmail.com

© Springer Nature Singapore Pte Ltd. 2018
M. Pant et al. (eds.), *Soft Computing: Theories and Applications*,
Advances in Intelligent Systems and Computing 584,
https://doi.org/10.1007/978-981-10-5699-4_67

predict the overall behavior and trends from the data which can ascertain help experts from different field to make accurate and efficient decision making.

Data mining and knowledge discovery from databases (KDD) are potentially known and widespread technology. The major fact is to constantly pertaining data mining technology is due to substantial outcomes from significant domains which are applied for future decision making [4–6]. The data mining technology is gaining new insights by merging itself with new advanced areas such as statistics, pattern recognition and others to develop new computation technology to benefit end users. Hence, data mining tends to be an important step in healthcare, business, web based, academia and others [7–10].

Although, data mining with healthcare is discussed as the complicate process as the results retrieved should be significant and accurate which probe benefit to patient care system [11–14]. Hence, the healthcare data mining is applied in different diseases which include cancer, diabetes, hypertension and others to solve real world problem. For example, if the patient is suffering from breast cancer at specific age then predictive analytics can be probe [6, 15]. The patients chances of survivability could be chased keeping the past referenced data and analysis conducted. The data mining technology can be forfeited applied to create models and predict the analytics for future decision making [16, 17].

Data mining provides a new technological advancement to effectively and efficiently measure the accuracy of data [18–20]. However, accuracy can be discussed as probability of data which is evidently true. In this paper we have utilized three classifiers decision tree, random tree, and random forest to evaluate which classifier performs more accurately. The contributed paper has been discussed with varied sections as below. Section 2 will have related work with some set of proposed solution. Section 3 algorithmic discussions with contributed datasets. In Sect. 4 overall representation of output results are discussed. Last section has conclusion and future works.

2 Related Works

An approach of neural network for heart disease was accomplished by Vazirani et al. [21]. He discussed about the diagnosis method manual and automatic to detect prognosis of disease. The study utilized both neural network and intelligent expert system to accomplish diagnosis of disease.

Another peculated work for mining medical data was conducted to detect hidden and unknown patterns while conducting series of experiments [11–13]. The approach was to discover that traditional technology is unable to handle large explosion of data, tends to be insufficient, hence data mining tends to be potentially benefited for the same. The focus was to discuss different data mining techniques in corresponding relevance to healthcare or clinical databases [2].

Aslandogan and Mahajani [3] worked on combinatorial effect with the focus on data mining with implications on medical databases. The approach of study was to discuss different classifiers such as K-Nearest Neighbor (K-NN) which works on criterion of nearest neighbor to measure the similarity among the data, Naïve Bayesian that is Bayesian probabilistic study to measure data with probability among the attributes and Decision Tree to discover a tree like structure where relevant data is kept at same level with labeled classes. The study was corresponded to discuss to measure the accuracy among the classifier using 10 fold cross validation and determine which classifier suited best for the study. The classifier was integrated together to determine the potential nature and improved accuracy. The study was referenced with Dempster's combination of study where both differential based uncertainties was coagulated with performance based linear and other combinations models to finally discover the improved performances for classification of data.

An approach to discover rules for heart disease prediction was conducted to improve diagnosis of disease [7]. The study was discovering novel rules from data with application to heart disease datasets. The paper discussed about the various aspects people suffering from heart disease with relevance to the risk factors during the prognosis of heart profusion and narrowing of arteries. The utmost three constraints were generated in relevance to study where the first constraint discuss that the attributes should be placed on the left side of rule, secondly the rule should be able to distinguish attributes in form of uninteresting groups, finally the number of rules generated inversely depends upon the size of data. The work has proven to be beneficially with help of constraint to generalize the rules and reducing them for better prediction of data for future outcomes of disease.

3   Research Methodology

The approach is to design an effective and efficient modeling technique that can characterize the data of class for future decision making [9, 10, 16, 17]. In this paper the focus is to build a predictive model for the relevant class and determine which classifier supports the same. The classifier utilized in our paper is random tree, decision rule and random forest; heart dataset is analyzed among the classifier while determining which is best suited for retrieval of effective and efficient patterns.

The proposed work focuses to retrieve effective patterns which can generalize the heart attack dataset to benefit patient care for future prognosis of disease. The discovery of pattern tends to be an automated process where classification of data is the perspired technology utilized to classify the data. Further, the classification techniques are evaluated among others and determine which classifier determines the maximum accuracy for the diagnosis of disease.

Table 1 Attributes of heart disease dataset

Attributes	Description	Values	Type
Age	Patient's age in years		Numeric_typ
Chest_pain	Pain observed in patient's chest	Typical Angina	Nominal_typ
		Atypical angina	
		Non-angina pain	
		Asymptomatic	
Rest_bpress	Blood pressure in resting phases (in mm Hg on admission to the hospital)		Numeric_typ
Blood_sugar	Blood sugar in fasting phase > 120 mg/dl	1 = true	Nominal_typ
		0 = false	
Rest_elecrto	Electrocardiographic results in rest state	1. Normal	Nominal_typ
		2. Hyper_left_vein	
		3. Wave_abnormility_st	

3.1 Patient Dataset

The datasets was collected from online repository of University of Lyon where the data is available with referred link [22]. There are 209 cases of heart patients. Description of datasets with their relative parameters is discussed in Table 1.

3.2 Tools

The rapid miner is a tool utilized it provides an integrated approach of utilization of various data mining algorithms which includes association rule [8, 10], decision tree, rule induction method, clustering and others for future decision making. It is a tool formulated on the concept of business model to detect pattern which can benefit different application domains [21, 23, 24].

The rapid miner is codified in java programming language [17, 24]. It provides graphical user interface (GUI) environment to interact with different analytical workflows. The work flows discussed are known as "Process" and they interact with each other through operators. Each operator is assigned a specific task to perform within a process and the output is input to the next process linked.

4 Results

The algorithms decision tree, random tree and random forest are implemented on heart disease dataset with cross validation using ten folds and determine different accuracy among the classifiers which suited best for dataset. The accuracy of each classifier is discussed with varied measures as below.

The overall comparison is analyzed among the classifier to detect which tends to be more effective and efficient for dataset. The decision tree shows the accuracy of 75.10% in Table 2, random tree represents the accuracy of 69.90% in Table 3, whereas random forest has accuracy of 75.60%. Further, the Area under curve (AUC) is also embarked to be considered for measuring accuracy (Figs. 1, 2, and 3) where we found that random forest (in Table 4) is best suited classifier with maximum accuracy then secondly decision tree and random tree has the least accuracy among them.

Table 2 Decision tree accuracy measure

	True_positive	True_negative	Class_precision (%)
Pred. negative	73	33	68.87
Pred. positive	19	84	81.85
Class recall (%)	79.35	71.79	

Table 3 Random tree accuracy measure

	True_positive	True_negative	Class_precision (%)
Pred. negative	45	16	73.77
Pred. positive	47	101	68.24
Class recall (%)	48.91	86.32	

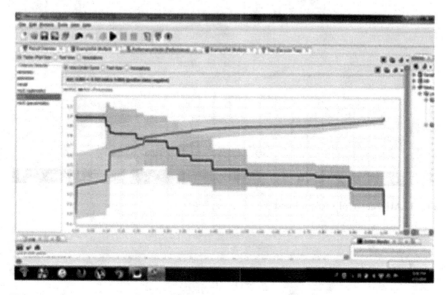

Fig. 1 AUC curve of decision tree validation

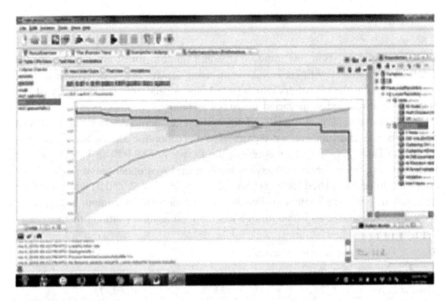

Fig. 2 AUC curve of random tree validation

Fig. 3 AUC curve of random forest

Table 4 Random forest accuracy measure

	True_positive	True_negative	Class_precision (%)
Pred. negative	54	13	80.60
Pred. positive	38	104	73.24
Class recall (%)	58.70	88.89	

5 Conclusions

In the proposed study varied data mining technique are employed for patients suffering from heart disease for future prognosis of disease. The work proposed in this paper discusses on varied classification techniques for prediction of heart attacks and determine best classifier among them. The results retried prove to be effective and efficient for discovery of knowledge among the data classified. The results can be fruitful for discovery of patterns among the healthcare practioners to identify which classifier utilized during a prediction process can benefit. The accuracy measured through recall, precision and AUC tends to be significant factors to determine which classifier is best suited for databases.

References

1. Gopal, S., Woodcock, C.: Theory and methods for accuracy assessment of thematic maps using fuzzy sets. Photogram. Eng. Remote Sens. **60**, 181–188 (1994)
2. Tsumoto, S.: Problems with mining medical data. 0-7695-0792-1 I00@ 2000 IEEE (2000)
3. Aslandogan, Y.A., Mahajani, G.A.: Evidence combination in medical data mining. In: Proceedings of the International Conference on Information Technology: Coding and Computing (ITCC'04) 0-7695-2108-8/04©2004 IEEE (2004)
4. Chauhan, R., Jangade, R.: A robust model for big healthcare analytics. Confluence (2016)
5. Jangade, R., Chauhan, R.: Big data with integrated cloud computing for healthcare analytics. INDIACOM (2016)
6. Glymour, C., Madigan, D., Pregibon, D., Smyth, P.: Statistical inference and data mining. Commu. ACM **39**(11), 35–41 (1996)
7. Ordonez, C.: Improving heart disease prediction using constrained association rules. Seminar Presentation at University of Tokyo (2004)
8. Le Duff, F., Muntean, C., Cuggia, M., Mabo, P.: Predicting survival causes after out of hospital cardiac arrest using data mining method. Stud. Health Technol. Inform. **107**(2), 1256–1259 (2004)
9. Szymanski, B., Han, L., Embrechts, M., Ross, A., Sternickel, K., Zhu, L.: Using efficient supanova kernel for heart disease diagnosis. In: Proceedings of ANNIE 06, Intelligent Engineering Systems Through Artificial Neural Networks, vol. 16, pp. 305–310 (2006)
10. Noh, K., Lee, H.G., Shon, H.S., Lee, B.J., Ryu, K.H.: Associative classification approach for diagnosing cardiovascular disease. Springer, Berlin, vol. 345, pp. 721–727 (2006)
11. Kaur, H., Chauhan, R., Alam, M.A., Aljunid, S., Salleh, M.: SPAGRID: a spatial grid framework for medical high dimensional databases. In: Proceedings of International Conference on Hybrid Artificial Intelligence Systems, HAIS 2012, Springer, Berlin, vol. 1, pp. 690–704 (2012)

12. Kaur, H., Chauhan, R., Aljunid, S.: Data mining cluster analysis on the influence of health factors in Casemix data. BMC J. Health Serv. Res. **12**(Suppl. 1), O3 (2012)
13. Chauhan, R., Kaur, H.: Predictive analytics and data mining: a framework for optimizing decisions with R tool. In: Tripathy, B., Acharjya, D. (eds.) Advances in Secure Computing, Internet Services, and Applications (pp. 73–88). Information Science Reference, Hershey. doi:10.4018/978-1-4666-4940-8.ch004
14. Chauhan, R., Kaur, H., Alam, A.: Data clustering method for discovering clusters in spatial cancer databases. Int. J. Comput. Appl. **10**(6), 9–14 (2010)
15. Han, J., Dong, G., Yin, Y.: Efficient mining of partial periodic patterns in time series database. In: IEEE Transaction on Knowledge and Data Engineering (1998)
16. Lee, H., Noh, K., Ryu, K.: Mining biosignal data: coronary artery disease diagnosis using linear and nonlinear features of HRV. In: LNAI 4819: Emerging Technologies in Knowledge Discovery and Data Mining, May 2007, pp. 56–66
17. Guru, N., Dahiya, A., Rajpal, N.: Decision support system for heart disease diagnosis using neural network. Delhi Bus. Rev. **8**(1), (2007)
18. Wang, H.: Medical knowledge acquisition through data mining. In: Proceedings of 2008 IEEE International Symposium on IT in Medicine and Education. 978-1-4244-2511-2/08©2008 Crown (2008)
19. Palaniappan, S., Awang, R.: Intelligent heart disease prediction system using data mining techniques. IJCSNS **8** (8), (2008)
20. Parthiban, L., Subramanian, R.: Intelligent heart disease prediction system using CANFIS and genetic algorithm. Int. J. Biol. Biomed. Med. Sci. **3**, 3 (2008)
21. Vazirani, H., Kala, R., Shukla, A., Tiwari, R.: Use of modular neural network for heart disease. IJCCT **1**(2, 3, 4), (2010). International Conference [ACCTA-2010], 3–5 Aug 2010, pp. 88–93 (Special Issue)
22. eric.univ-lyon2.fr/~ricco/tanagra/fichiers/heart_disease_male.xls
23. Dangare, C.S., Apte, S.S.: Improved study of heart disease prediction system using data mining classification techniques. IJCA **47**(10), 44–48 (2012) (0975—8887)
24. Vijiyarani, S., Sudha, S.: An efficient classification tree technique for heart disease prediction. In: ICRTCT—2013, Proceedings Published in IJCA (0975—8887), pp. 6–9 (2013)

Determination and Segmentation of Brain Tumor Using Threshold Segmentation with Morphological Operations

Natthan Singh and Shivani Goyal

Abstract Brain tumor detection and studying its behavior are very difficult processes in the field of medical image processing due to the nature of brain image, and its formation is very complex, which can only be studied by experienced person of the field. Segmentation acts as leading role in the field of medical image processing. Medical resonance (MR) image is a diagnostic tool for the diagnosis of brain and related images of medical field. The purpose of this paper is to show segmentation and determination of brain tumor in MR images. We are applying threshold segmentation with morphological operation with reconstruction. In this paper, we are also showing comparison analysis between watershed segmentation with thresholding.

Keywords MRI · Brain tumor · Segmentation · Watershed segmentation
Thresholding · Morphological operation

1 Introduction

The brain is very significant component of our nervous system. The researchers and doctors used MR images to study the complicated structure and function of the brain noninvasively, and MR imaging techniques are very helpful in this. The body of a person has various cells. Every cell is associated with a special kind of functionality. Cells are frequently divided, without any specific order, when they lost their capability to control the growth. These additional cells from a mass of tissue are known as tumor. The doctors used MR imaging as a supporting diag-

N. Singh (✉)
Computer Science and Engineering Department, IET Lucknow,
Lucknow 226021, Uttar Pradesh, India
e-mail: nsingh.iet@gmail.com

S. Goyal
Master of Technology, Computer Science Department, Banasthali University,
Jaipur 302001, Rajasthan, India
e-mail: shivanigoyal16jun@gmail.com

© Springer Nature Singapore Pte Ltd. 2018　　　　　　　　　　　　　　715
M. Pant et al. (eds.), *Soft Computing: Theories and Applications*,
Advances in Intelligent Systems and Computing 584,
https://doi.org/10.1007/978-981-10-5699-4_68

nostic tool for diagnosis of disease and their treatment. Soft tissues are generated through this modality of imaging. The founded medical images demonstrate the interior structure; however, peer images are required by the doctors need to know more, for example emphasizing the abnormal tissues, quantification of their size and portraying its shape. If the doctors consider these tasks themselves, it might not be satisfactory, taking a lot of time and overburdening them.

1.1 Operations and Types of Tumor

In processing of medical images, 3D segmentation of images plays a crucial role in stages which happen before implementing object recognition. With the help of 3D image segmentation, we diagnose the different diseases of brain automatically, and we take help of 3D segmentation for automated and quantitative study of images, for example measuring accurate size and volume of recognized part of the image. It is very challenging to measure the accurate diagnosis of brain due to assorted shape, sizes and visibility of tumors. Tumors have abruptly growing tendency due to this nearby tissues also get infected and in the result give a general unusual structure for healthy tissues too. We are proposing 3D segmentation technique for brain tumor with morphological operations by utilizing segmentation as a part of conjunction.

1.2 Tumor

Tumor is a word used in place of word neoplasm. Neoplasm is framed by abnormal development of cells. Tumor is a disease that is completely differed from cancer.

1.2.1 Types of Tumor

Generally, we put tumor in three categories: (1) benign; (2) premalignant; and (3) malignant.

1. Benign tumor: The behavior of benign tumor does not grow in an unexpected way. This tumor also does not infect its nearby healthy tissues and furthermore does not expand to non-nearby tissues. Moles are the familiar example of benign tumors. In short, if it can be detected in early stage, it can be cured easily.
2. Premalignant tumor: Pre-stage of cancer is known as premalignant tumor. Considering it as a disease, if it is not appropriately cured, it might prompt to cancer.
3. Malignant tumor: Malignancy is very dangerous kind of tumor, it expands very fast as time passed and at last, person, which has this type of tumor, died.

Malignant is fundamentally a medical term used for representing the serious type of progressing disease. Typically, to describe the cancer, the commonly used term by doctors is malignant tumor.

1.3 Magnetic Resonance (MR) Imaging

Generally, MR imaging is used to study the internal formation of body in detail in the field of biomedical. It is used to study the structure of the body in finer way for detection and visualization. The technique of MR imaging is generally used to determine the irregularities in the tissues, which have a much better procedure in comparison with computed tomography. That is why this procedure is very effective for the brain tumor determination and identification of cancer imaging.

Strong magnetic field is used in MR imaging in place of ionizing radiation that is used in CT scan. Magnetic field does the adjustment of nuclear magnetization; then, alignment of the magnetization is changed by radio frequencies, which can be identified by scanner. To get the additional information of the body, this signal can be further examined.

1.4 Image Segmentation

Segmentation of images is very crucial and, maybe, the most challenging job in image processing. Segmentation means to club the elements of an image that have the similar attributes, i.e., an image is divided into subparts to constitute regions or objects. All forthcoming tasks were to be analyzed; for example, object recognition and classification depend vigorously on the quality of the segmentation process.

2 Literature Review and Related Work

Various research papers concerning with segmentation techniques of medical images were studied. We are presenting a brief summary of literature survey here:

Zhang et al. [1] proposed segmentation of brain MRI images via a hidden Markov random field model and the expectation-maximization algorithm. This developed algorithm has the capacity to conceal the spatial properties of the image as well as factual properties [1]. Ahmed et al. [2] presented a changed fuzzy C-mean algorithm for influenced field estimation and segmentation of MR image information. This developed algorithm shows how fast the results we can generate [2]. Tolba et al. [3]

proposed MR brain image segmentation with Gaussian multi-resolution analyze and the expectation-maximization algorithm. This presented method shows how the results are less sensitive to noise [3]. Sing et al. [4] presented the segmentation of MRI of human brain utilizing fuzzy adaptational radial basic function neural network [4]. This presented technique indicates how we can preserve the sharpness of an image. SasiKala et al. [5] introduced an automatic segmentation of malignant tumor in MR images of brain with optimal texture features. Texture elements are removed from ordinary and tumor area in the brain images under study utilizing the spatial gray-level dependence technique and transformation of wavelet. Abdal-Maksood [6] proposed brain tumor segmentation technique relied on a hybrid clustering. This proposed method shows minimization of time for calculating the results [6]. Kailash Sinha et al. [7] presented proficient segmentation techniques for the determination of tumor in MRI images. It shows convergence and computing time reduced [7]. Anandgaonkar et al. [8] gave a method for detection and identification brain tumor in brain MR images utilizing fuzzy c-mean segmentation. This proposed work indicates whether the tumor is benign or malignant [8].

3 Segmentation Techniques

The main focus of this section is to briefly describe about the segmentation strategies, which are useful and helpful in determination and recognition of brain tumor in MR images.

3.1 K-means Clustering

The n numbers of findings are divided into k clusters in K-means clustering technique, in which every pixel has a place in the clusters. This is performed through the minimization of an objective function in such a manner that total number of squares is minimized within the cluster. Minimization initially begins within k cluster centers. The observations are reassigned to clusters on the basis of the commonness between the observations and cluster center. K-means clustering is a two-step process. In the initial step, a skull mask is produced to perform skull stripping via the MR image. In the next step, a propelled K-means algorithm is improved by two-level granularity arrangements of grids, and segmentation of images into gray matter, white matter and tumor region is performed by localization process depending on standard local deviation, and afterward length and breadth of the tumor are evaluated [9].

3.2   Region Growing

In this segmentation method, images are separated by arranging the closest pixel of same type. It begins with a pixel which has common properties. On the basis of similar properties, the nearby pixels are grouped continuously to the seed. In the process of splitting, regions are divided into subregions that do not satisfy a given similarity criteria. Splitting and conquering can be utilized together. Mostly, the performance relies on the chosen similarity basis. The technique of seeded region development is controlled through various starting seeds without tuning the similarity parameters.

3.3   Soft Computing

Self-Organizing Map or SOM is a kind of artificial neural network, which is used for unsupervised learning. Self-Organizing Map works in training and mapping modes. The vector quantization process is used for building training process map and mapping consequent orders of another input vector. SOM delineates neurons and nodes. SOM maps every neuron, which are connected with a weight vector map, data input vectors and location in the map space. The input space with higher dimension is mapped with the input space with lower dimension map space by the SOM. Mean, median, entropy, energy, correlation, contrast, variance, maximum and minimum intensity values are utilized for getting clear nature of the tumor [9].

3.4   Fuzzy C-mean Clustering

Fuzzy C-mean clustering is an information bundle technique in which every data point has a place toward a cluster to a degree indicated by a membership value. A set of n vectors is partitioned into c fuzzy groups through fuzzy C-mean clustering, and a cluster center is found for every group such that a cost function is minimized for measure of dissimilarities [9].

3.5   Image/Symmetry Analysis

This examination is an intelligent segmentation technique that notwithstanding region of the area and edge data utilizes earlier data, likewise its symmetry investigation which is more reliable in diagnosis cases of pathology. An adroitly simple directed piece-based, shape surface, content-based system has been broke down MR brain images with generally bringdown computational prerequisites. Ordering areas by method for their multi-parameter values makes the investigation of the locales of physiological and neurotic premium less demanding and more determinable.

3.6 Thresholding

Thresholding is a standout among the most well-known and most established method for image segmentation. In this method, image should be made out of regions and these have different ranges of gray scale. The histogram of an image has many peaks and valleys, in which every peak considers as one region and between the peaks, the valleys show a threshold value. Through histogram thresholding strategy, the image is partitioned into two equivalent parts, and histograms are contrasted to recognize the tumor, and for finding a legitimate physical dimension of brain tumor, cropping method is used. Based on local raw pixel data, the threshold method made a decision which helps in removing the elementary structure of an image, examining title superfluous subtle elements [9].

4 Existing Methodology

4.1 Image Acquisition

Images are acquired by MRI scan. The two-dimensional matrices are used to represent the scanned images, and these matrices have pixels as their element. MATLAB is used to store the images and showed as a grayscale image of 256 * 256. The images are indexed in gray scale from 0 to 255, where 0 indicates total black color and 255 shows immaculate white color. The intensity of elements of this range varies from black to white color.

4.2 Preprocessing

In this stage, image is upgraded in the way that better points of interest are enhanced and noise is extracted from the image. Generally, up-gradation and noise minimization methods are used, and these methods are implemented for better results. Improvement will bring about some eminent edges and a better image is acquired, noise will be minimized, and thus, the effect of blurry from the image is minimized. Besides, the up-gradation of image, as well as segmentation of image, will be adapted. Due to the enhancement and up-gradation of image, distinguishing edges will be detected easily, and therefore, overall quality of the whole image is improved. Edge identification will prompt to report the correct area of tumor. The steps that will be considered in the preprocessing stage are:

(a) The image obtained from MR image scan, stored in database, is reshaped to size 255 * 255 of grayscale image.

(b) Any noise element is removed from the image. This will be applied when visual quality of the image would not be up to the mark.

(c) A high-pass filter will be used for enhancing and edge determination on the noise-free, better quality image.

(d) The acquired enhanced image is then merged to the main image for better results.

4.3   Processing

In this stage, we apply segmentation techniques for group of areas.

Segmentation: The segmentation of image depends on dividing the image into regions. Segmentation is performed on the base of common characteristics. Common characteristics are keeping apart into groups. Fundamental objective behind segmentation is to taken out the main features from the image, and data can be easily understandable. Yet it is very difficult job in the area of medical image processing to segment brain tumor from MRI images (Fig. 1).

4.4   Post-processing

The methods used in the segmentation during the post-processing are as follows:

1. Threshold segmentation: The simplest method of image segmentation is the threshold segmentation. In this method, format of image is transformed from input grayscale image to binary. Based on the threshold value, conversion of format of image from gray scale to binary is take place. The key feature of this method is the criterion of selection of a threshold value. Some popular techniques taken in this segmentation incorporate maximum entropy method and K-means clustering technique for segmentation.

2. Watershed segmentation: This method is based on the intensity of pixels of an image, and it is one of the best techniques under this category. Pixels having the same intensities are grouped together. Watershed segmentation is very effective method for segmenting an image to extract tumor part from the image. Watershed is a numerical morphological working tool. Due Watershed is typically used for checking outcome instead of using as an input segmentation method since it for the most part experiences over segmentation and under segmentation explicitly by a user or they can be detected automatically by utilizing morphological tool.

3. Morphological operations: First, we convert the image into binary format and then applied some morphological operations on the converted image. The morphological operation is used to extract tumor portion of the image. Presently just the tumor part of the image is displayed as white color. This part of the image has the most intensity than other parts of the image.

Fig. 1 Steps of the
watershed segmentation with
thresholding

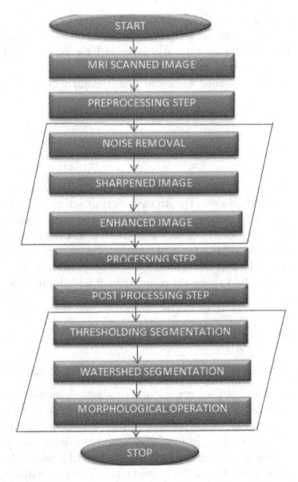

5 Results of Existing Method

The images of brain we are considering were taken from MIDAS: Community-designed database of MR brain images of healthy volunteers, [10] and we are using these images for study purpose only (Fig. 2).

6 Proposed Methodology

6.1 Image Acquisition

In the first step, we take an MRI scan image. This image is a two-dimensional image of 255 * 255 size.

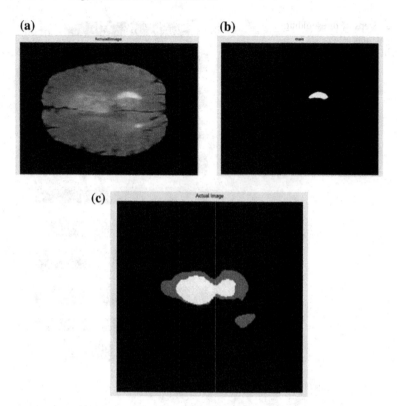

Fig. 2 Images of brain tumor (*Source* MIDAS: community-designed database of MR brain images of healthy volunteers) [10]

6.2 Gradient Operator

In the second step, we apply gradient operator on the MRI scan image, so that we can get the edges of the particular image (Fig. 3).

6.3 Morphological Operation

In this step, we apply morphological operations, so that we can get better results. These operations are as follows:

1. Opening: In this morphological operation, we get the image with large areas, so that we can get better results of tumor and can easily get the area where tumor lies exactly.
2. Opening by Reconstruction: In this step, we get the image after reconstruction, so that we can see the changes clearly.

Fig. 3 Steps of thresholding
segmentation using
morphological operators

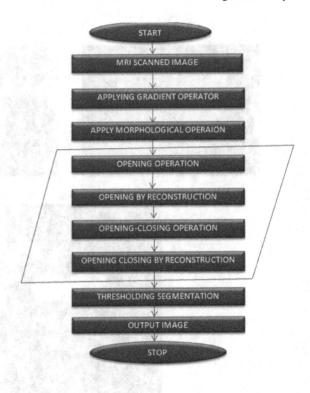

3. Opening–Closing: In this step, we apply both the morphological operations, so that those areas, which exactly lie in the tumor region, can see more clearly, and other area, which is not so important, can be avoided.
4. Opening–Closing by Reconstruction: In this step, we get the image after opening–closing reconstruction. In this step, we can see the changes immediately, and we can almost see the area which is affected by the tumor.

6.4 Thresholding Segmentation

In this final step, we get the final result. In this image, we covered the area in which exactly tumor lies. In the final step, we see the area which shows the clear image of tumor. From previous, we get the area in which tumor lies, but when we apply the final step we get the neighborhood area also. The benefit of this step is that we can also cure the neighboring area of the tumor area (Fig. 4).

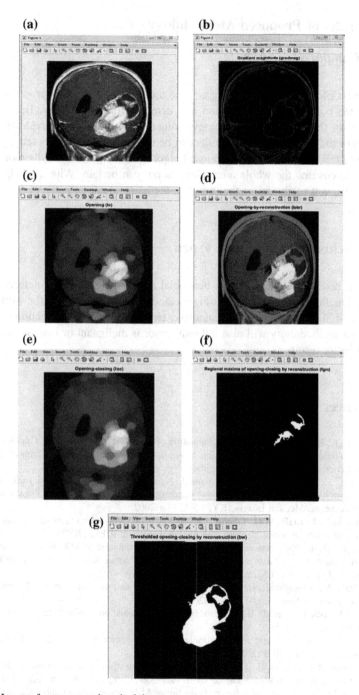

Fig. 4 Images from proposed methodology

7 Results of Proposed Methodology

7.1 Comparison Analysis

The methodology which is existing and the methodology which is proposed, both, are for the purpose of segmentation and detection of brain tumor. But both are different. In the existing methodology, sometimes it does not provide the exact location of the tumor, since the results can be negative, but the proposed algorithm is solving this issue. From the proposed algorithm, we get the results more clear, and it also covered the whole area where tumor can be lain. With the help of this method, we can easily detect the tumor in the exact location.

8 Conclusion and Future Work

From this research work, we get a beneficial result for brain tumor detection. The proposed method is better than existing method. We can also extend the work in the direction of 3D MRI images. We can also extend our work in the direction of that in which our methodology will also tell that tumor is malignant or benign so that we can get better results.

References

1. Zhang, Y., Brady, M., Smith, S.: Segmentation of brain MR images through a hidden Markov random field model and the expectation-maximization algorithm. In: Proceedings of the IEEE Transaction on Medical Images, Jan 2001
2. Ahmed, M.N., Yamany, S.M., Mohamed, N., Moriarty, T.: A modified fuzzy c-means algorithm for bias field estimation and segmentation of MRI data. In: Proceedings of the IEEE Transaction on Medical Images, KY, USA, Mar 2002
3. Tolba, M.F., Mostafa, M.G., Gharib, T.F., Salem, M.A.: MR-brain image segmentation using Gaussian multi resolution analysis and the em algorithm. ICEIS (2003)
4. Sing, K., Basu, D.K., Nasipuri, M., Kundu, M.: Segmentation of MR images of the human brain using fuzzy adaptive radial basis function neural network. In: Pattern Recognition and Machine Intelligence. LNCS. Springer, Berlin (2005)
5. Sasikala, M., Kumaravel, N. : A wavelet-based optimal texture feature set for classification of brain tumors. J. Med. Eng. Technol. (2009)
6. Abdal Maksood, E., et al.: Brain tumor segmentation based on a hybrid clustering technique. Egyptian Inf. J. (2015)
7. Sinha, K., et al.: Efficient segmentation methods for tumor detection in MRI images. In: Conference on Electrical (2014)
8. Anandgaonkar, G.P., et al.: Detection and identification of brain tumor in brain MR images using Fuzzy C-means segmentation. Int. J. Adv. Res. Comput. Commun. Eng. (2013)
9. Deshmukh, R.D.: Study of different brain tumor MRI image segmentation techniques. Int. J. Comput. Sci. Technol. (2014)
10. www.insight-journal.org/midas/gallery

A Brief Overview of Firefly Algorithm

Bilal and Millie Pant

Abstract Past one decade has seen a tremendous increase in the development of nature-inspired metaheuristics (NIM). In this paper, we provide a brief review on firefly algorithm (FA), one of the recent developments in the field of NIM for solving optimization problems. FA first proposed by Xin-She Yang in 2008 is a swarm intelligence-based algorithm slowly gaining popularity for solving continuous, multi-objective, dynamic and noisy, discrete, nonlinear, and multi-dimensional problem. It has a simple structure and is easy to apply on a wide range of problems as shown by different researchers.

Keywords Firefly algorithm · Nature-inspired metaheuristics (NIM) Swarm intelligence

1 Introduction

Situations arising in almost all spheres of human activities can be designed mathematically as optimization problems. Mathematically, the optimization problems can be categorized in several ways; for example, as continuous problems (dealing with real functions [1, 2]), constrained problems (having limitations or bounds [3, 4]), combinational problems where solution is determined from finite or infinite sets [5, 6–8] multimodal problems having large number of local solutions better than all neighboring solutions, but not good as the globally optimal solution [9], multi-objective problems, where the fineness of a solution is defined by its performance relative to different objectives [10, 11], and non-stationary problems which are dynamic and noisy [12].

Bilal (✉) · M. Pant
Department of Applied Science and Engineering, IIT Roorkee, Roorkee, India
e-mail: bilal25iitr@gmail.com

M. Pant
e-mail: millifpt@iitr.ac.in

© Springer Nature Singapore Pte Ltd. 2018
M. Pant et al. (eds.), *Soft Computing: Theories and Applications*,
Advances in Intelligent Systems and Computing 584,
https://doi.org/10.1007/978-981-10-5699-4_69

Often these problems are challenging and require a suitable method for their solutions. Classical methods are usually problem specific and depend on a certain set of rules to be followed before these methods can be applied. For example, Newton's type methods can be applied only to differentiable functions; likewise, simplex method can be applied only to problems which are strictly linear in nature and so on.

Consequently, in the past few decades, emphasis is being laid on techniques which are generic in nature and can be applied to a wide range of problems. Past few decades have witnessed the emergence of optimization techniques following based on some natural phenomena like theory of evolution or socio cooperative behavior displayed by various species including humans. These are stochastic search techniques and are collectively called nature-inspired metaheuristics (NIM).

In the present article, the search technique taken is firefly algorithm (FA), a comparatively newer addition to the class of NIM. FA belongs to a class of swarm intelligence (SI)-based algorithms. SI takes inspiration from the collective behavior of insects or animals. This term was first used by Beni and Wang [13] in 1989. SI is mainly based on the philosophy that by working together, species can solve the most complex task (e.g., mound building of termites). Various species follow simple interaction rules which they follow for communication and thus form a self-organized multi-agent system. A relation between SI-based algorithms, bio-inspired algorithms, and nature-inspired algorithms can be given as follows:

$$\text{SI-based} \subset \text{bio-inspired} \subset \text{nature-inspired}$$

SI-based $\subset$ bio-inspired $\subset$ nature-inspired

Most common swarm intelligence optimization techniques available in nature include particle swarm optimization (PSO) [14, 15], ant colony optimization (ACO) [16], artificial bee colony (ABC) [17, 18], firefly algorithm (FA) [19, 20], differential evolution (DE) [21, 22], bat algorithm [23], glow swarms [24], krill herd bio-inspired algorithms [25], cuckoo search [26], fish optimization [27], bacterial foraging optimization [28]; most of these algorithms are population based and have a natural advantage over single-point search algorithms [29].

The objective of the present study is to introduce the researchers with a brief overview of FA. The article consists of five sections including introduction. In the second section, we define the working, structure of firefly. In the third section, we classify the firefly algorithm. In the fourth section, we discuss the application of firefly algorithm, and the fifth section is for conclusion.

2 Introduction of Firefly Algorithm

A firefly algorithm was developed by Yang [19]. Fireflies are quite fascinating insects found typically in warm environment and are normally active during night when they look mesmerizing. Some interesting facts about fireflies are as follows:

- More than two thousand species of firefly can be found in nature, and mostly all fireflies produce tiny, periodic flashes. The pattern of flashing is incomparable for a particular species.
- The function of flashing light is either to attract partners (communication) or is a warning toward the predator.
- Each firefly can glow with a different intensity.

Many researchers have discussed firefly behavior in nature. For reference, the interested reader may see [30–32].

In FA algorithm, better fireflies have a smaller error and have a higher intensity. The flashing light can be expressed in such a way that it can be linked with the objective function which is to be optimized. In this paper, firstly, we discuss the structure of the firefly algorithm (FA), classification, and then move toward their applications.

The basic idea of FA is that a firefly will be attracted to any other firefly that has a higher intensity, and that attractiveness (the distance moved toward a more intense firefly) is stronger if the distance between the two fireflies is smaller. Firefly algorithm is based on a physical formulation of light intensity, which is varied at the distance from the eyes of beholder. In general, as the distance between two fireflies' increases, then light intensity decreases rapidly which means light intensity is inversely proportional to the distance between two fireflies [19].

The two major issues in FA are intensity and attractiveness. Since FA is a metaheuristic, one is free to define intensity in any manner while keeping in mind that higher intensity is associated with a better solution/position. The next major issue is to define attractiveness so that closer fireflies will move toward a more intense target. Thus, the brightness phenomenon is associated with objective function to be optimized.

Firefly algorithms use the following three idealized rules:

1. Fireflies are unisex so that they are attracted to each other regardless of their gender.
2. The attractiveness is proportional to the brightness, and both attraction and brightness decrease as the distance between consecutive fireflies increases. This implies that the firefly having lesser brightness will move toward the brighter firefly. However, if the brightness is same, the firefly will have a random movement.
3. For a maximization problem, brightness is proportional to the objective function.

2.1 Structure of Firefly

As already mentioned, in FA, there are two important parameters, intensity and attractiveness. A firefly will move toward the other firefly which is brighter than itself and the attractiveness depends on light intensity.

The light intensity and attractiveness both are inversely proportional to the distance d between attracted fireflies which implies that light intensity and attractiveness decrease as the value of d increases. These two parameters are defined below.

Intensity:

$$I = I_0 * e^{-\gamma d^2},$$

where

I light intensity
I_0 original light intensity or initial light intensity
D distance between firefly i and j
λ the light absorption coefficient controls the light intensity

Attractiveness

Attractiveness is proportional to the light intensity, and attractiveness β is defined as follows:

$$\beta = \beta_0 * e^{-\gamma d^2},$$

where

B_0 attractiveness when d equals to zero

The distance between two fireflies can be defined as follows:

$$d_{ij} = \|u_i - u_j\| = \sqrt{\sum_{k=1}^{n} (u_{i,k} - u_{j,k})^2}.$$

Movement of firefly:

When a firefly i is attracted toward the brighter firefly j, its movement is defined as follows:

$$u_i^{t+1} = u_i^t + \beta_0 * e^{-\gamma d_{ij}^2} * (u_j^t - u_i^t) + \alpha * \varepsilon_i.$$

The second term shows attraction, and γ is limitation. If γ tends to zero or is too small, then attractiveness and brightness become constant, $\beta = \beta_0$ implies that a firefly can be seen in any place, and this shows the firefly will also search global firefly. If γ tends to infinity or too large then attractiveness and brightness will decreases near to zero which means firefly will move randomly and the third term is for randomization here α is a parameter called random parameter.

ε_i can be replaced by (rand − 1/2), where rand is a random number generated between 0 and 1.

The pseudocode of basic FA is given below:

3 Classification

Many variants in firefly algorithms have been proposed in the literature. Attempts have been made to find the best possible combination of parameters through tuning [33] or by controlling the parameters adaptively.

Research has also been done on the movement of fireflies using different probability distributions Levy, uniform, Gaussian etc. or by using chaotic sequences. Binary and real coded, both versions of FA, are available in the literature. In general, more than 20 variants of firefly exist in the literature. Fig. 1 gives a pictorial representation of modified and hybridized variants of FA, while Tables 1 and 2 list the various modified and hybridized FA (Fig. 2).

Begin FA
Define objective function $f(x)$, where $x = (x_i, ,,, x_n)$
Generate an initial population
Define I, the light intensity
Define γ, absorption coefficient
While$(t < Max\_Generation)$
For $i = 1$ to $n(all\ n\ fireflies)$
  For $j = 1$ to $n(all\ n\ fireflies)$
   If$(I_i > I_j)$, move firefly i towards firefly j
   End if
   Evaluate new solutions and update light intensity;
  End for j
End for i
Rank the fireflies and find the current best
End while
End FA

Fig. 1 Pseudocode of basic FA

Table 1 Modified firefly algorithms

Topic	References
Elitist FA	[57]
Quaternion FA	[58]
Lévy flights randomized FA	[1, 59]
Gaussian randomized FA	[1, 2]
Harmonic clustering	[60]
Multi-population	[61]
Parallel firefly algorithm	[62]
Chaos randomized FA	[63]
Binary represented FA	[64]

Table 2 Hybrid firefly algorithms

Topic	References
Eagle strategy using Lévy walk	[65]
Ant colony	[66]
Neural network	[67]
Memetic algorithm	[5]
Differential evolution	[68]
Cellular learning automata	[69]
Genetic algorithm	[70]
Evolutionary strategies	[71]
Simulated annealing	[72]

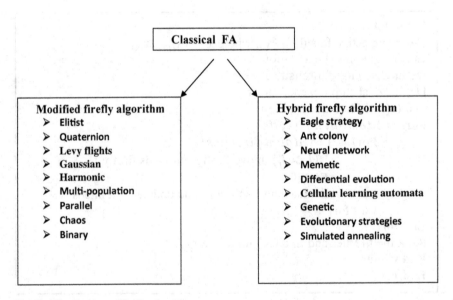

Fig. 2 Classification of firefly algorithm

4 Application

FA has been applied to several fields like economic emissions load dispatch problem, multi-level thresholding selection [34], vector quantization for image compression [35], finding optimal test sequence generation [36], object tracking [37]. Various optimization problems that have been dealt with FA are (Fig. 3) as follows:

- Continuous,
- Constrained,
- Combinatorial,
- Multimodal,
- Multi-objective,
- Dynamic and noisy.

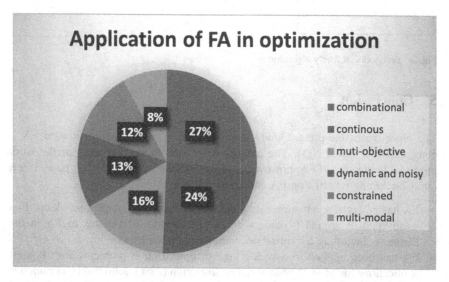

Pie-chart. Uses FA in optimization

Firefly algorithms have also been applied to various real-life applications problems like industrial optimization [38–40], image processing [35, 41], business optimization [42], antenna design [43], civil engineering [44], semantic web [45], robotics [46], chemistry [47], meteorology [48] and wireless sensor networks [49]; optimal placement and sizing of voltage-controlled distributed generators in unbalanced distribution networks [50], rescheduling of real power for congestion management with integration of pumped storage hydro-unit [51], for multi-stage transmission expansion planning with adequacy-security considerations in deregulated environments [52], optimizing real power loss and voltage stability limit of a large transmission network [53], for congestion management in deregulated environment [54].

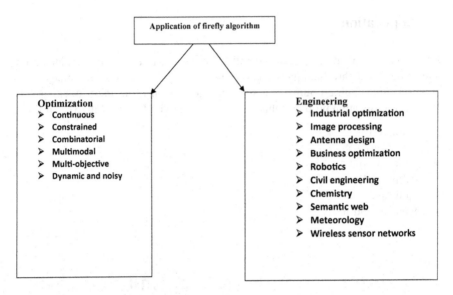

Fig. 3 Taxonomy of firefly algorithm

5 Conclusion

The objective of the present study is to familiarize the researchers toward FA, a comparatively newer NIM. The authors have tried to include most of the work proposed in the area of FA; however, it is possible that some important paper is left out. Concluding remarks on FA can be as follows:

- FA has emerged as an attractive alternative for solving complex problems. In a small span of around 6 years, we can find around 20 variants of FA in the literature, including hybridization, variation in parameters.
- FA has been applied to a wide range of optimization problems with highest applications found in combinatorial optimization (27%), followed by continuous optimization (24%). FA has also been applied to multi-objective, noisy, and dynamic optimization problems as well.
- FA has also been applied to different real-life problems like image processing, robotics, business optimization.
- FA can be combined with other soft computing techniques (such as neural network [55], genetic algorithm [56] and other NIAs [14–29]) to get better performance.

References

1. Yang, X.S.: Metaheuristic optimization: algorithm analysis and open problems. In: Pardalos, P.M., Rebennack, S. (eds.) Experimental Algorithms, vol. 6630, pp. 21–32. Lecture Notes in Computer Science. Springer, Berlin (2011)
2. Farahani, S.M., Abshouri, A.A., Nasiri, B., Meybodi, M.R.: A gaussian firefly algorithm. Int. J. Mach. Learn. Comput. 1(5), 448–454 (2011)
3. Gomes, H.M.: A firefly metaheuristic structural size and shape optimisation with natural frequency constraints. Int. J. Metaheuristics 2(1), 38–55 (2012)
4. Łukasik, S., Zak, S.: Firefly algorithm for continuous constrained optimization tasks. In: Computational Collective Intelligence. Semantic Web, Social Networks and Multiagent Systems, pp. 97–106. Springer, Berlin (2009)
5. Fister, I., Jr., Yang, X.-S., Fister, I., Brest, J.: Memetic Firefly Algorithm for Combinatorial Optimization, pp. 75–86. Jožef Stefan Institute (2012)
6. Jati, G.: Evolutionary discrete firefly algorithm for travelling salesman problem. In: Adaptive and Intelligent Systems, pp. 393–403 (2011)
7. Khadwilard, A., Chansombat, S., Thepphakorn, T., Thapatsuwan, P., Chainate, W., Pongcharoen, P.: Application of firefly algorithm and its parameter setting for job shop scheduling. In: 1st Symposium on Hands-On Research and, Development, pp. 1–10 (2011)
8. Kwiecień, J., Filipowicz, B.: Firefly algorithm in optimization of queueing systems. Tech. Sci. 60(2), 363–368 (2012)
9. Miguel, L.F.F., Lopez, R.H., Miguel, L.F.F.: Multimodal size, shape, and topology optimisation of truss structures using the firefly algorithm. Adv. Eng. Softw. 56, 23–37 (2013)
10. Abedinia, O., Amjady, N., Naderi, M.S.: Multi-objective environmental/economic dispatch using firefly technique. In: 11th International Conference on Environment and Electrical Engineering (EEEIC), pp. 461–466. IEEE (2012)
11. Amiri, B.K, Hossain, L., Crawford, J.W., Wigand, R.T.: Community detection in complex networks: Multi-objective enhanced firefly algorithm. Knowl. Based Syst. 46, 1–11 (2013)
12. Abshouri, A.A., Meybodi, M.R., Bakhtiary, A.: New firefly algorithm based on multi swarm & learning automata in dynamic environments. In: IEEE proceedings, pp. 73–77 (2011)
13. Beni, G., Wang, J.: Swarm intelligence in cellular robotic systems. In: Proceedings of NATO Advanced Workshop on Robots and Biological Systems, Tuscany, Italy, pp. 26–30 (1989)
14. Kennedy, J., Eberhart, R.: The particle swarm optimization: Social adaptation in information processing. In: Corne, D., Dorigo, M., Glover, F. (eds.) New Ideas in Optimization, pp. 379–387. McGraw Hill, London (1999)
15. Ansari, I.A., Pant, M., Ahn, C.W., Jeong, J.: PSO optimized multipurpose image watermarking using SVD and chaotic sequence. bio-inspired computing-theories and applications. In: Proceedings of the 10th International Conference, BIC-TA 2015 Hefei, China, vol. 562, 25–28 Sept 2015. Springer, Berlin (2016)
16. Dorigo, M., Di Caro, G.: The ant colony optimization meta-heuristic. In: Corne, D., Dorigo, M., Glover, F. (eds.) New Ideas in Optimization, pp. 11–32. McGraw-Hill, London (1999)
17. Karaboga, D., Basturk, B.: A powerful and efficient algorithm for numerical function optimization: artificial bee colony (ABC) algorithm. J. Global Optim. 39(3), 459–471 (2007)
18. Ansari, I.A., Pant, M., Ahn, C.W.: ABC optimized secured image watermarking scheme to find out the rightful ownership. Optik-Int. J. Light Electron Opt. 127(14), 5711–5721 (2016)
19. Yang, X.S.: Fire fly algorithm. Nat. Inspir. Metaheuristic Algorithms 79–90 (2008)
20. Gandomi, A., Yang, X.-S., Talatahari, S., Alavi, A.: Metaheuristic in modeling and optimization. In: Structures and Infrastructures, pp. 1–24. Elsevier, Waltham (2013)
21. Jauhar, S.K., Pant, M., Deep, A.: An approach to solve multi-criteria supplier selection while considering environmental aspects using differential evolution. In: International Conference on Swarm, Evolutionary, and Memetic Computing, pp. 199–208. Springer, Berlin (2013)

22. Kumar Jauhar, S., Pant, M., Nagar, M.C.: Differential evolution for sustainable supplier selection in pulp and paper industry: a DEA based approach. Comput. Methods Mater. Sci. **15** (2015)

23. Yang, X.S.: A new metaheuristic bat-inspired algorithm. In: Cruz, C., Gonzlez, J., Krasnogor, G.T.N., Pelta, D.A. (eds.) Nature Inspired Cooperative Strategies for Optimization (NISCO 2010), Studies in Computational Inteligence, vol. 284, pp. 65–74. Springer, Berlin (2010)

24. Krishnanand, K.N., Ghose, D.: Glowworm swarm optimization for simultaneous capture of multiple local optima of multimodal functions. Swarm intell. **3**(2) 87–124 (2009)

25. Gandomi, A.H., Alavi, A.H.: Krill herd: a new bio-inspired optimization algorithm. Commun. Nonlinear Sci. Numer. Simul. **17**(12) 4831–4845 (2012). doi:10.1016/j.cnsns.2012.05.010

26. Yang, X.S., Deb, S.: Cuckoo search via levy flights. In: World Congress on Nature & Biologically Inspired Computing (NaBIC 2009), IEEE Publications, pp. 210–214 (2009)

27. Li, X.-L., Qian, J.-X.: Studies on artificial fish swarm optimization algorithm based on decomposition and coordination techniques. J. Circuits Syst. **1**, 1–6 (2003)

28. Passino, K.M.: Bacterial foraging optimization. Innov. Dev. Swarm Intell. Appl. 219–233 (2012)

29. Prugel-Bennett, A.: Benefits of a population: five mechanisms that advantage population-based algorithms. IEEE Trans. Evol. Comput. **14**(4), 500–517 (2010)

30. Brasier, A., Tate, J., Habener, J., et al.: Optimized use of the firefly Luciferase assay as a reporter gene in mammalian cell lines. BioTechniques **7**(10), 11–16 (1989)

31. Strehler, B.L., Totter, J.R.: Fire fly luminescence in the study of energy transfer mechanisms. I. Substrate and enzyme determination. Arch. Biochem. Biophys. **40**(1), 28–41 (1952)

32. Deluca, M.: Firefly luciferase. Adv. Enzymol. Relat. Areas Mol. Biol. **40**, 37–68 (2006)

33. Eiben, A., Smit, S.: Parameter tuning for configuring and analysing evolutionary algorithms. Swarm Evol. Comput. **1**(1), 19–31 (2011)

34. Apostolopoulos, T., Vlachos, A.: Application of the firefly algorithm for solving the economic emissions load dispatch problem. Int. J. Comb. **2011** (2010)

35. Horng, M.-H.: Vector quantization using the firefly algorithm for image compression. Expert Syst. Appl. **39**(1), 1078–1091 (2012)

36. Srivatsava, P.R., Mallikarjun, B., Yang, X.-S.: Optimal test sequence generation using firefly algorithm. Swarm Evol. Comput. **8**, 44–53 (2013)

37. Gao, M.-L., et al.: Object tracking using firefly algorithm. IET Comput. Vis. **7**(4), 227–237 (2013)

38. Yang, X.S., Hosseini, S.S.S., Gandomi, A.H.: Firefly algorithm for solvingnon-convex economic dispatch problems with valve loading effect. Appl. Soft Comput. **12**(3), 1180–1186 (2011)

39. Mauder, T., Sandera, C., Stetina, J., Seda, M.: Optimization of the quality of continuously cast steel slabs using the firefly algorithm. Mater. Tehnol. **45**(4), 347–350 (2011)

40. Chatterjee, A., Mahanti, G., Chatterjee, A.: Design of a fully digital controlled reconfigurable switched beam concentric ring array antenna using firefly and particle swarm optimization algorithm. Prog. Electromagnet. Res. B **36**, 113–131 (2012)

41. Horng, M., Jiang, T.: The codebook design of image vector quantization based on the firefly algorithm. Comput. Collect. Intell. Technol. Appl. 438–447 (2010)

42. Yang, X.-S., Deb, S., Fong, S.: Accelerated particle swarm optimization and support vector machine for business optimization and applications. In: International Conference on Networked Digital Technologies, Springer, Berlin (2011)

43. Basu, Banani, Mahanti, Gautam Kumar: Fire fly and artificial bees colony algorithm for synthesis of scanned and broadside linear array antenna. Prog. Electromagnet. Res. B **32**, 169–190 (2011)

44. Gholizadeh, S., Barati, H.: A comparative study of three metaheuristics for optimum design of trusses. Int. J. Optim. Civil Eng. **3**(3), 423–441 (2012)

45. Pop, C.B., et al.: A hybrid firefly-inspired approach for optimal semantic web service composition. Scalable Comput. Pract. Exp. **12**(3), 363–370 (2011)

46. Severin, S., Rossmann, J.: A comparison of different metaheuristic algorithms for optimizing blended PTP movements for industrial robots. In: International Conference on Intelligent Robotics and Applications. Springer, Berlin (2012)
47. Fateen, S.E.K., Bonilla-Petriciolet, A., Rangaiah, G.P.: Evaluation of covariance matrix adaptation evolution strategy, shuffled complex evolution and firefly algorithms for phase stability, phase equilibrium and chemical equilibrium problems. Chem. Eng. Res. Des. **90** (12), 2051–2071 (2012)
48. Santos, A.F.D, et al.: Firefly optimization to determine the precipitation field on South America. Inverse Prob. Sci. Eng. **21**(3), 451–466 (2013)
49. Breza, M., McCann, J.A.: Lessons in implementing bio-inspired algorithms on wireless sensor networks. NASA/ESA Conference on Adaptive Hardware and Systems. AHS'08. IEEE (2008)
50. Othman, M.M., et al.: Optimal placement and sizing of voltage controlled distributed generators in unbalanced distribution networks using supervised firefly algorithm. Int. J. Electr. Power Energy Syst. **82**, 105–113 (2016)
51. Gope, S., et al.: Rescheduling of real power for congestion management with integration of pumped storage hydro unit using firefly algorithm. Int. J. Electr. Power Energy Syst. **83**, 434–442 (2016)
52. Rastgou, A., Moshtagh, J.: Application of firefly algorithm for multi-stage transmission expansion planning with adequacy-security considerations in deregulated environments. Appl. Soft Comput. **41**, 373–389 (2016)
53. Balachennaiah, P., Suryakalavathi, M., Nagendra, Palukuru: Optimizing real power loss and voltage stability limit of a large transmission network using firefly algorithm. Eng. Sci. Technol. Int J **19**(2), 800–810 (2016)
54. Verma, S., Mukherjee, V.: Firefly algorithm for congestion management in deregulated environment. Eng. Sci. Technol. Int. J. (2016)
55. Ansari, I.A., Singla, R.: BCI: an optimised speller using SSVEP. Int. J. Biomed. Eng. Technol. **22**(1), 31–46 (2016)
56. Jauhar, S.K., Pant, M.: Genetic algorithms in supply chain management: a critical analysis of the literature. Sādhanā **41**(9), 993–1017 (2016)
57. Ong, H.C., Tilahun, S.L.: Modified firefly algorithm. J. Appl. Math. **2012**, 12 (2012)
58. Fister, I., Yang, X.-S., Brest, J., Fister Jr., I.: Modified firefly algoirthm using quaternion representation. Expert Syst. Appl. (2013). doi:10.1016/j.eswa.2013.06.070
59. Yang, X.S.: Efficiency analysis of swarm intelligence and randomization techniques. J. Comput. Theor. Nanosci. **9**(2), 189–198 (2012)
60. Adaniya, M.H.A.C., Lima, F.M., Rodrigues, J.J.P.C., Abrao, T., Proenca, M.L.: Anomaly detection using DSNS and firefly harmonic clustering algorithm. In: IEEE International Conference on Communications (ICC), pp. 1183–1187. IEEE (2012)
61. Liu, G.: A multipopulation firefly algorithm for correlated data routing in underwater wireless sensor networks. Int. J. Distrib. Sens., Netw (2013)
62. Husselmann, A.V., Hawick, K.A.: Parallel parametric optimisation with firefly algorithms on graphical processing units. Technical Report CSTN-141 (2012)
63. Gandomi, A.H., Yang, X.-S., Talatahari, S., Alavi, A.H.: Firefly algorithm with chaos. Commun. Nonlinear Sci. Numer. Simul. **18**(1), 89–98 (2013)
64. Chandrasekaran, K., Simon, S.P., Padhy, N.P.: Binary real coded firefly algorithm for solving unit commitment problem. Inf. Sci. (2013). doi:10.1016/j.ins.2013.06.022
65. Yang, X.S., Deb, S.: Eagle strategy using levy walk and firefly algorithms for stochastic optimization. In: Nature Inspired Cooperative Strategies for Optimization (NICSO 2010), pp. 101–111 (2010)
66. Aruchamy, R., Vasantha, K.D.D.: A comparative performance study on hybrid swarm model for micro array data. Int. J. Comput. Appl. **30**(6), 10–14 (2011)
67. Hassanzadeh, T., Faez, K., Seyfi, G.: A speech recognition system based on structure equivalent fuzzy neural network trained by firefly algorithm. In: International Conference on Biomedical Engineering (ICoBE), pp. 63–67. IEEE (2012)

68. Abdullah, A., Deris, S., Mohamad, M., Hashim, S.: Anewhybrid firefly algorithm for complex and nonlinear problem. In: Omatu, S., et al. (eds.) Distributed Computing and Artificial Intelligence, vol. 151, pp. 673–680. Springer, Berlin (2012)
69. Hassanzadeh, T., Meybodi, M.R.: A new hybrid algorithm based on firefly algorithm and cellular learning automata. In: 20th Iranian Conference on Electrical Engineering, pp. 628–633. IEEE (2012)
70. Luthra, J., Pal, S.K.: A hybrid firefly algorithm using genetic operators for the crypt analysis of a monoalphabetic substitution cipher. In: World Congress on Information and Communication Technologies (WICT), pp. 202–206. IEEE (2011)
71. LulesegedTilahun, S., Ong, H.C.: Vector optimisation using fuzzy preference in evolutionary strategy based firefly algorithm. Int. J. Oper. Res. **16**(1), 81–95 (2013)
72. Vahedi Nouri, B., Fattahi, P., Ramezanian, R.: Hybrid firefly-simulated annealing algorithm for the flow shop problem with learning effects and flexible maintenance activities. Int. J. Prod. Res. (ahead-of-print), 1–15 (2013)

Analysis of Indian and Indian Politicians News in the New York Times

Irshad Ahmad Ansari and Suryakant

Abstract The news coverage of any country's media directly reflects the mind-set and interest of that particular country's citizens. Data mining is a very famous and old tool, which is used to understand the trend and major component of any raw data. In the present study, the effect of India and Indian politician is analyzed in the New York Times, a leading newspaper of the USA (United States of America). This analysis is trying to provide a comparative effect on India on the global level during the different government tenures (UPA-I, UPA-II, and NDA). To be fair with the analysis, only first two years of each government are considered during this study. Along with that, this study also provides a comparative effect on Indian politicians on the international level (especially in the USA).

Keywords Popularity analysis · Indian news · Data mining · Indian politicians India popularity in USA

1 Introduction

Present age of media and technology made the communication and information exchange between different countries very easy and fast [1, 2]. The information form one end of the world now reaches to another end faster than the blink of eye. Print and electronic media plays a very important role in this communication and works like a bridge in between different countries [3]. The growth and importance

I.A. Ansari (✉)
Electronics and Communication Engineering, PDPM Indian Institute of Information Technology Design and Manufacturing, Jabalpur, MP 482005, India
e-mail: 01.irshad@gmail.com; irshad@iiitdmj.ac.in

Suryakant
Department of PPE, Indian Institute of Technology Roorkee, Roorkee, India
e-mail: suryak111@gmail.com

© Springer Nature Singapore Pte Ltd. 2018
M. Pant et al. (eds.), *Soft Computing: Theories and Applications*,
Advances in Intelligent Systems and Computing 584,
https://doi.org/10.1007/978-981-10-5699-4_70

of any country can be analyzed by its international media coverage [4–6], because the importance can be directly related to the number of media reports (international level).

Generally, the news of any country comes into picture because of its political and economic conditions [5]. These types of news are directly affected by the governing government. If the government has policies or leaders that have the capability to affect the international market (either in good way or bad way), then they simply get more coverage in news.

It is a big issue for a developing country to be in the news of developed country's media, and it can be directly seen as its growing effect of that particular country on the international level [7, 8].

The development of Internet-based communication significantly influenced and changed the conventional media (press and electronic) world and has become a real competitor to them [9, 10]. The reach of Internet-based news articles and information is much deeper as compared to print media because of its flexible nature. Moreover, size limit and space is never a constraint for the Internet-based media, which provides it the ability to cover larger volume of news considering the vast variety of its user. This helps them to cover even very small of interest. In addition to this, Internet-based media can also fold and expand their news-covering strategies much faster than the conventional media because of the direct user integration and feedback over the Internet. Internet provides a greater level of audience participation and interactive feature adds to the popularity of the online newspapers. This feature enhances the ability of the public to actively participate in the discussion and control of content. Modification, immediacy, accessibility, easy and fast publication, user-friendly search, and no storage limitations make the Internet-based article much more powerful than the conventional media [9, 10]. These features make the online media for a powerful resource to check the public opinion. Because of which, the present study is also utilizing the online newspaper (The New York Times) in order to analyzing the effect of India on the international level.

2 Tools and Methods

Data mining is most sought after technique for data analysis [11–13], and same is utilized in this study too. Data mining is used to extract some useful information from the raw data [11]. The extraction of information from the raw data remains totally on the user's approach, and different people can interpret different useful information from the raw data [13].

2.1 Application Programming Interface

The basic tool used for this study is the application programming interface [14]. It is basically a set of subroutine protocols, definitions, and tools to provide an interface between web server and user-developed program [14, 15]. The API is provided by the Web site developer itself. Some famous APIs [16] are YouTube API, Flickr API, Google Maps API, Amazon Product Advertising API, and Twitter API.

The use of API makes the interaction between data mining program and web server very efficient and fast, because it is designed by the web owner itself so they are designed in the Web site architecture's efficient way.

The basic building blocks of good API are provided by the Web site owner, which can be used by the programmers in which so ever way they like. The APIs can be designed in hardware or software specific way too, but most commonly APIs are designed in operating system specific way. Figure 1 shows the basic architecture of API system. In this study, the API [17] of leading US online newspaper (The New York Times) is used.

2.2 User-Designed Code

Python language is used to design the user-specific code as well as the extraction of information from the API. The basic property of API is utilized in this code development, which enable the user to input/output user-specific function in order to copy files from one location to another without the requirement of understanding the operations occurring behind the architecture.

The user-designed code sends the request to make a connection with New York Times's server in order to fetch the specific data like news headlines related to India. This request asks for multiple parameters and same is supplied by the web server to the program. If no keyword is passed during the request, then sever sends the overall data available in that specific time frame, which is been scrutinized.

Fig. 1 Basic architecture of API system

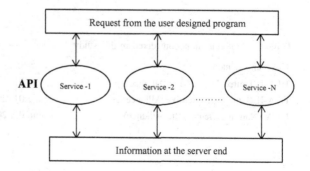

3   Research Problem Formulation and Test Parameters

The study is performed to recognize the importance of India in the international scenario with the help of media coverage of online newspaper (The New York Times). In order to do so, first of all the frequency of new article is searched related to concept "India" during the last three Indian governments and compared with each other. To be fair with the comparison, only the first two years are taken into consideration as the present Indian government completed its two-year tenure only.

Then after, the media coverage of India, Pakistan, China, and Sri Lanka is compared with each other during these three tenures of Indian government.

In addition to that, this study also provided the comparative popularity analysis of Indian politicians in the world scenario, and same is checked by fetching the news related to them in the different tenures of Indian governments because different tenures might result into different popularity levels.

Finally, the analysis of present government's schemes and its US visit is performed on the basis of their announcement time and visit duration, respectively. This analysis is done to see the effect (good or bad) of present government in the international domain.

3.1   Target Web site

A leading international news Web site of the USA, The New York Times—http://www.nytimes.com/.

3.2   Duration of Analysis

As this analysis is mainly focused on the examination of India's effect on international media, the time duration of analysis is taken from the Indian government's point of view and their tenure. Table 1 shows the three different time durations (different government tenures) considered for this study.

Table 1 Time durations considered in this study

Government name	Duration of analysis
UPA-I (United Progressive Alliance-1)	June 01, 2004–June 01, 2006
UPA-II (United Progressive Alliance-2)	June 01, 2009–June 01, 2011
NDA (National Democratic Alliance)	June 01, 2014–June 01, 2016

3.3 Countries Used for Comparison

In this study, three nearby countries of India are considered because they are all developing countries and situated in the same continent. The countries are China, Pakistan, and Sri Lanka.

3.4 Indian Politicians Used for Popularity Analysis

Some famous Indian politicians are considered in this study to see their global reach and effect on international news. The considered politicians are as follows:

1. Mr. Narendra Modi (Current prime minister (PM) of India and politically affiliated to Bharatiya Janata Party)
2. Dr. Manmohan Singh (Ex-Prime minister (PM) of India and politically affiliated to Indian National Congress)
3. Mrs. Sonia Gandhi (Vice-President and politically affiliated to Indian National Congress)
4. Mr. Arvind Kejriwal (Current Chief Minister of New Delhi and politically affiliated to Aam Aadmi Party)
5. Mr. Rahul Gandhi (Vice-President and politically affiliated to Indian National Congress)
6. Mr. Amit Shah (President and politically affiliated to Bharatiya Janata Party)

3.5 Indian Prime Minister's (Current) Visit of USA and Duration

Table 2 shows the different visit durations of Indian PM to USA

Table 2 Time durations of the US visits of Indian prime minister

Visit number	Government name (PM name)	Duration of visit
Visit—01	NDA (Mr. Modi)	September 26–30, 2014
Visit—02	NDA (Mr. Modi)	September 24–30, 2015
Visit—03	NDA (Mr. Modi)	September 31–01, 2016
Visit—04	NDA (Mr. Modi)	September 06–08, 2016

3.6 Indian Government Scheme's (Current Government) Used for Popularity Analysis

The following schemes (Swachh Bharat Abhiyan, Make in India, Digital India) of current Indian government are taken into account while checking India-related news popularity in international media. This analysis is done to see the Indian scheme's importance from the point of view of international media.

4 Analysis and Discussion of Obtained Results

Figure 2 shows the news frequency of concept (word) "India" during different Indian government's tenures with respect to different months. It can be clearly seen from Fig. 2 that India remain in more news during the NDA government than the earlier two governments (UPA-I and UPA-II) during its first two years. Though, in few months, UPA-II government also able to get more international media attention as compared to NDA.

Figure 3 provides the news frequencies of four developing counties of Asian continent (including India) with respect to three government tenures in India. It is quite clear from Fig. 3 that China is most dominating country in terms of international coverage. In addition to that, India-related news coverage was better in NDA government as compared to other last two tenures of UPA government.

Figures 4 and 5 are showing the popularity level of the news related to Dr. Manmohan Singh (Ex-PM of India) and Mr. Narendra Modi (current PM of India) in the international media, respectively. It is quite clear that the news frequency of current prime minister is quite high in his own tenure, but he was not so popular in rest two tenures of Ex-PM. In addition to that, the popularity of current PM is declined over the time in international media. Contrary to current PM, Ex-PM's popularity was very low irrespective of the tenure; that is, even when he was the prime minister of India, he got very less international media coverage.

Figure 6 is shows the popularity comparison of different Indian politicians in the international media. The PMs are popular in their own tenures, which is quite common but during UPA-II, president of congress (Mrs. Sonia Gandhi) got more attention by the international media than the Ex-PM himself, which is quite strange thing. UPA-II tenure also shows the rise of Mr. Arvind Kejriwal very strongly, which again fell down in NDA government. Mr. Modi is obviously more popular than earlier PM Dr. Singh. Mr. Rahul Gandhi also got popularity in UPA-II but it fell down drastically in NDA government. The popularity level of Mr. Amit Shah is almost similar in UPA-II and NDA. Though, he first gets international media attention during UPA-II tenure only.

Figure 7 shows the popularity of different Indian government (NDA government) campaigns, and it is quite visible the Make in INDIA gets the highest attention because it affects the global economy.

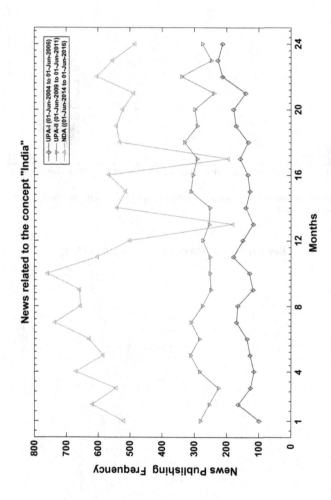

Fig. 2 The news frequency of concept (word) "India" during different Indian government's tenure

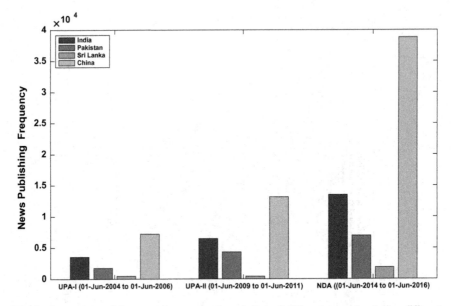

Fig. 3 Comparison of the news frequencies (popularity) of different countries during different Indian government's tenure

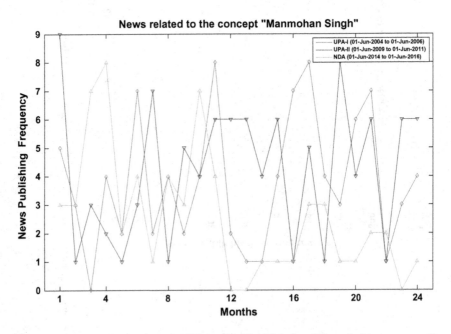

Fig. 4 The news frequency of concept "Manmohan Singh" during different Indian government's tenure

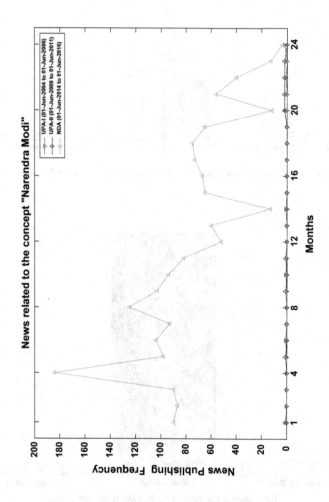

Fig. 5 The news frequency of concept "Narendra Modi" during different Indian government's tenure

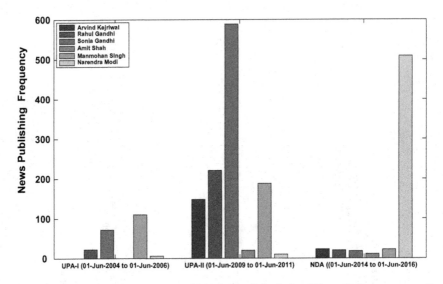

Fig. 6 Comparison of different Indian politician's popularity during different Indian government's tenure

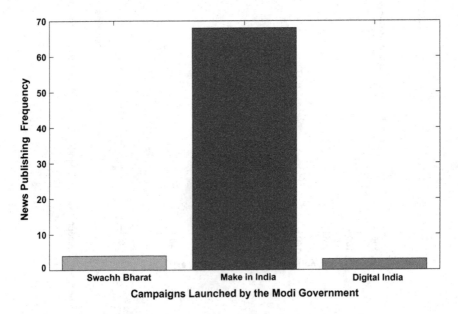

Fig. 7 Popularity analysis of different campaigns launched by NDA government

Figures 8 and 9 show the popularity level of India and Mr. Modi during his visits to the USA. Both the figures are showing the average popularity (per day coverage). Figure 8 shows an increasing interest of media, and then a decline as the

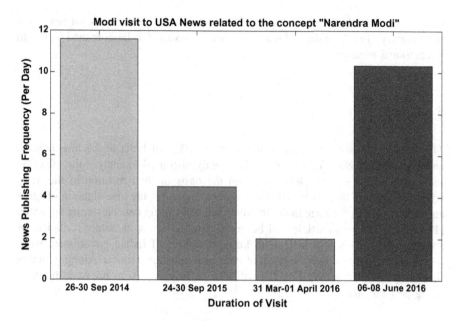

Fig. 8 Popularity analysis of Indian prime minister during US visit

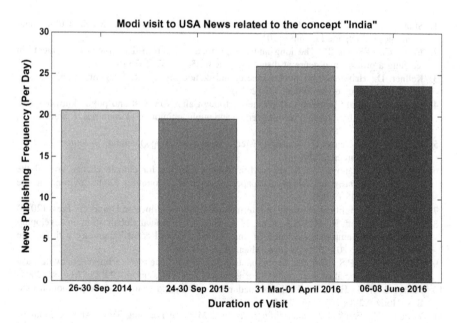

Fig. 9 Popularity analysis of India during US visits of Indian prime minister

same can be concluded using the Mr. Modi occurrence in international news. On the contrary, Fig. 9 shows almost constant behavior for India-related news in international media.

5 Conclusion

This study was focused on the analysis of the effect of India in the international news (using The New York Times). This study also tried to analyze the effect on Indian politicians on the global level on the basis of their mention in the news articles. In future, this study will further be extended for the investigation of India and Indian politician image in the international news using data clustering approach [18]; that is, the news article will be investigated to know its nature (criticism or appreciation), which will help us to know the effect of Indian government work (positive or negative) from the point of view of international media. Along with that soft computing techniques [19–22] will also be investigated to improve the clustering performance.

References

1. Straubhaar, J., LaRose, R., Davenport, L.: Media Now: Understanding Media, Culture, and Technology. Cengage Learning (2013)
2. Jones, C., O'Brien, T.: The long and bumpy road to multi-media: hi-tech experiments in teaching a professional genre at distance. System 25(2), 157–167 (1997)
3. Kellner, D.: Habermas, the public sphere, and democracy. In: Re-Imagining Public Space, pp. 19–43. Palgrave Macmillan, US (2014)
4. Aalberg, T., et al.: International TV news, foreign affairs interest and public knowledge: a comparative study of foreign news coverage and public opinion in 11 countries. J. Stud. 14(3), 387–406 (2013)
5. Riff, D., Lacy, S., Fico, F.: Analyzing Media Messages: Using Quantitative Content Analysis in Research. Routledge (2014)
6. Schmidt, A., Ivanova, A., Schäfer, M.S.: Media attention for climate change around the world: a comparative analysis of newspaper coverage in 27 countries. Glob. Environ. Change 23(5), 1233–1248 (2013)
7. Gilpin, R.: The political economy of international relations. Princeton University Press (2016)
8. Prasad, E., Rogoff, K., Wei, S.J., Kose, M.A.: Effects of financial globalization on developing countries: some empirical evidence. In: India's and China's Recent Experience with Reform and Growth, pp. 201–228. Palgrave Macmillan, UK (2005)
9. Leung, L., Lee, P.S.: Multiple determinants of life quality: The roles of Internet activities, use of new media, social support, and leisure activities. Telematics Inform. 22(3), 161–180 (2005)
10. Mangold, W.G., Faulds, D.J.: Social media: the new hybrid element of the promotion mix. Bus. Horiz. 52(4), 357–365 (2009)
11. Witten, I.H., Frank, E.: Data Mining: Practical Machine Learning Tools And Techniques. Morgan Kaufmann (2005)

12. Srivastava, J., Cooley, R., Deshpande, M., Tan, P.N.: Web usage mining: discovery and applications of usage patterns from web data. ACM SIGKDD Explor. Newsl. 1(2), 12–23 (2000)
13. Berkhin, P.: A survey of clustering data mining techniques. In: Grouping Multidimensional Data, pp. 25–71. Springer, Berlin (2006)
14. Hornick, M., Dadashev, D.: U.S. Patent No. 6,865,573. U.S. Patent and Trademark Office, Washington, DC (2005)
15. Stevens, W.R., Fenner, B., Rudoff, A.M.: UNIX Network Programming: The Sockets Networking API, vol. 1. Addison-Wesley Professional (2004)
16. http://www.computersciencezone.org/50-most-useful-apis-for-developers/
17. https://developer.nytimes.com/
18. Kant, S., Ansari, I.A.: An improved K means clustering with Atkinson index to classify liver patient dataset. Int. J. Syst. Assur. Eng. Manag. 1–7 (2015). 10.1007/s13198-015-0365-3
19. Jauhar, S.K., Pant, M.: Genetic algorithms in supply chain management: a critical analysis of the literature. Sādhanā 41(9), 993–1017 (2016)
20. Ansari, I.A., Pant, M., Ahn, C.W.: ABC optimized secured image watermarking scheme to find out the rightful ownership. Optik-Int. J. Light Electron Opt. 127(14), 5711–5721 (2016)
21. Zaheer, H., Pant, M., Monakhov, O., Monakhova, E.: A portfolio analysis of ten national banks through differential evolution. In: Proceedings of Fifth International Conference on Soft Computing for Problem Solving, pp. 851–862. Springer, Singapore (2016)
22. Jauhar, S.K., Pant, M., Nagar, A.K.: Sustainable educational supply chain performance measurement through DEA and differential evolution: a case on Indian HEI. J. Comput. Sci. (2016). doi:10.1016/j.jocs.2016.10.007

Comparative Analysis of Clustering Techniques for Customer Behaviour

Shalini and Deepika Singh

Abstract In real world, customer behaviour is changing and evolving over time. Business in each industry endeavours to expand client esteem and accomplish a high level state of consumer loyalty. Sells group concentrate on open doors for cross-sell and up-sell while client mind concentrates on certain key measurements, for example first call determination, snappy determination to client's issues, best service levels and quality scores. Customer clustering is utilized to know about behavioural patterns of customers so that industry or organization can make their marketing strategies according to the customers' preferences and retain them. This paper shows the performance review of clustering data mining techniques to know which technique is more suitable to identify customer profile and patterns for a retail store, to improve better customer satisfaction and retention.

Keywords Customer behaviour · Clustering · Data mining · Satisfaction

1 Introduction

The competitive business environment is consistently changing after some time. In this progressively evolving circumstance, customer behaviour is regularly complicating to anticipate and indeterminate. Customer analysis is crucial stage for organizations to make new battle for their current customers. On the off chance that an organization can comprehend customer elements and attempt endeavours to satisfy their needs and give friendly service, then the customer will be stronger to the undertaking. The motive of this study was to build up a strategy to recognize the characteristics of customers. Data mining strategies include the analysis of data put away in data warehouse. There are principally two data mining techniques proposed

Shalini (✉) · D. Singh
Teerthanker Mahaveer University, Moradabad, Uttar Pradesh, India
e-mail: shalini15.pundir@gmail.com

D. Singh
e-mail: deep.16feb84@gmail.com

© Springer Nature Singapore Pte Ltd. 2018
M. Pant et al. (eds.), *Soft Computing: Theories and Applications*,
Advances in Intelligent Systems and Computing 584,
https://doi.org/10.1007/978-981-10-5699-4_71

in the literature: classification and clustering. In this research paper, we have done a comparative study of clustering-based customer behaviour analysis strategies with the utilization of various data sets. Our primary objective is to demonstrate the comparison of various types of clustering algorithm of utilizing the WEKA tool.

Cluster analysis is a helpful technique for distinguishing homogeneous grouping of objects called groups. Observations in a particular cluster share numerous characteristics and, however, are exceptionally not at all like objects not having a place with that cluster. The goal of cluster analysis is to recognize groups of objects (in this case, customers) that are fundamentally the same as to their price consciousness and brand loyalty and relegate them into clusters. Subsequent to having settled on the clustering variables, we have to decide on the clustering technique to frame our groups of objects. This progression is urgent for the analysis, as various methods require different decisions before analysis.

In the literature, a number of clustering techniques have been proposed by different researchers; a few of them are: partitioning technique, hierarchical technique, density-based technique, grid-based technique, model-based technique.

2   State of Art of Clustering Techniques Used for CBA

In this section, the clustering techniques which are used for customer behaviour analysis (CBA) have been discussed.

(a) **K-Means Clustering**—An imperative group of clustering technique is dividing methods. The k-means method is the most imperative one for market research. This algorithm is not in view of distance measures, for example Euclidean distance or city–block distance, however, utilizes the inside group variety as a measure to frame homogeneous clusters. In particular, the method goes for dividing the information in a manner that the inside cluster variety is minimized.

The clustering procedure begins by arbitrarily allocating objects to various groups. The items are then progressively reassigned to different clusters to minimize the inside bunch variety, which is fundamentally the (squared) distance from every perception to the centre point of the related cluster. In the event that the reallocation of an object to another group diminishes the inside cluster variation, this object is reassigned to that cluster.

$$J = \sum_{j-1}^{k} \sum_{i-1}^{x} \left\| x_i^{(j)} - c_j \right\|^2 \tag{1}$$

where $\left\|x_i^{(j)} - c_j\right\|^2$ is a picked distance measure between a data point $x_i^{(j)}$ and the cluster centre c_j, and an indicator of the distance of the n data points from their individual cluster centres.

(b) **Hierarchical Clustering**—Hierarchical clustering techniques are portrayed by the tree-like structure built up over the span of the analysis. Most hierarchical strategies fall into a class called agglomerative clustering. For customer behaviour analysis, only agglomerative clustering is easy in use. In this category, clusters are sequentially framed from objects. At first, this kind of technique begins with every object presenting an individual cluster. These clusters are then consecutively combined as indicated by their closeness. In the first place, the two most comparable clusters are merged to shape a new cluster at the base of the chain of hierarchy. In the next step, another combination of clusters is combined and connected to a higher level of the hierarchy, etc. This permits a hierarchy of clusters to be built up from the bottom up.

(c) **Density-based Clustering**—Density-based clusters are characterized as clusters which are separated from different clusters by varying densities that means a group which has dense region of objects might be encompassed by low-density regions. DBSCAN clusters are recognized by watching at the density of points. Regions with a high density of points delineate the presence of cluster, while regions with a low density of points demonstrate clusters of noise or clusters of anomalies. This algorithm is especially suited to manage large data sets, with noise, and can distinguish clusters with various sizes and shapes.

(d) **EM (expectation maximization)**—the expectation–maximization (EM) calculation is created for incomplete data (Dempster and Laird 1977). It can be utilized to run maximum likelihood parameter expectation for mix models. It applies the rule of maximum likelihood to locate the model parameters. The E–M calculation rehashes the expectation (E) and maximization (M) steps iteratively after haphazardly initializing the mix demonstrate parameters. The E and M steps are iterated until a proposed merging is acquired.

3   Background Work

In the exploration of client behaviour analysis utilizing data mining procedures, we concentrated on different research papers and articles to assess the execution of various clustering algorithms. By using data mining method, we can analyse data of customer and can create group of proper data. The analysis of the person who needs to buy certain thing from shopping place, at the time of purchasing thing, why he needs to buy it? This is imperative idea. To consider his mental thought and changing over this into statical configuration and see that is there any specific format by which we can distinguish his obtaining pattern.

Hussain [1] has proposed the approach which is utilized to produce a high number of object classes. This kind of querying the object searches all sorts of object and data connected with it. It gives the yield in view of the rerank of picture and its object class. To start with it, download all the important pictures and the website pages. At that point on separating highlights, it researches about the downloaded page. Furthermore, put it in the database and after that, positioning is done in view of content encompassing and metadata highlights.

Osama Abu [2] has talked about k-means for clustering of object that has a place among a k-cluster. It is used on both small and huge data sets. It has less poor clustering for small data sets. Utilization was done by using LNK net package.

Tian et al. [3] proposed a methodical strategy for finding the initial centroids. This technique's result gives better impacts and less iterative time than the current k-means algorithm. This approach adds almost no work to the framework. This strategy will diminish the iterative time of the k-means algorithm, making the cluster analysis more proficient. The outcome for the small data set is not exceptionally remarkable when the refinement algorithm works over a small subset of an entirely substantial data set.

Tikmani and Tiwari [4] have examined the paper that considers a data set of 80 clients which won't not be the situation, in actuality, situations as the quantity of clients may be enormous and k-means won't not give the proposed result when the data set is extensive. As clarified, k-means algorithm would not end up being as powerful while handling huge data as it is while handling of small data sets. This is because of the reason that large size of data sets affects the clustering quality and the clustering time effectiveness.

Kim and Hong [5] proposed a methodology that engages online stores to offer customized marketing by segmenting their customer in view of customers' psycho-realistic data. We have developed an exploration model to disclose customers' desire to purchase in online stores and attempted the model with SEM. In perspective of research model, we clustered by organizing the k-means algorithm and SOM for more complex division of customers. At since quite a while ago last, applied the k-nearest neighbours technique for ordering clients that are excluded in the data set for showing and clustering. The predictive accuracy of k-nearest neighbours in the cases was sufficient for applying our philosophy in promoting.

4   Performance Analysis

Data sets: We used three data sets for performance analysis of clustering algorithm on WEKA. All data sets are in arff format. Super market data set has 4627 instances and 217 attributes, yeast data set has 445 instances and 9 attributes, and car has 518 instances and 7 attributes. Car and German credit are inbuilt in WEKA (3-6-6) (Table 1).

Table 1 Details of data sets

Name of data set	Type of file	No. of attribute	No. of instances	Attribute characteristics	Data set characteristics	Missing value
Super market	ARFF (attribute relation file format)	217	4627	Categorical integer	Multivariate	No
Yeast	ARFF (attribute relation file format)	9	445	Categorical integer	Multivariate	No
Car	ARFF (attribute relation file format)	7	518	Categorical integer	Multivariate	No

Evaluation of clustering on different data sets.

4.1 Clustering on Super Market Data Set

See Table 2.

4.2 Clustering on Yeast Data set

See Table 3.

4.3 Clustering on Car Data Set

See Table 4.

4.4 Compare the Time Taken to Built the Model of Clustering Algorithm in Different Data sets

4.4.1 Data set Super Market

See Fig. 1.

Table 2 Evaluation of clustering on super market data set with full training data test mode

Clustering algorithm	No. of instances	Test mode	No. of clusters generated	Clustered instances	Time taken to build the model (s)	Unclustered	Unlikelihood
EM	4627	Full training data	1	0 (100%)	201.6	0	−0.65 505
K-means	4627	Full training data	2	0 (36%) 1 (64%)	1.28	0	–
Hierarchical	4627	Full training data	2	0 (98%) 1 (2%)	261	0	–
Density-based clustering	4627	Full training data	2	0 (36%) 1 (64%)	3.55	0	−0.65 505

Table 3 Evaluation of clustering on yeast data set with full training data test mode

Clustering algorithm	No. of instances	Test mode	No. of clusters generated	Clustered instances	Time taken to build the model (s)	Unclustered	Unlikelihood
EM	445	Full training data	2	0 (28%) 1 (72%)	4.45	0	8.30211
K-means	445	Full training data	2	0 (34%) 1 (66%)	0.03	0	–
Hierarchical	445	Full training data	2	0 (100%) 1 (0%)	0.77	0	–
Density-based clustering	445	Full training data	2	0 (33%) 1 (67%)	0.03	0	6.75402

Table 4 Evaluation of clustering on car data set with full training data test mode

Clustering algorithm	No. of instances	Test mode	No. of clusters generated	Clustered instances	Time taken to build the model (s)	Unclustered	Unlikelihood
EM	518	Full training data	4	0 (16%), 1 (33%), 2 29%), 3 (22%)	15.6	0	7.7925
K-means	518	Full training data	2	0 (56%), 1 (44%)	0.02	0	–
Hierarchical	518	Full training data	2	0 (100%), 1 (0%)	0.34	0	–
Density-based clustering	518	Full training data	2	0 (57%), 1 (43%)	0.02	0	–8.27594

Fig. 1 Bar chart for super market data set to view the performance of different algorithms

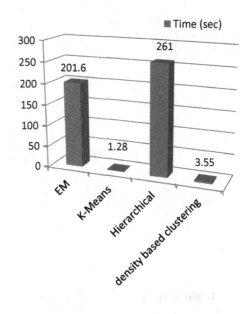

Fig. 2 Bar chart for yeast data set to view the performance of different algorithms

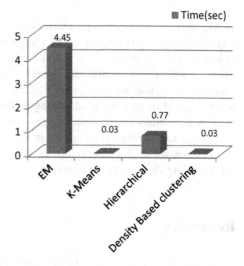

4.4.2 Data set Yeast

See Fig. 2.

4.4.3 Data set Car

See Fig. 3.

Fig. 3 Bar chart for car data
set to view the performance of
different algorithms

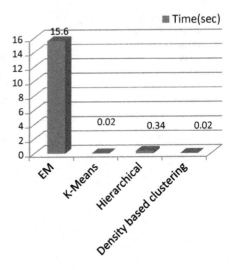

5   Conclusion

In this work, we study different clustering strategies on the market sells information. Clustering is the procedure for collecting the data into classes or clusters, so that objects inside a group have a high similarity in contrast to each other and extremely not at all like objects in different clusters. This research paper presents a report on customer behaviour analysis using EM algorithm, k-means algorithm, hierarchical algorithm and density-based clustering algorithm and has presented a comparative study which shows k-means outperforms other algorithm according to time which is taken to create clusters and performance.

In future, we shall try to explore the aspects of fuzzy-based clustering techniques for analysing the customer behaviour.

References

1. Hussain, S.T., Kanya, B.N.: Extracting images from the web using data mining technique. Int. J. Adv. Technol. Eng. Res. (2012)
2. Osama Abu, A.: Comparison between data clustering algorithms. Int. Arab J. Inf. Technol. **5** (3), 320–325 (2008)
3. Farajian, M.A., Mohammadi, S.: Mining the banking customer behavior using clustering and association rules methods. Int. J. Ind. Eng. **21**(4) (2010)
4. Tikmani, J., Tiwari, S.: Int. J. Innov. Res. Comput. Commun. Eng. 3(11) 2015 (An ISO 3297: 2007 Certified Organization)
5. Kim, E., Hong, T.: Segmenting customers in online stores from factors that affect the customer's intention to purchase. In: 2010 International Conference on Information Society (i-Society) (2010)
6. Jain, M., Verma, C.: Adapting k means for clustering in big data. IJCA 19–24 (2014)

7. Tiwari, M., et al.: Performance analysis of Data Mining algorithms in Weka. IOSR J. Comput. Eng. (IOSRJCE) **6**(3) 32–41 (2012). www.iosrjournals.org. ISSN 2278-0661, ISBN 2278-8727
8. Shrivastava, V.: Int. J. Comput. Inf. Technol. Bioinform. (IJCITB) **1**(3). ISSN 2278-7593
9. Singh, G.: Int. J. Adv. Res. Comput. Sci. Softw. Eng. Res. Pap. **4**(7) (2014). Available online at: www.ijarcsse.com. ISSN 2277
10. Zhang, H., Chai, J., Wang, Y., An, M., Li, B., Shen, Q.: Application of clustering algorithm on TV programmes preference grouping of subscribers. In: 2015 IEEE International Conference on Computer and Communications (ICCC) (2015)
11. Noughabi, E.A.Z., Albadvi, A., Far, B.H.: How can we explore patterns of customer segments' structural changes? In: 2015 IEEE International Conference on A Sequential Rule Mining Approach Information Reuse and Integration (IRI) (2015)
12. Victor, H.A.: Am. J. Comput. Commun. Control **1**(4), 66–74 (2014). ISSN 2375-3943
13. Sharma, N.: Comparison the various clustering algorithms of weka tools. Int. J. Emerg. Technol. Adv. Eng. **2**(5) (2012). Website www.ijetae.com ISSN 2250-2459
14. Rajagopal, S.: Customer data clustering using data mining technique. Int. J. Database Manag. IJDMS **3**(4) (2011) (Enterprise DW/BI Consultant Tata Consultancy Services, Newark, DE, USA)

Real-time Sentiment Analysis of Big Data Applications Using Twitter Data with Hadoop Framework

Divya Sehgal and Ambuj Kumar Agarwal

Abstract Twitter and other social networking sites generate huge amount of data on a daily basis. Frequent interactions can be witnessed with data generated through Linked-in, facebook, and gmail. Twitter being the largest social networking site generates data of very large amount because of millions of tweets and followers which are increasing per day. This imposes a big problem of processing and analyzing the data. As it is a case of handling big data, the technology of Hadoop comes into picture. Using Hadoop eases the process of analyzing the data. The work of analyzing twitter data is undertaken in the paper.

Keywords Big data · Hadoop · HDFS · Map reduce

1 Introduction

Today, the data generated through social networking sites is increasing day by day. Twitter one of the most popular social networking platforms deals with both structured and unstructured format of data. However, Twitter data is mostly in unstructured format like followers, tweets, likes, and expressions, etc. It is very difficult to process the data easily. All kinds of industries and companies are using this type of data for the future development and advertising work.

D. Sehgal (✉) · A.K. Agarwal
CCSIT, Teerthanker Mahaveer University, Moradabad, India
e-mail: sonasahgal199@gmail.com

A.K. Agarwal
e-mail: ambuj4u@gmail.com

© Springer Nature Singapore Pte Ltd. 2018 765
M. Pant et al. (eds.), *Soft Computing: Theories and Applications*,
Advances in Intelligent Systems and Computing 584,
https://doi.org/10.1007/978-981-10-5699-4_72

2 Big Data

Big data comprises of both structured and unstructured data, which includes video, audio, data generated through emails, etc. Social sites generate vast amount of data through daily activities. It should be understood that the historical data maintained in industries and business sector help them to thrive in this competitive world. This can be termed as big data. It is very difficult to analyze and process the Big data. However, with the help of Hadoop technology, the complexity in analyzing and processing of data can be reduced substantially. Big data is a heterogeneous collection of complex data sets and produced by varied sources like television, mobile, etc.

There are three characteristics of big data—

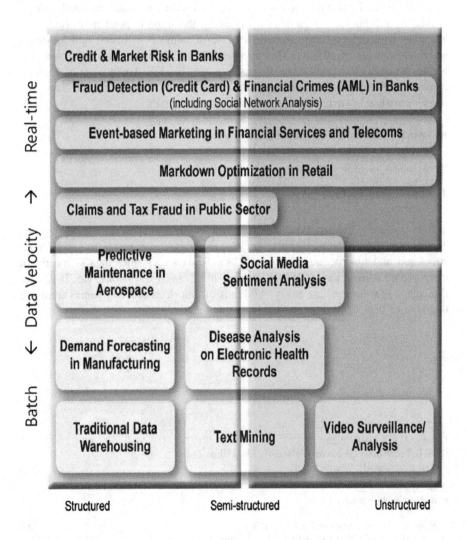

1. volume
2. velocity
3. variety

Structure of the big data—

1. **Volume**—Volume mainly defined in terms of amount of data. Volume of data is growing exponentially. Management of high volume of data in reference with processing and storing is always complex.
2. **Velocity**—Social sites are constantly generating complex data in unstructured and semi-structured form. Increasing the collection of big data with the help of mobile, televisions, and more advance technologies as Internet mainly influences high velocity.
3. **Variety**—Variety of big data exists in structured and unstructured format. Structured data are always of fixed format and there is no possibility of changes of this data like tabular data, ERP, etc.

3 Related Work

Sentiment analysis is very popular technology in today's world. Vast amount of work has been done in this field. Mostly, work in this area relates to storage of the data.

(i) Semantic analysis assumes much importance: also it deals with document and word type of the data and is mostly dependent on NLP processing techniques.
(ii) It deals with point-wise data and information and is mathematical in nature.

4 Hadoop

Hadoop is an open source framework which is freely available for every user. It is based on the Java programming framework. Hadoop is a project of apache. Hadoop is a framework which is available to support for the reliable and scalable distributed computing system. Hadoop framework was designed for solving the problems like processing the data and analysis the big data (Fig. 1).

1. Execution Engine (Map Reduce)
2. Hadoop distributed file system (HDFS) (Fig. 2)

Fig. 1 Data replication [1]

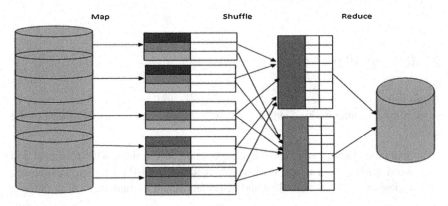

Fig. 2 Map reduce [2]

5 Hadoop Distributed File System (HDFS)

It is mainly handling large amount of big data. Big data stores the blocks in HDFS. It is client server architecture. HDFS comprises of name node and many data nodes. It stores the data for the named nodes that is known as name node. The Name node searches to track the data node positions. Also, it is responsible to support the file system operations. If the name node fails in running the operation, then Hadoop does not support and recovers any data node state. Data replication is done for achieving targets of fault tolerance. In HDFS, the large sized data cluster is stored as a parallel sequence of blocks.

Using Our Approach

In this paper, we focus on mainly speed performing analysis and also accuracy. What removes the various problems in big data technology? Like part of speech and tagging using opennlp, it is easy to solve the problem. It is mainly known as tagging and used for following purposes.

(i) Firstly: Usage of words like a, an can be stopped. It is not useful for the real-time sentiment analysis.
(ii) Second approach is unstructured to structured: The twitter messages and the comments are mostly unstructured i.e. comment of "on bajrange bhai jaan" "favorite" is written "favorable," "God" is written "good," "aswm" is written for "awesome," "bd" is written for "bad."
(iii) Thirdly, emoticons: It is most expressive approach available on ideas and opinion. It is symbolic representation converted to words at this stage.

1. Data and Real-time data features

In this paper, the real time is very important. It is obtained from the data streaming API's available from twitter. It uses keywords like we are using in the movie bajrange bhai jaan. Objects that we use to perform the sentiment analysis are submitted to the twitter APIs. This provides Twitter, the tweets that are related to only that object.

Twitter data mostly used unstructured data. A tweet mostly consists of maximum 140 characters and likes. The messages' comments consist of a user name and timestamp. Mostly, timestamp is useful for the future development in our project. It is also helpful for different geographical regions (Fig. 3).

2. Data defined part of speech

The file contains the obtained tweets.

3. Data in Root form

It is widely used program to increase the overall efficiency and lowers the time access of the system. Root forms the word on twitter for the tweet are changed to their root form and split all that word which is unwanted and extra storage of the derived word's sentiment analysis.

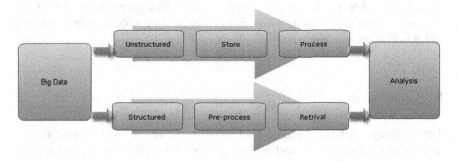

Fig. 3 Data processing [1]

4. Sentiment Data Directory

Now, the Real Time Data Directory is making and using standard directory for the big data sentiment, word net, and uses all condition for this word i.e. "good," "bad." The overall data is used to store the sentiment data in standard directories. It is local to the program, it is only primary memory. We can utilize our time in searching the main word in primary memory.

5. Map Reduce Algorithm

This is mapped reduce algorithm used for the tweets of the data. The sentiment values can be obtained from standard algorithm which we are using.

6 Final Data Accuracy

The complete overall accuracy in some twitters data i.e. accessing the data from sentiwordnet, opennlp, and wordnet is shown. The data are bajrange bhai jaan comments like negative word and positive word and neutral word. The data are compared by the help of movie bajrange bhai jaan, tweets like good, favorite, and negative words. The final special data available on web like following http://www. cs.tau.ac.il/~kfirbar/mlproject/twitter.data, now the checked data are as follows—

Sentiment	Count	Correct	%	Tolerance
Positive	739	520	72.22	−0.01
Negative	637	399	61.67	+0.05
Neutral	86	43	73.42	±0.003

The total accuracy of this project is 72.22. It is the mean of the total accuracy.

7 Total Time Efficiency

In our project, a necessary aspect is efficiency, that is why our project working well. Also reduces the time from hard disk that is only possible with the help of Hadoop and it is also using a lower time.

8 Conclusion

The Sentiment analysis is widely used at this time. The research is used for the analysis of the data. This project is also expanded to the social media platform and movie reviews like blogs and comment and likes per day. The accuracy is totally

value following. Also the use of hashtags and emoticons is very useful and necessary for social media data for this project. In our project, use of emoticons and hash tags like comment and tweets analyzed per day data. The total accuracy found is 72.22%.

References

1. Saleem, A., Agarwal, A.K.: Analysis and design of secure web services. In: Proceedings of Fifth International Conference on Soft Computing for Problem Solving, Springer Singapore (2016)
2. Shukla, S., Lakhmani, A., Agarwal, A.K.: Approaches of artificial intelligence in biomedical image processing: a leading tool between computer vision & biological vision. In: 2016 International Conference on Advances in Computing, Communication, & Automation (ICACCA) (Spring), Dehradun, 2016, pp. 1–6. doi:10.1109/ICACCA.2016.7578900
3. Beaver, D., Kumar, S., Li, H.C., Sobel, J., Vajgel, P.: Finding a needle in haystack: Facebook's photo storage. In: Proceedings of the nineth USENIX Conference on Operating Systems Design and Implementation, Berkeley, CA, USA, USENIX Association, pp. 1–8 (2010)
4. Rupanagunta, K., Zakkam, D., Rao, H.: How to mine unstructured data. Artic. Inf. Manag. (2012)
5. Marche, S.: Is Facebook making us lonely. Atlantic **309**(4), 60–69 (2012)
6. IBM What Is Big Data: Bring Big Data to the Enterprise, IBM (2012). Available: http://www.01.ibm.com/software/data/bigdata/
7. Duggal, R., Shukla, B., Khatri, S.K.: Big data analytics in Indian healthcare system—opportunities and challenges. In: Research Paper Accepted at National Conference on Computing, Communication and Information Processing (NCCCIP-2015), May 2015, pp. 92–104 (2015)
8. Chih-Wei, L., Chih-Ming, H., Chih-Hung, C., Chao-Tung, Y.: An improvement to data service in cloud computing with content sensitive transaction analysis and adaptation. In: Computer Software and Applications Conference Workshops (COMPSACW), 2013 IEEE 37th Annual, 2013, pp. 463–468
9. Purcell, B.: The emergence of "big data" technology and analytics. J. Technol. Res. (2013). Garlasu, D., Sandulescu, V., Halcu, I., Neculoiu, G.: A big data implementation based on grid computing. Grid Comput. (2013)
10. Agarwal, A.: Implementation of cylomatrix complexity matrix. J. Nat. Inspir. Comput. **1** (2013)
11. Agrawal, T., Agarwal, A.K., Singh, S.K.: Study of cloud computing and its security approaches
12. Saxena, A.K., Agarwal, A.K., Ather, D.: How to secure design using threat modeling
13. Agarwal, A.K., Katiyar, V.: A study of software matrix systems: a comparative study of existing software matrix systems
14. Fatima, S., Agarwal, A., Gupta, P.: Different approaches to convert speech into sign language. In: 2016 3rd International Conference on Computing for Sustainable Global Development (INDIACom), New Delhi, India, pp. 180–183 (2016)
15. Shukla, S., Agarwal, A.K., Lakhmani, A.: MICROCHIPS: a leading innovation in medicine. In: 2016 3rd International Conference on Computing for Sustainable Global Development (INDIACom), New Delhi, India, pp. 205–210 (2016)
16. Grosso, P., de Laat, C., Membrey, P.: Addressing big data issues in scientific data infrastructure, 20–24 May 2013
17. Lin, J.: MapReduce is good enough? The control project. IEEE Comput. **32** (2013)

18. Agarwal, S., Mozafari, B., Panda, A., Milner, H., Madden, S., Stoica, I.: BlinkDB: Queries with bounded errors and bounded response times on very large data (2013)
19. Chih-Wei, L., Chih-Ming, H., Chih-Hung, C., Chao-Tung, Y.: An improvement to data service in cloud computing with content sensitive transaction analysis and adaptation. In: Computer Software and Applications Conference Workshops (COMPSACW), 2013 IEEE 37th Annual, pp. 463–4681 (2013)
20. Sagiroglu, S., Sinance, D.: Big data: a review, 20–24 May 2013
21. Zhao, Y., Wu, J.: Dacha: a data aware caching for big-data applications using the Map Reduce framework. In: INFOCOM, 2013 Proceedings IEEE, Turin (2013)
22. Garlasu, D., Sandulescu, V., Halcu, I., Neculoiu, G.: A big data implementation, 17–19 Jan 2013
23. Zhang, X., Xu, F.: Survey of research on big data storage, 2–4 Sept 2013
24. Mukherjee, A., Datta, J., Jorapur, R., Singhvi, R., Haloi, S., Akram, W.: Shared disk big data analytics with Apache Hadoop, 18–22 Dec 2012
25. Bifet, A.: Mining big data in real time. Informatica 37, 15–20 (2013)

Addressing Security Concerns for Infrastructure of Cloud Computing

Shweta Gaur Sharma and Lakshmi Ahuja

Abstract Although cloud computing has emerged as a dominant paradigm in the era of Internet-based computing, it is still vulnerable to security loopholes and attacks. Cloud computing is a set of data and services available remotely via Internet. Data centers and physical existence of the server play a vital role in providing the services to the end user. The cloud service providers may be a third-party liaison with the data center managing companies, but they need to follow the protocol of connectivity. This paper focuses on security measures for infrastructure of cloud computing architecture. We have focused on the security concerns faced and the solution currently implemented. The process helped to analyze the gap in the security systems and understand the importance of infrastructural resources.

Keywords Data centers · CSP · Infrastructure · IaaS · Security measure

1 Introduction

Cloud computing in today's world has become indispensable yet faces major security challenges. Cloud computing is not just confined to delivering services but has expanded its horizon into all over Internet and the hardware and system software as well. The hardware and software come together in form of data center and that is termed as cloud. Now, cloud computing is simply making data, platform and services available to the client by the means of varied interconnectivity modes. We can further categorize it as "public cloud" which is based on pay as you go model and "private cloud" refer to internal data center of an organization. [1] A "hybrid

S.G. Sharma (✉) · L. Ahuja
Amity University, Noida, India
e-mail: Myself.sh@gmail.com

L. Ahuja
e-mail: lahuja@amity.edu

© Springer Nature Singapore Pte Ltd. 2018
M. Pant et al. (eds.), *Soft Computing: Theories and Applications*,
Advances in Intelligent Systems and Computing 584,
https://doi.org/10.1007/978-981-10-5699-4_73

Fig. 1 Service stack of cloud computing

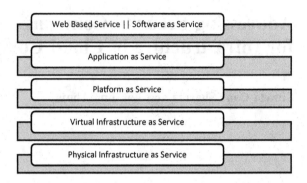

Web Based Service || Software as Service

Application as Service

Platform as Service

Virtual Infrastructure as Service

Physical Infrastructure as Service

cloud" combines multiple private and public clouds. In "community cloud," cloud infrastructure is shared by specific communities with common concerns.

Cloud computing as a collection of services [2] which can be represented as a layered cloud computing architecture is represented in Fig. 1.

Cloud computing provides opportunity to share huge amount of distributed resources to different geographical locations. It makes it difficult to implement traditional security measures on the cloud model. It includes members from different environments and varied local systems. And hence it makes the system more vulnerable. Cloud computing security has evolved as a major concern for the users as well as vendors. Moving to private cloud infrastructure is also not the way out as interconnectivity issue, data tampering, malicious insiders, and many such security concerns are grooming. In this paper, we have reviewed the major security concerns [2] for operating in the infrastructure of cloud environment across all the service stacks of cloud. The prime objective is to illustrate the concerns raised, researched, and discussed, and the solutions are proposed. For gap analysis, further these issues are addressed [3] in form of solutions available.

2 Addressing the Security Concerns for IaaS

Security in a cloud environment requires a systematic point of view, out of which security will be formulated by extenuating protection to a trusted third party. The prime threats deal with data integrity, confidentiality, authenticity, and availability [3] (Fig. 2).

Cloud security has emerged as the prime concern over decades, and infrastructure security needs to be modeled; the prime security concerns for infrastructure as a service can be listed as under:

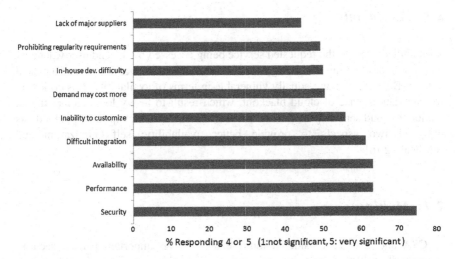

Fig. 2 Challenges in considering cloud computing (*Source* New IDC it cloud services survey: top benefits and challenges, 2009 [4])

2.1 Privacy Issues

Privacy in cloud differs according to the cloud utility model. Some cloud security models hold a high security threat in comparison to others like if you are gathering, transferring, or processing someone's personal information. And if the process is designed to fetch and process the personal information dynamically, then the risk is even higher. Most of the processing on public clouds has low risk in comparison, but on the contrary if the public cloud service provider holds and manages your information, then it rises the risk factor. Legal laws have come into action for controlling data leakage or breach of privacy.

2.2 Assess Control

Assess is another major factor, as geographical location of data center can be across any data center where government interventions also take place. Every location has their own norms on data supervision which leads to threats like data tampering. Legal rights of the area lead to processing and service unavailability. Moreover, the risk of data theft by internal employees or by data thieves from the physical location is another threat [5].

2.3   Availability

Availability refers to the requested service being accessed by the end user whenever and wherever the request is being raised. It has twofold impact in terms of data and service. Both of them are equally vulnerable in terms of availability. It is critical to recover data in case of cloud blackout, which result to heavy losses for business. Although cloud service providers do provide unified data backup facility, cloud has designed two aspects to provide better availability: self-optimization and self-healing [6].

2.4   Multi-tenancy

In CSA and ENSIA considered multi-tenancy as an important part of security measure. It enables to create an environment where single instance of software runs on multiple client organization simultaneously. It is a substitute for independent job scheduling as well as resource management. In contrast, the same may also be achieved by virtualization which introduces new security vulnerabilities. Therefore, virtualization requires special connect on security measures.

2.5   Trust Issues

Trust is established between users and cloud service providers. Trust management system manages the level of trust in compliance, governance, risk management, availability, integrity, confidentiality, and privacy. [7] Lack of trust leads in migration of the end user across the platforms which highly impacts business processes.

2.6   Data Security

Varied encryption techniques are used for securing sensitive information. Inverse Caesar Cipher; Playfair Cipher; fully homomorphic encryption; block based symmetric cryptography; classical encryption, DES algorithm, and many more [8].

2.7 Others

The other threats also include session hijacking; traffic flow analysis; exposure in network; defacement; connection flooding; DDOS; impersonation; disrupting communication, and many more.

3 Current Solutions

3.1 Service-Level Agreements

In cloud computing environment, the SLA is the formal contract between the user and the vendor. Since there is minimal physical interaction between the two, this commitment is made in form of the contract to ensure security measures are met. It is treated as a legal document between the two. SLA is the future perspective of trust between the cloud service provider and the user.

3.2 Vendor Viability

If your vendor goes out of business, it can be tremendously affect your business performance. Checking out their financials references and bibliographies will help you sense your business longevity.

3.3 Client–Server Authentication

A certification authority is involved in cloud environment who certifies entities involved in the transactions or processes. The PKI certification authority is responsible for such certifications. We have SAML [3], an XML-based standard for exchanging authentication and authorization of data between security domains. The Shibboleth and SAML design process have been coupled to ensure Shibboleth is in standards based. Shibboleth's added value lies in support for privacy, and business process reengineering via user attributes, extensive policy controls, and large-scale support systems.

3.4   Network Virtualization

IBM originally pioneered the concept of virtualization in 1960s with M44/44X systems. [9] The Amazon Elastic Compute cloud (EC2) is the most popular virtual network in current scenario. Nimbus and Eucalyptus are popular private IaaS platforms [10]. Other virtualization technologies include Microsoft Azure Platform, Open Stack, Open VZ, Oracle VM, VMware, and many more.

3.5   Inter-cloud Security (PKI)

With the increased number of Internet users, the data center network topology becomes more complex as huge number of tenants save their data at the common server, and the same tenant may also store their data at differently located geographical locations or servers. It results into multiple backup series, as a result the network boundaries tend to be blurred. The network security in cloud is currently governed by dynamic and virtual boundaries. The complete system of inter-cloud security is ensured by varied algorithms and processes designed [11].

4   Conclusion

In this paper, we have summarized the vulnerability of cloud computing and the emerging role of security measures in the area of infrastructure of cloud computing. We have discussed the prime issues faced in establishing infrastructure of cloud computing structure. The various researches done in this area are summed along with the solution proposed. In the gap analysis, it was found that virtualization of cloud systems for higher availability is growing but is the most vulnerable aspect. Moreover, geographical location of data center also contributes to the availability and trust worthiness of the cloud service provider.

References

1. Armbrust, M., et al.,: A view of cloud computing. Communications of the ACM, vol. 53, no. 4, pp. 52–55. Jan. 2010 [Online]. Available: http://dl.acm.org/citation.cfm?id=1721672. Accessed Nov. 18 2016
2. Shen, Z., Li, L., Yan, F., Wu, J.: Cloud computing system based on trusted computing platform. In: 2010 International Conference on Intelligent Computation Technology and Automation, China: IEEE Computer Society, 2010, pp. 942–945 [Online]. Available: http://acpce.yolasite.com/resources/05522632.pdf. Accessed: Sep. 18, 2016
3. Zissis, D., Lekkas, D.: Addressing cloud computing security issues. Future Gen. Comput. Syst. **28**(3), 583–592 (2012)

4. Gens, F.: New IDC it cloud services survey: top benefits and challenges (2009). http://blogs. idc.com/ie/?p=730
5. Zissis, D., Lekkas, D.: Addressing cloud computing security issues. Future Gen. Comput. Syst. **28**(3), 583–592 (2012)
6. Mckinley, P., Samimi, F., Shapiro, J., Tang, C.: Service clouds: a distributed infrastructure for constructing autonomic communication services. In: 2006 2nd IEEE International Symposium on Dependable, Autonomic and Secure Computing (2006)
7. Prasann, P., Vikas, K.: Trust management issues in cloud computing environment, in leveraging information technology for competitive advantage. Enriched Publication **2013**, 71–79 (2013)
8. Hemalatha, N., Jenis, A., Donald, A.C., Arockiam, L.: A comparative analysis of encryption techniques and data security issues in cloud computing. Int. J. Comput. Appl. IJCA **96**(16), 1–6 (2014)
9. Creasy, R.: The origin of VM/370 time sharing system. IBM J. Res. Dev. **25**(5), 483–490 (1981)
10. Younge, J., Henschel, R., Brown, J.T., Laszewski, G.V., Qiu, J., Fox, G.C.: Analysis of Virtualization Technologies for High Performance Computing Environments. In: 2011 IEEE 4th International Conference on Cloud Computing, Oct. 2011
11. Chen, Z., Dong, W., Li, H., Zhang, P., Chen, X., Cao, J.: Collaborative network security in multi-tenant data center for cloud computing. Tsinghua Sci. Technol. **19**(1), 82–94 (2014)

Author Index

A

Abdul, Azeem Mohammed, 147, 157
Abraham, Ajith, 267
Agarwal, Ambuj Kumar, 765
Agrawal, Lokesh, 343
Ahuja, Lakshmi, 773
Ali, Imran, 405
Amin, Sobia, 309
Ansari, Irshad Ahmad, 739
Arif-Ul-Islam, 67
Arora, Shaina, 51
Audu, Eliazar Elisha, 233

B

Bala, Manju, 1
Bandyopadhyay, A., 201, 213, 323, 343
Banerjee, Arpita, 651, 659
Banerjee, C., 651, 659
Banwari, Anamika, 585
Bateja, Ritika, 255
Bhardwaj, Tushar, 485
Bhattacharjee, Kamanasish, 617, 625
Bhatt, Anurag, 289
Bhatt, Ashutosh Kumar, 255, 289
Bhowmik, Chiranjib, 31
Bhowmik, Sumit, 31
Bijawat, Prashant, 373
Bilal, 727
Bonello, James, 233
Buhlan, Rakesh, 419

C

Chaturvedi, Rekha, 697
Chaudhari, Sharad S., 525
Chauhan, Ritu, 707
Chaurasia, Amit, 687
Cherukuvada, Srikanth, 147, 157
Chhajed, Rutuja, 343
Chhipa, Indu, 473

Chopra, Devansh, 577

D

Dave, Devanshi D., 181
Dey, Vidyut, 79
Doti, R., 201, 213, 323, 333
Dubey, Sanjay Kumar, 255, 289, 595

F

Faubert, J., 201, 213, 323, 333
Fujita, Daisuke, 343

G

Gangwar, Sachin, 31
Garg, Anchal, 577
Garg, Gaurav, 233
Garg, Lalit, 233
Garg, Shivani, 133
Gaurav, Vivek Kumar, 419
Ghosh, Batu, 343
Ghosh, S., 201, 213, 323, 343
Gotra, Shailza, 585
Goyal, Shivani, 715
Gupta, Preeti, 279
Gupta, Raj Kumar, 687
Gupta, Richa, 309
Gupta, Surbhi, 537
Gupta, Vikas, 433

H

Hashim, Zain, 585
Hussain, Athar, 495

I

Ishan, Lav, 453

J

Jangade, Rajesh, 707
Jauhar, Sunil Kumar, 405

© Springer Nature Singapore Pte Ltd. 2018
M. Pant et al. (eds.), *Soft Computing: Theories and Applications*,
Advances in Intelligent Systems and Computing 584,
https://doi.org/10.1007/978-981-10-5699-4

Jha, Brajesh Kumar, 181
Jha, P.C., 505
Johari, Nancy, 11

K
Kadarla, Kavitha, 567
Kamble, Prashant D., 525
Kaul, Arshia, 505
Keshwala, Ushaben, 225
Khare, Sudhir, 461
Khurana, Palak, 547, 555
Kumar, Alok, 387
Kumar, Amit, 387
Kumar, Archek Praveen, 687, 697
Kumar, Cheruku Sandesh, 697
Kumari, Lalita, 79
Kumari, Pinki, 113
Kumar, Mohit, 485
Kumar, Parveen, 605
Kumar, Pravesh, 21, 245
Kumar, Raushan, 41
Kumar, Surendra, 21
Kumar, Sushil, 547, 555, 577, 617, 625
Kunroo, Mohd Hussain, 387

L
Lone, Showkat Ahmad, 67
Lugo, J.E., 201, 213, 323, 333

M
Mahajan, Akanshu, 21
Mahara, Tripti, 515
Mann, Mukesh, 679
Manoharan, R., 167
Mehrotra, Deepti, 1, 279, 309
Mehta, Rupa, 89
Mishra, Nidhi, 51, 133
Mogha, Sandeep Kumar, 387, 405
Mudgal, Neha, 59
Mukherjee, Tanushri, 373

N
Nautiyal, Chanda Thapliyal, 123

P
Pandey, Pallavi, 41
Pandey, S.K., 667
Pant, Millie, 191, 245, 405, 461, 505, 617, 727
Poonia, Ajeet Singh, 651, 659
Prajapati, Pankaj P., 301
Prakash, Divya, 397
Prasad, M., 167

R
Raheja, J.L., 79
Rahman, Ahmadur, 67
Rajpurohit, Jitendra, 267
Rajput, Nitesh Singh, 453
Rana, U.S., 123, 405
Rani, Kajal, 637
Rathi, Priya, 101
Rathore, Vijay Singh, 605
Rawat, S., 113, 201, 213, 225, 323, 373, 687,
    697
Ray, Amitava, 31
Ray, K., 113, 201, 213, 225, 323, 343, 373
Rekapally, Ruchita, 707
Roy, Ratnadeep, 687, 697

S
Sagar, Raj Kumar, 637
Sahare, Shilpa B., 525
Sahu, Satyajit, 343
Saklani, Yogita, 461
Sangwan, Om Prakash, 679
Saroliya, Anil, 41
Satija, Ajay, 495
Saxena, Sanjeev, 585
Saxena, Shweta, 59
Sehgal, Divya, 765
Sehgal, Rajni, 1
Sethi, Sushanta K., 419
Shah, Mihir V., 301
Shalini, 753
Sharawat, Kirti, 595
Sharma, Abhay, 697
Sharma, Anshika, 547, 555
Sharma, Chhaya, 419
Sharma, Ekta, 577
Sharma, S.C., 379, 485, 567
Sharma, Shreta, 667
Sharma, Shweta Gaur, 773
Sharma, S.K., 651, 659
Sharma, Tarun Kumar, 267, 279, 397, 453
Shrivastava, R.L., 525
Shukla, Dipesh Dilipbhai, 453
Singhal, Abhishek, 537
Singhal, Trapti, 473
Singh, Brahmjit, 433
Singh, Deepika, 753
Singh, Dipti, 495
Singh, Himanshu, 461
Singh, H.P., 21, 245
Singh, Meenu, 505
Singh, Natthan, 11, 715

Singh, P., 113, 201, 213, 323
Singh, Pallavi, 59
Singh, Raghvendra, 113
Singh, Santar Pal, 379
Singh, Shailendra Narayan, 547, 555
Singh, Sunita, 123
Somwanshi, Devendra, 473
Soni, Abhishek, 441
Sonika, 101
Soujanya, Annaram, 147, 157
Sridevi, G., 147, 157
Suryakant, 739

T
Tidke, Bharat, 89
Tiwari, Arti, 617, 625
Tomar, Pradeep, 679
Toshniwal, Sandeep, 373, 441

Tripathi, Anurag, 397

U
Uday Kanth Reddy, K., 567
Umar, Syed, 147, 157
Untawale, Sachin P., 525

V
Vaishali, 267

Y
Yadav, Ashwani, 473
Yadav, Ashwani Kumar, 687
Yadav, Rohit, 515

Z
Zaheer, Hira, 191
Zaidi, Nabeel, 605

Printed in the United States
By Bookmasters